T0299180

Wave Theory of Information

Understand the relationship between information theory and the physics of wave propagation with this expert guide. Balancing fundamental theory with engineering applications, it describes the mechanism and limits for the representation and communication of information using electromagnetic waves. Information-theoretic laws relating functional approximation and quantum uncertainty principles to entropy, capacity, mutual information, rate–distortion, and degrees of freedom of bandlimited radiation are derived and explained. Both stochastic and deterministic approaches are explored, and applications for remote sensing and signal reconstruction, wireless communication, and networks of multiple transmitters and receivers are reviewed. With end-of-chapter exercises and suggestions for further reading enabling in-depth understanding of key concepts, it is the ideal resource for researchers and graduate students in electrical engineering, physics, and applied mathematics looking for a fresh perspective on information theory.

Massimo Franceschetti is a Professor in the Department of Electrical and Computer Engineering at the University of California, San Diego, and a Research Affiliate of the California Institute of Telecommunications and Information Technology. He is the coauthor of *Random Networks for Communication* (Cambridge, 2008).

"This is an excellent textbook that ties together information theory and wave theory in a very insightful and understandable way. It is of great value and highly recommended for students, researchers and practitioners. Professor Franceschetti brings a highly valuable textbook based on many years of teaching and research."
Charles Elachi, ***California Institute of Technology and Director Emeritus of the Jet Propulsion Laboratory (NASA)***

"This book is about the physics of information and communication. It could be considered to be an exposition of Shannon information theory, where information is transmitted via electromagnetic waves. Surely Shannon would approve of it."
Sanjoy K. Mitter, ***Massachusetts Institute of Technology***

"Communication and information are inherently physical. Most of the literature, however, abstracts out the physics, treating them as mathematical or engineering disciplines. Although abstractions are necessary in the design of systems, much is lost in understanding the fundamental limits and how these disciplines fit together with the underlying physics. Franceschetti breaks the disciplinary boundaries, presenting communication and information as physical phenomena in a coherent, mathematically sophisticated, and lucid manner."
Abbas El Gamal, ***Stanford University***

Wave Theory of Information

MASSIMO FRANCESCHETTI
University of California, San Diego

Shaftesbury Road, Cambridge CB2 8EA, United Kingdom

One Liberty Plaza, 20th Floor, New York, NY 10006, USA

477 Williamstown Road, Port Melbourne, VIC 3207, Australia

314–321, 3rd Floor, Plot 3, Splendor Forum, Jasola District Centre, New Delhi – 110025, India

103 Penang Road, #05–06/07, Visioncrest Commercial, Singapore 238467

Cambridge University Press is part of Cambridge University Press & Assessment, a department of the University of Cambridge.

We share the University's mission to contribute to society through the pursuit of education, learning and research at the highest international levels of excellence.

www.cambridge.org
Information on this title: www.cambridge.org/9781107022317

DOI: 10.1017/9781108165020

First published 2018

A catalogue record for this publication is available from the British Library

Library of Congress Cataloging-in-Publication data
Names: Franceschetti, Massimo, author.
Title: Wave theory of information / Massimo Franceschetti, University of California, San Diego.
Description: Cambridge : Cambridge University Press, 2017. |
Includes bibliographical references.
Identifiers: LCCN 2017032961 | ISBN 9781107022317 (hardback)
Subjects: LCSH: Information theory. | Electromagnetic waves. |
Wave-motion, Theory of.
Classification: LCC Q360.F73 2017 | DDC 003/.54–dc23
LC record available at https://lccn.loc.gov/2017032961

ISBN 978-1-107-02231-7 Hardback

About the Cover

The picture represents the electromagnetic emission from stellar dust pervading our galaxy, measured by the Planck satellite of the European Space Agency. The colors represent the intensity, while the texture reflects the orientation of the field. The intensity of radiation peaks along the galactic plane, at the center of the image, where the field is aligned along almost parallel lines following the spiral structure of the Milky Way. Cloud formations are visible immediately above and below the plane, where the field's structure becomes less regular. The emission carries information regarding the evolution of our galaxy, as the turbulent structure of the field is related to the processes taking place when stars are born.

Acknowledgment: M.-A. Miville-Deschênes, CNRS, Institut d'Astrophysique Spatiale, Université Paris-XI, Orsay, France.

To my wife Isabella,
opera in my head.

Contents

Preface *page* xvii
Notation xx

1 Introduction 1
1.1 The Physics of Information 1
1.1.1 Shannon's Laws 1
1.1.2 Concentration Behaviors 2
1.1.3 Applications 4
1.2 The Dimensionality of the Space 5
1.2.1 Bandlimitation Filtering 5
1.2.2 The Number of Degrees of Freedom 7
1.2.3 Space–Time Fields 9
1.2.4 Super-resolution 11
1.3 Deterministic Information Measures 13
1.3.1 Kolmogorov Entropy 14
1.3.2 Kolmogorov Capacity 14
1.3.3 Quantized Unit of Information 17
1.4 Probabilistic Information Measures 18
1.4.1 Statistical Entropy 19
1.4.2 Differential Entropy 21
1.4.3 Typical Waveforms 22
1.4.4 Quantized Typical Waveforms 23
1.4.5 Mutual Information 25
1.4.6 Shannon Capacity 26
1.4.7 Gaussian Noise 29
1.4.8 Capacity with Gaussian Noise 30
1.5 Energy Limits 33
1.5.1 The Low-Energy Regime 33
1.5.2 The High-Energy Regime 34
1.5.3 Quantized Radiation 35
1.5.4 Universal Limits 37
1.6 *Tour d'Horizon* 41

1.7 Summary and Further Reading 43
1.8 Test Your Understanding 44

2 **Signals** 48
2.1 Representations 48
2.2 Information Content 49
2.2.1 Bandlimited Signals 50
2.2.2 Timelimited Signals 51
2.2.3 Impossibility of Time–Frequency Limiting 52
2.2.4 Shannon's Program 53
2.3 Heisenberg's Uncertainty Principle 55
2.3.1 The Uncertainty Principle for Signals 55
2.3.2 The Uncertainty Principle in Quantum Mechanics 56
2.3.3 Entropic Uncertainty Principle 58
2.3.4 Uncertainty Principle Over Arbitrary Measurable Sets 58
2.3.5 Converse to the Uncertainty Principle 59
2.4 The Folk Theorem 60
2.4.1 Problems with the Folk Theorem 61
2.5 Slepian's Concentration Problem 63
2.5.1 A "Lucky Accident" 65
2.5.2 Most Concentrated Functions 66
2.5.3 Geometric View of Concentration 68
2.6 Spheroidal Wave Functions 70
2.6.1 The Wave Equation 71
2.6.2 The Helmholtz Equation 72
2.7 Series Representations 75
2.7.1 Prolate Spheroidal Orthogonal Representation 76
2.7.2 Other Orthogonal Representations 78
2.7.3 Minimum Energy Error 79
2.8 Summary and Further Reading 81
2.9 Test Your Understanding 83

3 **Functional Approximation** 87
3.1 Signals and Functional Spaces 87
3.2 Kolmogorov N-Width 88
3.3 Degrees of Freedom of Bandlimited Signals 90
3.3.1 Computation of the N-Widths 91
3.4 Hilbert–Schmidt Integral Operators 95
3.4.1 Timelimiting and Bandlimiting Operators 98
3.4.2 Hilbert–Schmidt Decomposition 100
3.4.3 Singular Value Decomposition 102
3.5 Extensions 104
3.5.1 Approximately Bandlimited Signals 104
3.5.2 Multi-band Signals 105

3.5.3 Signals of Multiple Variables 108
3.5.4 Hybrid Scaling Regimes 111
3.6 Blind Sensing 113
3.6.1 Robustness of Blind Sensing 115
3.6.2 Fractal Dimension 116
3.7 Compressed Sensing 118
3.7.1 Robustness of Compressed Sensing 119
3.7.2 Probabilistic Reconstruction 121
3.7.3 Information Dimension 122
3.8 Summary and Further Reading 123
3.9 Test Your Understanding 124

4 Electromagnetic Propagation 130
4.1 Maxwell's Equations 130
4.2 Propagation Media 132
4.2.1 Perfectly Conductive Media 134
4.2.2 Dielectric Media 137
4.3 Conservation of Power 138
4.4 Plane Wave Propagation 139
4.4.1 Lossless Case 140
4.4.2 Lossy Case 141
4.4.3 Boundary Effects 142
4.4.4 Evanescent Waves 143
4.5 The Wave Equation for the Potentials 144
4.6 Radiation 146
4.6.1 The Far-Field Region 149
4.6.2 The Fraunhofer Region 151
4.7 Equivalence and Uniqueness 152
4.8 Summary and Further Reading 153
4.9 Test Your Understanding 154

5 Deterministic Representations 157
5.1 The Spectral Domains 157
5.1.1 Four Field Representations 157
5.1.2 The Space–Frequency Spectral Domain 158
5.2 System Representations 159
5.2.1 Linear, Time-Invariant Systems 159
5.2.2 Linear, Time-Invariant, Homogeneous Media 160
5.2.3 Green's Function in Free Space for the Potential 160
5.2.4 Green's Function in Free Space for the Field 161
5.2.5 Green's Function for Cylindrical Propagation 162
5.3 Discrete Radiating Elements 164
5.3.1 Single Transmitter–Receiver Pair 164
5.3.2 Multiple Transmitters and Receivers 166

5.3.3 Singular Value Decomposition 166
5.4 Communication Systems: Arbitrary Radiating Elements 167
5.4.1 Hilbert–Schmidt Decomposition 168
5.4.2 Optimal Communication Architecture 170
5.5 Summary and Further Reading 171
5.6 Test Your Understanding 172

6 Stochastic Representations 173
6.1 Stochastic Models 173
6.2 Green's Function for a Random Environment 174
6.2.1 Linear, Time-Varying Systems 174
6.2.2 Linear, Space–Time-Varying Systems 176
6.3 Multi-path 176
6.3.1 Frequency-Varying Green's Function: Coherence Bandwidth 179
6.3.2 Time-Varying Green's Function: Coherence Time 181
6.3.3 Mutual Coherence Function 183
6.3.4 Spatially Varying Green's Function: Coherence Distance 186
6.4 Karhunen–Loève Representation 186
6.4.1 Time-Varying Green's Function 187
6.4.2 Optimality of the Karhunen–Loève Representation 190
6.4.3 Stochastic Diversity 191
6.4.4 Constant Power Spectral Density 193
6.4.5 Frequency-Varying Green's Function 194
6.4.6 Spatially Varying Green's Function 195
6.5 Summary and Further Reading 196
6.6 Test Your Understanding 197

7 Communication Technologies 200
7.1 Applications 200
7.2 Propagation Effects 200
7.2.1 Multiplexing 203
7.2.2 Diversity 204
7.3 Overview of Current Technologies 205
7.3.1 OFDM 205
7.3.2 MC-CDMA 205
7.3.3 GSM 206
7.3.4 DS-CDMA 206
7.3.5 MIMO 207
7.4 Principles of Operation 208
7.4.1 Orthogonal Spectrum Division 209
7.4.2 Orthogonal Code Division 212
7.4.3 Exploiting Diversity 216
7.4.4 Orthogonal Spatial Division 217

7.5 Network Strategies 219
7.5.1 Multi-hop 219
7.5.2 Hierarchical Cooperation 221
7.5.3 Interference Alignment 222
7.5.4 A Layered View 224
7.5.5 Degrees of Freedom 225
7.6 Summary and Further Reading 226
7.7 Test Your Understanding 227

8 The Space–Wavenumber Domain 230
8.1 Spatial Configurations 230
8.2 Radiation Model 231
8.3 The Field's Functional Space 232
8.4 Spatial Bandwidth 233
8.4.1 Bandlimitation Error 234
8.4.2 Phase Transition of the Bandlimitation Error 236
8.4.3 Asymptotic Evaluation 238
8.4.4 Critical Bandwidth 240
8.4.5 Size of the Transition Window 243
8.5 Degrees of Freedom 244
8.5.1 Hilbert–Schmidt Decomposition 246
8.5.2 Sampling 248
8.6 Cut-Set Integrals 250
8.6.1 Linear Cut-Set Integral 251
8.6.2 Surface Cut-Set Integral 253
8.6.3 Applications to Canonical Geometries 256
8.7 Backscattering 258
8.8 Summary and Further Reading 261
8.9 Test Your Understanding 261

9 The Time–Frequency Domain 265
9.1 Frequency-Bandlimited Signals 265
9.2 Radiation with Arbitrary Multiple Scattering 266
9.2.1 Two-Dimensional Circular Domains 267
9.2.2 Three-Dimensional Spherical Domains 269
9.2.3 General Rotationally Symmetric Domains 270
9.3 Modulated Signals 272
9.4 Alternative Derivations 273
9.5 Summary and Further Reading 274
9.6 Test Your Understanding 274

10 Multiple Scattering Theory 275
10.1 Radiation with Multiple Scattering 275
10.1.1 The Basic Equation 276

10.1.2 Multi-path Propagation 277
10.2 Multiple Scattering in Random Media 278
10.2.1 Born Approximation 281
10.2.2 Complete Solutions 281
10.2.3 Cross Sections 283
10.3 Random Walk Theory 284
10.3.1 Radiated Power Density 289
10.3.2 Full Power Density 289
10.3.3 Diffusive Regime 290
10.3.4 Transport Theory 290
10.4 Path Loss Measurements 291
10.5 Pulse Propagation in Random Media 292
10.5.1 Expected Space–Time Power Response 292
10.5.2 Random Walk Interpretation 296
10.5.3 Expected Space–Frequency Power Response 297
10.5.4 Correlation Functions 298
10.6 Power Delay Profile Measurements 299
10.7 Summary and Further Reading 300
10.8 Test Your Understanding 301

11 Noise Processes 303
11.1 Measurement Uncertainty 303
11.1.1 Thermal Noise 303
11.1.2 Shot Noise 304
11.1.3 Quantum Noise 306
11.1.4 Radiation Noise 306
11.2 The Black Body 307
11.2.1 Radiation Law, Classical Derivation 307
11.2.2 Thermal Noise, Classical Derivation 311
11.2.3 Quantum Mechanical Correction 312
11.3 Equilibrium Configurations 314
11.3.1 Statistical Entropy 316
11.3.2 Thermodynamic Entropy 317
11.3.3 The Second Law of Thermodynamics 318
11.3.4 Probabilistic Interpretation 319
11.3.5 Asymptotic Equipartition Property 320
11.3.6 Entropy and Noise 321
11.4 Relative Entropy 322
11.5 The Microwave Window 323
11.6 Quantum Complementarity 325
11.7 Entropy of a Black Body 326
11.7.1 Total Energy 326
11.7.2 Thermodynamic Entropy 327
11.7.3 The Planck Length 328

11.7.4 Gravitational Limits 329
11.8 Entropy of Arbitrary Systems 329
11.8.1 The Holographic Bound 330
11.8.2 The Universal Entropy Bound 330
11.9 Entropy of Black Holes 331
11.10 Maximum Entropy Distributions 333
11.11 Summary and Further Reading 336
11.12 Test Your Understanding 338

12 Information-Theoretic Quantities 343
12.1 Communication Using Signals 343
12.2 Shannon Capacity 344
12.2.1 Sphere Packing 347
12.2.2 Random Coding 349
12.2.3 Capacity and Mutual Information 350
12.2.4 Limiting Regimes 352
12.2.5 Quantum Constraints 354
12.2.6 Capacity of the Noiseless Photon Channel 355
12.2.7 Colored Gaussian Noise 356
12.2.8 Minimum Energy Transmission 357
12.3 A More Rigorous Formulation 358
12.3.1 Timelimited Signals 359
12.3.2 Bandlimited Signals 360
12.3.3 Refined Noise Models 362
12.4 Shannon Entropy 364
12.4.1 Rate–Distortion Function 364
12.4.2 Rate–Distortion and Mutual Information 366
12.5 Kolmogorov's Deterministic Quantities 368
12.5.1 ϵ-Coverings, ϵ-Nets, and ϵ-Entropy 369
12.5.2 ϵ-Distinguishable Sets and ϵ-Capacity 370
12.5.3 Relation Between ϵ-Entropy and ϵ-Capacity 370
12.6 Basic Deterministic–Stochastic Model Relations 371
12.6.1 Capacity 371
12.6.2 Rate–Distortion 373
12.7 Information Dimensionality 374
12.7.1 Metric Dimension 374
12.7.2 Functional Dimension and Metric Order 376
12.7.3 Infinite-Dimensional Spaces 377
12.8 Bandlimited Signals 378
12.8.1 Capacity and Packing 378
12.8.2 Entropy and Covering 378
12.8.3 ϵ-Capacity of Bandlimited Signals 379
12.8.4 (ϵ, δ)-Capacity of Bandlimited Signals 381

12.8.5 ϵ-Entropy of Bandlimited Signals 382
12.8.6 Comparison with Stochastic Quantities 383
12.9 Spatially Distributed Systems 384
12.9.1 Capacity with Channel State Information 384
12.9.2 Capacity without Channel State Information 387
12.10 Summary and Further Reading 388
12.11 Test Your Understanding 389

13 Universal Entropy Bounds 391
13.1 Bandlimited Radiation 391
13.2 Deterministic Signals 392
13.2.1 Quantization Error 393
13.2.2 Kolmogorov Entropy Bound 394
13.2.3 Saturating the Bound 396
13.3 Stochastic Signals 397
13.3.1 Shannon Rate–Distortion Bound 398
13.3.2 Shannon Entropy Bound 398
13.4 One-Dimensional Radiation 400
13.5 Applications 401
13.5.1 High-Energy Limits 401
13.5.2 Relation to Current Technologies 402
13.6 On Models and Reality 403
13.7 Summary and Further Reading 405
13.8 Test Your Understanding 406

Appendix A Elements of Functional Analysis 407

Appendix B Vector Calculus 422

Appendix C Methods for Asymptotic Evaluation of Integrals 428

Appendix D Stochastic Integration 433

Appendix E Special Functions 434

Appendix F Electromagnetic Spectrum 437

Bibliography 438
Index 447

Preface

Claude Elwood Shannon, the giant who ignited the digital revolution, is the father of information theory and a hero for many engineers and scientists. There are many excellent textbooks describing the many facets of his work, so why add another one? The ambitious goal is to provide a completely different perspective. The writing reflects my desire to abhor duplication and to attempt to break through the compartmentalized walls of several disciplines. Rather than copying a Picasso, I have tried to frame it and place it in a broader context.

The motivation also came from my experience as a teacher. The Electrical and Computer Engineering Department of the University of California at San Diego, in the spotlight of its annual workshop on information theory and applications, attracting several hundred participants from around the world, may be considered a holy destination for graduate students in information theory. Many gifted young minds join our department every year with the ultimate goal of earning a PhD in this venerable subject. Here, thanks to the work of many esteemed colleagues, they can become experts in coding and communication theories, point-to-point and network information theories, and wired and wireless information systems. Over my years of teaching, however, I have noticed that sometimes students are missing the master plan for how these topics are tied together and how are they related to the fundamental sciences. Some questions that may catch them off guard are: How much information can be radiated by a waveform at the most fundamental level? How is the physical entropy related to the information-theoretic limits of communication? How does the energy and the quantum nature of radiation limit information? How is information theory related to other branches of mathematics besides probability theory, like functional analysis and approximation theory? On top of these, there is the overarching question, of paramount importance for the engineer, of how communication technologies are influenced by fundamental limits in a practical setting. To fill these gaps, this book focuses on information theory from the point of view of wave theory, and describes connections with different branches of physics and mathematics.

David Hilbert, studying functional representations in terms of orthogonal basis sets in early twentieth-century Germany, contributed to underpinning the mathematical concept of information associated with a waveform. After Shannon's breakthrough work, his approach was later followed by the Soviet mathematician Andrey Kolmogorov, who developed information-theoretic concepts in a purely deterministic setting, and by Shannon's colleagues at Bell Laboratories: Henry Landau, Henry Pollack, and David

Slepian. Their mathematical works are the basis of the wave theory of information presented in this book. We expand upon them, and place them in the context of communication with electromagnetic waves.

From the physics perspective, much has been written on the relationship between thermodynamics and information theory. Parallels between the statistical mechanics of Boltzmann and Gibbs and Shannon's definition of entropy led to many important advancements in the analysis and design of complex engineering systems. Once again, *repetita iuvant, sed continuata secant.* We briefly touch upon these topics, but focus on the less beaten path, uncovering the relationship between information theory and the physics of Heisenberg, Maxwell, and Planck. We describe how information physically propagates using waves, and what the limitations are for this process. What was first addressed in the pioneering works of Dennis Gabor and Giuliano Toraldo di Francia is revisited here in the rigorous setting of the theory of functional approximation. Using these tools we provide, for the first time in a book, a complete derivation of the information-theoretic notion of degrees of freedom of a wave starting from the Maxwell equations, and relate it to the concept of entropy, and to the principles of quantized radiation and of quantum indeterminacy. We also provide analogous derivations for stochastic processes and discuss communication technologies from the point of view of functional representations, which turns out to be very useful to uncover the core architectural ideas fundamental to communication systems. When these are viewed in the context of physical limits, one realizes that there still is "plenty of room at the bottom." Engineers are far from reaching the limits that nature imposes on communication: our students have a bright future in front of them!

Now, a word on style and organization. Although the treatment requires a great deal of mathematics and assumes that the reader has some familiarity with probability theory, stochastic processes, and real analysis, this is not a mathematics book. From the outset, I have made the decision to avoid writing a text as a sequence of theorems and proofs. Instead, I focus on describing the ideas that are behind the results, the relevant mathematical techniques, and the philosophy behind their arguments. When not given, rigorous proofs should follow easily once these basic concepts are grasped. This approach is also reflected in the small set of exercises provided at the end of each chapter. They are designed to complement the text, and to provide a more in-depth understanding of the material. When given, solutions are often sketchy, emphasize intuition over rigor, and encourage the reader to fill in the details. Pointers to research papers for further reading are also provided at the end of each chapter.

A grouping into an introductory sequence of topics (Chapters 1–6), central results (Chapters 8 and 9), and an in-depth sequence (Chapters 10–13) is the most natural for using the book to teach a two-quarter graduate course. The demarcation line between these topics can be somewhat shifted, based on the taste of the instructor. The book could also be used in a one-semester course with a selection of the in-depth topics, and limiting the exposition of some of the details of the central results. Within this organization, Chapter 7 is an *intermezzo*, focusing on wireless communication technologies, and on how they exploit information-theoretic representations. Of course, this can only scratch the surface of a large field of study, and the interested reader should refer to the wide range of literature for a more in-depth account. A *tour d'horizon* of the book's content is provided at the end of the first chapter.

The material has been tested over the course of seven years in the annual graduate course I have taught at the University of California at San Diego. I wish to thank the many students who attended the course and provided feedback on the lecture notes, which were early incarnations of this book, especially the students in my research group, Taehyung Jay Lim and Hamed Omidvar, who read many sections in detail and provided invaluable comments. I also enjoyed interactions with my colleague Young-Han Kim, who read parts of the manuscript and provided detailed feedback. The presentation of blind sensing and compressed sensing in Chapter 3 has been enriched by conversations with my colleague Rayan Saab. Recurrent visits to the group led by Bernard Fleury at Aalborg University, Denmark, influenced the presentation of the material on stochastic models and their relationship to communication systems presented in Chapters 6 and 10. Sergio Verdú of Princeton University kindly offered some stylistic suggestions to improve the presentation of the material in Chapters 1 and 12. Many exchanges with Edward Lee of the University of California at Berkeley and with my colleague George Papen on the physical meaning of information helped to shape the presentation in Chapter 13. Interactions with Sanjoy Mitter of the Massachusetts Institute of Technology also stimulated many of the physical questions addressed in the book. My editors Phil Meyler and Julie Lancashire at Cambridge University Press were very patient with my eternal postponement of manuscript delivery, and provided excellent professional advice throughout.

A final "thank you" goes to my family, who patiently accepted, with "minimal" complaint, my lack of presence, due to the long retreats in my downstairs hideout.

Massimo Franceschetti

Notation

Asymptotics

$f(x) \sim g(x)$ as $x \to x_0$	$\Longleftrightarrow$	$\lim_{x \to x_0} f(x)/g(x) = 1$
$f(x) = o(g(x))$ as $x \to x_0$	$\Longleftrightarrow$	$\lim_{x \to x_0} f(x)/g(x) = 0$
$f(x) = O(g(x))$ as $x \to x_0$	$\Longleftrightarrow$	$\lim_{x \to x_0} \lvert f(x)/g(x) \rvert < \infty$

Approximations

$f(x) \simeq g(x)$	$g(x)$ is a finite-degree Taylor polynomial of $f(x)$
$f(x) \approx g(x)$	$f(x)$ is approximately equal to $g(x)$ in some numerical sense
$f(x) \gg g(x)$	$f(x)$ is much greater than $g(x)$ in some numerical sense
$f(x) \ll g(x)$	$f(x)$ is much smaller than $g(x)$ in some numerical sense

Domains

$t \in \mathbb{R}$	time
$\omega \in \mathbb{R}$	angular frequency
$\mathbf{r} \in \mathbb{R}^3$	spatial
$\mathbf{k} \in \mathbb{R}^3$	wavenumber
$\phi \in [0, 2\pi]$	angular
$\lambda \in \mathbb{R}^+$	wavelength
$w \in \mathbb{R}^+$	scalar wavenumber
$f(t) \leftrightarrow F(\omega)$	time–angular frequency Fourier transform pairs
$f(\mathbf{r}) \leftrightarrow \widehat{f}(\mathbf{k})$	space–wavenumber Fourier transform pairs
$f(\phi) \leftrightarrow \widehat{f}(w)$	angle–wavenumber Fourier transform pairs

Signals

Ω	angular frequency bandwidth
W	wavenumber bandwidth
$\text{sinc}\,(t)$	waveform $(\sin t)/t$
$\text{rect}\,(t/T)$	rectangular waveform of support T and unitary amplitude
$U(t)$	Heaviside's step function: $U(t) = 0$ for $t < 0$, $U(t) = 1$ for $x \geq 0$
$\delta(t)$	Dirac's impulse distribution

Complex Numbers

j	imaginary unit
$f^*(\cdot)$	conjugate of complex signal f
$\Re f(\cdot)$	real part of complex signal f
$\Im f(\cdot)$	imaginary part of complex signal f
$\mathrm{M}^\dagger$	conjugate transpose of matrix M

Functional Spaces

L^2	square-integrable signals
$\mathscr{B}_\Omega$	bandlimited signals of spectral support $[-\Omega, \Omega]$
$\mathscr{T}_T$	timelimited signals of time support $[-T/2, T/2]$
N_0	Nyquist number, $N_0 = \Omega T/\pi$
$\alpha^2(T)$	fraction of a signal's energy in $[-T/2, T/2]$
$\beta^2(\Omega)$	fraction of a signal's energy in $[-\Omega, \Omega]$
$\mathscr{E}(\epsilon_T)$	the set of ϵ_T-concentrated, bandlimited signals
$\lVert \cdot \rVert$	norm
$\langle \cdot \rangle$	inner product
$\mathscr{S}_n \subset \mathscr{S}$	an n-dimensional subspace of the space $\mathscr{S}$
$D_{\mathscr{S}_n}(\mathscr{A})$	deviation of the set $\mathscr{A}$ from $\mathscr{S}_n$
$d_n(\mathscr{A}, \mathscr{S})$	Kolmogorov n-width of the set $\mathscr{A}$ in $\mathscr{S}$
$N_\epsilon(\mathscr{A})$	number of degrees of freedom at level ϵ of the set $\mathscr{A}$

Fields

$\bar{\mathbf{x}}, \bar{\mathbf{y}}, \bar{\mathbf{z}}$	unit vectors along the coordinate axes
$f(x,y,z)$	a scalar field
$\mathbf{f}(x,y,z)$	a vector field: $f_x(x,y,z)\,\bar{\mathbf{x}} + f_y(x,y,z)\,\bar{\mathbf{y}} + f_z(x,y,z)\,\bar{\mathbf{z}}$
∇f	gradient
$\nabla \cdot \mathbf{f}$	divergence
$\nabla \times \mathbf{f}$	curl
$\nabla^2 f$	scalar Laplacian
$\nabla^2 \mathbf{f}$	vector Laplacian
$\underline{\mathbf{g}}(\mathbf{r},t)$	dyadic space–time Green's function
$\underline{\mathbf{G}}(\mathbf{r},\omega)$	dyadic space–frequency Green's function

Physical Constants

ℓ_{p}	Planck's length [m]
$\hbar$	reduced Planck's constant [J s]
k_{B}	Boltzmann's constant [J K^{-1}]
ϵ_0	permittivity of the vacuum [F m^{-1}]
μ_0	permeability of the vacuum [H m^{-1}]
$\epsilon = \epsilon_{\mathrm{r}}\epsilon_0$	permittivity of the medium [F m^{-1}]
$\mu = \mu_{\mathrm{r}}\mu_0$	permeability of the medium [H m^{-1}]
σ	conductivity of the medium [S m^{-1}]
$c = 1/\sqrt{\epsilon\mu}$	propagation speed of electromagnetic wave [m s^{-1}]
$\beta = 2\pi/\lambda = \omega/c$	propagation coefficient [m^{-1}]

Probability

$\mathscr{A}$	a set
$\lvert\mathscr{A}\rvert$	cardinality of the set $\mathscr{A}$
Z	a random variable
$\mathbb{P}(\mathsf{Z} \in \mathscr{A})$	probability that realization of random variable Z is in $\mathscr{A}$
$\mathbb{E}(\mathsf{Z})$	expected value of Z
$f_{\mathsf{Z}}(z)$	probability density function of random variable Z
$\mathsf{Z}(t)$	a random process
$s_{\mathsf{Z}}(t,t')$	autocorrelation of $\mathsf{Z}(t)$
$s_{\mathsf{Z}}(\tau)$	autocorrelation of wide-sense stationary process $\mathsf{Z}(t)$
$S_{\mathsf{Z}}(\omega)$	power spectral density of wide-sense stationary process $\mathsf{Z}(t)$

Entropy and Capacity

H_C	Thermodynamic entropy [J K^{-1}]
H_B	Boltzmann entropy [J K^{-1}]
H_G	Gibbs entropy [J K^{-1}]
H	Shannon entropy [bits]
H_ϵ	Kolmogorov ϵ-entropy [bits]
$\bar{H}_\epsilon$	Kolmogorov ϵ-entropy per unit time [bits s^{-1}]
R_N	Shannon rate distortion function [bits s^{-1}]
C	Shannon capacity [bits s^{-1}]
C_ϵ	Kolmogorov ϵ-capacity [bits]
$\bar{C}_\epsilon$	Kolmogorov ϵ-capacity per unit time [bits s^{-1}]
C_ϵ^δ	ϵ-delta capacity [bits]
$\bar{C}_\epsilon^\delta$	ϵ-delta capacity per unit time [bits s^{-1}]
$h(f)$	differential entropy of probability density function f
$D(p\|\|q)$	relative entropy between probability mass functions p and q
$I(\mathsf{X};\mathsf{Y})$	mutual information between random variables X and Y

1 Introduction

Nihil est in intellectu quod non prius fuerit in sensu.[1]

1.1 The Physics of Information

This book describes the limits for the communication of information with waves. How many ideas can we communicate by writing on a sheet of paper? How well can we hear a concert? How many details can we distinguish in an image? How much data can we get from our internet connection? These are all questions related to the transport of information by waves. Our sensing ability to capture the differences between distinct waveforms dictates the limits of the amount of information that is delivered by a propagating wave. The problem of quantifying this precisely requires a mathematical description and a physical understanding of both the propagation and the communication processes.

We focus on the propagation of electromagnetic waves as described by Maxwell's theory of electromagnetism, and on communication as described by Shannon's theory of information. Although our treatment is mostly based on classical field theory, we also consider limiting regimes where the classical theory must give way to discrete quantum formulations. The old question of whether information is physics or mathematics resounds here. Information is certainly described mathematically, but we argue that it also has a definite physical structure. The central theme of this book is that Shannon's information-theoretic limits are natural. They are revealed by observing physical quantities at certain asymptotic scales where finite dimensionality emerges and observational uncertainties are averaged out. These limits are also rigorous, and obey the mathematical rules that govern the model of reality on which the physical theories are based.

1.1.1 Shannon's Laws

Originally introduced by Claude Elwood Shannon in 1948, and continuing up to its latest developments in multi-user communication networks, information theory describes

[1] Empiricist claim adopted from the Peripatetic school. A principle subscribed to by Aristotle, St. Thomas, and Locke; opposed by Plato, St. Augustine, and Leibniz.

in the language of mathematics the limits for the communication of information. These limits are operational, of real engineering significance, and independent of the semantic aspects associated with the communication process. They arise from the following set of constraints expressing our inability to appreciate the *infinitum*:

- *Finite energy*: Communication occurs by a finite expenditure of energy.
- *Finite dimensionality*: Communication occurs by selecting among a range of possible choices, each identified by a finite number of attributes.
- *Finite resolution*: Each attribute can be observed with limited precision.

According to Shannon, reliable communication of information occurs if the probability of miscommunication *tends* to zero in appropriate limiting regimes. In these regimes, some interesting physical phenomena also occur: the space–time fields used to convey information become amenable to a discrete representation, and the finite dimensionality of the physical world is revealed. This allows us to view information-theoretic results as being imposed by the laws of nature. Information theory classically considers time asymptotics, used to describe point-to-point communication. Spatial asymptotics are their natural counterparts, used to extend the description to communication between multiple transmitters and receivers, and to the remote sensing of the world around us. There is a beautiful duality between the two, and this book attempts to capture it by providing a unified treatment of space–time waveforms.

1.1.2 Concentration Behaviors

At the basis of the asymptotic arguments leading to information-theoretic limits is the notion of *concentration*.

Consider a space–time waveform $f(x,y,z,t)$ of finite energy, transmitted for T seconds. As $T \to \infty$, we can define the effective frequency bandwidth of the waveform as the effective spectral support in the Fourier-transformed angular frequency domain – see Figure 1.1. This definition is made possible by the mathematics at the basis of wave theory that predict *spectral concentration*. As the time domain support is stretched, the signal, when viewed in the frequency domain, can be more and more concentrated inside the bandwidth. Thanks to this phenomenon, electromagnetic signals can be considered, for large T, as occupying an essentially finite bandwidth. Signals of finite energy and finite bandwidth enjoy another important mathematical property. They exhibit a limit on the amount of variation they can undergo in any given time interval and thus, when viewed at finite resolution, on the amount of information they can carry over time. The same limitation also applies to the spatial domain. As the region where the signal is observed is stretched by scaling all of its coordinates, spectral concentration occurs, and this allows the definition of the effective bandwidth in the wavenumber domain, that is the Fourier transform of the spatial domain. This limits the number of spatial configurations of the waveform, and thus, when viewed at finite resolution,

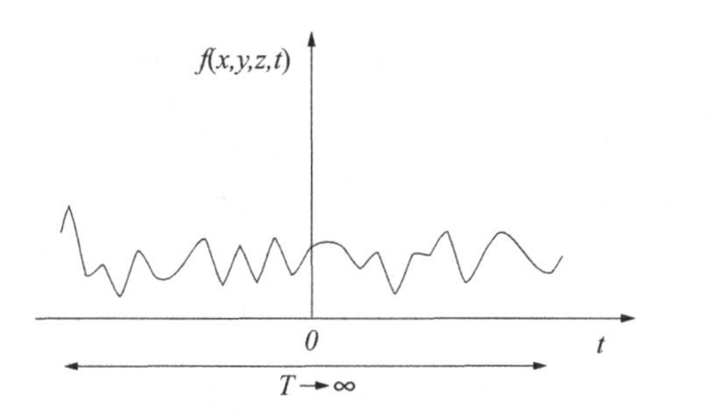

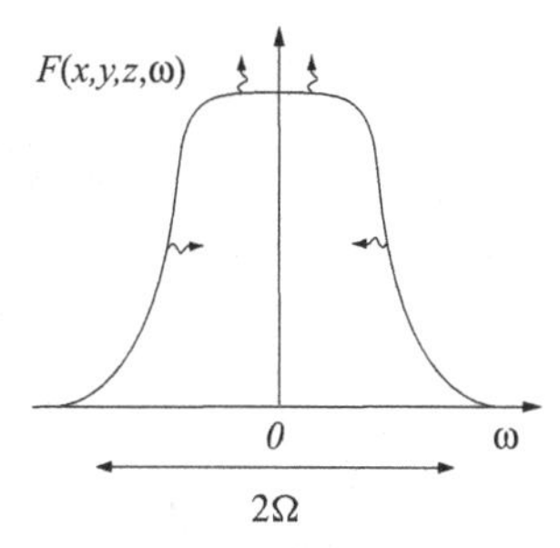

Fig. 1.1 Spectral concentration.

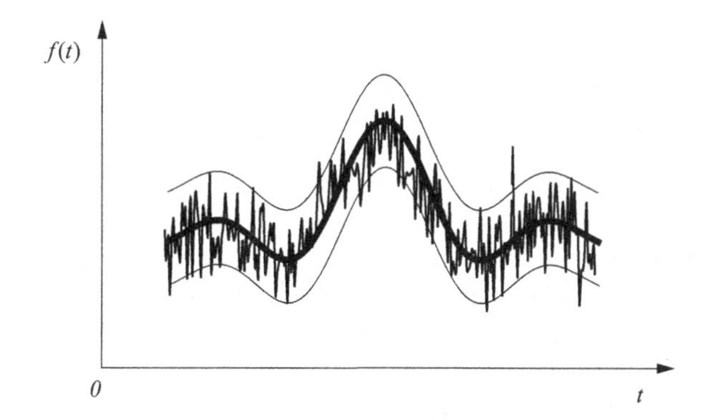

Fig. 1.2 Probabilistic concentration.

the amount of information it can carry over space. This limitation is important in the context of network information theory, when multiple transmitters and receivers in a communication system are distributed in space. It is also important in the context of imaging systems, where it leads to spatial resolution limits of the constructed image.

When considering space and time asymptotics, another kind of concentration phenomenon also occurs. The precision level at which the signal can be observed *probabilistically concentrates* around its typical value. Every physical apparatus measuring a signal is affected by a measurement error: repeated measurements appear to fluctuate randomly by a small amount. This is a consequence of the quantized nature of the world observed at the microscopic scale. Over many repetitions, the uncertainty with which the signal is observed is typically contained within its standard deviation – see Figure 1.2. This allows us to view the uncertainty of the observation as concentrated around its typical value and determines a resolution limit at which the signal can be observed. Combined with the constraints on the form of the signal due to spectral concentration mentioned above, it poses an ultimate limit on the amount of information that can be transported by waves in time and space.

The same concentration behaviors leading to information-theoretic limits are also at the basis of quantum mechanics and statistical mechanics. Spectral concentration is at the basis of Heisenberg's uncertainty principle, stating that physical quantities related by

Fourier transforms cannot be determined simultaneously, as pinpointing one precisely always implies the smearing of the other. On the other hand, probabilistic concentration is at the heart of statistical mechanics. Due to probabilistic concentration, only some realizations of a stochastic process have non-negligible probability of occurrence, and these typical outcomes are nearly equally probable. Based on this premise, statistical mechanics explains the thermodynamic behavior of large systems in terms of typical outcomes of random microscopic events. The information-theoretic approach is another instance of this method, as it exploits probabilistic concentration to describe the typical behavior of encoding and decoding systems formed by large ensembles of random variables.

1.1.3 Applications

When discussing information carried by electromagnetic waves, it is also natural to mention practical applications. In the context of electromagnetics, the operational aspects of information theory have found vast applications ranging from remote sensing and imaging to communication systems. During the wireless revolution of the turn of the millennium, the physical layer of digital communications has successfully been abstracted using approximate probabilistic models of the propagation medium. Information theory has fruitfully developed in this framework, and important theoretical contributions have inspired creative minds, who improved engineering designs. Many practical advancements have been influenced by a clearer understanding of the information-theoretic limits of different channel models. As theoretical studies found practical applications, industry flourished. Sometimes, however, this has come at the expense of casting a shadow on the physical limits of communication. The fundamental question posed by Shannon regarding the ultimate limits of communication has been obfuscated by the myriad results regarding different approximate models of physical reality.

Research in wireless communication has expanded the knowledge tree into an intricate forest of narrow cases, more of interest to the practitioner than to the scientist seeking a fundamental understanding. These cases have provided useful design guidelines, but they have also somewhat hidden the fundamental limits. This situation is somehow natural: as a field becomes more mature, improvements tend to become more sectorized, and technology, rather than basic advancements, becomes the main driver for progress. Maturity, however, should also open up the opportunity of revisiting, reordering, reinterpreting, and pruning the knowledge tree, revealing its basic skeleton. With it, we also wish to reveal the misconception that a rigorous physical treatment is too complex to provide valuable engineering insights. A physical treatment not only shows that Shannon's theory is part of the fundamental description of our physical universe, but it also provides insights into the design and operation of real engineering systems. After all, if, as Alfréd Rényi (1984) put it, information theory came out of the realization that the flow of information, like other physical quantities, can be expressed numerically, then our engineering designs should best exploit its physical nature.

For this reason, we devote Chapter 7 to discussing communication technologies, and we revisit these technologies in the light of our physical treatment of information. Of course, we can only discuss some of the fundamental principles; for a more in-depth perspective the reader should refer to the wide range of available engineering literature.

1.2 The Dimensionality of the Space

To quantify the amount of information carried by electromagnetic waves of finite energy, we represent the possible messages that the transmitter may send by an ensemble of waveforms, and then attempt to quantify the amount of information transferred by the correct selection at the receiver of one element from this ensemble, according to Shannon's theory. This selection implies that a certain amount of information has been transported between the two. To realize this program, we are faced with a first fundamental question:

> How many distinct waveforms can we possibly select from?

The answer depends on the size of the signals' space, or the *number of degrees of freedom*, available for communication and on the resolution at which each degree of freedom can be observed. The number of degrees of freedom is limited by the nature of the wave propagation process, while the resolution is limited by the uncertainty associated with the observation process.

1.2.1 Bandlimitation Filtering

Any radiated waveform has an essentially finite bandwidth. Due to the interaction with the propagation medium and with the measurement apparatus used to detect the signal, the high-frequency components are cut off, and any signal appears as the output of a linear filter. Typically this filter also shapes the frequency profile, distorting the transmitted waveform. A simple example of this phenomenon occurs in a scattering environment. Due to multiple scattering, multiple copies of the transmitted waveform, carrying different delays and attenuations while traveling on different paths, may overlap at the receiver, creating interference and distorting the original signal – see Figure 1.3. While in ideal free space the magnitude of the frequency response of the propagation environment would be flat across all frequencies, and the phase of the response would be proportional to the distance between transmitter and receiver, in the presence of a large amount of scattering the response is a highly varying signal, due to the interference over the multiple scattered paths.

An analogous effect occurs in the spatial domain. The signal observed along any spatial cut-set that separates transmitters and receivers is the image of all the elements radiating from one side of the cut to the other. These radiating waveforms may interfere along the cut, leading to spatial filtering – see Figure 1.4.

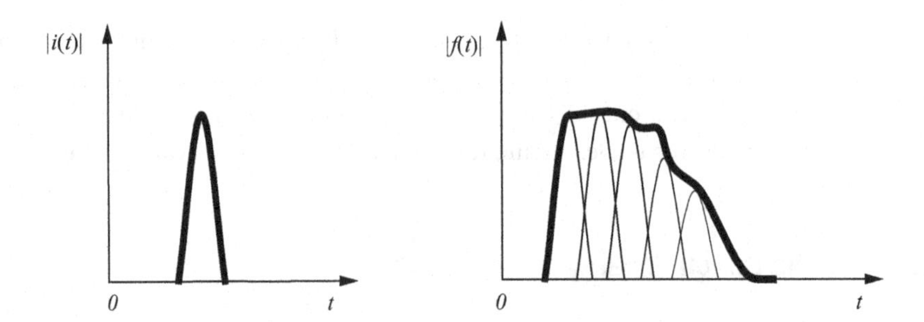

Fig. 1.3 Spreading of the signal in the time domain due to multiple scattering. Transmitted signal: $i(t)$; received signal: $f(t)$.

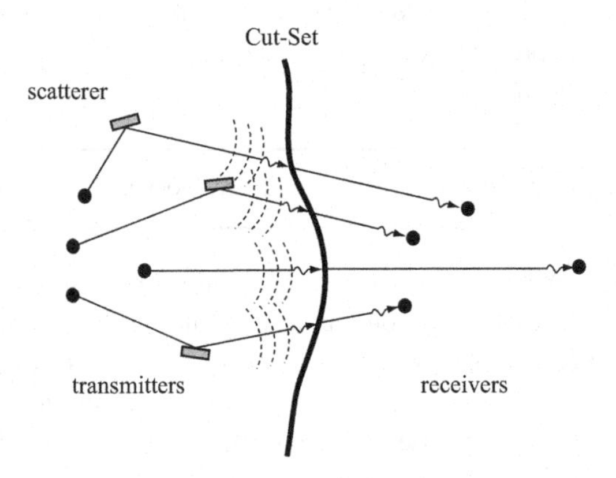

Fig. 1.4 Multiple signals overlap over the cut-set boundary.

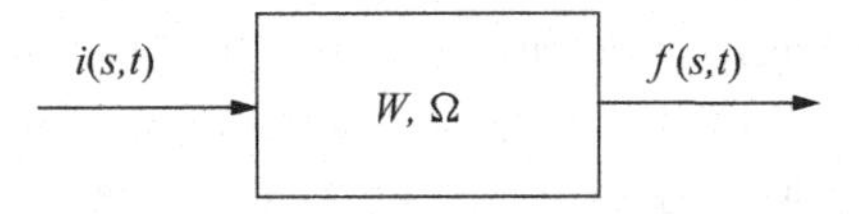

Fig. 1.5 Propagation filtering.

A block diagram of the propagation filtering effect that occurs when a scalar source current $i(s,t)$ of one spatial and one temporal variable produces a scalar electromagnetic field $f(s,t)$ observed along a given cut-set boundary is depicted in Figure 1.5. The figure shows that the effect of propagation is analogous to that of a linear filter of frequency cut-off Ω and of wavenumber cut-off W. The form of the transfer function depends on the features of the environment where propagation occurs and is studied in Chapters 8 and 9. This filtering operation limits the number of distinct space–time waveforms that can be observed at the receiver, and makes the space of electromagnetic signals essentially bandlimited and suitable for an information-theoretic analysis.

1.2.2 The Number of Degrees of Freedom

The physics of propagation dictate that any observed electromagnetic field is an essentially bandlimited function. This basic property allows us to define the size of the signals' space in terms of the number of degrees of freedom. Consider a one-dimensional, real, scalar waveform f of a single scalar variable t. We assume that f is square-integrable, and

$$\int_{-\infty}^{\infty} f^2(t)dt \leq E. \tag{1.1}$$

This ensures that the waveform can be expanded in a series of, possibly complex, orthonormal basis functions $\{\psi_n\}$,

$$f(t) = \sum_{n=1}^{\infty} a_n \psi_n(t), \tag{1.2}$$

where

$$a_n = \int_{-\infty}^{\infty} f(t)\psi_n^*(t)dt. \tag{1.3}$$

The equality in (1.2) is intended in the "energy" sense:

$$\lim_{N\to\infty} \int_{-\infty}^{\infty} [f(t) - f_N(t)]^2 dt = 0, \tag{1.4}$$

where

$$f_N(t) = \sum_{n=1}^{N} a_n \psi_n(t). \tag{1.5}$$

In the language of mathematics, f is in $L^2(-\infty,\infty)$, and it can be viewed as a point in an infinite-dimensional space of coordinates given by the coefficients $\{a_n\}$ in (1.3). By varying the values of these coefficients, we can create distinct waveforms and use them to communicate information. If the orthonormal set of basis functions $\{\psi_n\}$ is complete, then using (1.2) we can construct any element in the space of signals defined by (1.1). By associating a waveform in this space with a given message that the transmitter wishes to communicate, the correct selection of the same waveform at the receiver implies that a certain amount of information is transferred between the two. One may reasonably expect that only a finite number of coefficients is in practice needed to specify the waveform up to any given accuracy, while using a larger number does not significantly improve the resolution at the receiver. It turns out that the question of what the smallest N is beyond which varying higher-order coefficients does not change the form of the waveform significantly has a remarkably precise answer.

Determining this number over all possible choices of basis functions is a question first posed by Kolmogorov in 1936, and corresponds to determining the *number of degrees of freedom* of the waveform. Consider an observation interval $[-T/2, T/2]$, and introduce the norm

$$\|f\| = \left(\int_{-T/2}^{T/2} f^2(t)dt\right)^{1/2}. \tag{1.6}$$

The problem amounts to determining the interval of values of N for which the approximation error for any signal f,

$$\|e_N\| = \|f(t) - f_N(t)\|, \tag{1.7}$$

transitions from being close to its maximum to being close to zero.

For signals of spectral support in an interval of size 2Ω, and observed over a time interval of size T, the angular frequency bandwidth Ω and the size of the observation interval play a key role in determining the number of degrees of freedom, as the approximation error undergoes a *phase transition* at the scale of the product ΩT.

For any $\epsilon > 0$, letting N_ϵ be the minimum number of basis functions for which the normalized energy of the error

$$\frac{\|e_{N_\epsilon}\|^2}{E} \leq \epsilon, \tag{1.8}$$

and letting

$$N_0 = \frac{\Omega T}{\pi}, \tag{1.9}$$

we have

$$\lim_{N_0 \to \infty} \frac{N_\epsilon}{N_0} = 1. \tag{1.10}$$

This result was a crowning achievement of three scientists, Henry Landau, Henry Pollak, and David Slepian, working at Bell Laboratories in the 1960s and 70s, and we discuss it in detail in Chapters 2 and 3. It identifies the number of degrees of freedom with the time-bandwidth product $N_0 = \Omega T/\pi$, and shows that this number is, up to first order, independent of the level of approximation ϵ. This is evident by rewriting (1.10) as

$$N_\epsilon = N_0 + o(N_0) \quad \text{as } N_0 \to \infty, \tag{1.11}$$

where the dependence on ϵ appears hidden as a pre-constant of the second-order term $o(N_0)$ in the phase transition of the number of degrees of freedom. Varying the approximation level does not affect the first-order scaling of the result. Figure 1.6 shows the transition of the approximation error for the optimal choice of basis functions. According to (1.10), the transition occurs in a small interval and tends to become a step function when viewed at the scale of N_0.

A widely used representation for bandlimited waveforms is the cardinal series that uses sampled values of the waveform as coefficients $\{a_n\}$ and real functions of the form $\text{sinc}\,(t) = (\sin t)/t$ as the basis set $\{\psi_n\}$, yielding the Kotelnikov–Shannon–Whittaker sampling representation

$$f(t) = \sum_{n=-\infty}^{\infty} f(n\pi/\Omega)\,\text{sinc}\,(\Omega t - n\pi), \tag{1.12}$$

which interpolates the signal from regularly spaced samples at frequency Ω/π. This representation is suboptimal in terms of the approximation error (1.7). The optimal interpolating functions, called prolate spheroidal wave functions, that achieve the

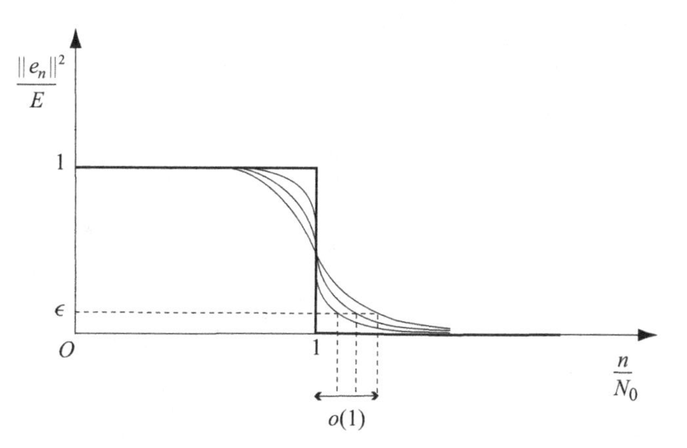

Fig. 1.6 Phase transition of the approximation error. The transition becomes sharper, and its width, viewed at the scale of N_0, shrinks to zero as $N_0 \to \infty$.

smallest approximation error are obtained by solving an eigenvalue problem that we study in Chapters 2 and 3.

1.2.3 Space–Time Fields

The electromagnetic field is in general a function of four scalar variables: three spatial and one temporal. It follows that in order to appreciate the total field's informational content in terms of degrees of freedom, we need to extend the treatment above to higher dimensions.

Let us first consider the canonical case of a two-dimensional domain of cylindrical symmetry, in which an electromagnetic field is radiated by current sources located inside a circular domain of radius r, and oriented perpendicular to the domain. The sources can also be induced by multiple scattering inside the domain. In any case, the radiated field away from the sources is completely determined by the field on the cut-set boundary surrounding the sources and through which it propagates – see Figure 1.7. On this boundary, we can refer to a scalar field $f(\phi,t)$ that is a function of only two scalar variables: one angular and one temporal. The corresponding four representations, linked by Fourier transforms, are depicted in Figure 1.8, where ω indicates the transformed coordinate of the time variable t and w indicates the wavenumber that is the transformed coordinate of the angular variable ϕ.

Letting Ω be the angular frequency bandwidth and W be the wavenumber bandwidth, we now wish to determine the total number of degrees of freedom of the space–time field $f(\phi,t)$. To visualize the phase transition, we fix the bandwidth Ω and the size of the angular observation interval $S = 2\pi$, and scale the time support where the signal is observed $T \to \infty$ and the wavenumber bandwidth $W \to \infty$. Using the results of the monodimensional case, we have that as $T \to \infty$ the number of time–frequency degrees of freedom is of the order of

$$N_0 = \frac{\Omega T}{\pi}. \tag{1.13}$$

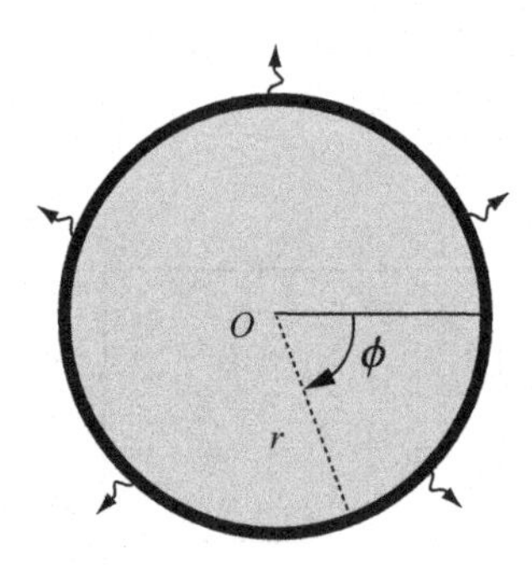

Fig. 1.7 Cylindrical propagation, cut-set boundary.

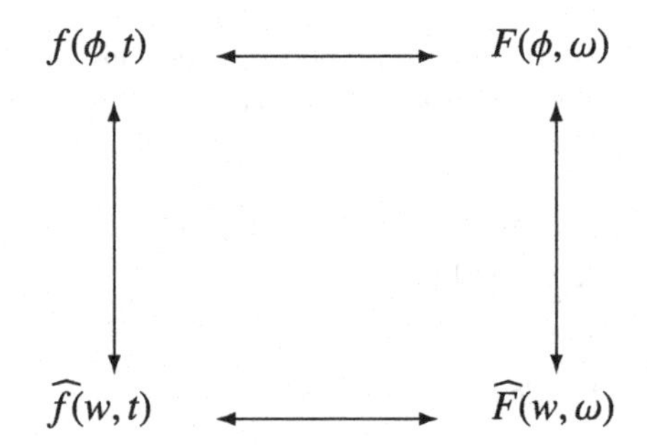

Fig. 1.8 Four field representations linked by Fourier transforms.

In a symmetric fashion, letting the wavenumber bandwidth $W \to \infty$, we have that the number of space–wavenumber degrees of freedom over an observation interval $S = 2\pi$ is of the order of

$$N_0 = \frac{W 2\pi}{\pi}. \tag{1.14}$$

The wavenumber bandwidth is related to the frequency of transmission. As we shall see in Chapter 8, every positive frequency component $\omega > 0$ of the signal has a wavenumber bandwidth that is, for any possible configuration of sources and scatterers inside the circular radiating domain, at most $W = \omega r / c$. It follows that the appropriate asymptotic regime $W \to \infty$ can be obtained by letting $r \to \infty$, so that by (1.14) the number of space–wavenumber degrees of freedom at angular frequency ω becomes of the order of

$$N_0(\omega) = \frac{2\pi r \omega}{c\pi} = \frac{2\pi r}{\lambda/2}, \tag{1.15}$$

where $\lambda = 2\pi c/\omega$ is the radiated wavelength, and c is the propagation speed of the field. The total number of degrees of freedom can now be obtained by integrating (1.15) over the bandwidth, and multiplying the result by T/π. It follows that as $r, T \to \infty$ the total number of degrees of freedom is of the order of

$$N_0 = \frac{2\pi r T}{c\pi^2} \int_0^{\Omega} \omega d\omega = \frac{2\pi r T \Omega^2}{2c\pi^2}. \tag{1.16}$$

Rearranging terms, we have

$$N_0 = \frac{\Omega T}{\pi} \frac{2\pi r \Omega}{2\pi c}. \tag{1.17}$$

Equation (1.17) shows that the number of degrees of freedom is the product of two factors, each viewed in an appropriate asymptotic regime: one accounting for the number of degrees of freedom in the time–frequency domain, $\Omega T/\pi$, as $T \to \infty$, and the other accounting for the number of degrees of freedom in the space–wavenumber domain, $2\pi r\Omega/(2\pi c)$, as $r \to \infty$. The second factor corresponds to the perimeter of the disc of radius r normalized by an interval of wavelengths $2\pi c/\Omega$. Its physical interpretation is that of a *spatial cut-set* imposing a limit on the amount of information, per time–frequency degree of freedom, that can flow from the interior of the domain to its exterior.

An analogous computation yields the number of degrees of freedom of the field radiated by arbitrary sources and scatterers in three-dimensional space. For spherical systems of radius r, we have the number of space–wavenumber degrees of freedom at frequency ω,

$$N_0(\omega) = \frac{4\pi r^2 \omega^2}{(c\pi)^2} = \frac{4\pi r^2}{(\lambda/2)^2}, \tag{1.18}$$

and the total number of degrees of freedom:

$$N_0 = \frac{4\pi r^2 T}{c^2 \pi^3} \int_0^{\Omega} \omega^2 d\omega = \frac{4\pi r^2 T \Omega^3}{3c^2 \pi^3}. \tag{1.19}$$

Once again, rearranging terms we have

$$N_0 = \frac{\Omega T}{\pi} \frac{4\pi r^2 \Omega^2}{3c^2 \pi^2}. \tag{1.20}$$

The number of degrees of freedom is the product of two factors, each viewed in an appropriate asymptotic regime: one accounting for the number of degrees of freedom in the time–frequency domain, $\Omega T/\pi$, as $T \to \infty$, and the other accounting for the number of degrees of freedom in the space–wavenumber domain, $4\pi r^2 \Omega^2/(3c^2\pi^2)$, as $r \to \infty$. The second factor corresponds to the spatial cut-set represented by the normalized surface area of the sphere of radius r through which the information must flow. This imposes a limit on the amount of information, per time–frequency degree of freedom, that can flow across the cut.

The results above lead to an interpretation of information as a conservative quantity that is radiated from a volume to the outside space. This interpretation is supported by a rigorous theory, which is presented in Chapters 8 and 9.

1.2.4 Super-resolution

We have argued that the number of degrees of freedom of square-integrable, bandlimited signals undergoes a phase transition. For signals of a single variable, this number quickly drops to zero in correspondence with a critical value N_0, given by the product of the size of the observation interval in the natural time and the size of the support set of

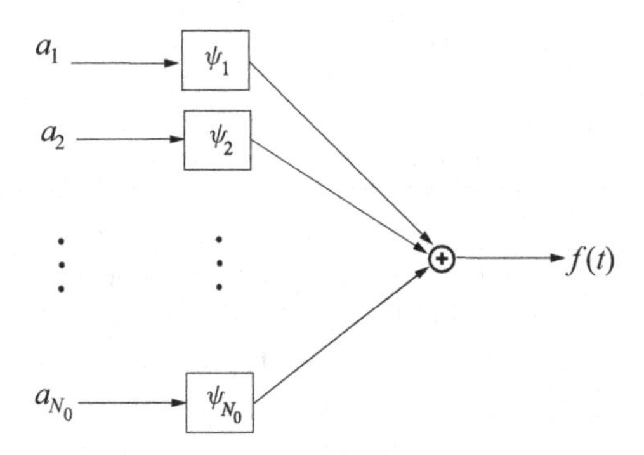

Fig. 1.9 Multiple channels resulting from a signal's decomposition.

the signal in the transformed frequency domain. The number of degrees of freedom of the signal is in principle infinite, but the phase transition dictates that no improvement of the accuracy of the observation can go substantially beyond a given characteristic scale. It follows that any waveform in the space can be identified, up to arbitrary accuracy, by essentially $N_0 = \Omega T/\pi$ real numbers. Modulating the signal's coefficients in excess of N_0 does not change the shape of the waveform significantly, so that these higher-order coefficients cannot be used to communicate any additional information. On the other hand, the most significant N_0 coefficients essentially determine the shape of the waveform, and we can view them as independent channels that can be used to multiplex different streams of information from a transmitter to a receiver. Any assignment of coefficients in the signal's expansion (1.2) compatible with the constraint (1.1) identifies one element of the space, and its correct selection at the receiver amounts to communicating N_0 real numbers – see Figure 1.9.

One could argue that by reducing the value of ϵ we can increase the number of degrees of freedom within the transition window. This amounts to having detectors capable of capturing the decaying tail of the approximation error, so that they are able to distinguish between signals that differ in their higher-order coefficients. These detectors can in principle achieve unbounded *super-resolution* – see Figure 1.10.

In imaging systems operating in the space–wavenumber domain, a number of technologies have been proposed to achieve super-resolution, including near-field microscopy and the use of meta-materials that facilitate beamforming. The steepness of the phase transition, however, imposes an exponential cost on all of these technologies. The same mathematics at the basis of basic quantum indeterminacy principles ensures that all efforts aimed at breaking the informational limit on the number of degrees of freedom are eventually doomed when viewed at the scale of N_0, when the transition in the approximation error becomes a step function. Chapter 12 further discusses super-resolution limits. The conclusion is that the rules of spectral concentration dictate an *asymptotic limit* independent of the technology employed:

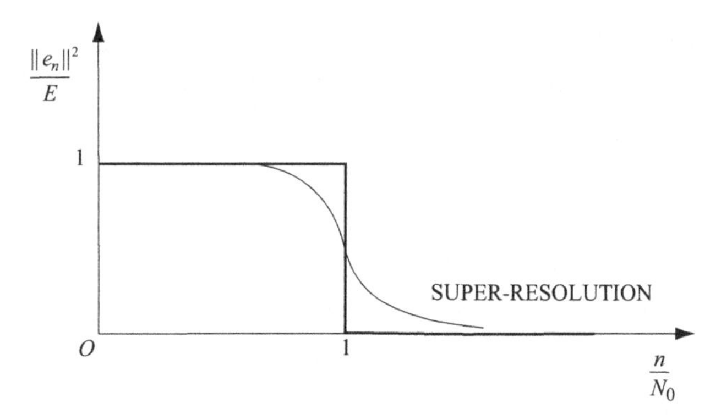

Fig. 1.10 Super-resolving detectors aim to capture the decaying tail of the number of degrees of freedom of the signal.

> It is impossible to resolve a number of dimensions larger than is naturally imposed by propagation constraints and expressed *asymptotically* by the number of degrees of freedom of the field.

1.3 Deterministic Information Measures

Every electromagnetic waveform is essentially identified by N_0 real numbers. This limitation is due to the laws of propagation that impose bandlimitation and hence a certain amount of smoothness of the corresponding waveform. The mathematics at the basis of this result are the rules of spectral concentration. The result, however, does not pose a limit on the amount of information carried by the signal. The N_0 real numbers identifying the waveform can be specified up to arbitrary precision, and this results in an infinite number of possible waveforms that can be used for communication. Even if the rules of spectral concentration pose a limit on the effective dimensionality of the signals' space in terms of degrees of freedom, the amount of information associated with the selection of one waveform in the space is still unbounded if each coordinate can be specified with arbitrary accuracy. Some kind of indeterminacy must be introduced to limit the resolution at which the signal can be observed. Such indeterminacy follows from the intrinsic uncertainty associated with the observation process.

Waveforms that are distinct as mathematical objects cannot be resolved as distinct physical quantities at a scale smaller than the noise affecting the measurement process. For this reason, we may consider two waveforms *distinguishable* only if, for a given $\epsilon > 0$,

$$\|f_1 - f_2\| > \epsilon. \tag{1.21}$$

This point of view yields a discretization of the space consistent with the quantized nature of the observable world, and allows an information-theoretic description of the space of signals in terms of the universal unit of information: the *bit*.

1.3.1 Kolmogorov Entropy

Following up on his 1930's investigations on the number of degrees of freedom of functions, and while working under the influence of Shannon's 1948 work, in 1956 Andrey Nikolaevich Kolmogorov introduced the notion of ϵ-*entropy*. This is defined as:

> The minimum number H_ϵ of bits that can represent any signal in the space within accuracy ϵ.

Viewing bandlimited signals observed over a large time interval and subject to the constraint (1.1) as points in a space of N_0 dimensions and of radius

$$\left(\int_{-\infty}^{\infty} f^2(t)dt\right)^{1/2} = \left(\sum_{n=1}^{N_0} a_n^2\right)^{1/2} \leq \sqrt{E}, \tag{1.22}$$

the ϵ-entropy geometrically corresponds to the logarithm base two of the *covering number* $Q_\epsilon(E)$, that is, the minimum number of balls of radius ϵ such that their union contains the whole space, and we have

$$H_\epsilon = \log Q_\epsilon(E) \text{ bits.} \tag{1.23}$$

In this way, any point in the space can be identified, in the sense of (1.21), with the center of a corresponding ϵ-ball covering it – see Figure 1.11 – and any signal can be represented by essentially H_ϵ bits.

The ϵ-entropy is related to the number of degrees of freedom. Given the interpretation of the number of degrees of freedom as the effective dimensionality of an infinite-dimensional functional space, and since we may expect the covering number of this space to be roughly of the order of the volume of the space divided by the volume of the ϵ-ball, a reasonable guess may be that the entropy is roughly

$$H_\epsilon = \log(\sqrt{E}/\epsilon)^{N_0} = N_0 \log(\sqrt{E}/\epsilon). \tag{1.24}$$

As we shall see in Chapter 12, this intuition is essentially correct; the entropy grows linearly with the number of degrees of freedom, and logarithmically with the ratio of the maximum energy of the signal to the energy of the noise. This latter term, called the *signal-to-noise ratio*, can be geometrically viewed as imposing a quantization of the space at level ϵ due to the noise inherent in the measurement process.

1.3.2 Kolmogorov Capacity

Another information-theoretic notion introduced by Kolmogorov is the ϵ-*capacity*. This is defined as:

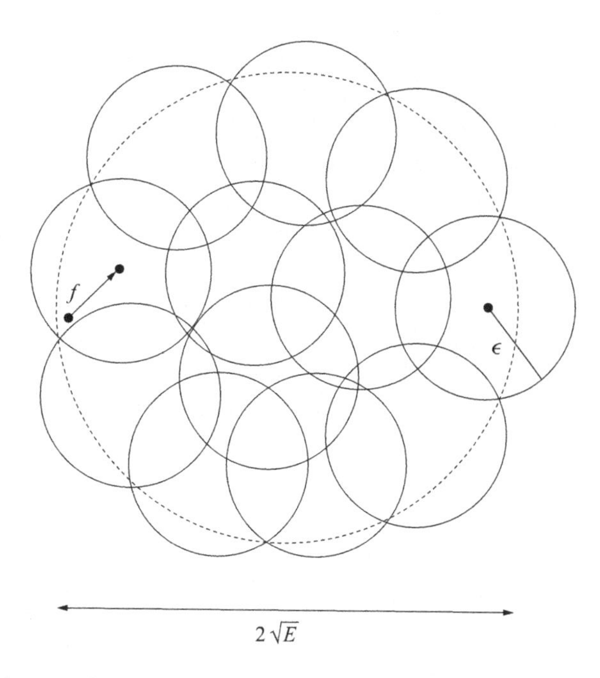

Fig. 1.11 Every signal in the space is within ϵ of the center of a covering ball.

> The maximum number C_ϵ of bits that can be communicated by selecting one signal at the transmitter, and observing it at the receiver with perturbation at most $\epsilon/2$.

Geometrically, the ϵ-capacity corresponds to the logarithm base two of the *packing number* $M_\epsilon(E)$, that is, the maximum number of disjoint balls of radius $\epsilon/2$ with their centers situated inside the signals' space – see Figure 1.12. Since any two signals in the packing are at least ϵ apart, they cannot be confused when each of them is observed with perturbation at most $\epsilon/2$, and we have

$$C_\epsilon = \log M_\epsilon(E) \text{ bits.} \tag{1.25}$$

An upper bound on the ϵ-capacity can be computed in terms of ϵ-entropy and using (1.24), yielding

$$C_\epsilon \leq H_{\epsilon/2} = N_0 \log(2\sqrt{E}/\epsilon). \tag{1.26}$$

The inequality in (1.26) follows from the observation that if, by contradiction, $M_\epsilon > Q_{\epsilon/2}$, then there must be one ϵ-ball that covers at least two points that are more than ϵ apart, which is impossible – see Figure 1.13.

A lower bound on the ϵ-capacity can be obtained by constructing a *codebook* composed of a subset of signals in the space, each corresponding to a given message. A transmitter can select any one of these signals, which is observed at the receiver with perturbation at most $\epsilon/2$. By choosing signals in the codebook to be at distance at least ϵ from each other, the receiver can decode the message without error. Counting the signals in the codebook, and taking the logarithm, gives a lower bound on the capacity. A simple codebook follows from the lattice packing depicted in Figure 1.14. This codebook is, however, far from being optimal. In higher dimensions, the volume of the hypersphere tends to concentrate on its boundary, a phenomenon called *sphere hardening*. The packing in the inscribed hypercube then captures only a vanishing fraction of the volume available in the hypersphere, leading to an extremely inefficient packing of the space by balls of fixed radii. A random codebook argument, although not constructive, leads to a tighter capacity bound than the one obtained with the simple packing depicted in Figure 1.14, and shows that the capacity can scale linearly with the number of degrees of freedom, and logarithmically with the signal-to-noise ratio. These results are described in Chapter 12.

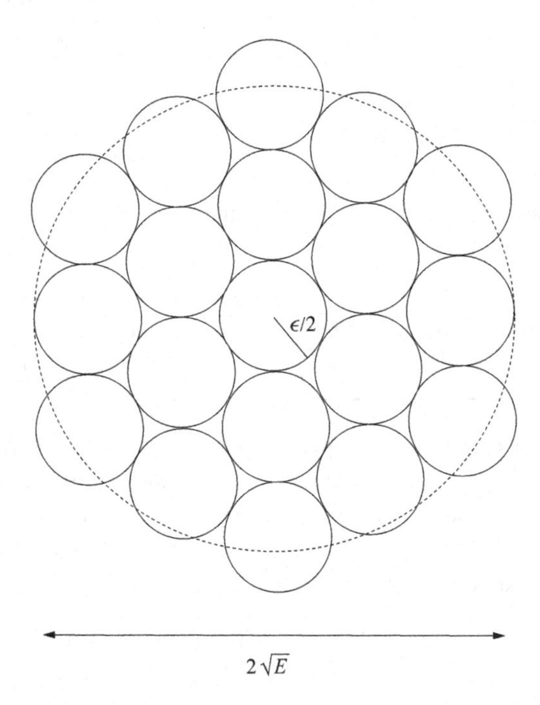

Fig. 1.12 Any two signals identified by the centers of the packing balls are at a distance of at least ϵ from each other.

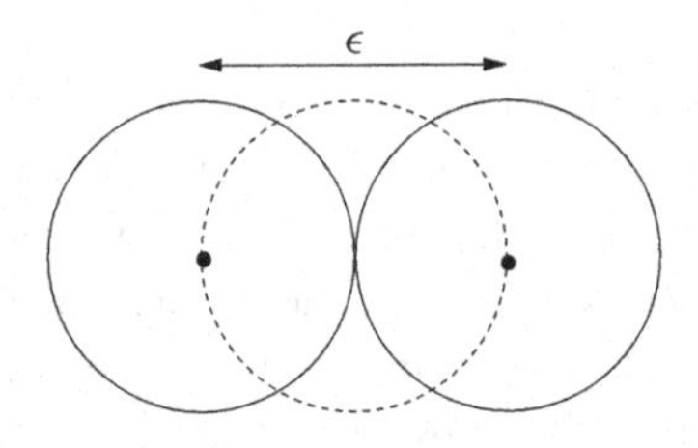

Fig. 1.13 Illustration of the bound $C_\epsilon \leq H_{\epsilon/2}$.

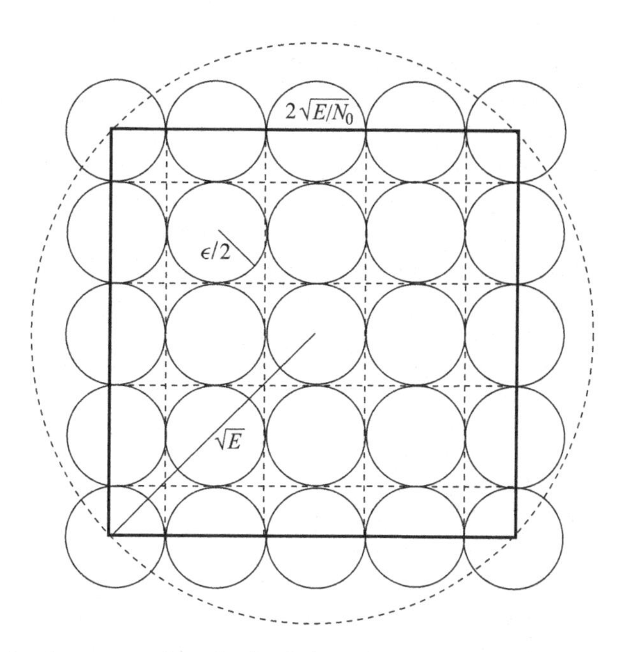

Fig. 1.14 Lattice packing for the computation of a simple lower bound on the ϵ-capacity.

1.3.3 Quantized Unit of Information

We now summarize. Introducing a resolution limit in the signals' space leads to two mathematical notions that quantify information in a discrete way, in terms of the universal unit of information: the bit. The entropy represents the smallest number of bits needed to identify any signal in the space, and is geometrically represented in terms of sphere covering at resolution ϵ, while the capacity represents the largest number of bits carried by any signal in the space, and is geometrically represented in terms of sphere packing at resolution $\epsilon/2$. Both entropy and capacity are linearly related to the number of degrees of freedom and logarithmically related to the ratio of the maximum norm of the signal to the maximum norm of the perturbation with which the signal is observed.

The mathematical definitions of entropy and capacity are used to quantify information in terms of degrees of freedom, corresponding to the "massiveness" or effective dimensionality of the space, and in terms of resolution uncertainty, corresponding to the signal-to-noise ratio. These quantities are determined by the physics. Wave propagation ensures that the observed field is a filtered version of the transmitted one through a propagation operator that projects it onto the space of effectively bandlimited signals of essentially N_0 dimensions. The signal-to-noise ratio is determined by the signal's energy and by the intrinsic uncertainty associated to the measurement process that is imposed by the quantized nature of the world. Even if fields are continuous quantities, they can only be observed in a quantized fashion, dictated by the measurement uncertainty.

Kolmogorov's ϵ-entropy and capacity are also closely related to the probabilistic notions of entropy and capacity used in information theory and statistical mechanics. In this context, Shannon's entropy represents the smallest number of bits needed to identify any typical realization of a continuous stochastic process, quantized at level ϵ. Shannon's capacity represents the largest number of bits that can be transported by any signal in the space, when this is subject to a stochastic perturbation of standard deviation ϵ. In this setting, a geometric picture completely analogous to the one described in the deterministic case arises, where the value of ϵ is replaced by the statistical dispersion of the noise associated with the measurement process, which probabilistically concentrates around its typical value.

1.4 Probabilistic Information Measures

The functional theory of information, based on the notions of ϵ-entropy and ϵ-capacity and developed by the Russian school of Kolmogorov and his disciples, was strongly influenced by the probabilistic theory of information developed by Shannon and his disciples in the West. The probabilistic point of view is based on the observation that highly unpredictable signals maximize the "surprise" of observing a given outcome. This surprise is identified with the amount of information that is transported by the signal. It follows that a highly unpredictable signal is highly informative, in the sense that the message it represents maximally alters the state of the receiver, and thus it must be described by a large number of bits.

To make these considerations precise, we model the coefficients in the signal's representation as random variables described by a probability distribution, and introduce the concept of statistical dispersion, or entropy, associated with this distribution. Informationally rich signals, having long bit representations, correspond to distributions with large entropy. For signals having the maximum entropy distribution, compatible with the signals' energy constraint, the typical values of the coefficients are maximally dispersed in a space of N_0 dimensions, and they are all roughly equally probable. This is the asymptotic equipartition property, expressed rigorously by the Shannon–McMillan–Breiman theorem, and ensures that the surprise associated with a typical realization is always maximized.

When signals are subject to noisy observations, the Shannon capacity is defined in terms of the largest number of bits that can be transported from transmitter to receiver by any one signal in the space, with negligible probability of error. This can be described in terms of another probabilistic quantity, the *mutual information*, which depends only on the joint distribution of the observed and transmitted signals. The supremum of the mutual information, over all possible distributions for the transmitted signals and compatible with the energy constraint, is the Shannon capacity. This statement is Shannon's coding theorem, which relates the operational definition of capacity to the maximization of a given information functional.

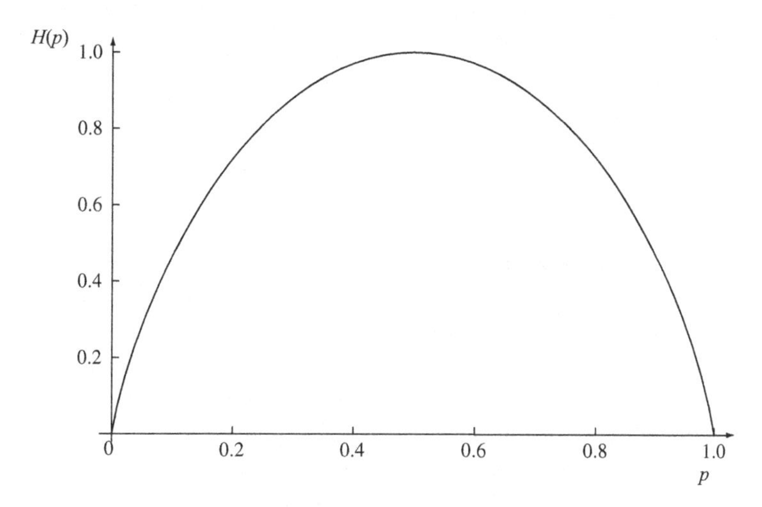

Fig. 1.15 Entropy of a Bernoulli random variable of parameter p.

1.4.1 Statistical Entropy

The probabilistic quantity used to describe the extent of the random fluctuations of the signal's coefficients is the *statistical entropy*. This arises in thermodynamics, statistical mechanics, and information theory with slightly different definitions that are all related to each other.

The Shannon (1948) entropy of a discrete random variable X taking values in a set $\mathscr{X}$ with probability mass function $p_{\mathsf{X}}(x)$ is

$$\begin{aligned} H &= -\sum_{x\in\mathscr{X}} p(x)\log p(x) \\ &= -\mathbb{E}_{\mathsf{X}} \log p(\mathsf{X}), \end{aligned} \tag{1.27}$$

where we define, by continuity, $0\log 0 = 0$. The logarithm is taken base two, and this only amounts to choosing the unit of measure to be bits. A change of base simply corresponds to multiplication by a constant factor and thus to a change of units, as $\log_b p = \log_b a \log_a p$.

The Shannon entropy represents a measure of the uncertainty of the random variable. Viewing this uncertainty as the "surprise" of observing a given outcome, it also measures the amount of information transferred in the observation process. When the random variable takes only one value with probability one, there is no surprise in the realization and the entropy is zero. When the outcome is unknown, the probability mass has a wider distribution and the entropy becomes positive. When the distribution becomes uniform among all possible outcomes, the entropy is maximized and it corresponds to the logarithm of the number of possible outcomes. A plot of the entropy for the simplest case of a Bernoulli random variable that takes values with probability p and $1-p$ is depicted in Figure 1.15.

When our random variable can assume any one of N values, it can be viewed as a statistical mechanical system, where each state $n \in \{1,2,\ldots,N\}$ occurs with probability p_n. The Gibbs (1902) entropy of such a system is

$$H_{\mathrm{G}} = -k_{\mathrm{B}} \sum_{n=1}^{N} p_n \log p_n, \tag{1.28}$$

where k_{B} is Boltzmann's constant, which has the units of energy divided by temperature. This normalization constant makes the definition consistent with the phenomenological notion of entropy developed by Clausius (1850–65) in a series of papers on the mechanical theory of heat. According to Clausius, a variation of thermodynamic entropy is defined by a corresponding variation of energy, in terms of heat absorbed by the system, at a given absolute temperature T_{K}, namely

$$dH_{\mathrm{C}} = \frac{dE}{T_{\mathrm{K}}}. \tag{1.29}$$

To see the correspondence between Gibbs' and Clausius' definitions, consider an isolated system in thermal equilibrium, where all states are equally probable. In this case, letting $p_n = 1/N$, the Gibbs entropy (1.28) achieves its maximum value and reduces to the Boltzmann (1896–8) form[2]

$$H_{\mathrm{B}} = k_{\mathrm{B}} \log N. \tag{1.30}$$

Assuming that each state of the system is to be defined by an ensemble of M elementary modes that take binary values, we have $N = 2^M$ and, by (1.30),

$$H_{\mathrm{B}} = k_{\mathrm{B}} M. \tag{1.31}$$

Assuming that at equilibrium each elementary mode, due to thermal agitations, has average energy proportional to the temperature and equal to $k_{\mathrm{B}} T_{\mathrm{K}}$, then the total average energy of the system is

$$E = k_{\mathrm{B}} T_{\mathrm{K}} M. \tag{1.32}$$

Differentiating both sides, dividing by T_{K}, and using (1.31), it follows that the phenomenological definition of the thermodynamic entropy of an isolated system in thermal equilibrium and subject to an average energy constraint coincides with the statistical mechanical definition, and corresponds to the configuration in which all states are equally probable, so that the Gibbs entropy is maximized and attains Boltzmann's form. The conclusion is that at equilibrium statistical and thermodynamic entropies coincide, and Shannon's definition is simply a dimensionless version of the thermodynamic entropy.

We further examine the connection between statistical and thermodynamic entropies in Chapter 11.

[2] Both the Gibbs and Boltzmann entropies are usually expressed in natural units, or nats, taking the logarithm base e. In this book we fix the units of information to be bits and take the logarithm base two, to favor the comparison with the Shannon entropy. The equivalent expression in nats can easily be obtained by dividing the value in bits by $\log_2 e$.

1.4.2 Differential Entropy

To appreciate the relationship between statistical entropy and the number of degrees of freedom of an electromagnetic waveform, we introduce the *differential entropy* of a continuous random variable X of probability density function $g_{\mathsf{X}}(x)$. This is also due to Shannon (1948), and is given by

$$\begin{aligned} h_{\mathsf{X}}(g) &= -\int_{\mathscr{X}} g(x) \log g(x) dx \\ &= -\mathbb{E}_{\mathsf{X}} \log g(\mathsf{X}), \end{aligned} \tag{1.33}$$

where $\mathscr{X}$ is the support set of the random variable. The joint differential entropy of two random variables is

$$\begin{aligned} h_{\mathsf{X},\mathsf{Y}}(g) &= -\int_{\mathscr{X}} \int_{\mathscr{Y}} g(x,y) \log g(x,y) dx dy \\ &= -\mathbb{E}_{\mathsf{X},\mathsf{Y}} \log g(\mathsf{X},\mathsf{Y}), \end{aligned} \tag{1.34}$$

and when the two variables are independent, we have

$$h_{\mathsf{X},\mathsf{Y}} = h_{\mathsf{X}} + h_{\mathsf{Y}}. \tag{1.35}$$

The definition naturally extends to random vectors. If $\mathsf{X}_1, \mathsf{X}_2, \ldots, \mathsf{X}_{N_0}$ are mutually independent, we have

$$h_{(\mathsf{X}_1,\mathsf{X}_2,\ldots,\mathsf{X}_{N_0})} = \sum_{n=1}^{N_0} h_{\mathsf{X}_n}, \tag{1.36}$$

and if they also have the same distribution, we have

$$h_{(\mathsf{X}_1,\mathsf{X}_2,\ldots,\mathsf{X}_{N_0})} = N_0 h, \tag{1.37}$$

where $h = h_{\mathsf{X}_1}(f)$ is the differential entropy of a single random variable.

Although each random variable in the set $\{\mathsf{X}_n\}$ can take values over an arbitrary, possibly infinite, support, for any finite value of h the rules of probabilistic concentration ensure that the typical realizations occur over a much smaller set. More precisely, the asymptotic equipartition property ensures that as $N_0 \to \infty$ the volume of the support set of the random variables is only roughly $2^{N_0 h}$ and all realizations inside this volume are roughly equally probable. This situation can be visualized by having realizations inside an equivalent hypercube in a space of N_0 dimensions, with each side length of range 2^h, and it provides an interpretation of the differential entropy as the logarithm of the equivalent side length of the smallest volume that contains most of the probability – see Figure 1.16.

The precise statement is given by defining the typical support set $V_\epsilon(N_0)$ composed of sequences $(x_1, x_2, \ldots, x_{N_0}) \in \mathbb{R}^{N_0}$ such that, for all $\epsilon > 0$,

$$2^{-N_0(h+\epsilon)} \leq g(x_1, x_2, \ldots, x_{N_0}) \leq 2^{-N_0(h-\epsilon)}. \tag{1.38}$$

The aggregate probability of the typical set for N_0 sufficiently large is

$$\int_{V_\epsilon(N_0)} g(x_1, x_2, \ldots, x_{N_0}) dx_1 dx_2 \cdots dx_{N_0} > 1 - \epsilon, \tag{1.39}$$

and its volume is

$$(1-\epsilon)2^{N_0(h-\epsilon)} \le \int_{V_\epsilon(N_0)} dx_1 dx_2 \cdots dx_{N_0} \le 2^{N_0(h+\epsilon)}. \tag{1.40}$$

1.4.3 Typical Waveforms

By identifying the sequence of random variables $\{\mathsf{X}_n\}$ with the coefficients representing a given waveform, we can view each possible waveform as a random point in a space of N_0 dimensions. It follows that a large typical volume corresponds to signals that are widely dispersed, while a small typical volume means that most signals are concentrated in a small portion of the space and asymptotic equipartition can be expressed by the statement:

> Almost all observed signals are almost equally probable.

To obtain an information-theoretic description of the typical waveforms, we proceed in an analogous way to the one followed in the deterministic setting: to bound the size of the typical volume we introduce an energy constraint that restricts the possible signal's configurations. Then, in order to obtain a bit representation, we introduce a resolution limit at which each waveform can be observed.

To realize this program, we consider random waveforms of the form

$$f(t) = \sum_{n=1}^{N_0} a_n \psi_n(t), \tag{1.41}$$

where $\{a_n\}$ are realizations of an independent and identically distributed (i.i.d.) real-valued random process $\{\mathsf{A}_n\}$, and ψ_n are deterministic, orthonormal, real basis functions. Each random variable in the process is distributed as A and satisfies the expected squared norm per degree of freedom constraint,

$$\mathbb{E}(\mathsf{A}^2) \le P. \tag{1.42}$$

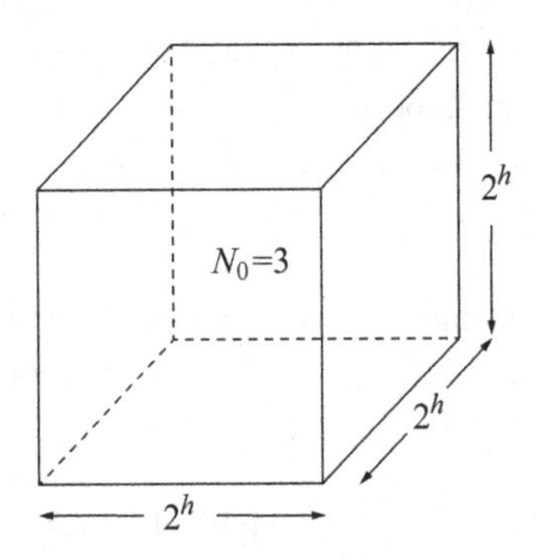

Fig. 1.16 The volume of the typical set coincides with the volume of an equivalent hypercube of side length 2^h.

This constraint should be compared with (1.22). While in the deterministic setting we imposed an energy constraint on the whole signal, here we impose an average constraint separately on each dimension of the space.

Maximizing the differential entropy subject to (1.42), we obtain the upper bound

$$h_{\mathsf{A}} \leq \frac{1}{2}\log(2\pi eP). \tag{1.43}$$

The maximizing density that achieves the bound in (1.43) is a zero-mean Gaussian of variance P – see Problem 1.15. By asymptotic equipartition, the volume of the typical set is then roughly at most

$$2^{N_0 h_{\mathsf{A}}} \leq (2\pi eP)^{N_0/2}. \tag{1.44}$$

It follows that whatever distribution we choose for the signal's coefficients, compatible with the constraint on the expected squared norm per degree of freedom (1.42), the signals' space is always confined within an effective volume at most of the order of the square root of the imposed constraint raised to the number of degrees of freedom. The largest effective volume is achieved by a Gaussian probability density function, which provides the largest possible statistical dispersion of the observed waveform.

1.4.4 Quantized Typical Waveforms

The amount of information of our continuum state space of signals depends not only on its statistical dispersion, but also on the level of quantization at which each signal is observed. If we quantize each dimension of the signals' space at level ϵ, then the number of quantized typical signals is roughly

$$N = \frac{2^{N_0 h_{\mathsf{A}}}}{\epsilon^{N_0}} \leq \frac{(2\pi eP)^{N_0/2}}{\epsilon^{N_0}} = \left(\frac{\sqrt{2\pi eP}}{\epsilon}\right)^{N_0}, \tag{1.45}$$

and since these are roughly equiprobable, the corresponding statistical entropy of the equipartitioned system of quantized signals is

$$H = \log N \leq N_0 \log\left(\frac{\sqrt{2\pi eP}}{\epsilon}\right), \tag{1.46}$$

where the equality is achieved by a zero-mean Gaussian of variance P. The statistical entropy in (1.46) also corresponds to the number of bits needed on average to represent any quantized signal in the space. This follows from the observation that there are N quantized typical signals, and each can be identified by a sequence of $H = \log N$ bits.

The bound (1.46) should be compared with (1.24). In the deterministic case the Kolmogorov ϵ-entropy of the signals' space grows at most linearly with the number of degrees of freedom and at most logarithmically with the signal-to-noise ratio $\sqrt{E}/\epsilon$. In the stochastic case, the Shannon entropy of the space of ϵ-quantized signals grows at most linearly with the number of degrees of freedom and at most logarithmically with the expected signal-to-noise per degree of freedom ratio $\sqrt{P}/\epsilon$. While in the deterministic model we bounded the energy of the signal and introduced quantization

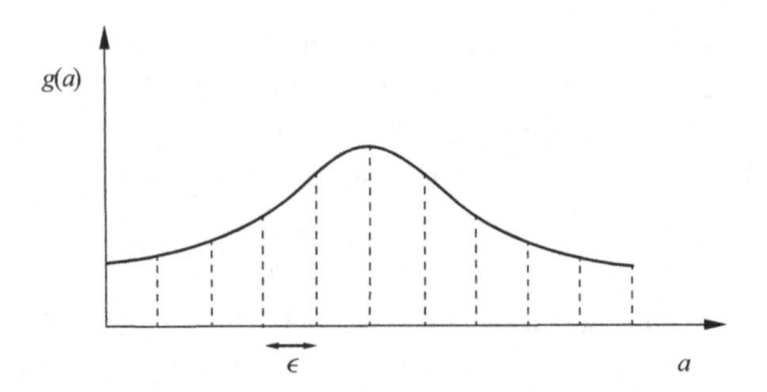

Fig. 1.17 Quantization of a continuous random variable of probability density function $g(a)$.

in terms of Euclidean distances, in the stochastic model we introduced an average constraint and a quantization on each dimension of the space.

The result (1.46) can be derived by writing the statistical entropy of the space of ϵ-quantized signals in terms of the entropy of a single quantized coefficient. Consider a coefficient A distributed according to a continuous density $g(a)$. We perform quantization as indicated in Figure 1.17. By continuity, for all k we let $a(k)$ be a value such that

$$g(a(k))\epsilon = \int_{k\epsilon}^{(k+1)\epsilon} g(a)da. \tag{1.47}$$

The quantized random variable A^ϵ is defined as

$$\mathsf{A}^\epsilon = a(k) \;\text{ if }\; k\epsilon \leq \mathsf{A} < (k+1)\epsilon. \tag{1.48}$$

We then have

$$\lim_{\epsilon \to 0}(H_{\mathsf{A}^\epsilon} + \log\epsilon) = h_{\mathsf{A}}. \tag{1.49}$$

This result can be derived by performing the following computation:

$$\begin{aligned} H_{\mathsf{A}^\epsilon} &= -\sum_{k=-\infty}^{\infty} g(a(k))\epsilon\log[g(a(k))\epsilon] \\ &= -\sum_{k=-\infty}^{\infty} g(a(k))\epsilon\log[g(a(k))] - \sum_{k=-\infty}^{\infty} g(a(k))\epsilon\log\epsilon \\ &= -\sum_{k=-\infty}^{\infty} g(a(k))\epsilon\log[g(a(k))] - \log\epsilon\sum_{k=-\infty}^{\infty}\int_{k\epsilon}^{(k+1)\epsilon} g(a)da \\ &= -\sum_{k=-\infty}^{\infty} g(a(k))\epsilon\log[g(a(k))] - \log\epsilon. \end{aligned} \tag{1.50}$$

As $\epsilon \to 0$, the first term approaches the differential entropy of A by definition of Riemann integrability, establishing (1.49).

It now follows that for small values of ϵ, by drawing N_0 independent random coefficients subject to the constraint (1.42), the entropy of the quantized sequence is approximately

$$\begin{aligned} H &= N_0 H_{\mathsf{A}^\epsilon} \\ &= N_0(h_{\mathsf{A}} - \log \epsilon) \\ &\leq N_0 \log\left(\frac{\sqrt{2\pi eP}}{\epsilon}\right), \end{aligned} \tag{1.51}$$

establishing (1.46).

1.4.5 Mutual Information

Shannon's entropy is a probabilistic quantity that measures information in terms of the number of bits needed to identify any typical realization of a random waveform, quantized at level ϵ on each dimension of the space, and subject to an expected norm per degree of freedom constraint. It is the stochastic analog of Kolmogorov's entropy, which counts the number of bits needed to identify any waveform in the space, quantized at level ϵ and subject to an energy constraint on the whole signal.

Another information-theoretic quantity, analogous to Kolmogorov's capacity, is used to quantify in a stochastic setting the largest amount of information that can be transported by any waveform in the space. In this case, the stochastic model assumes that observations are random variables governed by a probability distribution. It follows that the signals' coefficients that are selected randomly at the transmitter and observed at the receiver have a joint probability density $g(x,y)$. Given any transmitted random coefficient X, the observed coefficient Y has a conditional distribution $g(y|x)$, and we say that X is observed through the *probabilistic channel* depicted in Figure 1.18.

The *conditional differential entropy* represents the remaining uncertainty about the outcome of Y given the outcome of X,

$$\begin{aligned} h_{\mathsf{Y}|\mathsf{X}} &= -\int_{\mathcal{X}}\int_{\mathcal{Y}} g(x,y)\log g(y|x)dxdy \\ &= -\mathbb{E}_{\mathsf{X},\mathsf{Y}} \log g(\mathsf{Y}|\mathsf{X}). \end{aligned} \tag{1.52}$$

A simple calculation yields the following chain rule:

$$h_{\mathsf{X},\mathsf{Y}} = h_{\mathsf{X}} + h_{\mathsf{Y}|\mathsf{X}}, \tag{1.53}$$

and when the two variables are independent, we have

$$h_{\mathsf{Y}|\mathsf{X}} = h_{\mathsf{Y}}, \tag{1.54}$$

so that (1.53) reduces to (1.35).

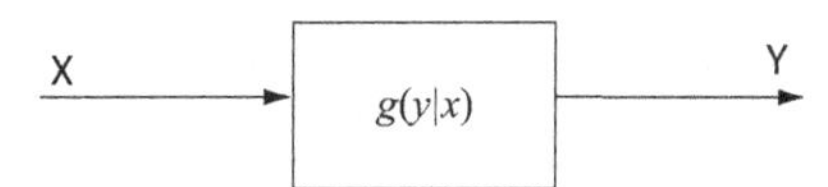

Fig. 1.18 Probabilistic channel.

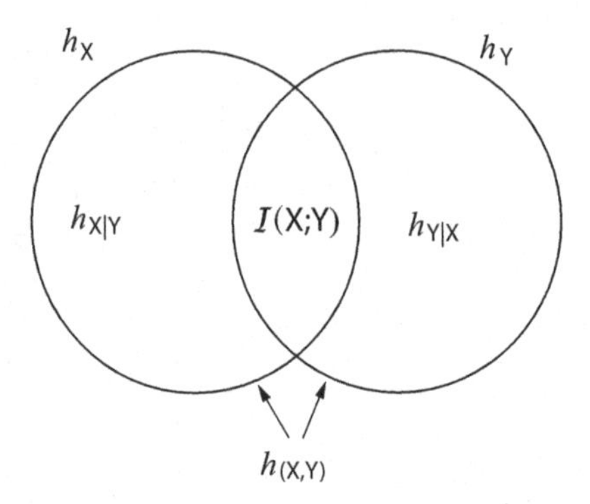

Fig. 1.19 Diagram describing the mutual information.

The *mutual information* between random variables X and Y is

$$\mathcal{I}(X;Y) = \int_{\mathcal{X}} \int_{\mathcal{Y}} g(x,y) \log \frac{g(x,y)}{g(x)g(y)} dxdy. \tag{1.55}$$

This can be interpreted as the reduction in the uncertainty of Y due the knowledge of the outcome of X, or vice versa, and can also be written as

$$\begin{aligned}\mathcal{I}(X;Y) &= h_Y - h_{Y|X} \\ &= h_X - h_{X|Y} \\ &= h_X + h_Y - h_{X,Y} \\ &= h_{X,Y} - h_{X|Y} - h_{Y|X}.\end{aligned} \tag{1.56}$$

Figure 1.19 gives a graphical interpretation of these results.

Using (1.56) and (1.49), we have that

$$\begin{aligned}\mathcal{I}(X;Y) &= h_Y - h_{Y|X} \\ &= \lim_{\epsilon\to 0}(H_{Y_\epsilon} - \log\epsilon - H_{Y_\epsilon|X_\epsilon} + \log\epsilon) \\ &= \lim_{\epsilon\to 0}(H_{Y_\epsilon} - H_{Y_\epsilon|X_\epsilon}),\end{aligned} \tag{1.57}$$

showing that the mutual information between two random variables is the limit of the mutual information in the discrete setting between their quantized versions.

1.4.6 Shannon Capacity

We now consider the supremum of the mutual information over all possible distributions for the channel's input, namely

$$C = \sup_{g_X} \mathcal{I}(X;Y), \tag{1.58}$$

where g_X varies over the set of admissible distributions for the random variable X. In the case of the stochastic model of signals considered so far, these would be all distributions

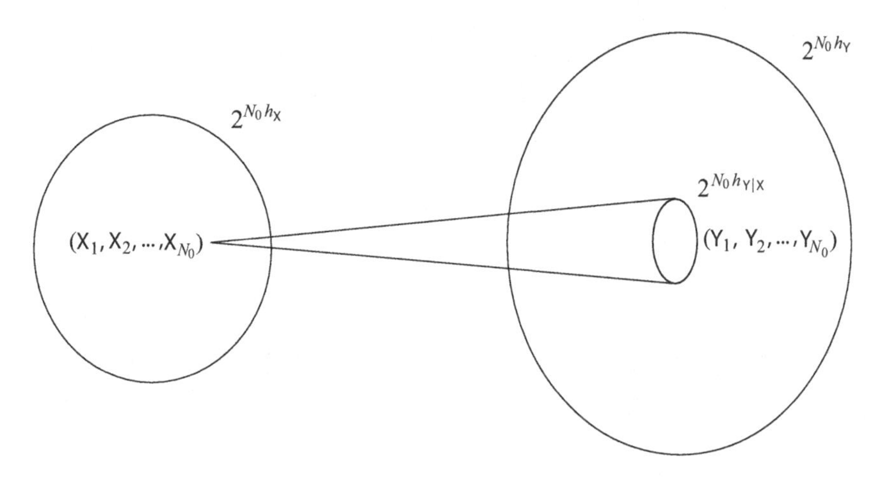

Fig. 1.20 Uncertainty spread due to noise.

satisfying the average constraint

$$\mathbb{E}(X^2) \leq P. \tag{1.59}$$

The quantity in (1.58) can be given an operational interpretation in terms of the largest number of bits per degree of freedom that can be transported by N_0 usages of the channel with negligible probability of error, namely the capacity of the channel. To view such a correspondence, consider a signal whose coefficients are drawn independently and satisfy (1.59), and use it to represent a given message that a transmitter wishes to communicate to the receiver. This signal is selected at the transmitter and observed at the receiver through a noisy channel. Due the noise, given the knowledge of the N_0 coefficients there may be a residual uncertainty about the observed signal. It follows that the coefficients of the received signal can be modeled as a random vector $(Y_1, Y_2, \dots, Y_{N_0})$ of conditional probability $g(y_1, y_2, \dots, y_{N_0} | x_1, x_2, \dots, x_{N_0})$. Assuming the noise acts independently on all signals' coefficients, using X and Y as a shorthand notation for X_1 and Y_1, we also have

$$h_{(Y_1,Y_2,\dots,Y_{N_0} | X_1,X_2,\dots,X_{N_0})} = N_0 h_{Y|X}. \tag{1.60}$$

The volume of the state space of the typical outputs given the input signal is $2^{N_0 h_{Y|X}}$, and all signals within this volume are roughly equally probable. Thus, each of them is an equally good candidate for the received signal, given the transmitted one. Figure 1.20 gives a visual representation of this situation. Since the total volume of the typical outputs is $2^{N_0 h_Y}$, and each signal carries an uncertainty volume of $2^{N_0 h_{Y|X}}$, it follows that we can distinguish without confusion among at most a number of signals

$$N \leq 2^{N_0 (h_Y - h_{Y|X})}. \tag{1.61}$$

The correct selection at the receiver of one among these signals corresponds to the act of communicating information, and the amount of information communicated is at most $\log N$ bits, corresponding to the number of binary digits that identify the signal.

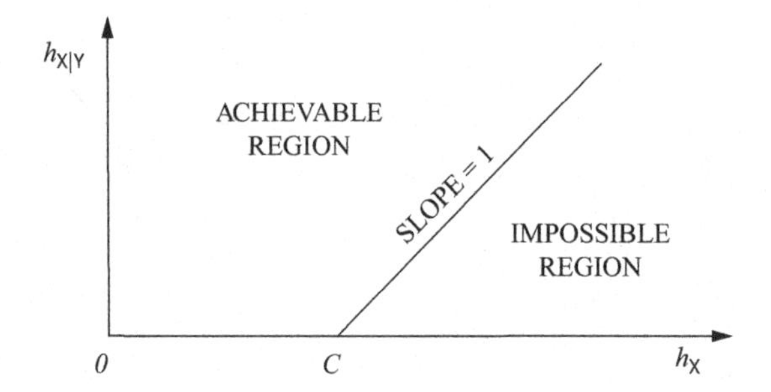

Fig. 1.21 Residual uncertainty as a function of the entropy of the input.

By (1.61), and using the definition of mutual information (1.56), an upper bound on the number of bits per degree of freedom transported by the signal is

$$\begin{aligned}\frac{\log N}{N_0} &\leq h_{\mathsf{Y}} - h_{\mathsf{Y}|\mathsf{X}} \\ &= h_{\mathsf{X}} - h_{\mathsf{X}|\mathsf{Y}} \\ &= \mathcal{I}(\mathsf{X};\mathsf{Y}) \\ &\leq \sup_{g_{\mathsf{X}}} \mathcal{I}(\mathsf{X};\mathsf{Y}) \text{ bits per degree of freedom.}\end{aligned} \tag{1.62}$$

It follows that the largest number of bits that can be reliably transported by the waveform, namely the capacity of the channel, is upper bounded by the quantity in (1.58). The term "reliable" here is intended in the sense that the probability of identifying the wrong signal tends to zero as $N_0 \to \infty$, and probabilistic concentration of signals around their typical values occurs. A more detailed argument provides a matching lower bound, and shows that the rate in (1.58) is achievable, with vanishing probability of error. As in the deterministic case, this requires the construction of a codebook composed of a subset of signals in the space, each corresponding to a given message. A transmitter can select any one of these signals, whose noisy version is observed at the receiver through the probabilistic channel. By an appropriate choice of the codebook and of the decoding function applied to the received signal, the receiver can then reliably identify the correct message.

The bound in (1.62) can also be written as

$$h_{\mathsf{X}|\mathsf{Y}} \geq h_{\mathsf{X}} - C, \tag{1.63}$$

providing a lower bound on the residual uncertainty of the variable X given the observation of Y as a function of the entropy of the input variable – see Figure 1.21. The residual uncertainty is zero, provided that every coefficient of the signal carries at most C bits.

The operational setting we have just described is depicted in Figure 1.22. One among N possible messages is selected at the transmitter and *encoded* into N_0 random coefficients $\{\mathsf{X}_n\}$, using an encoding function $\mathsf{X}^{N_0} : \{1,2,\ldots,N\} \to \mathbb{R}^{N_0}$. Each coefficient goes through a channel governed by a conditional distribution $g(y|x)$, forming an

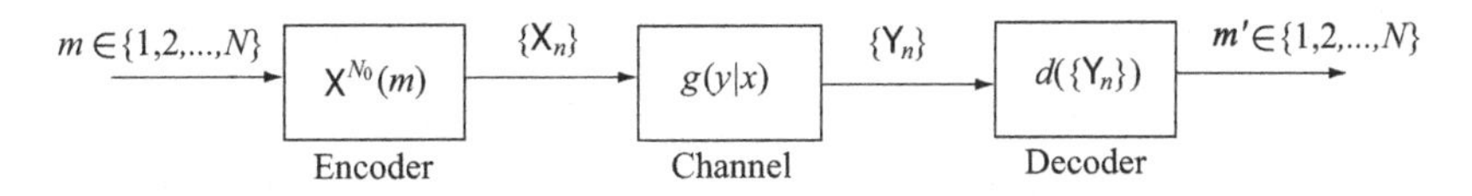

Fig. 1.22 Communication system.

output waveform represented by random coefficients $\{Y_n\}$ that are observed at the receiver. The *decoding* function $d : \mathbb{R}^{N_0} \to \{1,2,\dots N\}$ selects a "guess" $d(\{Y_n\})$ for the possible transmitted message, given the observed coefficients. The rate of this (N, N_0) communication system is

$$R = \frac{\log N}{N_0} \text{ bits per degree of freedom,} \tag{1.64}$$

and the probability of error for message m is

$$p_e(m) = \mathbb{P}(d(\{Y_n\}) \neq m \,|\, \{x_n\} = X^{N_0}(m)). \tag{1.65}$$

A rate $R > 0$ is achievable if there exist a sequence of $(\lceil 2^{N_0 R} \rceil, N_0)$ communication systems with encoding and decoding functions over blocks of N_0 coefficients, whose rate is at least R, and such that

$$\lim_{N_0 \to \infty} \max_{m \in \{1,2,\dots,N\}} p_e(m) = 0. \tag{1.66}$$

The supremum of the achievable rates is the Shannon capacity of the channel, and Shannon's channel coding theorem is stated as follows:

> The supremum of the mutual information between input and output coincides with the Shannon capacity of the channel.

1.4.7 Gaussian Noise

The Shannon capacity can be viewed as a limit due to an *effective resolution level* imposed by the noise inherent in the observation of the waveform. This makes the results above consistent with the deterministic view expressed by (1.26), showing that the capacity grows at most linearly with the number of degrees of freedom, but only logarithmically with the signal-to-noise ratio. We provide this interpretation in the context of Gaussian uncertainty, for which the Shannon capacity is expressed in terms of number of degrees of freedom, signal's energy, and noise constraints.

The uncertainty in the signal's observation is due to a variety of causes, but all of them can be traced back to the quantized nature of the world arising at the microscopic scale. At a very small scale, continuum fields are described by quantum particles whose configurations are uncertain, and repeated measurements appear to fluctuate randomly around their average values. A mathematical model for this situation adds a zero

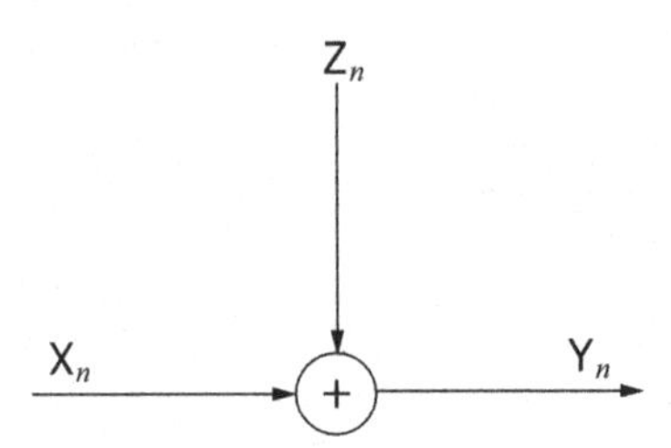

Fig. 1.23 The additive Gaussian channel.

mean Gaussian random variable of standard deviation ϵ independently to each field's coefficient, obtaining the channel depicted in Figure 1.23:

$$\mathsf{Y}_n = \mathsf{X}_n + \mathsf{Z}_n. \tag{1.67}$$

The model is justified as follows. The entropy associated with the noise measures its statistical dispersion: low differential entropy implies that noise realizations are confined to a small effective volume, while high differential entropy indicates that outcomes are widely dispersed. The probability density function that maximizes the uncertainty of the observation, subject to the average constraint

$$\mathbb{E}(\mathsf{Z}_n^2) \leq \epsilon^2, \tag{1.68}$$

is the Gaussian one that achieves the maximum differential entropy

$$h_{\mathsf{Z}_n} = \frac{1}{2} \log(2\pi e \epsilon^2). \tag{1.69}$$

This distribution provides the most surprising observation, for the given moment constraint. In addition, by the central limit theorem, the Gaussian assumption is valid in a large number of practical situations where the noise models the cumulative effect of a variety of random effects. A treatment of maximum entropy distributions and their relationship with noise modeling, statistical mechanics, and the second law is given in Chapter 11.

1.4.8 Capacity with Gaussian Noise

In Chapter 12, we provide a derivation of the capacity of the additive Gaussian noise channel considering communication with waveforms of N_0 degrees of freedom, subject to the constraint

$$\left(\sum_{n=1}^{N_0} x_n^2 \right)^{1/2} \leq \sqrt{PN_0}. \tag{1.70}$$

Compared to (1.59), the more stringent constraint (1.70) is an empirical average rather than a statistical one, since from (1.70) we have

$$\frac{1}{N_0} \sum_{n=0}^{N_0} x_n^2 \leq P. \tag{1.71}$$

This ensures that *all* transmitted signals are inside a hypersphere of radius $\sqrt{PN_0}$, and requires that any realization of a random waveform must draw N_0 coefficients in a dependent fashion. Geometrically, since we are only allowed to pick signals at random inside the hypersphere of radius $\sqrt{PN_0}$, signals that are close to the boundary along one dimension must necessarily be far from it along the other dimensions. In contrast, in Section 1.4.6 we drew coefficients independently, subject to the weaker constraint (1.59). Fortunately, however, probabilistic concentration ensures that for any independent construction,

$$\lim_{N_0 \to \infty} \frac{1}{N_0} \sum_{n=1}^{N_0} x_n^2 = \mathbb{E}(\mathsf{X}^2) \leq P, \tag{1.72}$$

and the probability of constructing a random signal that violates the deterministic constraint (1.70) is arbitrarily small as $N_0 \to \infty$. The geometric interpretation of (1.72) is that by drawing independent coefficients subject to the constraint (1.59), signals are contained with high probability inside the high-dimensional sphere of radius $\sqrt{PN_0}$, as $N_0 \to \infty$.

The capacity of the additive Gaussian channel subject to (1.70) is arguably the most notorious formula in information theory:

$$C = \frac{1}{2} \log \left(1 + \frac{P}{\epsilon^2} \right) \text{ bits per degree of freedom.} \tag{1.73}$$

Consistent with the deterministic setting, this result shows that the largest number of bits carried by any signal in the space scales linearly with the number of degrees of freedom and logarithmically with the signal-to-noise ratio.

The result in (1.73) can be viewed in a geometric setting by considering the standard deviation ϵ as the uncertainty level around each coordinate, and $\sqrt{PN_0}$ as the effective radius of the signals' space. In this way, a picture of Shannon's capacity that is completely analogous to the one of Kolmogorov's 2ϵ-capacity arises. The geometric insight based on sphere packing of the space on which the two models are built is the same. However, while in Kolmogorov's deterministic model packing is performed with "hard" spheres of radius ϵ and communication in the presence of arbitrarily distributed noise over a bounded support is performed without error, in Shannon's stochastic model packing is performed with "soft" spheres of effective radius $\sqrt{N_0}\epsilon$ and communication in the presence of Gaussian noise of unbounded support is performed with arbitrarily low probability of error. In both cases, spectral concentration ensures that the size of the signals' space is essentially of N_0 dimensions. Probabilistic concentration ensures that the noise in Shannon's model concentrates around its standard deviation, so that results are similar in the two cases.

In Shannon's model, when Gaussian noise is added to each signal's dimension, it creates a region of uncertainty around the signal's value. In high dimensions this region is of radius $\sqrt{N_0}\epsilon$, and has a spherical shape. The shape of the region can be visualized

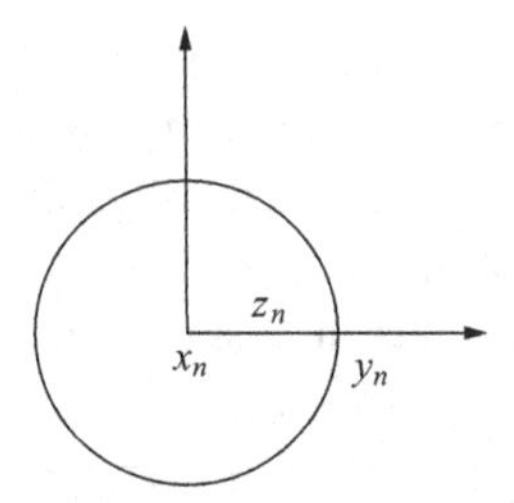

Fig. 1.24 Cross section along the nth dimension of the uncertainty sphere.

as follows. By (1.70), signals correspond to points located inside a high-dimensional sphere of radius $\sqrt{PN_0}$. Consider the difference vector of the coordinates of the points corresponding to transmitted and received signals,

$$(y_1 - x_1, y_2 - y_2, \ldots, y_n - x_n) = (z_1, z_2, \ldots, z_{N_0}). \tag{1.74}$$

According to our channel model, this vector is composed of N_0 independent Gaussian variables, whose joint probability density function factorizes into the product of the individual densities, which depends only on the sum $\sum_{n=1}^{N_0} z_n^2$. This shows that the region of uncertainty around each received point is spherical – see Figure 1.24. If each signal has a typical range that is within the ball of radius $\sqrt{PN_0}$, when it is observed it appears as having a range of radius $\sqrt{N_0(P+\epsilon^2)}$. Comparing the volume of the observed signal space to the one of the uncertainty ball defined by the noise, it follows that the number of distinguishable signals is roughly given by

$$N = \left(\sqrt{\frac{N_0(P+\epsilon^2)}{N_0\epsilon^2}}\right)^{N_0} = \left(\sqrt{1+\frac{P}{\epsilon^2}}\right)^{N_0}. \tag{1.75}$$

By taking the logarithm and dividing by N_0, we have that the correct selection at the receiver of one signal in this ensemble corresponds to communicating a number of bits per degree of freedom given by (1.73).

Finally, compared to (1.22) the constraint (1.70) allows the norm of the signal to scale with $\sqrt{N_0}$, rather than being a constant. The reason for this should be clear: since the noise is assumed to act independently on each signal's coefficient, the statistical spread of the output, given the input signal, corresponds to an uncertainty ball of radius $\sqrt{N_0}\epsilon$. It follows that the norm of the signal should also be proportional to $\sqrt{N_0}$, to avoid a vanishing signal-to-noise ratio as $N_0 \to \infty$. In contrast, in the case of (1.22) the capacity is computed assuming an uncertainty ball of fixed radius ϵ.

A detailed comparison between the deterministic and the stochastic model is given in Chapter 12.

1.5 Energy Limits

Electromagnetic waveforms can transport an amount of information that scales linearly with the number of degrees of freedom, representing the dimensionality of the space, and logarithmically with the ratio of the maximum energy of the signal to the energy of the noise. In the deterministic model of Kolmogorov, in which a bounded amount of noise of norm at most ϵ is added to the signal, we can communicate a number of bits proportional to the number of degrees of freedom. In the stochastic model of Shannon, in which random noise of standard deviation at most ϵ is added independently on each coordinate of the signal's space and the signal's energy is proportional to the number of coordinates, we can communicate a number of bits proportional to the number of degrees of freedom, with arbitrarily low probability of error.

We now explore two additional limiting regimes. In a low signal-to-noise ratio regime, where the energy of the signal vanishes compared to the energy of the noise, the capacity is directly proportional to the energy of the signal and inversely proportional to the number of degrees of freedom. On the other hand, if we keep a constant signal-to-noise ratio and increase the number of degrees of freedom, the capacity increases. However, this also requires energy expenditure, and it turns out that the number of degrees of freedom is ultimately limited by the laws of high-energy physics, and cannot grow beyond what is imposed by general relativity to keep the radiating system gravitationally stable.

1.5.1 The Low-Energy Regime

To explore the low-energy regime, we consider bandlimited signals with N_0 degrees of freedom subject to the fixed energy constraint (1.22), and assume the addition of random Gaussian noise independently to each degree of freedom, subject to (1.68). In this case, the energy of the signal is bounded, while the total amount of noise is proportional to N_0, and we have

$$\frac{1}{N_0}\sum_{n=1}^{N_0} x_n^2 \leq \frac{E}{N_0}, \tag{1.76}$$

which tends to zero as $N_0 \to \infty$. Substituting E/N_0 for P into (1.73) and using a first-order Taylor expansion of the logarithmic function, we have

$$\begin{aligned} C &= \frac{1}{2}\log\left(1+\frac{E}{N_0\epsilon^2}\right) \\ &\simeq \frac{E}{2N_0\epsilon^2}\log e \ \text{ bits per degree of freedom.} \end{aligned} \tag{1.77}$$

It follows that in a regime where the energy of the signal is negligible compared to the energy of the noise, the total amount of information carried by any one signal in the space and expressed in bits is proportional to the energy of the signal, and remains bounded even if the number of degrees of freedom tends to infinity. On the other hand, the capacity per degree of freedom in (1.77) vanishes as $N_0 \to \infty$. This is due to the

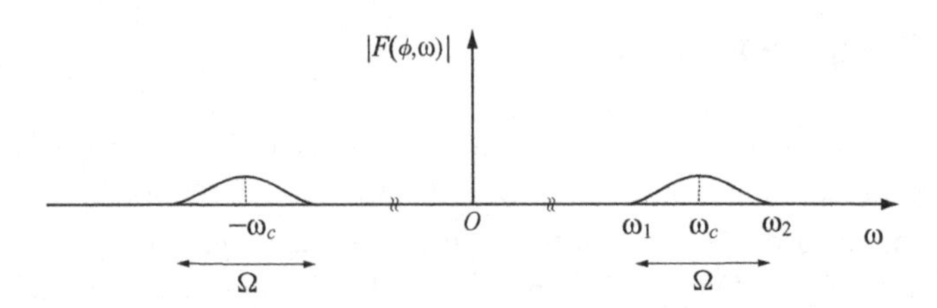

Fig. 1.25 Radiation over a bandwidth Ω centered around ω_c.

signal being spread over a large number of degrees of freedom, while a constant amount of noise is added to each degree of freedom.

1.5.2 The High-Energy Regime

In both the deterministic model of Kolmogorov and the stochastic model of Shannon, we can increase the amount of information associated with the waveforms in the signals' space by increasing the signal-to-noise ratio. By (1.24), (1.26), (1.46), and (1.73), this increases entropy and capacity by a logarithmic factor. We now ask whether we can also spend energy to obtain a linear increase of the amount of information, keeping a fixed signal-to-noise ratio. A possible strategy seems to be to increase the number of degrees of freedom, since this increases entropy and capacity linearly, and by (1.15) and (1.18) it can be accomplished by increasing the frequency of radiation. It turns out, however, that high-frequency signals are also observed at a coarser resolution, so that increasing the frequency while keeping the signal-to-noise ratio constant requires a corresponding increase of the energy per degree of freedom of the radiated signal, and an ultimate limit to the amount of information is imposed by the laws of high-energy physics.

To view these effects in more detail, let us have a closer look at the quantities determining the number of degrees of freedom. By (1.15), in a two-dimensional setting the number of space–wavenumber degrees of freedom at every frequency ω depends on size of the cut-set boundary and on the frequency of radiation. For any arbitrary configuration of sources and scatterers, we can increase the number of space–wavenumber degrees of freedom by transmitting at higher and higher frequencies. This improves the spatial resolution of the received waveform on the cut-set boundary. Similarly, in a three-dimensional setting (1.18) shows that the number of spatial degrees of freedom at each frequency ω increases with the frequency of radiation.

When radiation occurs over a range of frequencies of support 2Ω centered around the origin, the total number of degrees of freedom is given by (1.16) and (1.19), in two and three dimensions respectively. These equations show that the number of degrees of freedom grows with the largest frequency Ω of the radiated signal.

Finally, when radiation occurs over a bandwidth Ω centered around a *carrier frequency* $\omega_c \gg \Omega$, as depicted in Figure 1.25, a computation analogous to (1.16) gives the following total number of degrees of freedom in the two-dimensional setting:

$$N_0 = \frac{T}{\pi} \frac{2\pi r}{c\pi} \int_{\omega_1}^{\omega_2} \omega d\omega$$

$$
\begin{aligned}
&= \frac{T}{\pi}\frac{2\pi r}{c\pi}\frac{(\omega_2^2-\omega_1^2)}{2} \\
&= \frac{\Omega T}{\pi}\frac{2\pi r\omega_c}{c\pi}. \qquad (1.78)
\end{aligned}
$$

Similarly, a computation analogous to (1.19) gives the following total number of degrees of freedom in the three-dimensional setting:

$$
\begin{aligned}
N_0 &= \frac{T}{\pi}\frac{4\pi r^2}{c^2\pi^2}\int_{\omega_1}^{\omega_2}\omega^2 d\omega \\
&= \frac{T}{\pi}\frac{4\pi r^2}{c^2\pi^2}\frac{(\omega_2^3-\omega_1^3)}{3} \\
&= \frac{\Omega T}{\pi}\frac{4\pi r^2}{c^2\pi^2}\frac{(\omega_2^2+\omega_1^2+\omega_1\omega_2)}{3} \\
&= \frac{\Omega T}{\pi}\frac{4\pi r^2\omega_c^2}{(c\pi)^2}(1+\kappa), \qquad (1.79)
\end{aligned}
$$

where the constant $\kappa \approx 0$ for $\omega_c \gg \Omega$. It follows that the total number of degrees of freedom is essentially given by

$$
N_0 = \frac{\Omega T}{\pi}\frac{4\pi r^2\omega_c^2}{(c\pi)^2} = \frac{\Omega T}{\pi}\frac{4\pi r^2}{(\lambda_c/2)^2}, \qquad (1.80)
$$

where $\lambda_c = 2\pi c/\omega_c$ is the *carrier wavelength* of the radiated waveform. These results show that information increases with the carrier frequency, which can be considered the representative frequency of the radiated signal, or equivalently with the inverse of the carrier wavelength.

The information gain obtained at high frequency, however, comes at a price: transmitting at higher and higher frequencies while maintaining a constant signal-to-noise ratio requires a corresponding increase of the energy per degree of freedom of the radiated signal. To illustrate this point in more detail, we need to take a closer look at the quantum mechanical nature of the radiation process.

1.5.3 Quantized Radiation

Consider a space–time waveform of fixed angular frequency bandwidth Ω, centered around ω_c, and radiated in the two-dimensional geometric setting described in Section 1.2.3. This waveform has a total number of degrees of freedom in the space–wavenumber and time–frequency domains given by (1.78), and it can be expanded in terms of orthonormal, real basis functions as

$$
f(\phi,t) = \sum_{n=1}^{N_0} a_n\psi_n(\phi,t). \qquad (1.81)
$$

Electromagnetic radiation occurs in discrete quanta of energy called *photons*. The square of each term in the series (1.81) can be interpreted as an energy density that is proportional to the number of photons observed at space–time point (ϕ,t) along the

nth dimension of the signal's space. Integration of $[a_n\psi_n(\phi,t)]^2$ over a space–time observation window gives the energy detected in a typical observation, over the given window, and along the nth dimension of the space spanned by the given basis representation. By orthonormality, it follows that the total amount of energy that can be detected along the nth dimension of the signal's space is given by

$$E_n = \int_{-\pi}^{\pi}\int_{-\infty}^{\infty}[a_n\psi_n(\phi,t)]^2 dtd\phi = a_n^2, \tag{1.82}$$

and we also have

$$a_n = \int_{-\pi}^{\pi}\int_{-\infty}^{\infty} f(\phi,t)\psi_n(\phi,t)dtd\phi. \tag{1.83}$$

The total energy of the signal, proportional to the number of photons detected over all dimensions of the space, is then given by

$$\begin{aligned}\int_{-\pi}^{\pi}\int_{-\infty}^{\infty} f^2(\phi,t)dtd\phi &= \int_{-\pi}^{\pi}\int_{-\infty}^{\infty}\sum_{n=1}^{N_0}[a_n\psi_n(\phi,t)]^2 dtd\phi \\ &\quad + \int_{-\pi}^{\pi}\int_{-\infty}^{\infty}\sum_{\substack{n,m=1\\ n\neq m}}^{N_0} a_n a_m \psi_n(\phi,t)\psi_m(\phi,t)dtd\phi \\ &= \sum_{n=1}^{N_0} a_n^2.\end{aligned} \tag{1.84}$$

The photons in our signal have a range of frequencies of bandwidth Ω centered around $\pm\omega_c$, as depicted in Figure 1.25. Their energy, measured in joules, is proportional to their positive frequency of radiation according to Planck's equation,

$$E_p = \hbar\omega\ [J], \tag{1.85}$$

where $\hbar$ is the reduced Planck's constant and $\omega > 0$. For $\omega_c \gg \Omega$, the energy of each radiated photon is essentially $\hbar\omega_c$ and an increase of the carrier frequency yields a corresponding increase of the energy of all the photons composing the signal. It follows that maintaining a constant signal-to-noise ratio at increasing frequencies of radiation requires an increase in the energy of the signal along each dimension of the space, since any given level of uncertainty in the number of detected photons corresponds to wider energy fluctuations in the observed signal level. Figure 1.26 illustrates this point, showing the quantized energy levels along the nth dimension of the signal's space for two different values of the carrier frequency $\omega_c^{(1)} < \omega_c^{(2)}$. At the lower frequency, a given number of detected photons on any dimension corresponds to a lower energy compared to the higher frequency level.

We now summarize. The observed waveform is the image of the sources and scatterers through a propagation operator. This operator essentially behaves as a filter, projecting the number of observable waveform configurations onto a lower-dimensional space with a number of degrees of freedom that critically depends on the spatial extension of the cut-set boundary of the radiating system, and on the frequency of radiation. It follows that we can increase the number of degrees of freedom by increasing the radiated frequency. This, however, comes at the price of increasing the radiated energy per degree of freedom, due to the coarser resolution at which signals are observed that is imposed by the more energetic photons composing the signal.

A natural question regarding the high-energy limit of information transmission then arises:

Can we increase information indefinitely by spending more and more energy?

It turns out that even with no constraint on the energy expenditure, nature precludes information growing arbitrarily large. In the high-energy regime we are faced with yet another basic physical constraint: a volume packed with high-energy configurations of radiating elements can become gravitationally unstable, and as it collapses it reaches its ultimate informational limit.

1.5.4 Universal Limits

Information is ultimately limited by the laws of high-energy physics, which preclude it increasing indefinitely. In order to increase the amount of information transported by waves, we need to increase the radiated energy, and this requires packing more and more energy at the source of the radiation. Eventually, this strategy comes to an abrupt stop, and a limiting value for the amount of information that can be transported is reached. This limit is well beyond the capability of any practical communication system and is of purely theoretical interest.

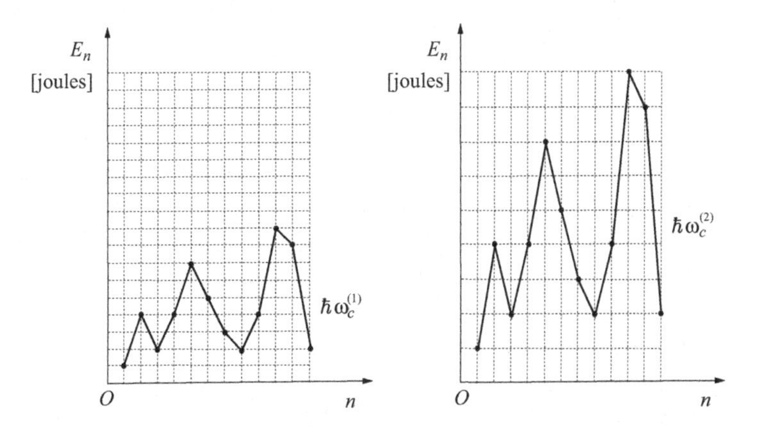

Fig. 1.26 A given number of photons along the nth dimension of the signals' space yields a higher energy value when photons are radiated at higher frequencies.

To describe this information bound we recall that according to general relativity any three-dimensional spherical region is characterized by a critical Schwarzschild radius

$$r_S = \frac{2Gm}{c^2}, \tag{1.86}$$

where G is Newton's gravitational constant and m is the mass of the region. If the radius of the region is smaller than its Schwarzschild radius it will collapse into a black hole. The surface at the Schwazschild radius is called the *event horizon* and it represents the boundary of the black hole. By Einstein's relation

$$E = mc^2, \tag{1.87}$$

increasing the energy inside a volume effectively increases its mass, and eventually causes it to collapse into a black hole.

To appreciate how this phenomenon limits the amount of information inside any volume, consider a spherical region of space of radius r and volume V, without restrictions on its content. We discretize the space into a grid where the cell spacing is given by the Planck length ℓ_P. We shall define this length precisely in Chapter 11 based on quantum indeterminacy principles; here, we simply posit it to be the smallest length scale at which any physical object can be localized. In this first-order calculation we assume cubic cells and do not worry about modifying the shape of the cells near the boundary of the region. We place one quantum state per cell. We assume that each quantum state has two possible energy levels, and that the energy of each state is bounded above by the Planck energy, corresponding to the largest amount of energy that can be localized to a Planck-sized cell without producing a tiny black hole there. It follows that the size of the state space of this system is roughly

$$N = 2^{V/\ell_P^3}, \tag{1.88}$$

and assuming the system is in equilibrium and all states are equally probable, the Shannon entropy is

$$H = V/\ell_P^3. \tag{1.89}$$

The problem with this estimate is that most of the states are too massive to be gravitationally stable. The gravitational limit imposed by the Planck spacing is too weak to prevent black hole formation at larger scales. Even if we have been careful to limit the energy in each Planck cell to avoid its collapse, it is still possible that the union of the cells composing our volume has configurations that lead to gravitational collapse of the whole volume. It is then natural to ask, how many states does a volume that is gravitationally stable have?

The *holographic information bound*, originally proposed in 1993 by the Dutch physicist Gerard 't Hooft and the American physicist Leonard Susskind, states that this number must be less than the number of states of a black hole of the same size as our region, and this black hole has a much smaller number of states than our original estimate, given by

$$N = 2^H, \tag{1.90}$$

where

$$H = \frac{4\pi r^2}{4\ell_\mathrm{p}^2} \log e \text{ bits} \tag{1.91}$$

is the black hole's entropy. We derive the Bekenstein–Hawking entropy formula for black holes (1.91), named after Israeli physicist Jacob Bekenstein and British physicist Stephen Hawking, in Chapter 11. The holographic bound then follows from the basic observation that if the entropy of our system exceeds (1.91), then one could add mass to it and turn it into a black hole having a lower entropy, violating the second law; or more precisely, the generalized second law that includes black hole entropy.

Consider an isolated system of mass m and entropy H'. Let r be the radius of the smallest sphere that surrounds the system and S its surface area. We assume that $r > r_\mathrm{S}$, so that the system is gravitationally stable. The system can be converted into a black hole by collapsing a spherical shell of mass $m - m_\mathrm{BH}$ and entropy H'' onto itself, where m_BH is the mass of a black hole of the same surface area S. The collapsing process preserves spherical symmetry and ensures that the new system has radius $r < r_\mathrm{S}$, so that it collapses into a black hole. The initial entropy in this thermodynamic process is

$$H = H' + H''. \tag{1.92}$$

The final entropy is given by (1.91). Since by the discussion in Section 1.4.1 the Shannon entropy times the Boltzmann constant k_B coincides with the thermodynamic entropy of an equipartitioned system at thermal equilibrium, by the second law we then have

$$H' + H'' \leq \frac{4\pi r^2}{4\ell_\mathrm{p}^2} \log e. \tag{1.93}$$

Since the entropy of the shell must be non-negative, the result follows.

The conclusion is that the most informational object that fits in our region is a black hole with entropy proportional to its event horizon, and any quantum mechanical configuration inside a spherical region of space of surface boundary S can be represented by at most $\log e$ bits of information per four Planck areas on the boundary of the region – see Figure 1.27.

Consider now a waveform observed on a spherical boundary surrounding our region, radiated from sources all internal to the region. Viewing the given configuration of sources as a point in the states' space defined by the quantum cells producing a corresponding point in the signals' space, and letting the number of possible source points be N and the number of possible signal points be M, the number of waveforms that can be generated cannot exceed the total number of source configurations inside the region, and we have $M \leq N$.

An important consequence of this limitation is that we cannot increase the number of degrees of freedom of the waveform arbitrarily: this requires increasing the number of

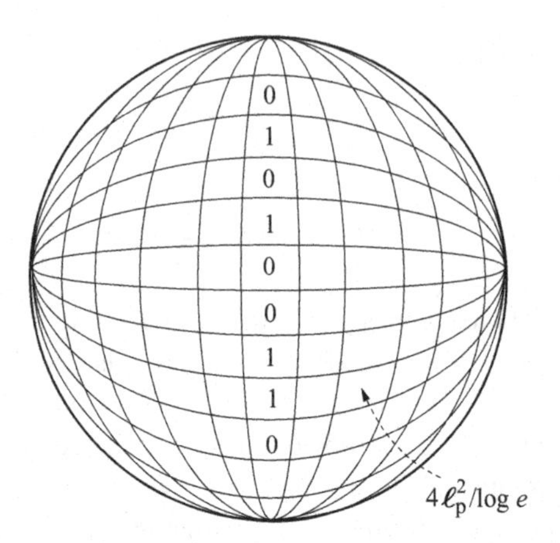

Fig. 1.27 Illustration of the holographic information bound. Similar to a hologram, the whole three-dimensional information can be encoded on the volume's surface area.

degrees of freedom of the source of radiation, and it eventually leads to its gravitational collapse. Gravity imposes that the radiation wavelength can only be decreased up to the Planck scale. Past this limit, the radiated waveform requires excessive energy and the radiating system collapses into a black hole, with entropy proportional to the area of its cut-set boundary expressed in Planck-scale units.

We now summarize. In classical field theory, a number of possible field configurations is projected onto a lower-dimensional space, of dimension proportional to the cut-set area viewed at the wavelength scale. When quantization of energy and gravity are taken into account, the dimension can be at most proportional to the cut-set area viewed at the Planck scale. These constraints are due to very different physical reasons: namely, propagation filtering and gravitational collapse, and occur at very different scales.

To have an idea of the very different scales involved, according to the holographic bound the information contained in an object of size of the order of one meter, and so energetic as to be on the verge of gravitational collapse, would be of the order of 10^{70} bits. To recover the whole information on the cut-set boundary surrounding the object through electromagnetic radiation we would need to convert the whole mass of the object into radiation of wavelength of the order of 10^{-35} meters. In comparison, microwaves for radio communication have wavelengths of the order of a centimeter, optical wavelengths are around 10^{-7} meters, and gamma rays start at a mere 10^{-12} meters – see Appendix F. In nature, most high-frequency detections come from astronomical sources, and the lowest wavelength of gamma rays that has been detected is around 10^{-20} meters, a long way from our hypothetical, highly informative field. Holographic bounds are clearly not going to matter for any practical system. Nevertheless, from a physical perspective they are very relevant to identify the ultimate information-theoretic limits imposed by nature. In Chapters 11 and 13 we take a closer look at these limitations, considering universal entropy bounds for noise signals and bandlimited signals.

1.6 *Tour d'Horizon*

This chapter has provided a roadmap for the topics addressed in the remainder of the book. A central result is the cut-set area bound on the number of degrees of freedom per radiated frequency, leading to the total number of degrees of freedom of electromagnetic signals, (1.19) and (1.80). A complete derivation of these results appears in Chapters 8 and 9, where they are also extended to more general cut-set surfaces with rotational symmetry. Chapters 2, 3, 4, and 5 build all the necessary background for this derivation and take us on a journey through the theory of functional approximation, decomposition of operators in infinite-dimensional Hilbert spaces, and electromagnetic wave propagation. Chapter 6 provides an analogous description of signals from a stochastic perspective, and Chapter 7 is an *intermezzo* that precedes the more technically demanding content of Chapters 8 and 9; Chapter 7 describes how the degrees of freedom and the stochastic diversity of electromagnetic waveforms are exploited in current communication technologies. It discusses the principles behind orthogonal frequency division, code division, time division, and multiple-antenna systems, viewing all of these technologies through the lens of the orthogonal representations examined in the previous chapters. It also gives an overview of the methods that have been proposed to operate next-generation communication systems arising in a network setting. The remaining chapters provide an additional in-depth look at some selected topics in wave theory and at their relationship with information theory.

We now provide a brief summary of the contents of the single chapters. In Chapter 2 we introduce the communication problem, define the signals' space, introduce Slepian's concentration problem, and discuss how this is related to the number of degrees of freedom of bandlimited functions. We show that the prolate spheroidal wave functions, solving the concentration problem and serving as the optimal representation basis for bandlimited signals, also arise in the context of wave propagation. We also discuss how Slepian's problem is related to the impossibility of simultaneous localization of signals in time and frequency, which provides the mathematical justification for Heisenberg's uncertainty principle in quantum mechanics. Thus, the same mathematics of spectral concentration at the basis of information-theoretic results is at the basis of the observational limits of our world.

In Chapter 3 we revisit these concepts in the more rigorous setting of approximation theory, giving the precise definition of the number of degrees of freedom through the notion of Kolmogorov's N-width. We compute the number of degrees of freedom of bandlimited signals in terms of eigenvalues of a certain Fredholm integral equation that arises from Slepian's concentration problem, and place the problem in the more general setting of the Hilbert–Schmidt decomposition of operators in infinite-dimensional spaces. Slepian's problem corresponds to the decomposition of one such operator, that can be performed explicitly. We also provide the generalization of the problem for multi-band signals and for signals of multiple variables, presenting basic results in analysis that are used in subsequent chapters to determine the number of degrees of freedom of electromagnetic space–time signals. Finally, we introduce the problem of signal reconstruction from blind measurements and relate it to the analogous problem of compressed sensing in a discrete setting.

Having developed all of this theory to determine the information associated with bandlimited signals, we need to prepare the stage for the application to the space of electromagnetic signals. Chapters 4 and 5 provide all the necessary ingredients for this. Electromagnetic signals are effectively bandlimited functions that are the image of the sources through a propagation operator. To study this bandlimitation property, in Chapter 4 we review the basic concepts of Maxwell's theory of electromagnetic wave propagation and discuss the physics behind the radiation process. In Chapter 5 we introduce Green's propagation operator, which effectively filters the field onto the space of bandlimited functions. The Hilbert–Schmidt decomposition of this operator leads to the information-theoretic optimal representation of the field in the time–frequency and space–wavenumber domains, and provides the number of parallel channels that can be used for communication, corresponding to the number of degrees of freedom of the field. This representation can be used to multiplex information in space, using multiple antennas, and in time–frequency domains, using appropriate modulation techniques. We also show in Chapter 6 that an analogous representation arises when the Green's kernel is modeled as a stochastic process. In this case, the analog of the Hilbert–Schmidt decomposition is the Karhunen–Loève decomposition, leading to the notion of "richness" of stochastic variations of the Green's function in the different domains. These representations are applied in Chapter 7, which is devoted to the description of communication technologies.

In Chapters 8 and 9, we establish results on the number of degrees of freedom by first showing that the field radiated in an arbitrary scattering environment and filtered by the Green's operator is an effectively bandlimited function, and then applying the theory of spectral concentration in the time–frequency and space–wavenumber domains. We study the spatial bandlimitation property leading to (1.15) and (1.18) in different geometric settings, and introduce the notion of a cut-set integral, measuring the richness of the information content of the waveform with respect to the boundary surface through which the information must flow. In Chapter 9 we extend these results by considering the time–frequency domain in conjunction with the space–wavenumber domain, and establishing (1.16) and (1.19) using the functional analysis machinery developed in the previous chapters. In Chapter 10 we further discuss the stochastic diversity that arises from the physics of propagation, using multiple scattering theory.

Having established the results on number of degrees of freedom and stochastic diversity, we turn to the problem of communication in the presence of noise. Chapter 11 is devoted to exploring the connection between information theory and thermodynamics in the context of developing useful noise models. We show that these mathematical models are rooted in the quantum mechanical nature of the world. We also show that the second law of thermodynamics provides an interpretation of the noise as a signal of maximum uncertainty subject to imposed physical constraints. In this context, we explore the relationship between statistical entropy and thermodynamic entropy.

Chapter 12 considers the notions of entropy, capacity, and dimensionality in the deterministic model of Kolmogorov and in the stochastic model of Shannon, and explores relationships between the two. The book concludes with a discussion in Chapter 13 of universal information bounds, accounting for quantum uncertainty and energy limits.

1.7 Summary and Further Reading

The topics of this chapter are developed in the remainder of the book, and we provide more extensive references in the following chapters. Here we just list some key references. Shannon's (1948) original paper is still a "must read" for any information theory student. Shannon (1949) provides more of an engineering point of view, and focuses on continuous waveform signals. Classic texts in information theory are Cover and Thomas (2006) and Gallager (1968).

Slepian's concentration problem, along with a description of the main results on bandlimited signals obtained with his colleagues at Bell Laboratories, are surveyed in Slepian (1976, 1983) and Landau (1985). The monograph by Hogan and Lakey (2012) provides some additional updated material.

The problem of best approximating a functional space over all finite-dimensional subspaces was formulated by Kolmogorov (1936). Kolmogorov was acquainted with Shannon's work in the early 1950s and immediately recognized that "his mathematical intuition is remarkably precise." The definitions of ϵ-entropy and capacity appearing in Kolmogorov (1956) were certainly influenced by Shannon's work, as well as by his long-standing interest in the approximation of functions in metric spaces from the early 1930s. A review appears in Kolmogorov and Tikhomirov (1959). Standard references for these topics are the books by Pinkus (1985) and Lorentz (1986).

The thermodynamic definition of entropy was introduced by Clausius (1850–65). Statistical definitions are due to Boltzmann (1896–8) for equipartitioned systems, and Gibbs (1902) for non-equipartitioned ones. Jaynes (1965) explored connections between thermodynamic and statistical definitions via the asymptotic equipartition property. Cover (1994) studied the second law of thermodynamics from the point of view of stochastic processes and information theory. Further connections between statistical physics and information theory are discussed by Mézard and Montanari (2009), and Merhav (2010).

The question of how much information an electromagnetic waveform can carry in time and space was first posed in the works of Toraldo di Francia (1955, 1969) and Gabor (1953, 1961). The cut-set area bound on the number of spatial degrees of freedom of time-harmonic electromagnetic fields was derived by Bucci and Franceschetti (1987, 1989), and extended by Franceschetti (2016) to frequency-bandlimited waveforms.

More details on the quantum nature of radiation can be found in many classic texts, among which Heitler (1954) and Loudon (2000) are good representative examples. The holographic information bound was first proposed by 't Hooft (1993). Susskind (1995) gave it a string-theoretic formulation. A comprehensive review appears in Bousso (2002).

1.8 Test Your Understanding

Problems

1.1 There is no such thing as a "thin shell of small volume." Consider the N-dimensional box $C^N(1)$ of side length 1. Show that all the volume in high dimensions concentrates on the surface of the box; namely, for any $0 < \epsilon < 1$,

$$\lim_{N\to\infty} \text{Vol}(C^N(1) - C^N(1-\epsilon)) = 1. \tag{1.94}$$

1.2 Consider a shell that gets thinner as $N \to \infty$. Compute, for all $a > 0$,

$$\lim_{N\to\infty} \text{Vol}(C^N(1) - C^N(1-a/N)), \tag{1.95}$$

and notice that the choice of a determines the fraction of volume that is trapped inside the shell.

1.3 The volume of the N-dimensional ball $B^N(r)$ of radius r is

$$\text{Vol}(B^N(r)) = \frac{2\pi^{N/2}r^N}{\Gamma(N/2)N}, \tag{1.96}$$

where $\Gamma(\cdot)$ is Euler's Gamma function. Verify that for $r = 1$ this has a maximum at $N = 5$ and decreases to zero as $N \to \infty$.

1.4 Consider a box of side length two and a ball of radius one inscribed inside it. Show that the ball contains only a negligible fraction of the whole volume in high dimensions, namely

$$\lim_{N\to\infty} \frac{\text{Vol}(C^N(2))}{\text{Vol}(B^N(1))} = \infty. \tag{1.97}$$

1.5 Consider a ball of radius one. Compute the side length of the largest box inscribed inside it. Show that in this case the box contains only a negligible fraction of the whole volume in high dimensions.

1.6 Are the volumes in Problems 1.4 and 1.5 growing or shrinking as $N \to \infty$?

1.7 Show that the volume of an N-dimensional ball $B^N(1)$ of radius one in high dimensions is concentrated near the boundary; namely, for all $0 < \epsilon < 1$,

$$\lim_{N\to\infty} \frac{\text{Vol}(B^N(1) - B^N(1-\epsilon))}{\text{Vol}(B^N(1))} = 1. \tag{1.98}$$

Compare this result with the analogous one for the box.

1.8 Consider a box of side length $2R$ and the smallest ball containing it. Show that the radius of the ball in N dimensions is $R\sqrt{N}$.

1.9 Show that

$$\lim_{N\to\infty} \text{Vol}(B^N(R\sqrt{N})) = \begin{cases} 0 & \text{if } R \leq 1/\sqrt{2\pi e} \\ \infty & \text{if } R > 1/\sqrt{2\pi e}. \end{cases}$$

1.10 Compute a lower bound on the ϵ-capacity of bandlimited functions using the lattice packing in Figure 1.14 and assuming that the number of degrees of freedom is equal to N_0. Show that as $N_0 \to \infty$ the lower bound grows as $\sqrt{N_0}$ so that the corresponding bound on the number of bits per degree of freedom carried by any waveform in the space tends to zero.

1.11 Explain why the integration interval in (1.16) and (1.19) is $[0, \Omega]$ rather than $[-\Omega, \Omega]$.

1.12 Verify that (1.78) and (1.79) reduce to (1.16) and (1.19) when the signal has a spectral support of size 2Ω centered around the origin.

1.13 To compute the capacity of the additive white Gaussian noise channel, we have considered communication with waveforms of N_0 degrees of freedom, subject to the constraint

$$\int_{-\infty}^{\infty} f^2(t)dt = \sum_{n=1}^{N_0} a_n^2 \leq PN_0, \tag{1.99}$$

where $\{a_n\}$ are the coefficients of an orthonormal signal representation of f over the real line. In this way, signals correspond to points inside the high-dimensional sphere of radius $\sqrt{PN_0}$. In his original derivation, Shannon considered bandlimited signals and used the sampling representation (1.12). Show that in this case, we have

$$\int_{-\infty}^{\infty} f^2(t)dt = \frac{\pi}{\Omega}\sum_{n=1}^{N_0} a_n^2. \tag{1.100}$$

1.14 Given the result in Problem 1.13, what is the appropriate constraint that allows us to view each signal as a point inside a sphere of radius $\sqrt{PN_0}$, in Shannon's model?

1.15 Show that among all continuous random variables defined over the reals and having finite differential entropy and variance σ^2, the differential entropy is maximized only for Gaussians.

Solution

Let f and g be continuous probability density functions over the reals. By Jensen's inequality, we have

$$\int_{\mathbb{R}} f\log(g/f)dx \leq \log\int_{\mathbb{R}} f(g/f)dx = \log\int_{\mathbb{R}} g(x)dx = 0. \tag{1.101}$$

It follows that

$$-\int_{\mathbb{R}} f\log f dx \leq -\int_{\mathbb{R}} f\log g dx, \tag{1.102}$$

where equality holds if and only if $g = f$. We now let

$$g(x) = \frac{1}{\sqrt{2\pi\sigma^2}}\exp\left\{-(x-\mu)^2/2\sigma^2\right\}, \tag{1.103}$$

and we get

$$
\begin{aligned}
h(f) &\leq \log e \frac{\mathbb{E}[(X-\mu)^2]}{2\sigma^2} + \frac{1}{2}\log(2\pi\sigma^2) \\
&= \log e \frac{\sigma^2}{2\sigma^2} + \frac{1}{2}\log(2\pi\sigma^2) \\
&= \frac{1}{2}\log(2\pi e\sigma^2), \qquad (1.104)
\end{aligned}
$$

where the right-hand side can be verified to be the differential entropy of the chosen Gaussian distribution.

1.16 Assume the coefficients that represent a signal $f(t)$ are drawn independently from a distribution with bounded support of size S. The coefficients are quantized at level ϵ, and the entropy of a single quantized coefficient is H_{A^ϵ}. The entropy of the quantized signal is denoted by H. Show that the signal can be identified by a sequence of bits of average length at most $N_0 H_{A^\epsilon}$.

Solution

We give a sketch of the proof. We divide the space of quantized signals into the set of atypical signals $\mathscr{A}$ and the set of typical signals $\mathscr{A}^c$. By (1.46), there are roughly $N = 2^H$ typical quantized signals. On the other hand, the number of atypical quantized signals is $N' \leq (S/\epsilon)^{N_0}$. We order the quantized signals and identify them by their indexes. The indexing of the typical signals requires at most H bits, while that of the atypical signals requires at most $N_0 \log(S/\epsilon)$ bits. We consider a quantized signal q, of bit-length $\ell(q)$, obtained by the realization of N_0 random coefficients, each quantized at level ϵ. The expected length of the sequence of bits identifying this signal is

$$
\begin{aligned}
\mathbb{E}(\ell) &= \sum_{q\in\mathscr{A}^c} p(q)\ell(q) + \sum_{q\in\mathscr{A}} p(q)\ell(q) \\
&\leq \sum_{q\in\mathscr{A}^c} p(q)H + \sum_{q\in\mathscr{A}} p(q)N_0\log(S/\epsilon) \\
&= \mathbb{P}(\mathscr{A}^c)H + \mathbb{P}(\mathscr{A})N_0\log(S/\epsilon). \qquad (1.105)
\end{aligned}
$$

For small ϵ, from the results in Section 1.4.4 we have $H \approx N_0 H_{A^\epsilon}$, so that

$$
\mathbb{E}(\ell) \leq N_0[\mathbb{P}(\mathscr{A}^c)H_{A^\epsilon} + \mathbb{P}(\mathscr{A})\log(S/\epsilon)]. \qquad (1.106)
$$

Finally, by choosing N_0 large enough we have that $\mathbb{P}(\mathscr{A}^c)$ can be made arbitrarily close to one and $\mathbb{P}(\mathscr{A})$ arbitrarily close to zero, completing the proof sketch.

1.17 Consider N_0 i.i.d. Gaussian random variables of zero mean and variance P. Compute upper and lower bounds on the size of the typical set in terms of differential entropy as $N_0 \to \infty$, and compare them with the volume of the N_0-dimensional ball of radius $\sqrt{PN_0}$, that follows from (1.72).

Solution

From (1.40) and using the differential entropy of a Gaussian random variable (1.104),

we have

$$(1-\epsilon)\frac{(\sqrt{2\pi eP})^{N_0}}{2^{N_0\epsilon}} \le \int_{V_\epsilon(N_0)} dx_1 dx_2 \cdots dx_{N_0} \le (\sqrt{2\pi eP})^{N_0} 2^{N_0\epsilon}. \tag{1.107}$$

On the other hand, as $N_0 \to \infty$ the volume of the ball can be approximated by

$$\mathrm{Vol}(B^{N_0}) = \frac{(\sqrt{2\pi eP})^{N_0}}{\sqrt{N_0\pi}}, \tag{1.108}$$

which falls between the upper and lower bounds in (1.107).

2 Signals

For whatsoever is capable of sufficient differences, and those perceptible to the sense, is in nature competent to express cogitations.[1]

2.1 Representations

We consider a class of real functions that we call *signals* that are defined on a real line. Signals have important physical interpretations in terms of real, measured quantities. For example, $f(t)$ can represent the voltage measured at time t across two terminals of an antenna. In what follows, we do not deal directly with units of measure, however these units should be made explicit whenever a given physical interpretation is considered.

The instantaneous power of a signal $f(t)$ is defined by $f^2(t)$, and its energy by

$$\int_{-\infty}^{\infty} f^2(t)dt < \infty. \tag{2.1}$$

From (2.1) it follows that the space of signals for the communication engineer coincides with the $L^2(-\infty, \infty)$ space of the square-integrable functions for the mathematician.

An alternative representation of the signal $f(t)$ is given by the Fourier transform

$$F(\omega) = \int_{-\infty}^{\infty} f(t) \exp(-j\omega t)dt, \tag{2.2}$$

where ω is a real line called *angular frequency* and the engineering notation j is used to denote the imaginary unit. $F(\omega)$ is called the *spectrum* of the signal. For real $f(t)$, we have that $|F(\omega)| = |F^*(-\omega)|$, and the absolute value of the spectrum is an even function of the frequency. The inverse Fourier transform

$$f(t) = \frac{1}{2\pi} \int_{-\infty}^{\infty} F(\omega) \exp(j\omega t)d\omega \tag{2.3}$$

allows the modeling of the signal as the continuous superposition of different spectral components, each at angular frequency ω. The equality in the inversion formula (2.3) is

[1] F. Bacon (1605). *The Advancement of Learning*, reprinted 1869, Clarendon Press, p. 166.

intended here in the energy sense,[2]

$$\lim_{\Omega\to\infty}\int_{-\infty}^{\infty}|f(t)-\int_{-\Omega}^{\Omega}F(\omega)\exp(j\omega t)d\omega|^2dt=0. \tag{2.4}$$

Parseval's theorem states that the energy is conserved in the spectrum, so that the function $F(\omega)$ is also integrable in the absolute square, namely

$$\frac{1}{2\pi}\int_{-\infty}^{+\infty}|F(\omega)|^2d\omega=\int_{-\infty}^{\infty}f^2(t)dt<\infty. \tag{2.5}$$

In practical settings, the real, measured quantities that the signals represent exhibit amplitude spectra of finite support and are observed for a finite time. For example, electromagnetic signals used for communication are confined to frequency bands that are used for different applications (e.g., radio broadcast, cellular telephones, internet access, etc.). The set $\mathscr{B}_\Omega$ of signals *bandlimited* to Ω is defined as the set of all signals whose amplitude spectra vanish for $|\omega|>\Omega$. The set $\mathscr{T}_T$ of signals *timelimited* to T is defined as the set of all signals that vanish for $|t|>T/2$.

Sometimes, rather than signals in $L^2(-\infty,\infty)$ it is useful to consider periodic signals that are in $L^2(a,a+T_0)$, for arbitrary $a\in\mathbb{R}$ and fundamental period $T_0=2\pi/\omega_0$. These have the Fourier series representation

$$f(t)=\sum_{n=-\infty}^{\infty}c_n\exp(jn\omega_0t), \tag{2.6}$$

where convergence here is intended as

$$\lim_{N\to\infty}\int_a^{a+T}|f(t)-\sum_{n=-N}^{N}c_n\exp(jn\omega_0t)|^2dt=0. \tag{2.7}$$

The coefficients of the series are given by

$$c_n=\frac{1}{T_0}\int_a^{a+T_0}f(t)\exp(-jn\omega_0t)dt. \tag{2.8}$$

Parseval's theorem in this case states that the average power over a period is conserved in the spectrum, namely

$$\sum_{n=-\infty}^{\infty}|c_n|^2=\frac{1}{T_0}\int_a^{a+T_0}f^2(t)dt. \tag{2.9}$$

2.2 Information Content

The information a signal carries is its *form*, in the sense of Plato's *eidos*. Signals with different shapes can have different meanings, and choosing one among them implies

[2] A much stronger convergence result holds. Carleson's theorem states that the inverse Fourier transform of square-integrable functions converges point-wise almost everywhere. Throughout the book we are only concerned with convergence in the energy sense, which is the most relevant for physical applications.

that a certain amount of information is transmitted via the selection process. Using mathematics, this concept can be made rigorous by geometrically representing a signal as a point that can be uniquely identified by its coordinates in a high-dimensional space. These coordinates allow us to distinguish among different signals, numerically representing the amount of information that can be communicated by selecting one of them.

A bandlimitation of the spectral support of the signal reflects into a limitation on how diverse the corresponding waveform can be, and in turn on how many coordinates are required to identify it. The number of coordinates that uniquely identify any bandlimited signal undergoes a *phase transition* around a critical value representing the effective dimension of the signal's space. This means that a number of coordinates slightly below the critical value does not suffice to identify any signal, while a number slightly above it does suffice. The meaning of the word "slightly" here refers to the width of the transition around the critical point that can be characterized precisely, and the sharp cut-off occurring within this interval provides a fundamental limit on the amount of information that can be communicated by transmitting an arbitrary signal in the space.

2.2.1 Bandlimited Signals

The Kotelnikov–Shannon–Whittaker interpolation formula (cardinal series, for short) is widely used in engineering. It states that if $f(t)$ is bandlimited to Ω, then

$$f(t) = \sum_{n=-\infty}^{\infty} f(n\pi/\Omega)\,\mathrm{sinc}\,(\Omega t - n\pi), \tag{2.10}$$

with the conventional notation $\mathrm{sinc}\, t = (\sin t)/t$.

It follows that any bandlimited signal can be identified by specifying a discrete sequence of real numbers representing the sampled values of the signal spaced by π/Ω. The signal can be constructed using the interpolating functions $\mathrm{sinc}(\cdot)$, each centered at a sampled point and whose amplitude is adjusted to that of the corresponding sample. The interpolating functions have value zero at all sampled points except the one where they are centered.

In the context of linear filters, the interpolation can be viewed as applying an ideal low-pass filter of cut-off frequency Ω, whose impulse response is $\mathrm{sinc}(\Omega t)$, and whose input is the ideal δ-pulse train modulated by the samples $f(n\pi/\Omega)$, as shown in Figure 2.1. This interpretation makes the formula attractive and widely used in practical settings. Relaxing the condition on the ideal filter cut-off and on the ideal δ-pulses leads to a reconstruction error that can be minimized using higher sampling rates.

A mathematical derivation of the formula can be easily obtained by replicating the spectrum $F(\omega)$ along the ω axis at every interval of size 2Ω. This gives a periodic signal $F_s(\omega)$ of the fundamental period 2Ω. By applying (2.6) and (2.8) with the substitutions $t \to \omega$, $T_0 \to 2\Omega$, and $\omega_0 \to \pi/\Omega$, we obtain

$$F_s(\omega) = \sum_{n=-\infty}^{\infty} F(\omega - 2\Omega n) = \frac{\pi}{\Omega} \sum_{n=-\infty}^{\infty} f(n\pi/\Omega)\exp[-jn(\pi/\Omega)\omega]. \tag{2.11}$$

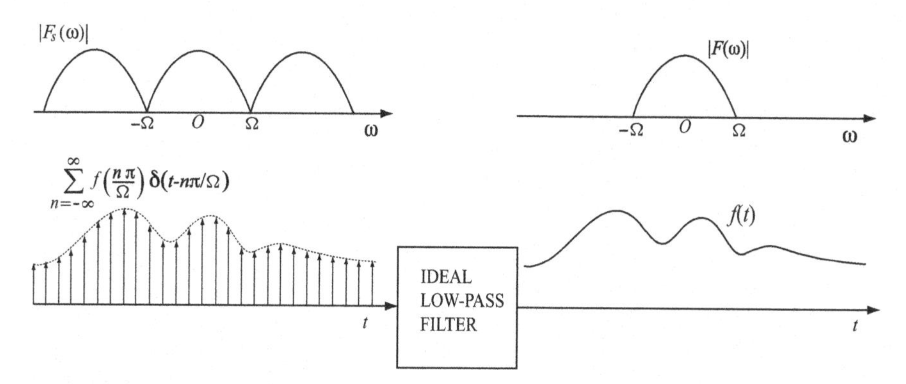

Fig. 2.1 Physical interpretation of the cardinal series interpolation formula.

The original signal is then given by the replicated signal times the rectangular window of support 2Ω, namely

$$
\begin{aligned}
F(\omega) &= \mathrm{rect}\left(\frac{\omega}{2\Omega}\right) F_s(\omega) \\
&= \mathrm{rect}\left(\frac{\omega}{2\Omega}\right) \frac{\pi}{\Omega} \sum_{n=-\infty}^{\infty} f(n\pi/\Omega) \exp[-jn(\pi/\Omega)\omega] \\
&= \sum_{n=-\infty}^{\infty} f(n\pi/\Omega) \frac{\pi}{\Omega} \mathrm{rect}\left(\frac{\omega}{2\Omega}\right) \exp[-jn(\pi/\Omega)\omega]. \qquad (2.12)
\end{aligned}
$$

Finally, taking the inverse transform of (2.12) and using the Fourier transform pair

$$\frac{\Omega}{\pi} \mathrm{sinc}\,(\Omega t) \longleftrightarrow \mathrm{rect}\left(\frac{\omega}{2\Omega}\right), \qquad (2.13)$$

the formula (2.10) follows. In this last step, we have assumed that the signal is sufficiently well behaved that term by term integration is allowed.

The intuitive justification for the cardinal series is that, if $f(t)$ contains no angular frequencies higher than Ω, then it cannot change to a substantially new value in a time less than one half-cycle of the highest frequency, corresponding to the sampling interval π/Ω, and all values in between samples can be perfectly reconstructed. In short, a bandwidth limitation translates into a limitation of the diversity of the signal, namely of the amount of variation that the signal can undergo in any given sampling interval.

2.2.2 Timelimited Signals

If $f(t)$ is also timelimited to $[-T/2, T/2]$, then the non-zero terms in (2.10) occur only for $|n| \leq \Omega T/(2\pi)$; the number of sinc$(\cdot)$ functions needed to interpolate any arbitrary bandlimited and timelimited signal is finite, and it is given by the Nyquist number[3]

$$N_0 = \Omega T/\pi. \qquad (2.14)$$

[3] We use the convention throughout that the Nyquist number is rounded to the largest integer no greater than $\Omega T/\pi + 1$.

It follows that a waveform $f(t)$, viewed in a simple environment such as the real plane, can be replaced by a simpler entity such as a point, viewed in a complex environment such as an N_0-dimensional space. In the context of communication, this means that using signals bandlimited to Ω one can transmit only N_0 real numbers in time T, and these numbers can be used to completely identify any one signal of duration T.

This remarkable intuition, due to Shannon (1948, 1949), allows engineers to treat continuous signals as discrete ensembles corresponding to their coordinates in an N_0-dimensional space, and is the essential idea supporting today's digital communication technology. As powerful as it is, the intuition is nevertheless imprecise, and it requires a great deal of mathematical argument to make it stand on solid ground.

2.2.3 Impossibility of Time–Frequency Limiting

The first problem that we encounter is that the only signal that is both bandlimited and timelimited is the trivial always-zero signal. A mathematical proof of the above statement is the following: consider the complex extension of $f(t)$ defined in (2.3) to the upper complex half-plane, by taking $\{t = x + jy : y > 0, x, y \in \mathbb{R}\}$. If $F(\omega)$ is bandlimited to Ω, then by the Paley–Wiener theorem stated in Appendix A.1, this extension is an entire function of exponential type Ω. Such a function is holomorphic over the whole complex plane and therefore it equals its own Taylor series everywhere. Now, if $f(t) = 0$ in an interval, then all of its derivatives would also be zero inside this interval and the Taylor expansion would require it to be zero everywhere. The above proof uses tools from analysis; the solution to Problem 2.10 provides a more elementary derivation.

It follows that the perfect reconstruction formula (2.10) is only valid if one observes bandlimited signals over an infinite time, and this is not possible in practice. Signals of finite time support cannot be rigorously identified by N_0 numbers, and if one attempts to use this finite number of samples then a certain reconstruction error must occur. This interpolation error is due to the missing samples of the tails of the signals in the time domain. On the other hand, a timelimited signal cannot be bandlimited and its reconstruction from N_0 samples has an aliasing error due to the overlaps of the tails of the replicas in the spectral domain. The situation is depicted in Figure 2.2, and creates a problem in defining precisely the amount of information that is carried by the transmitted signals.

One possible way around these issues is to argue that signals can be, if not perfectly timelimited and bandlimited, at least approximately so. The aliasing and interpolation errors can then be neglected, if the approximation is sufficiently accurate. The problem is then to appropriately define the meaning of the term "approximately" and to relate this definition to the quality of the reconstruction. After Shannon's breakthrough proposal, this objective kept communication engineers busy for a long time building rigorous foundations for their methods.

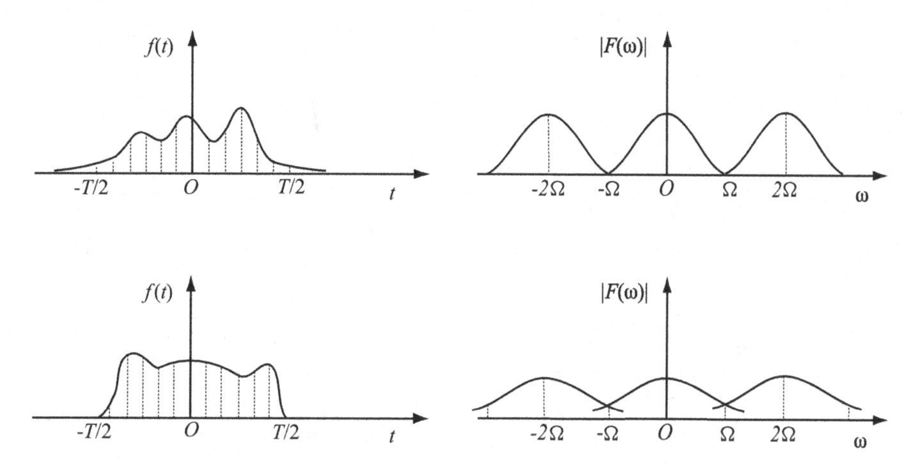

Fig. 2.2 It is impossible to achieve simultaneous concentration of signals in time and frequency. Sampling reconstruction carries either an interpolation error due to the missing samples of the tails of the signals in the time domain, or an aliasing error due to the overlap of the tails of the replicas in the spectral domain.

2.2.4 Shannon's Program

We first provide a non-rigorous, but intuitively appealing, argument for how bandlimited signals can be used for communication. In the late 1940s, this was Shannon's program, which sparked the digital communication revolution.

We view a set of N_0 numbers as coordinates of a point in a space of N_0 dimensions. Each selection of these numbers identifies a point in this space. Ignoring the problem of the impossibility of simultaneous concentration of signals in time and frequency, we assume for the moment that there is exactly one point corresponding to each signal approximately bandlimited to Ω and with approximate duration T. We make some comments on the geometric insights provided by such an approximate, discrete representation.

Consider a point $\mathbf{x} = (x_1, \ldots, x_{N_0})$, whose coordinates correspond to the samples of a signal $f_x(t)$. The distance of $\mathbf{x}$ from the origin of the space is given by

$$d(\mathbf{x}, 0) = \sqrt{\sum_{n=1}^{N_0} x_n^2}. \tag{2.15}$$

By (2.10), using Parseval's theorem and (2.13), it follows that the total energy of the signal is given by

$$\int_{-\infty}^{\infty} f_x^2(t)dt = \frac{\pi}{\Omega} \sum_{n=1}^{N_0} x_n^2. \tag{2.16}$$

Letting the average power over time T be

$$P = \frac{1}{T} \int_{-T/2}^{T/2} f_x^2(t)dt, \tag{2.17}$$

we have

$$
\begin{aligned}
d(\mathbf{x},0) &= \sqrt{\frac{\Omega}{\pi}\int_{-\infty}^{\infty} f_x^2(t)dt} \\
&= \sqrt{P\frac{\Omega T}{\pi}} \\
&= \sqrt{PN_0}.
\end{aligned} \tag{2.18}
$$

It follows that the space of signals whose average power is at most P corresponds to the set of points within the sphere of radius

$$r = \sqrt{PN_0}. \tag{2.19}$$

Similarly, one can compute the distance between any two signals $f_y(t)$ and $f_z(t)$ as the geometric distance between the corresponding points, namely

$$d(\mathbf{y},\mathbf{z}) = \sqrt{\sum_{n=1}^{N_0} |y_n - z_n|^2} = \sqrt{\frac{\Omega}{\pi}\int_{-T/2}^{T/2} [f_y(t) - f_z(t)]^2 dt}. \tag{2.20}$$

An additional quantity that has an important role in the space of signals is the *noise* that is inevitably added to every measured signal. When noise is added to the signal, the corresponding point is moved a certain random distance in the space, roughly proportional to the standard deviation of the noise. This produces a small region of uncertainty about each point in the space and provides a resolution limit at which one can distinguish different signals. Points that are too close together have overlapping regions of uncertainty and cannot be distinguished from one another with reasonable accuracy. With an obvious physical analogy, a set of signals in the space is called a *constellation*, an example of which is depicted in Figure 2.3.

We conclude that there are three main features that limit the amount of information that can be communicated by transmitting signals. One is the dimension of the space, which determines the number of coordinates that are communicated by transmitting the signal. Another is the noise, which limits the resolution at which we can observe each coordinate. Finally, the average power used for transmission limits the amount of separation between signals in the space.

In a digital communication system that uses continuous waveforms for communication, a transmitter can select any one of the signals in the space, subject to the geometric constraint (2.19), by specifying N_0 coordinate points. The signal is then observed at the receiver, with typical perturbation of the order of the standard deviation of the noise. By keeping the possible transmitted signals sufficiently well separated in the signal space, the receiver can likely correctly identify which one of them has been selected at the transmitter, so that an amount of information proportional to N_0 is communicated at each transmission event. This communication system relies on the basic idea of treating continuous signals as discrete ensembles corresponding to their coordinates in an N_0-dimensional space.

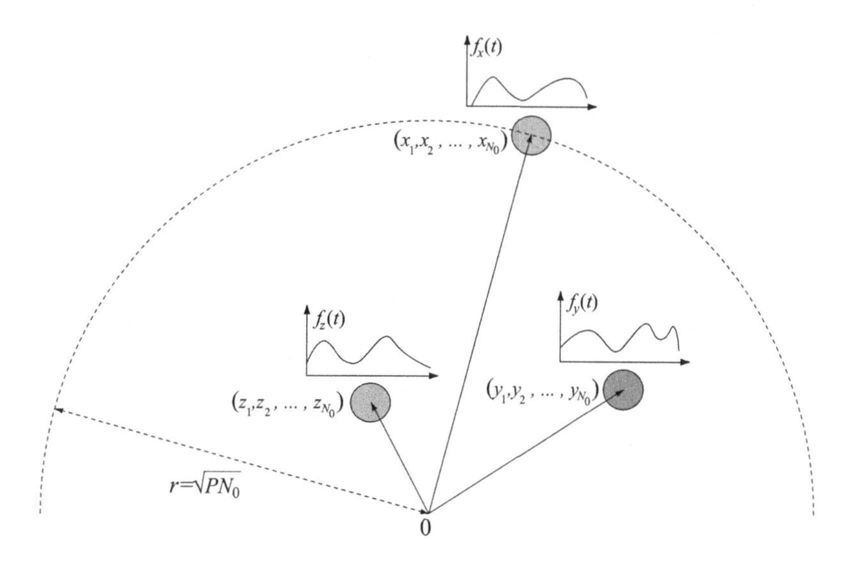

Fig. 2.3 Signal constellation. The uncertainty region about each signal point is depicted as a gray disc.

2.3 Heisenberg's Uncertainty Principle

A significant obstacle to Shannon's program of having a discrete representation of the signals' space is the impossibility of achieving simultaneous concentration of signals and their spectra. Any bandlimited signal cannot be timelimited, and so it cannot be determined by a finite number of samples. This mathematical result is also at the basis of Heisenberg's uncertainty principle. Published by Werner Heisenberg (1927), the principle states that it is impossible to determine both the position and the momentum of an elementary particle with an arbitrary degree of accuracy. In mathematical terms, it provides an upper bound on the amount of simultaneous concentration of a signal and its Fourier transform.

2.3.1 The Uncertainty Principle for Signals

To formulate the principle in the context of signal theory, let us consider a normalized signal $f(t)$ of unit energy, so that

$$\frac{1}{2\pi}\int_{-\infty}^{+\infty} |F(\omega)|^2 d\omega = \int_{-\infty}^{\infty} f^2(t)dt = 1. \tag{2.21}$$

In this way, we can interpret $f^2(t)$ and $|F(\omega)|^2/(2\pi)$ as probability densities, and their variances σ_t^2 and σ_ω^2 as measures of concentration of these densities. The uncertainty principle states that

$$\sigma_t^2 \sigma_\omega^2 \geq \frac{1}{4}. \tag{2.22}$$

In words, there is a uniform bound on the amount of simultaneous concentration that signals can have in the time and frequency domains.

In order to show that (2.22) holds, since the value of the variance is invariant to translation of the signal, we can assume that signals have zero mean time and zero mean frequency. In this case, we have

$$\sigma_t^2 = \int_{-\infty}^{\infty} t^2 f^2(t) dt, \tag{2.23}$$

$$\sigma_\omega^2 = \frac{1}{2\pi} \int_{-\infty}^{\infty} \omega^2 |F(\omega)|^2 d\omega. \tag{2.24}$$

The derivative property of the Fourier transform, in conjunction with Parseval's theorem, yields

$$\frac{1}{2\pi} \int_{-\infty}^{\infty} \omega^2 |F(\omega)|^2 d\omega = \frac{1}{2\pi} \int_{-\infty}^{\infty} |j\omega F(\omega)|^2 d\omega = \int_{-\infty}^{\infty} \left| \frac{d}{dt} f(t) \right|^2 dt. \tag{2.25}$$

Combining (2.23), (2.24), and (2.25), and using the Schwarz inequality, we have

$$\begin{aligned} \sigma_t^2 \sigma_\omega^2 &\geq \left| \int_{-\infty}^{\infty} t f(t) \frac{d}{dt} f(t) dt \right|^2 \\ &= \frac{1}{4} \left| \int_{-\infty}^{\infty} \frac{d}{dt} (t f^2(t)) dt - \int_{-\infty}^{\infty} f^2(t) dt \right|^2 \\ &= \frac{1}{4}, \end{aligned} \tag{2.26}$$

where the last equality holds since the first integral is zero and the second integral is one by virtue of (2.21).

2.3.2 The Uncertainty Principle in Quantum Mechanics

We now consider the physical implications of the above result. The uncertainty principle arises in the context of the wave–particle duality of electromagnetic waves. Heisenberg considered the following example to show the uncertainty principle in action. Consider illuminating an electron with a microscope. The interaction of the electron with the illuminating light corresponds to a collision of at least one photon with the electron. Classical optics predicts that larger wavenumbers (i.e., shorter wavelengths) allow higher observational resolutions. However, the collision of light with the electron disturbs the momentum of the electron. The larger the wavenumber of the illuminating light, and thus the precision of the determined position, the larger this unknown change in momentum can be. In Heisenberg own words:

> At the instant of time when the position is determined, that is, at the instant when the photon is scattered by the electron, the electron undergoes a discontinuous change in momentum. This change is the greater the smaller the wavelength of the light employed, i.e., the more exact the determination of the position. At the instant at which the position of the electron is known, its momentum therefore can be known only up to magnitudes which correspond to that discontinuous change; thus, the more precisely the position is determined, the less precisely the momentum is known, and conversely. (Heisenberg 1927, pp. 174–5).

A mathematical explanation of why the above attempt to determine accurately both the momentum and the position of the electron fails can be given as follows. A scalar electric field is a real function of space and time. Viewing the field at a given location as a time signal $f(t)$, we have the usual Fourier transform pair

$$f(t) \longleftrightarrow F(\omega), \tag{2.27}$$

and we have seen above that this signal cannot be simultaneously concentrated in time and frequency. Analogous considerations hold if instead we consider the field at a given time as a signal of space. Considering one-dimensional space for simplicity, the signal is characterized by the Fourier transform pair

$$g(z) \longleftrightarrow \widehat{g}(k_z), \tag{2.28}$$

where z is the spatial coordinate and $k_z = \omega/c$ is the wavenumber that constitutes the spectral variable corresponding to the spatial coordinate z. It follows that the field cannot be simultaneously concentrated in space and in the wavenumber spectrum: electromagnetic concentration in an arbitrarily small region of space results in a wavenumber amplitude spectrum of large support, and vice versa.

In relativistic mechanics, the momentum p of a single photon is proportional to the wavenumber of the associated electromagnetic field. This can be seen by combining Einstein's equation for the energy,

$$E = mc^2, \tag{2.29}$$

with the equation for the momentum,

$$p = mv, \tag{2.30}$$

and letting m be the relativistic mass

$$m = \frac{m_0}{\sqrt{1 - v^2/c^2}}, \tag{2.31}$$

where m_0 is the rest mass, v is the velocity, and c is the speed of light. By (2.30) and (2.31), we have

$$\begin{aligned} p^2c^2 &= \frac{m_0^2 v^2 c^2}{1 - v^2/c^2} \\ &= \frac{m_0^2 c^4 v^2/c^2}{1 - v^2/c^2} \\ &= \frac{m_0^2 c^4 \left(v^2/c^2 - 1\right)}{1 - v^2/c^2} + \frac{m_0^2 c^4}{1 - v^2/c^2} \\ &= -m_0^2 c^4 + (mc^2)^2; \end{aligned} \tag{2.32}$$

rearranging terms and using (2.29), we obtain the relativistic energy–momentum equation

$$\sqrt{p^2c^2 + m_0^2 c^4} = mc^2 = E. \tag{2.33}$$

Finally, letting the rest mass of the photon tend to zero and using Planck's formula for the energy of a photon, we have

$$p = \frac{E}{c} = \frac{\hbar\omega}{c} = \hbar k_z, \tag{2.34}$$

where ω is the angular frequency of the associated electromagnetic signal, k_z is the wavenumber, and $\hbar$ is the reduced Planck's constant.

We now conclude that the uncertainty on the wavenumber of the signal in the spectral domain translates into an uncertainty on the momentum of the corresponding photon particle. Heisenberg's principle in quantum mechanics immediately follows:

$$\sigma_p^2 \sigma_z^2 \geq \frac{\hbar^2}{4}. \tag{2.35}$$

The only kind of signal with a definite momentum is an infinite regular periodic oscillation over all space, which has no definite position. Conversely, the only kind of signal with a definite position is concentrated at one point, and such a signal has an indefinite momentum. In quantum mechanics, it is impossible to describe particles with a definite position and a definite momentum. The more precise the position, the less precise the momentum.

2.3.3 Entropic Uncertainty Principle

Heisenberg's uncertainty principle considers signals of unit energy and their variances as measures of concentration. An analogous result can be derived using the differential entropy defined in (1.33) as a concentration measure. Similar to the variance, this measures the statistical dispersion of a signal around its typical value. In this case, the uncertainty principle provides a lower bound on the sum of the temporal and spectral differential entropies of a unit energy signal satisfying (2.21), and we have

$$h(f^2) + h(|F|^2/(2\pi)) \geq \log(\pi e). \tag{2.36}$$

In both (2.22) and (2.36), the equality holds in the case of Gaussian signals

$$f^2(t) = \frac{1}{\sqrt{2\pi}\sigma} \exp[-t^2/(2\sigma^2)], \tag{2.37}$$

for which

$$\sigma_t^2 \sigma_\omega^2 = \frac{1}{4} \tag{2.38}$$

and

$$h(f^2) + h(|F|^2/(2\pi)) = \log(\pi e). \tag{2.39}$$

2.3.4 Uncertainty Principle Over Arbitrary Measurable Sets

Another uncertainty principle can be derived considering the portion of the signal's energy inside a given measurable set as a measure of concentration over this set. For example, consider signals of unit energy whose fraction of energy outside the interval $[-T/2, T/2]$ is at most ϵ_T^2, namely

$$1 - \int_{-T/2}^{T/2} f^2(t) dt \leq \epsilon_T^2, \tag{2.40}$$

and whose fraction of energy outside the interval $[-\Omega, \Omega]$ is at most ϵ_Ω^2, namely

$$1 - \frac{1}{2\pi} \int_{-\Omega}^{\Omega} |F(\omega)|^2 d\omega \leq \epsilon_\Omega^2. \tag{2.41}$$

In this case, we have the uncertainty principle

$$\frac{\Omega T}{\pi} \geq [1 - (\epsilon_T + \epsilon_\Omega)]^2, \tag{2.42}$$

showing that a signal and its Fourier transform cannot be both arbitrarily concentrated over intervals of size T and 2Ω. This result also holds for arbitrary measurable sets that are not necessarily intervals – see Appendix A.2.

2.3.5 Converse to the Uncertainty Principle

The uncertainty principle provides an upper bound on the amount of simultaneous concentration in time and frequency of any signal in the space. One can also ask about the converse, namely what the most concentrated signals are that satisfy (2.22), (2.36), or (2.42). For example, we may wish to determine signals of given frequency concentration that achieve the largest time concentration, or vice versa.

When concentration is measured in terms of variance, or entropy, then the Gaussian achieves the largest simultaneous concentration in time and frequency – see Problems 2.8 and 2.9. On the other hand, when concentration is measured in terms of the fraction of the signal's energy over given measurable sets in time and frequency, as in (2.40) and (2.41), then the most concentrated signals are somewhat more difficult to determine. These signals, however, have the additional remarkable property of providing an optimal orthogonal basis representation for any bandlimited signal. The error associated with this representation drops sharply to zero when slightly more than a critical number N_0 of basis functions are used for the approximation, and N_0 can be identified with the effective dimensionality of the space of bandlimited signals.

It turns out that these highly concentrated basis functions are the solutions of an integral equation defined on the sets of concentration. They can be obtained explicitly in the case of intervals, and the critical number of functions needed to represent any signal up to arbitrary accuracy can be determined in an asymptotic order sense.

It took communication engineers a great deal of effort to rigorously derive the above results. For a long time, the standard "hand waving" argument to determine the asymptotic dimensionality of the space of bandlimited signals using an orthogonal basis representation relied on the somewhat simpler, but sub-optimal, cardinal series sampling representation (2.10). The idea was to approximate bandlimited signals using a finite number $N_0 = \Omega T/\pi$ of terms of the cardinal series, corresponding to sampled signal values collected inside a time interval of size T. Then, noticing that as $T \to \infty$ a vanishing portion $\epsilon_T^2 \to 0$ of the signal's energy is neglected and a better and better approximation is achieved, one may consider N_0 as being the asymptotic dimensionality of the space. In this way, any real bandlimited signal of unbounded time support can be approximated by a finite number of samples collected inside a finite interval, and thus appears as a point in a high-dimensional space. Since Shannon's first outline of this argument in 1948, it has been the undisputed "folk theorem" of communication engineering.

2.4 The Folk Theorem

The communication engineer wants to work with signals that are somewhat concentrated in both the time and the frequency domains. These are the kind of signals that seem to be the most natural, because physical devices are characterized by a finite frequency response, and because signals are really observed for a finite time. Signals of this kind can be represented via the cardinal series by a discrete set of N_0 sampled points, and open the possibility of using a geometric approach to the design of communication systems.

On the other hand, the same mathematics that is at the basis of the fundamental physical indeterminacy laws of quantum mechanics seems to prohibit this. Signals that appear highly concentrated in frequency must be widely dispersed in time, and vice versa. For some time engineers ignored the issue. The dilemma was hand-waved as being only an apparent one, more of interest to the mathematician than the practitioner. They made the observation that as $N_0 = \Omega T/\pi \to \infty$, the number of samples grows and both the sampled values of the original signal and the reconstruction error of the cardinal series become negligible. So – they argued – if the error cut-off is sufficiently sharp, then N_0 can still be considered the asymptotic dimension of the signals' space. This led to the formulation of the following folk theorem that engineers have used with great success to design sophisticated, real communication systems:

> For large ΩT, the space of signals of approximate duration T and approximate bandwidth Ω has approximate dimension $N_0 = \Omega T/\pi$.

The "theorem" is far from being rigorous, and its imprecise statement is rightfully belittled by the analyst. Nevertheless, it gives the right intuition of representing signals as points in a space of N_0 dimensions, and allows communication engineers to use geometric arguments to design communication systems and study their performance. This geometric intuition creates a bridge between the continuous world of signals and Shannon's vision of a digital world of information.

2.4.1 Problems with the Folk Theorem

Let us have a closer look at the representation error of the cardinal series. We show that using the cardinal series, the reconstruction error decays slowly and makes it impossible to identify the desired sharp error cut-off required to define the effective dimensionality of the signals' space. This raises the question of whether there is a better, more concentrated basis than the one used in the cardinal series to optimally represent any bandlimited signal.

In the following discussion we fix the bandwidth Ω and consider a signal of unbounded time support. The treatment by fixing the time duration and considering unbounded bandwidth is completely analogous. We consider signals in the set $\mathscr{E}(\epsilon_T)$ of bandlimited signals subject to the energy constraint

$$\int_{-\infty}^{\infty} f^2(t)dt \leq 1, \tag{2.43}$$

and with at most a fraction of ϵ_T^2 energy outside the interval $[-T/2, T/2]$, namely

$$1 - \frac{\int_{-T/2}^{T/2} f^2(t)dt}{\int_{-\infty}^{\infty} f^2(t)dt} \leq \epsilon_T^2. \tag{2.44}$$

As $T \to \infty$ we can also let $\epsilon_T \to 0$, as the signal resembles more and more a timelimited one. Consider the truncation error,

$$\begin{aligned} e_N(t) &= f(t) - \sum_{n=-N}^{N} f(n\pi/\Omega)\, \text{sinc}\,(\Omega t - n\pi) \\ &= \sum_{|n|>N} f(n\pi/\Omega)\, \text{sinc}\,(\Omega t - n\pi). \end{aligned} \tag{2.45}$$

By Parseval's theorem and (2.13) it follows that the total energy of the error is given by

$$\int_{-\infty}^{\infty} (e_N(t))^2 dt = \frac{\pi}{\Omega} \sum_{|n|>N} f^2(n\pi/\Omega), \tag{2.46}$$

which tends to zero as $N \to \infty$, because

$$\frac{\pi}{\Omega} \sum_{n=-\infty}^{\infty} f^2(n\pi/\Omega) = \int_{-\infty}^{\infty} f^2(t)dt < \infty. \tag{2.47}$$

The value of N required to guarantee a small error depends on the signal we wish to approximate. Ideally, we would like to be able to represent any bandlimited signal using roughly $N_0 = \Omega T/\pi$ interpolating functions, and have the energy of the error be of the order of the energy of the signal outside the interval $[-T/2, T/2]$, as $T \to \infty$. In this way, all signals with negligible energy outside $[-T/2, T/2]$ would not present a significant error when represented using N_0 samples collected inside this interval. Letting $T \to \infty$, the energy of the error and ϵ_T would tend to zero at the same rate, and N_0 could be identified with the effective dimension of the signals' space, consistent with our version of the folk theorem.

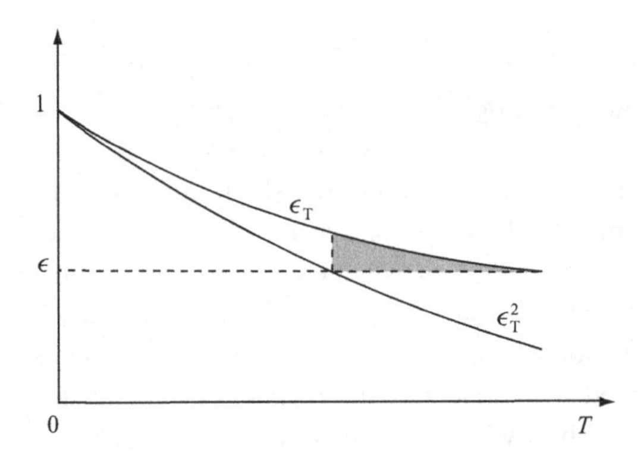

Fig. 2.4 Decay of the energy of the error using N_0 samples for reconstruction. When the tail of the signal's energy ϵ_T^2 is below ϵ, i.e., not detectable, the energy of the error can be within the highlighted area, i.e., detectable.

Unfortunately, the desire above cannot be fulfilled. It can be shown that for any arbitrarily large constants a and N', there exists a concentrated signal f with ϵ_T small enough such that

$$\int_{-\infty}^{\infty} (e_{N_0+N'}(t))^2 dt > a\epsilon_T^2. \tag{2.48}$$

Thus, retaining N_0 plus any constant number of terms of the cardinal series, we cannot approximate every concentrated signal to a degree proportional to its unconcentrated energy. There are signals for which, as $T \to \infty$, the energy of the approximation is greater than an arbitrarily large multiple of the unconcentrated energy.

Another negative result regarding the approximation using the cardinal series states that even an almost linear increase in the number of terms does not suffice for approximating f to within a constant factor of ϵ_T^2. Namely, for any $\gamma < 1$, there exists $\eta > 0$ and a signal f with ϵ_T small enough, such that

$$\int_{-\infty}^{\infty} (e_{N_0+N_0^\gamma}(t))^2 dt > (1+\eta)\epsilon_T^2. \tag{2.49}$$

On the other hand, for all f and ϵ_T, what holds is a weaker bound of the type

$$\int_{-\infty}^{\infty} (e_{N_0}(t))^2 dt \leq \pi\epsilon_T + \epsilon_T^2. \tag{2.50}$$

To see what problem this creates, assume that the accuracy of our measurement is a small $\epsilon > 0$. Since ϵ_T approaches zero as $T \to \infty$ at a slower rate than ϵ_T^2, we can choose T so large that

$$\epsilon_T^2 < \epsilon < \epsilon_T \tag{2.51}$$

(see Figure 2.4). Consider now a signal that is practically timelimited to T, in the sense that our measurement cannot detect it after T seconds because its energy outside the observation interval, which is bounded by ϵ_T^2, has vanished to a level below ϵ. According to (2.50), this signal can be reconstructed by collecting N_0 samples placed inside the detectable time window with error at most of the order of ϵ_T, strictly above ϵ. Thus, an undetectable portion of the signal may lead to a detectable error. This undesirable physical behavior is due to the signal's energy approaching zero as $T \to \infty$ at a faster rate than the reconstruction error, something our imprecise statement of the folk theorem did not take into account. By (2.49), using the cardinal series with N_0 plus any number of terms that grows sublinearly with N_0, we cannot reconstruct every bandlimited and almost timelimited signal to a degree proportional to the unconcentrated part ϵ_T^2 of its energy. The degree of the approximation can only be bounded by the square root of the unconcentrated energy.

This opens the question of whether a better representation than the cardinal series can achieve an error proportional to the unconcentrated part of the signal's energy using a number of terms only slightly larger than N_0. It turns out that this optimal representation is obtained by solving the problem of what are the most time-concentrated, orthogonal, signals of given frequency bandwidth. These are the most concentrated signals that still satisfy the uncertainty principle (2.42), and can be used as the optimal basis set to represent bandlimited signals.

2.5 Slepian's Concentration Problem

There are problems with the statement of the folk theorem and the way the approximation error decays. In the early days of communication theory these were overlooked for some time. The dimension of the space of signals the communication engineer worked with was set to approximately N_0, for large ΩT. The precise meaning of "approximately" and "large" was cautiously swept under the carpet.

In the 1960s, David Slepian and his colleagues Henry Landau and Henry Pollak at Bell Laboratories, moved by an effort to have rigorous engineering models that are practically relevant, finally asked the right question, came up with a precise answer, and provided a theory that has been refined over the span of roughly two decades. In doing so, the discrete geometrical approach of representing continuous signals as points in a finite-dimensional space, put forth by Shannon, was placed on solid mathematical ground.

Slepian's concentration problem can be stated as follows. For a given $T > 0$, define the concentration of a signal $f(t)$ as

$$\alpha^2(T) = \frac{\int_{-T/2}^{T/2} f^2(t)dt}{\int_{-\infty}^{\infty} f^2(t)dt}, \tag{2.52}$$

namely the fraction of the signal's energy that lies in a given time interval of width T centered at the origin. If $f(t)$ is timelimited to T, then the concentration has its largest value, one. Similarly, for a given $\Omega > 0$, define the concentration of the spectrum of $f(t)$ as

$$\beta^2(\Omega) = \frac{\int_{-\Omega}^{\Omega} |F(\omega)|^2 d\omega}{\int_{-\infty}^{\infty} |F(\omega)|^2 d\omega}, \tag{2.53}$$

namely the fraction of the signal's energy that lies in a given frequency band of width 2Ω centered at the origin. If $f(t)$ is bandlimited to Ω, then the spectral concentration has its largest value, one. The concentration problem is now stated:

> Determine how large $\alpha(T)$ can be for $f(t) \in \mathscr{B}_\Omega(t)$, and how large $\beta(\Omega)$ can be for $f(t) \in \mathscr{T}_T$.

Solving this question leads to the identification of a family of bandlimited signals highly concentrated in the interval $[-T/2, T/2]$ that can serve as an optimal basis to represent all bandlimited signals. The dimension of this basis exhibits a phase transition in the neighborhood of $N_0 = \Omega T/\pi$, whose threshold window can be characterized precisely, providing a rigorous justification for the engineering folk theorem, and a solid foundation for the geometric approach to the study of communication with signals.

To investigate the concentration problem, we write

$$\alpha^2(T) = \frac{\int_{-T/2}^{T/2} dt \, \frac{1}{2\pi} \int_{-\Omega}^{\Omega} F(\tilde{\omega}) \exp(j\tilde{\omega}t) d\tilde{\omega} \, \frac{1}{2\pi} \int_{-\Omega}^{\Omega} F^*(\omega) \exp(-j\omega t) d\omega}{\frac{1}{2\pi} \int_{-\infty}^{\infty} |F(\omega)|^2 d\omega}$$

$$= \frac{\int_{-\Omega}^{\Omega} d\omega \int_{-\Omega}^{\Omega} d\tilde{\omega} \, \frac{\sin(\omega - \tilde{\omega})T/2}{\pi(\omega - \tilde{\omega})} F(\tilde{\omega}) F^*(\omega)}{\int_{-\Omega}^{\Omega} F(\omega) F^*(\omega) d\omega}. \tag{2.54}$$

To maximize $\alpha^2(T)$ we can solve the eigenvalue equation (see Problem 2.1)

$$\alpha^2(T) F(\omega) = \int_{-\Omega}^{\Omega} \frac{\sin \frac{T}{2}(\omega - \tilde{\omega})}{\pi(\omega - \tilde{\omega})} F(\tilde{\omega}) \, d\tilde{\omega}. \tag{2.55}$$

An analogous derivation leads to

$$\beta^2(\Omega) f(t) = \int_{-T/2}^{T/2} \frac{\sin \Omega(t - \tilde{t})}{\pi(t - \tilde{t})} f(\tilde{t}) \, d\tilde{t}. \tag{2.56}$$

Equations (2.55) and (2.56) can be written as

$$\int_{-1}^{1} \frac{\sin c_0(x - y)}{\pi(x - y)} \varphi(y) dy = \lambda \varphi(x), \quad |x| < 1, \tag{2.57}$$

where (2.55) is obtained by letting

$$y = \frac{\tilde{\omega}}{\Omega}, \; x = \frac{\omega}{\Omega}, \; \varphi(y) = F(\Omega y), \; \lambda = \alpha^2(T), \; c_0 = \frac{\Omega T}{2}, \tag{2.58}$$

and (2.56) is obtained by letting

$$y = \tilde{t}\,\frac{2}{T}, \; x = t\,\frac{2}{T}, \; \varphi(y) = f\left(\frac{T}{2} y\right), \; \lambda = \beta^2(\Omega), \; c_0 = \frac{\Omega T}{2}. \tag{2.59}$$

Equation (2.57) is recognized as a homogeneous Fredholm integral equation of the second kind. The Fredholm operator is compact, and the solutions to (2.57) are a countable set of real eigenfunctions $\varphi_0(x), \varphi_1(x), \varphi_2(x), \ldots$, orthogonal, and complete in $L^2(-1,1)$, and the corresponding eigenvalues $\lambda_0 \geq \lambda_1 \geq \lambda_2 \geq \cdots$ are positive, real, and such that $\lim_{n\to\infty} \lambda_n = 0$.

It follows that any signal $f(t) \in \mathscr{B}_\Omega$ can have at most a λ_0-fraction of its energy inside a time interval $[-T/2, T/2]$. The signal that achieves such a concentration is the one whose spectral representation is given by the eigenfunction $\varphi_0(x)$, evaluated in the interval $(-1,1)$, with an appropriate change of scale given by (2.58). Similarly, any signal $f(t) \in \mathscr{T}_T$ can have at most a λ_0-fraction of its energy inside a spectral interval $(-\Omega, \Omega)$, and the one that achieves such a concentration has a time representation again given by $\varphi_0(x)$, for $x \in (-1,1)$, with the appropriate change of scale given by (2.59).

2.5.1 A "Lucky Accident"

The solutions $\{\varphi_n(x)\}$ of the integral equation (2.57) can be obtained by solving the differential equation

$$\frac{d}{dx}(1-x^2)\frac{d\varphi(x)}{dx} + \left(\chi - c_0^2 x^2\right)\varphi(x) = 0, \quad |x| < 1. \tag{2.60}$$

This follows by defining the differential and integral operators

$$\mathcal{P}\varphi = \frac{d}{dx}(1-x^2)\frac{d\varphi}{dx} - c_0^2 x^2 \varphi, \tag{2.61}$$

$$\mathcal{Q}\varphi = \int_{-1}^{1} \frac{\sin c_0(x-y)}{\pi(x-y)}\varphi(y)dy, \tag{2.62}$$

and noticing the commutative property

$$\mathcal{Q}\mathcal{P}\varphi = \mathcal{P}\mathcal{Q}\varphi. \tag{2.63}$$

Since commuting operators admitting a complete set of eigenfunctions share the same eigenfunctions, it follows that the solutions of (2.60) are also solutions of (2.57). Slepian refers to this as a "lucky accident" that allowed him and his collaborators to find the solution to the concentration problem in terms of that of a well-known equation in physics.

The solutions of (2.60) arise in the context of the wave equation, and are known as the prolate spheroidal wave functions of the first kind and of order zero. They are a set of solutions of the Helmholtz equation when this is expressed in a suitable coordinate system. The corresponding eigenvalues are positive, discrete reals $\chi_0 \leq \chi_1 \leq \chi_2 \leq \cdots$. When the eigenfunctions are indexed by increasing values of χ, they agree with the notation of indexing by decreasing values of λ used above. It follows that the prolate spheroidal wave functions are the most concentrated, orthogonal, bandlimited functions, and enjoy many properties useful for applications in mathematical physics. Among those, one key property is that their energy falls off sharply beyond a critical phase transition point corresponding to values of the indexes in the neighborhood of $N_0 = \Omega T/\pi$. This leads to the notion of the asymptotic dimensionality of the space of bandlimited signals.

2.5.2 Most Concentrated Functions

The eigenfunctions $\{\varphi_n(x)\}$ of (2.57) are also well defined for all $x \in \mathbb{R}$. It can be shown that they are orthonormal in $L^2(-\infty,\infty)$ and complete in $\mathscr{B}_1$ there, as well as orthogonal and complete in $L^2(-1,1)$, as already noted. They have exactly n zeros in $(-1,1)$, and they are even or odd as n is even or odd. We have

$$\int_{-\infty}^{\infty} \varphi_n(x)\varphi_m(x)dx = \begin{cases} 1 & \text{if } n = m, \\ 0 & \text{otherwise;} \end{cases} \tag{2.64}$$

$$\int_{-1}^{1} \varphi_n(x)\varphi_m(x)dx = \begin{cases} \lambda_n & \text{if } n = m, \\ 0 & \text{otherwise.} \end{cases} \tag{2.65}$$

The notation conceals the fact that both the $\{\varphi_n\}$ and the $\{\lambda_n\}$ depend on the parameter $c_0 = \Omega T/2$. When necessary, we write $\lambda_n = \lambda_n(c_0)$, $\varphi_n(x) = \varphi_n(c_0,x)$. It turns out that $\{\lambda_n(c_0)\}$ are continuous functions of c_0, and for fixed c_0 they fall off rapidly with increasing n, once n exceeds the Nyquist number $N_0 = 2c_0/\pi = \Omega T/\pi$. This is the phase transition behavior referred to above, which allows the determination of the exact dimension of the space of bandlimited signals spanned by the eigenfunctions $\{\varphi_n\}$. Another important property is that the Fourier transform of $\varphi_n(x)$ restricted to $|x| < 1$ has the same form as $\varphi_n(x)$, except for a scale factor (see also Section 2.6.2 below), namely

$$j^n \sqrt{\frac{\lambda_n}{2\pi c_0}} \varphi_n(x) = \frac{1}{2\pi} \int_{-1}^{1} \varphi_n(s) \exp(jc_0xs)ds. \tag{2.66}$$

We can now define the signals

$$\psi_n(t) = \sqrt{\frac{2}{T}} \varphi_n\left(\frac{2t}{T}\right), \tag{2.67}$$

$$\Xi_n(\omega) = \frac{1}{j^n} \sqrt{\frac{2\pi}{\Omega}} \varphi_n\left(\frac{\omega}{\Omega}\right), \tag{2.68}$$

and by virtue of (2.66), the corresponding direct and inverse transforms are

$$\Psi_n(\omega) = \frac{1}{\sqrt{\lambda_n}} \Xi_n(\omega) \mathrm{rect}\,(\omega/(2\Omega)), \tag{2.69}$$

$$\xi_n(t) = \frac{1}{\sqrt{\lambda_n}} \psi_n(t) \mathrm{rect}\,(t/T). \tag{2.70}$$

It can be readily verified that

$$\int_{-\infty}^{\infty} \psi_n(t)\psi_m(t)dt = \begin{cases} 1 & \text{if } n = m, \\ 0 & \text{otherwise;} \end{cases} \tag{2.71}$$

$$\int_{-T/2}^{T/2} \psi_n(t)\psi_m(t)dt = \begin{cases} \lambda_n & \text{if } n = m, \\ 0 & \text{otherwise;} \end{cases} \tag{2.72}$$

$$\psi_n(t) \in \mathscr{B}_\Omega, \tag{2.73}$$

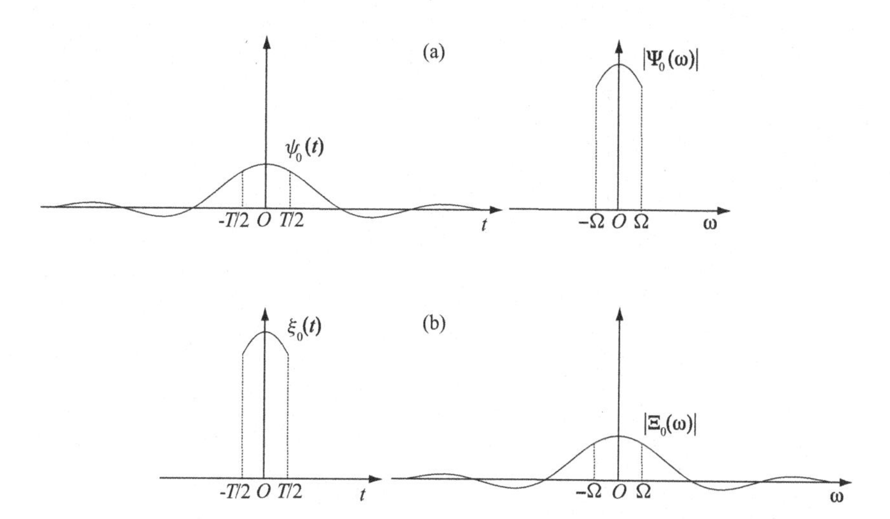

Fig. 2.5 Illustration of the minimum energy spill. In (a), $\psi_0(t)$ contains a λ_0-fraction of its energy in the interval $[-T/2, T/2]$. Its Fourier transform is $\Psi_0(\omega) \in \mathscr{B}_\Omega$. In (b), $\Xi_0(\omega)$ contains a λ_0-fraction of its energy in the interval $(-\Omega, \Omega)$. Its inverse Fourier transform is $\xi_0(t) \in \mathscr{T}_T$.

and

$$\frac{1}{2\pi}\int_{-\infty}^{\infty} \Xi_n(\omega)\Xi_m^*(\omega)d\omega = \begin{cases} 1 & \text{if } n = m, \\ 0 & \text{otherwise;} \end{cases} \tag{2.74}$$

$$\frac{1}{2\pi}\int_{-\Omega}^{\Omega} \Xi_n(\omega)\Xi_m^*(\omega)d\omega = \begin{cases} \lambda_n & \text{if } n = m, \\ 0 & \text{otherwise;} \end{cases} \tag{2.75}$$

$$\Xi_n(\omega) \in \mathscr{T}_T. \tag{2.76}$$

The signals $\psi_n(t) \in \mathscr{B}_\Omega$ viewed in the spectral domain have their whole energy inside the interval $(-\Omega, \Omega)$, while viewed in the time domain they have only a λ_n-fraction of their energy concentrated in the time interval $[-T/2, T/2]$. Their timelimited versions $\xi_n(t)$ have the opposite appearance: their whole energy is concentrated in the time interval $[-T/2, T/2]$, while only a λ_n-fraction of their energy lies inside the interval $(-\Omega, \Omega)$. Notice that $\xi_n(t)$ is obtained by timelimiting $\psi_n(t)$ and renormalizing by $1/\sqrt{\lambda_n}$ to have unit energy. The effect of the timelimitation operation thus appears as producing a corresponding energy spill in the spectrum. For $\psi_0(t)$ the timelimitation operation causes the minimum energy spill in the spectrum, as shown in Figure 2.5(a). A similar behavior is observed regarding the energy spill in time of $\psi_n(t)$ caused by the bandlimitation operation of $\Xi_n(\omega)$, as shown in Figure 2.5(b).

We conclude that $\psi_0(t)$ is the most concentrated signal in $[-T/2, T/2]$ among all signals in $\mathscr{B}_\Omega$, and its concentration is $\alpha^2(T) = \lambda_0$. Among signals in $\mathscr{B}_\Omega$ that are orthogonal to $\psi_0(t)$, $\psi_1(t)$ is the most concentrated; for it, $\alpha^2(T) = \lambda_1$. In general, $\psi_n(t)$, whose concentration is λ_n, is the most concentrated signal in $\mathscr{B}_\Omega$ that is orthogonal to $\psi_0, \psi_1, \ldots, \psi_{n-1}$. The same holds for signals $\Xi_n(\omega) \in \mathscr{T}_T$: the most concentrated is $\Xi_0(\omega)$, having concentration $\beta^2(\Omega) = \lambda_0$, followed by $\Xi_1(\omega)$, and so on.

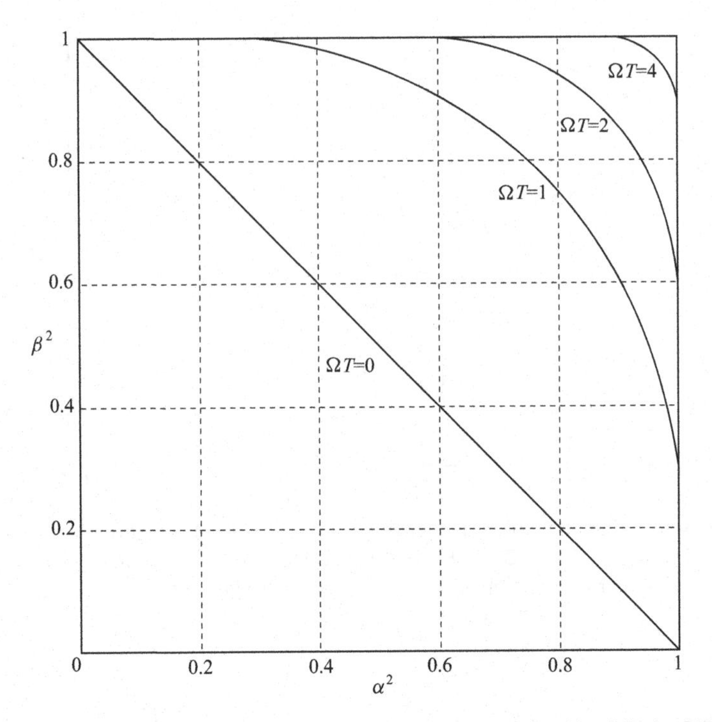

Fig. 2.6 Permissible regions of concentration lie below the curves parametrized by ΩT. As $\Omega T \to \infty$, higher concentrations are allowed.

The answer to the concentration problem is now stated:

$$\text{If } f(t) \in \mathscr{B}_\Omega \text{ then } \alpha(T) \le \sqrt{\lambda_0}; \tag{2.77}$$

$$\text{if } f(t) \in \mathscr{T}_T \text{ then } \beta(\Omega) \le \sqrt{\lambda_0}. \tag{2.78}$$

The permissible region of concentration can also be computed in greater generality as a function of α and β, and is depicted in Figure 2.6 for different values of ΩT. The excluded region of the square $0 \le \alpha^2 \le 1, 0 \le \beta^2 \le 1$ lies above the ellipse

$$\cos^{-1}\alpha + \cos^{-1}\beta = \cos^{-1}\sqrt{\lambda_0}. \tag{2.79}$$

Notice that as $\Omega T \to \infty$, higher concentrations are allowed.

2.5.3 Geometric View of Concentration

An alternative view of Slepian's concentration problem can be given in a geometric setting. In this context, Heisenberg's uncertainty principle corresponds to the impossibility of having an arbitrarily small angle between the subspaces corresponding to timelimited and bandlimited signals.

We begin by noticing that complex functions in $L^2(-\infty,\infty)$ form a Hilbert space with inner product

$$\langle f,g\rangle = \int_{-\infty}^{\infty} f(t)g^*(t)dt \tag{2.80}$$

and squared norm given by the energy,

$$\|f\|^2 = \langle f,f\rangle. \tag{2.81}$$

The signals in $\mathscr{T}_T$ and in $\mathscr{B}_\Omega$ form two linear subspaces of this Hilbert space. The observation made previously that a bandlimited function that vanishes for $|t| > T/2$ must vanish everywhere can be written as

$$\mathscr{T}_T \cap \mathscr{B}_\Omega = \emptyset, \tag{2.82}$$

namely the two subspaces have no function in common except the all-zero function.

The inner product allows us to define the angle between two functions f and g. First, notice that

$$-1 \le \frac{\Re\langle f,g\rangle}{\|f\|\cdot\|g\|} \le 1, \tag{2.83}$$

since

$$|\Re\langle f,g\rangle| \le |\langle f,g\rangle|, \tag{2.84}$$

and by the Schwarz inequality,

$$|\langle f,g\rangle| \le \|f\|\cdot\|g\|. \tag{2.85}$$

By (2.83), we can define the angle

$$\theta(f,g) = \cos^{-1}\frac{\Re\langle f,g\rangle}{\|f\|\cdot\|g\|}, \tag{2.86}$$

where the extreme values 0 and π for the angle $\theta(f,g)$ can be reached only if f and g are proportional (so that equality holds in the Schwarz inequality) and $\langle f,g\rangle$ is real. Now, define the projection operators that assign to every function its projection onto the subspace of timelimited and bandlimited functions:

$$\mathcal{T}f = \begin{cases} f & \text{if } |t| \le T/2, \\ 0 & \text{if } |t| > T/2; \end{cases} \tag{2.87}$$

$$\mathcal{B}f = \mathcal{F}^{-1}\widehat{\mathcal{B}}\mathcal{F}f, \tag{2.88}$$

where $\mathcal{F}$ indicates the Fourier transform, and

$$\widehat{\mathcal{B}}F = \begin{cases} F & \text{if } |\omega| \le \Omega, \\ 0 & \text{if } |\omega| > \Omega. \end{cases} \tag{2.89}$$

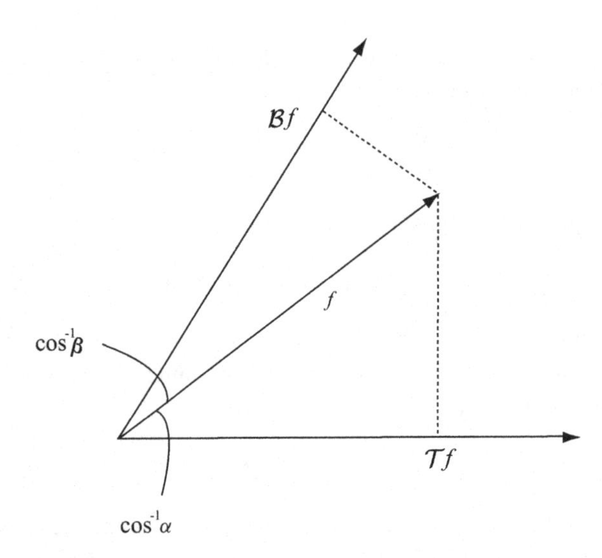

Fig. 2.7 Function f and its projection angles onto the timelimited and bandlimited subspaces. $\mathcal{F}f$ indicates the Fourier transform of f.

Recalling (2.52), (2.53), and Parseval's theorem, we have

$$\theta(f(t), \mathcal{T}f(t)) = \cos^{-1} \frac{\|\mathcal{T}f(t)\|}{\|f(t)\|} = \cos^{-1} \alpha, \tag{2.90}$$

$$\theta(f(t), \mathcal{B}f(t)) = \cos^{-1} \frac{\|\widehat{\mathcal{B}}F(\omega)\|}{\|F(\omega)\|} = \cos^{-1} \beta. \tag{2.91}$$

See Figure 2.7 for a representation.

The concentration result can now be viewed in terms of the existence of a least angle between the two subspaces $\mathscr{B}_\Omega$ and $\mathscr{T}_T$, so that a bandlimited function and a timelimited one cannot be very close together. If $f(t) \in \mathscr{B}_\Omega$ then $\beta = 1$, and by (2.77) this minimum angle equals $\cos^{-1}\sqrt{\lambda_0}$ and is achieved by $\varphi_0(2t/T) \in \mathscr{B}_\Omega$. On the other hand, if $f(t) \in \mathscr{T}_T$ then $\alpha = 1$ and, by (2.78), the minimum angle $\cos^{-1}\sqrt{\lambda_0}$ is achieved by $\varphi_0(\omega/\Omega) \in \mathscr{T}_T$ – see Figure 2.8. Finally, the requirement of being below the ellipses in Figure 2.6, namely

$$\cos^{-1} \alpha + \cos^{-1} \beta \geq \cos^{-1} \sqrt{\lambda_0}, \tag{2.92}$$

corresponds to the sum of the projection angles onto the timelimited and bandlimited subspaces being at least $\cos^{-1}\sqrt{\lambda_0}$.

2.6 Spheroidal Wave Functions

Many of the properties of the eigenfunctions and eigenvalues in (2.57) described in the previous section can be obtained through their interpretation as prolate spheroidal wave functions that are extensively studied in physics. These functions are used to describe

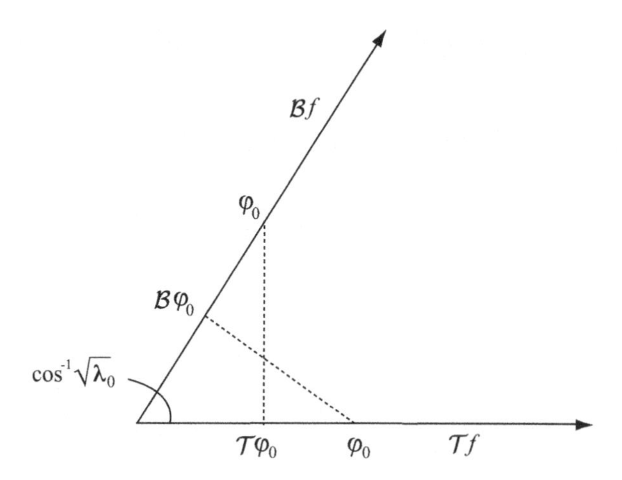

Fig. 2.8 The minimum angle between subspaces is achieved by the most concentrated function φ_0.

the spatial component of the wave propagation process, and we describe them in this context next.

2.6.1 The Wave Equation

The wave equation is a hyperbolic partial differential equation that describes the propagation of signals through a medium. It arises in a number of fields, including acoustics, electromagnetism, and fluid dynamics. In vector form, the homogeneous version is

$$\left(\nabla^2 - \frac{1}{v^2}\frac{\partial}{\partial t^2}\right)\mathbf{u}(\mathbf{r},t) = 0, \tag{2.93}$$

where $\mathbf{u}(\mathbf{r},t)$ is a space–time field, $\mathbf{r} = (x,y,z)$ is the spatial coordinate, ∇^2 is the spatial Laplacian, and v is the velocity of the wave. Here, we show an easy derivation of the scalar one-dimensional case, $\mathbf{u}(\mathbf{r},t) = u(x,t)$, obtained by considering the waves formed by a stretched elastic string. The analogous derivation for the vector case in three-dimensional space is given in Chapter 4, by considering the waves describing the propagation of electromagnetic signals.

With reference to Figure 2.9, consider the string to be initially horizontal. At time $t = 0$, we displace it in the vertical direction and then we release it, so that it starts oscillating. We want to write the equation that describes the vertical displacement $u(x,t)$ as a function of space x and time t. We assume that the density of the string ρ is constant and its displacement is small, so that the mass of a length $\Delta\ell$ of the string is $\rho\Delta\ell \approx \rho\Delta x$. Consider two points spaced by $\Delta\ell$ along the string. We write the tension forces there as F_1 and F_2 respectively. Since there is no horizontal motion, we have

$$F_1\cos\theta_1 = F_2\cos\theta_2 = F. \tag{2.94}$$

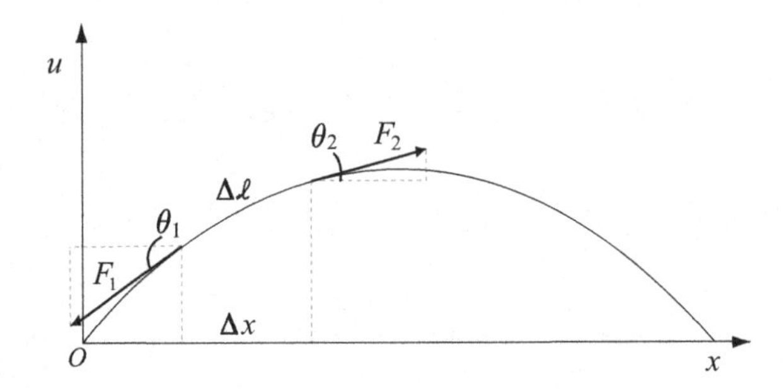

Fig. 2.9 Example of the one-dimensional, scalar wave equation: the vibrating string.

For the vertical components, we have

$$\rho \Delta x \frac{\partial^2 u(x,t)}{\partial t^2} = F_2 \sin\theta_2 - F_1 \sin\theta_1. \tag{2.95}$$

Dividing by $F\Delta x$ and using (2.94), we obtain

$$\begin{aligned} \frac{\rho}{F}\frac{\partial^2 u(x,t)}{\partial t^2} &= \frac{1}{\Delta x}(\tan\theta_2 - \tan\theta_1) \\ &= \frac{1}{\Delta x}\left(\frac{\partial u(x+\Delta x,t)}{\partial x} - \frac{\partial u(x,t)}{\partial x}\right), \end{aligned} \tag{2.96}$$

and letting $\Delta x \to 0$ we finally obtain

$$\frac{1}{v^2}\frac{\partial^2 u(x,t)}{\partial t^2} = \frac{\partial^2 u(x,t)}{\partial x^2}, \tag{2.97}$$

which is the one-dimensional, scalar version of (2.93), where $v^2 = F/\rho$ is the square of the propagation velocity of the vertical displacement of the string along the horizontal axis.

2.6.2 The Helmholtz Equation

We now look for a solution to the wave equation in the case of a scalar space–time field in three-dimensional space, $u(\mathbf{r},t)$, $\mathbf{r} = (x,y,z)$. Assuming $u(\mathbf{r},t)$ is separable, we write

$$u(\mathbf{r},t) = \gamma(\mathbf{r})\upsilon(t). \tag{2.98}$$

Substituting (2.98) into (2.93), we obtain

$$\frac{\nabla^2\gamma}{\gamma} = \frac{1}{v^2\upsilon}\frac{d^2\upsilon}{dt^2}. \tag{2.99}$$

Since the expression on the left-hand side depends only on $\mathbf{r}$, whereas the right-hand side depends only on t, both sides of the equation must be equal to a constant value.

Hence, we obtain the two coupled equations

$$\begin{cases} \dfrac{\nabla^2 \gamma}{\gamma} = -a^2, & (2.100a) \\ \dfrac{1}{v^2 \upsilon} \dfrac{d^2 \upsilon}{dt^2} = -a^2. & (2.100b) \end{cases}$$

We rewrite the time-independent component (2.100a) as

$$(\nabla^2 + a^2)\gamma(\mathbf{r}) = 0, \tag{2.101}$$

which is an elliptic partial differential equation called the scalar Helmholtz equation. This can also be solved by separation of variables, in a suitable coordinate system.

Two of the eleven coordinate systems in which the scalar Helmholtz equation (2.101) is separable are the prolate and oblate spheroidal coordinates.[4] These are formed by rotating the two-dimensional elliptic coordinate system, consisting of ellipses and hyperbolas, about the major and minor axes of the ellipses, respectively. Due to the geometrical symmetry, solutions obtained in the prolate and those in the oblate spheroidal system are related by simple transformations. We focus on solutions in the prolate case. The prolate coordinate system with interfocal distance d is depicted in Figure 2.10. The orthogonal curvilinear coordinates are (η, ξ, θ), with

$$x = d\xi\eta, \tag{2.102}$$

$$y = d\sqrt{(\xi^2 - 1)(1 - \eta^2)}\cos\theta, \tag{2.103}$$

$$z = d\sqrt{(\xi^2 - 1)(1 - \eta^2)}\sin\theta, \tag{2.104}$$

$$|\xi| > 1 \text{ and } |\eta| < 1. \tag{2.105}$$

In the limit of $d \to 0$ the system reduces to the spherical coordinate system. For d finite, the surface of constant ξ becomes spherical as $\xi \to \infty$.

Writing (2.101) in prolate spheroidal coordinates, and letting $c_0 = ad/2$, the separated solutions are

$$\begin{cases} \gamma_{n,m}(\eta, \xi, \theta) = S_{m,n}(c_0, \eta) R_{m,n}(c_0, \xi)\cos(m\theta), & (2.106a) \\ \gamma_{n,m}(\eta, \xi, \theta) = S_{m,n}(c_0, \eta) R_{m,n}(c_0, \xi)\sin(m\theta), & (2.106b) \end{cases}$$

where $\{S_{m,n}(c_0, \eta)\}$ and $\{R_{m,n}(c_0, \xi)\}$ are called angular prolate spheroidal functions of order m and degree n, and radial prolate spheroidal functions of order m and degree n, respectively.

[4] The complete set is composed of: Cartesian, confocal ellipsoidal, confocal paraboloidal, conical, cylindrical, elliptic cylindrical, paraboloidal, parabolic cylindrical, prolate spheroidal, oblate spheroidal, and spherical coordinates. Laplace's equation is obtained for $a = 0$ and is separable in the two additional bispherical coordinates and toroidal coordinates.

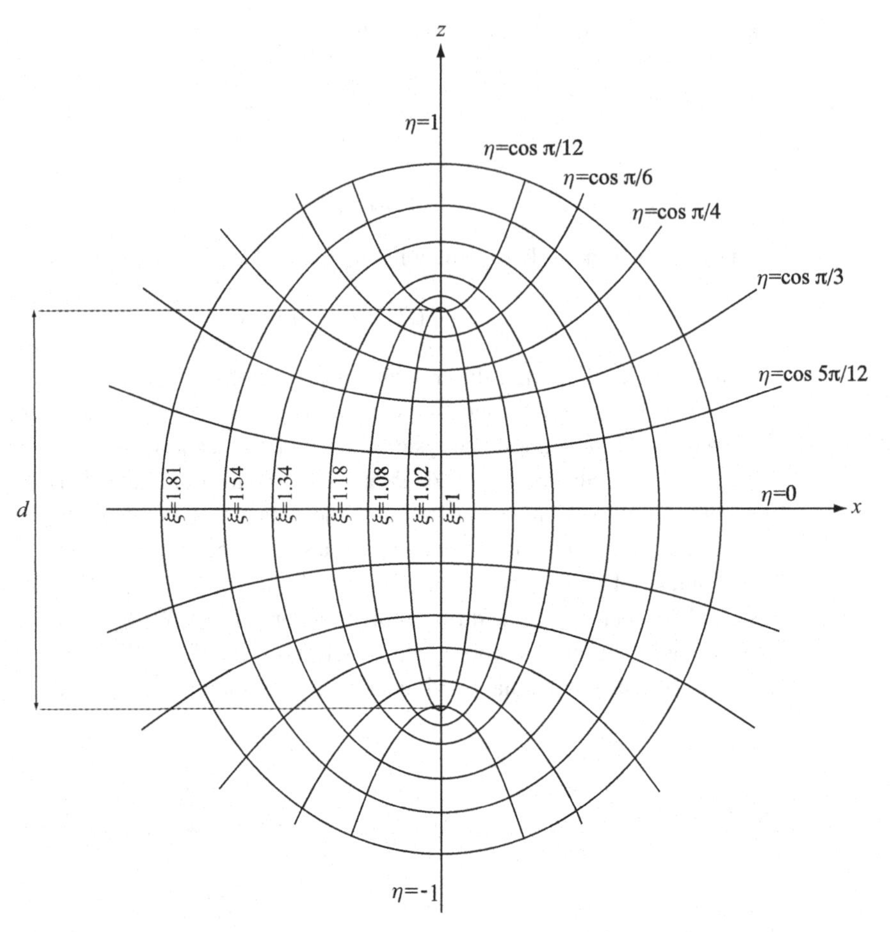

Fig. 2.10 The prolate spheroidal coordinate system. The coordinates η and ξ are depicted. The third coordinate θ corresponds to the rotation angle around the z axis, starting from the plane $y = 0$ and moving towards the oriented y axis.

Both angular and radial functions are solutions of the Sturm–Liouville eigenvalue problem

$$\frac{d}{dx}(1-x^2)\frac{df(c_0,x)}{dx} + \left(\chi_{m,n} - c_0^2 x^2 - \frac{m^2}{1-x^2}\right)f(c_0,x) = 0. \tag{2.107}$$

However, the range of the variable is different: for the radial wave function, $|x| > 1$; for the angular wave function, $|x| < 1$. Solutions that are finite at $x = \pm 1$ are called prolate spheroidal functions of the first kind. We focus on solutions of the first kind and of order zero. In this case, the ordinary differential equation (2.107) simplifies to

$$\frac{d}{dx}(1-x^2)\frac{df(c_0,x)}{dx} + \left(\chi_{0,n} - c_0^2 x^2\right)f(c_0,x) = 0, \tag{2.108}$$

and a first set of solutions obtained for $|x| < 1$ are the eigenfunctions $\{S_{0,n}(c_0,x)\}$, each corresponding to a real positive eigenvalue in the set $\chi_{0,0}(c_0) \leq \chi_{0,1}(c_0) \leq \chi_{0,2}(c_0) \leq \cdots$.

Each eigenfunction is real for real x, and can be extended to be an entire function on the whole complex plane. The eigenfunctions are orthogonal and complete in $L^2(-1,1)$. $S_{0,n}(c_0,t)$ has exactly n zeros in $(-1,1)$, and is even or odd as n is even or odd. The eigenvalues are continuous functions of c_0 and $\chi_{0,n}(0) = n(n+1)$, $n = 0,1,2,\ldots$

A second set of solutions, obtained for $|x| > 1$, are the radial functions, which differ from the angular functions only by a real scale factor, namely

$$S_{0,n}(c_0,x) = \kappa_n(c_0)R_{0,n}(c_0,x). \tag{2.109}$$

The functions $\{S_{0,n}(c_0,x)\}$ and $\{R_{0,n}(c_0,x)\}$ have a number of important properties that can be deduced from (2.108). For $c_0 = 0$, we have $S_{0,n}(0,x) = P_n(x)$, the nth Legendre polynomial.

There are several expansions, approximations, and integral relations for the prolate spheroidal wave functions in the literature. Among these, we have

$$\frac{2c_0}{\pi}R_{0,n}^2(c_0,1)S_{0,n}(c_0,x) = \int_{-1}^{1}\frac{\sin c_0(x-s)}{\pi(x-s)}S_{0,n}(c_0,s)ds. \tag{2.110}$$

From this, we obtain the integral equation (2.57) studied in the previous section, letting

$$\lambda_n(c_0) = \frac{2c_0}{\pi}R_{0,n}^2(c_0,1), \tag{2.111}$$

and using the normalization

$$\varphi_n(c_0,x) = \sqrt{\frac{\lambda_n(c_0)}{\int_{-1}^{1}S_{0,n}^2(c_0,x)dx}}\,S_{0,n}(c_0,x), \tag{2.112}$$

where the notation of indexing the eigenfunctions $\{S_{0,n}\}$ by increasing values of the eigenvalues $\{\chi_n\}$ agrees with the indexing of the eigenfunctions $\{\psi_n\}$ by decreasing values of the eigenvalues $\{\lambda_n\}$. Another useful integral relation is

$$2j^nR_{0,n}(c_0,1)S_{0,n}(c_0,x) = \int_{-1}^{1}\exp(jc_0xs)S_{0,n}(c_0,s)ds, \tag{2.113}$$

which leads to (2.66) and the Fourier transform pairs studied thereafter. Figure 2.11 puts (2.113) into action, as it shows the energy spill of $S_{0,0}(1,s)$ when it is restricted to the interval $|s| < 1$ and inverse Fourier transformed, obtaining the scaled signal $R_{0,0}(1,1)S_{0,0}(1,x)/\pi$. The result is analogous to that in Figure 2.5.

2.7 Series Representations

The orthogonality and concentration properties of the prolate spheroidal functions make them a natural candidate for the reconstruction of bandlimited signals observed over a finite domain. The possibility of extrapolating the whole signal from a local observation should not be surprising. Since $f \in \mathscr{B}_\Omega$ is an entire function, a familiar representation of f is in terms of its Taylor series, which must converge at any point. In principle, one could extrapolate the whole function by calculating successive derivatives at a single point. In some sense, the bandlimitation property makes the signal so smooth

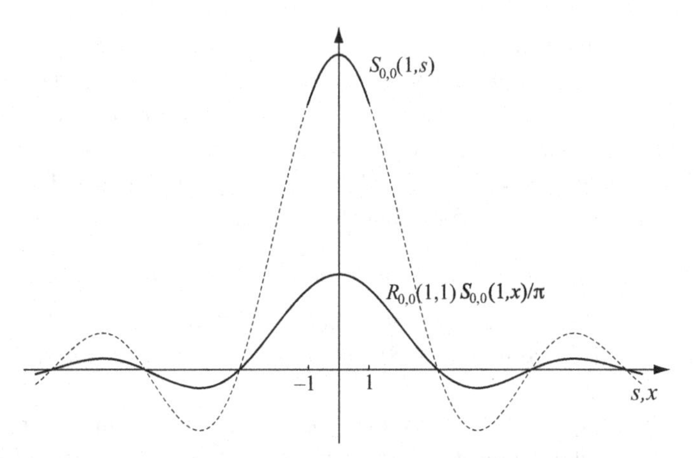

Fig. 2.11 The inverse transform of $S_{0,0}(1,s)$ restricted to $|s| < 1$ has the same form as $S_{0,0}(1,s)$, except for a scale factor. Both representations are shown on the same plot for comparison.

that knowing it with arbitrary precision at some point allows us to predict its behavior everywhere. In practice, however, the resulting finite-degree polynomial of a truncated Taylor series would give a poor approximation to the function sufficiently far from the point of expansion, no matter how large the number of terms retained.

The cardinal series formula provides a first alternative based on an orthogonal representation. Although not requiring knowledge of derivatives at any given point, it does require observing the signal at different sampling points over the whole real line. It then interpolates the signal using $\text{sinc}(\cdot)$ functions placed at those sampling points. Truncating the formula to a sufficiently large number of terms leads to a reconstruction error that remains small for all t. However, we have seen in Section 2.4.1 that the error decay is of the order of the square root of the unconcentrated part of the signal's energy. This makes the formula attractive mainly for the simplicity of the practical realization through sampling and low-pass filtering.

The prolate spheroidal functions provide a second, more attractive, alternative. Thanks to their double orthogonality property over both $(-\infty,\infty)$ and $[-T/2,T/2]$, they combine the advantages of the Taylor and the cardinal series. On the one hand, it is possible to extrapolate the signal everywhere using only local knowledge of the signal over $[-T/2,T/2]$. On the other hand, taking a sufficiently large number of interpolating functions, the reconstruction error remains small for all t. In addition to these two properties, the reconstruction is also optimal in a precise energy sense.

2.7.1 Prolate Spheroidal Orthogonal Representation

Let $f(t) \in \mathscr{B}_\Omega$; we wish to approximate f using the orthonormal prolate spheroidal basis functions $\{\psi_n\}$. We consider the N-dimensional approximation

$$f_N(t) = \sum_{n=0}^{N-1} a_n \psi_n(t), \tag{2.114}$$

where, since the functions $\{\psi_n\}$ are real, the coefficients are given by

$$a_n = \int_{-\infty}^{\infty} f(t)\psi_n(t)dt. \tag{2.115}$$

It is easy to show that this choice of coefficients, which is valid for any real orthonormal basis, minimizes the energy of the error associated with the approximation. This follows by letting

$$g_N = \sum_{n=0}^{N-1} b_n \psi_n(t), \tag{2.116}$$

computing the energy of the approximation error for g_N using the orthonormality (2.71) of ψ_n in $(-\infty, \infty)$,

$$\begin{aligned}
\int_{-\infty}^{\infty} (f(t) - g_N(t))^2 dt &= \int_{-\infty}^{\infty} (f(t) - \sum_{n=0}^{N-1} b_n \psi_n(t))^2 dt \\
&= \int_{-\infty}^{\infty} f^2(t)dt + \sum_{n=0}^{N-1} b_n^2 - 2\sum_{n=0}^{N-1} b_n \int_{-\infty}^{\infty} f(t)\psi_n(t)dt \\
&= \int_{-\infty}^{\infty} f^2(t)dt + \sum_{n=0}^{N-1} b_n^2 - 2\sum_{n=0}^{N-1} a_n b_n \\
&= \int_{-\infty}^{\infty} f^2(t)dt - \sum_{n=0}^{N-1} a_n^2 + \sum_{n=0}^{N-1} (a_n - b_n)^2,
\end{aligned} \tag{2.117}$$

and noticing that (2.117) is minimum when its last term is zero, namely when

$$b_n = a_n, \quad 0 \le n \le N-1. \tag{2.118}$$

It follows that the minimum energy of the error is

$$\begin{aligned}
\int_{-\infty}^{\infty} (e_N(t))^2 dt &= \int_{-\infty}^{\infty} (f(t) - f_N(t))^2 dt \\
&= \int_{-\infty}^{\infty} f^2(t)dt - \sum_{n=0}^{N-1} a_n^2.
\end{aligned} \tag{2.119}$$

Since the energy of the error is non-negative for all N, it follows that

$$\sum_{n=0}^{N-1} a_n^2 \le \int_{-\infty}^{\infty} f^2(t)dt, \tag{2.120}$$

and since N is arbitrary and the right-hand side does not depend on N, we also have

$$\sum_{n=0}^{\infty} a_n^2 \le \int_{-\infty}^{\infty} f^2(t)dt < \infty, \tag{2.121}$$

that is, *Bessel's inequality*.

At this point, it is not yet clear whether (2.119) converges to zero or not. Indeed, the *Riesz–Fischer theorem* ensures that

$$\lim_{N\to\infty}\int_{-\infty}^{\infty}(e_N(t))^2 dt = 0, \tag{2.122}$$

so that $f(t)$ can be represented in the $L^2(-\infty,\infty)$ sense by the series

$$f(t) = \sum_{n=0}^{\infty} a_n \psi_n(t). \tag{2.123}$$

It follows that in this case Bessel's inequality (2.121) assumes Parseval's form,

$$\sum_{n=0}^{\infty} a_n^2 = \int_{-\infty}^{\infty} f^2(t)dt. \tag{2.124}$$

Substituting (2.124) into (2.119), we have

$$\int_{-\infty}^{\infty}(e_N(t))^2 dt = \sum_{n=N}^{\infty} a_n^2, \tag{2.125}$$

which can be made as small as desired by taking N large, by virtue of (2.121).

By analogous reasoning, using the orthogonality (2.72) of the $\{\psi_n\}$ in $[-T/2, T/2]$, we have

$$a_n = \frac{1}{\lambda_n}\int_{-T/2}^{T/2} f(t)\psi_n(t)dt, \tag{2.126}$$

and the quality of the fit of f_N to f in the interval $[-T/2, T/2]$ is

$$\int_{-T/2}^{T/2}(e_N(t))^2 dt = \sum_{n=N}^{\infty} a_n^2 \lambda_n. \tag{2.127}$$

Notice that since $\lambda_n \to 0$, it is possible that (2.127) is small for values of N for which (2.125) is still large. Perhaps not surprisingly, the goodness of fit inside the interval is not an indication of the goodness of fit elsewhere.

2.7.2 Other Orthogonal Representations

The prolate spheroidal representation is only one example of the decomposition of a normed space over an orthogonal basis set. Another example commonly encountered is the Fourier series (2.6), where the basis functions are the complex exponentials $\{\exp(jn\omega_0 t)\}$ whose coefficients are given by (2.8), and Parseval's theorem is given by (2.9). This representation can be used to represent any square-integrable signal over the finite interval $[-T_0/2, T_0/2]$, where $T_0 = 2\pi/\omega_0$. A third, also common, example is the cardinal series representation (2.10) that uses the basis functions $\{\text{sinc}(\Omega t - n\pi)\}$ that are orthogonal over the whole real line, complete over the class of square-integrable signals that are bandlimited to the interval $[-\Omega, \Omega]$, and whose coefficients are the sampled signal's values taken every interval of length π/Ω. Thus, they can be used to represent any bandlimited signal over the whole real line. More examples of orthogonal decompositions are given in Appendix A.3.

Among the different representations, the prolate spheroidal one is optimal in a precise approximation-theoretic sense. It provides the minimum energy error for all functions in the space, among all possible basis representations. This result is described next, and leads to the notion of information associated with the waveform in terms of the asymptotic dimensionality of the optimal approximating subspace.

2.7.3 Minimum Energy Error

Let $\mathscr{E}(\epsilon_T)$ be the set of signals $f(t) \in \mathscr{B}_\Omega$ subject to the energy constraint (2.43) and concentration constraint (2.44). The best basis $\{u_n\}$ that, for any given N, achieves the minimum energy error for all signals in the space,

$$\inf_{\{u_n\}_0^{N-1}} \sup_{f\in\mathscr{E}(\epsilon_T)} \inf_{\{a_n\}_0^{N-1}} \int_{-\infty}^{\infty} \left(f(t) - \sum_{n=0}^{N-1} a_n u_n(t) \right)^2 dt, \tag{2.128}$$

is composed of the N linearly independent, most concentrated, bandlimited functions, which are the prolate spheroidal ones $\{\psi_n(t)\}$. In this case, we have

$$\int_{-\infty}^{\infty} (e_{N_0}(t))^2 dt = O(\epsilon_T^2) \text{ as } \epsilon_T \to 0, \tag{2.129}$$

namely, the N_0 best basis functions suffice to approximate any concentrated signal to a degree proportional to the "unconcentrated" part ϵ_T^2 of its energy. This main result, which we shall prove in the next chapter, should be contrasted with (2.50), mentioned earlier for the cardinal series.

We can then ask: what does it take to bring the proportionality constant in the order notation in (2.129) arbitrarily close to one? It turns out that for any $\epsilon > 0$, a number of functions $N = \Omega T/\pi + N'$ for any constant N' cannot suffice for approximating f to within $(1+\epsilon)\epsilon_T^2$, but a logarithmically growing extra number of terms does. Namely, there is a constant $C(\epsilon)$ such that, for $N = N_0 + C \log N_0$ and T large enough, we have

$$\int_{-\infty}^{\infty} (e_N(t))^2 dt \leq (1+\epsilon)\epsilon_T^2. \tag{2.130}$$

The two results in (2.129) and (2.130), the latter attributed by Landau and Pollak (1962) to Shannon,[5] are very much related to the width of the n-interval in which the eigenvalues $\{\lambda_n\}$ fall from one to zero. We examine this transition next, and return to the proof of (2.129) and (2.130) in the next chapter.

Figure 2.12 shows the behavior of the eigenvalues for different values of $N_0 = \Omega T/\pi$. It appears that for $n \ll N_0$ most of the eigenvalues are close to one, while for $n \gg N_0$

[5] In a conversation with the author, Landau remarked that the problem of determining the precise width of the transition window of the eigenvalues was posed to him by Shannon. Indeed, Landau and Pollak (1962) attribute Theorem 4 of their paper to Shannon. Landau solved the problem completely, together with Widom, in 1980.

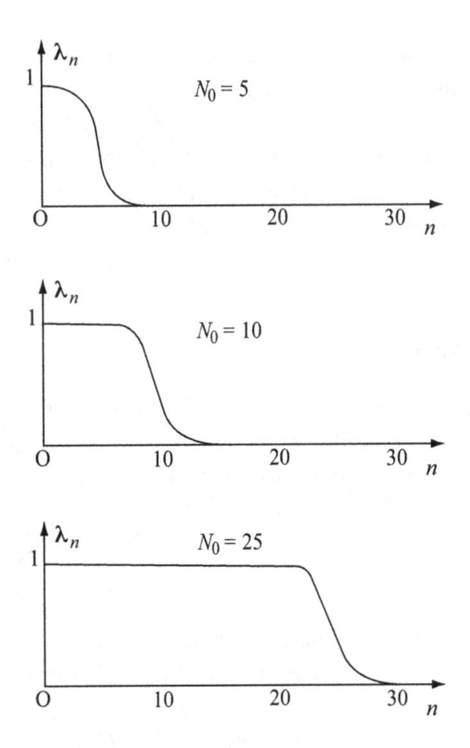

Fig. 2.12 Behavior of the eigenvalues for different values of N_0.

most of them are nearly zero. When $n \approx N_0$, $\lambda_n \approx 1/2$. As N_0 gets large, the width of the n-interval of the transition increases as $\log N_0$.

The exact width of the transition can be computed. It was first established non-rigorously by Slepian (1965), and later rigorously by Landau and Widom (1980). The result is as follows:

$$\lim_{N_0\to\infty} \lambda_n(N_0) = \begin{cases} 0 & \text{for } n = \lfloor (1+\nu)N_0 \rfloor, \\ (1+\exp(k\pi^2))^{-1} & \text{for } n = \lfloor N_0 + k\log(N_0\pi/2) \rfloor, \\ 1 & \text{for } n = \lfloor (1-\nu)N_0 \rfloor, \end{cases} \tag{2.131}$$

where ν is an arbitrarily small positive constant, and we have used the notation $\lfloor\cdot\rfloor$ to indicate the largest integer less than or equal to the argument inside the brackets.

Pictorially, the phase transition represented by (2.131) can be better viewed at the scale of N_0. When the eigenvalues are plotted on an index scale normalized by N_0, Figure 2.13 shows that the transition becomes sharper as $N_0 \to \infty$ and tends to become a step function. From (2.131), it follows that for all $0 < \epsilon < 1$, choosing

$$\begin{aligned} n &= N_0 + \frac{1}{\pi^2}\log\left(\frac{1-\epsilon}{\epsilon}\right)\log(N_0\pi/2) + o(\log N_0) \\ &= N_0 + O(\log N_0) \quad \text{as } N_0 \to \infty, \end{aligned} \tag{2.132}$$

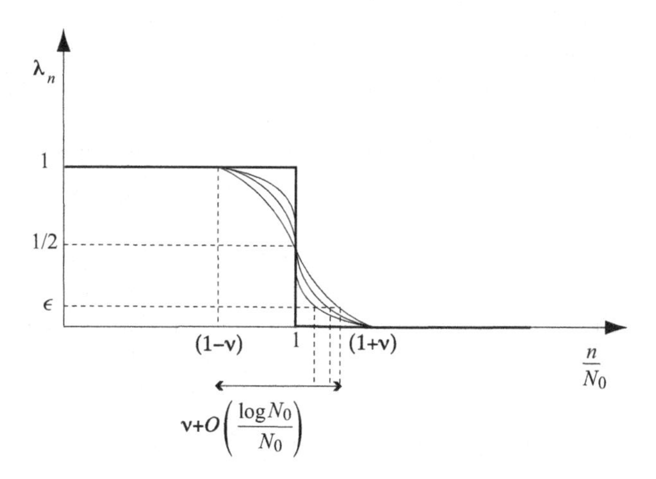

Fig. 2.13 Phase transition of the eigenvalues. Viewed at the scale of N_0, the transition occurs within an arbitrarily small interval of 2ν. The transition becomes sharper as $N_0 \to \infty$, and tends to become a step function.

we have $\lambda_n(N_0) < \epsilon$. This result suggests that as $T \to \infty$, the number of eigenfunctions that are useful to interpolate bandlimited signals within ϵ accuracy is only slightly larger than N_0, because all the remaining eigenfunctions have negligible energy inside the interval $[-T/2, T/2]$ and thus they provide, as $T \to \infty$, a negligible contribution to the reconstruction. We are then led to define the asymptotic dimension of the signals' space, as $T \to \infty$, to be the number in (2.132). The dimension of the signals' space defined in this way depends only logarithmically on the accuracy of the approximation ϵ, and only in its second-order term, which is a very desirable property in practice.

In view of the above arguments, Shannon's program of having a geometric theory of signals, based on finite ensembles of discrete numbers and supported by a rigorous version of the folk theorem, seems now to be within reach. To complete the picture, however, we need to make the above considerations rigorous, and provide the connection between the behavior of the eigenvalues (2.131) and the approximation error bounds (2.129) and (2.130). This can be done in the context of the Kolmogorv N-width, a topic discussed in the next chapter.

2.8 Summary and Further Reading

The leitmotiv of this chapter has been that bandlimited signals used for communication can be represented as points in high-dimensional space. This follows from the cardinal series sampling representation of signals that was introduced by Shannon (1948, 1949) in information theory, and by Kotelnikov (1933) in the Russian literature. In mathematics, it is known as the cardinal series and goes as far back as to Whittaker (1915), or even earlier to Borel (1897). An excellent historical account on the origin of the cardinal series is given in Higgins' (1985) paper. Extensive reviews on sampling theory are in the papers by Jerry (1977), Unser (2000), and Garcia (2000).

Shannon's (1949) landmark paper contains many of the fundamental ideas regarding the signals' space used for communication. It was perhaps written earlier than his most celebrated work, which appeared in 1948, adopting more of an engineering point of view than a mathematical one, and focusing on continuous-time waveforms. The exact time placement of Shannon's (1949) paper has often been a source of debate in the history of science. Since the original publication states a submission date of 1940, some argued that it was written in 1940 but remained classified until after the end of World War II. Shannon, however, stated in an interview that the 1940 date must have been a typographical error. Nevertheless, he cites this work in his earlier 1948 paper, leading to the belief that its contents were probably in circulation in the United States by 1948.

Shannon's work posed the fundamental problem of determining the interface between the continuous world of signals and the discrete world of information, namely how long a discrete sequence of real numbers must be to represent continuous signals. On the one hand, some non-rigorous arguments suggest that this should be close to $\Omega T/\pi$, where Ω is the bandwidth of the signal and T its time duration. We called this critical length the Nyquist number after Nyquist (1928), who pointed out the fundamental importance of the time interval π/Ω in connection with telegraphy. On the other hand, signals cannot be simultaneously timelimited and bandlimited, and soon one realizes that asymptotics must come into the picture.

In an analogy with physics, the indeterminacy principle of Heisenberg (1927) prevents signals being simultaneously concentrated in time and frequency, and leads to what we have called Slepian's concentration problem. This has been investigated for about twenty years and the main reference is a sequence of five papers by Slepian and Pollak (1961), Landau and Pollak (1961, 1962), and Slepian (1964, 1978). Two survey papers by Slepian (1976, 1983) and one by Landau (1985) also outline some of the main ideas. Details on the asymptotic analysis of the eigenvalues can be found in an additional paper by Slepian (1965) and Landau and Widom (1980). We also point out the monograph by Flammer (1957), which contains many details on prolate spheroidal functions, and the one by Hogan and Lakey (2012) containing some updated material. Our treatment of the problem loosely followed the references above, putting more emphasis on the physical insights.

The cardinal series provides a suboptimal representation of bandlimited signals in terms of discrete coefficients achieving a given reconstruction error that are the sampled values of the signal. An optimal representation is obtained by interpolating the prolate spheroidal functions that are solutions of Slepian's concentration problem. The coefficients of this orthogonal representation, however, are not the sampled signal's values. Alternative sampling representations with good truncation errors have been proposed in the literature, including the approximate prolate spheroidal sampling functions introduced by Knab (1979, 1983).

Rigorous derivations of Heisenberg's uncertainty principle are due to Kennard (1927) and Weyl (1928). The entropic uncertainty principle was proved by Beckner (1975), following a conjecture by Hirschman (1957), while the case of concentration over arbitrary measurable sets was proved by Donoho and Stark (1989). We provide their derivation in Appendix A.2. A mathematical survey of the uncertainty principle appears in Folland and Sitaram (1997). The concentration properties of Gaussian signals led Gabor (1946) to propose using shifted and modulated Gaussians as non-orthogonal basis functions to represent signals. He called these basis functions "logons," from the Greek λέγω, representing the minimal volume that any signal can occupy in the time–frequency plane according to the uncertainty principle. These kinds of representation are at the basis of wavelet analysis in signal processing. Vetterli, Kovačević, and Goyal (2014a, 2014b) provide an up to date overview of the field.

Finally, we wish to make some additional remarks on an important aspect of this chapter: the encounter with the concept of a phase transition, which arises repeatedly throughout the book. Phase transitions are commonly encountered in information theory and statistical physics when stochastic concentration around typical sets occurs. Here, we have instead encountered it in the deterministic setting of vector spaces subject to regularity constraints. The idea is that when certain asymptotic regimes are studied under a set of constraints, a concentration behavior occurs and a sharp transition can be identified. In our case, the transition is due to the bandlimitation constraint and reveals the limit on the amount of information associated with this class of signals.

2.9 Test Your Understanding

Problems

2.1 Show that $\alpha^2(T) \leq \lambda_0$, and that to achieve the highest concentration λ_0 we must solve (2.55).

Solution

We use the compact notation defined in Section 2.5.3, and also let $\widehat{\mathcal{T}}F = \mathcal{F}\mathcal{T}\mathcal{F}^{-1}F$. We let $\{\varphi_n\}$ be solutions of $\widehat{\mathcal{B}\mathcal{T}}\varphi_n = \lambda\varphi_n$. By the Schwarz inequality and the completeness of $\{\varphi_n\}$, we have:

$$\begin{aligned}
\alpha^2 &= \frac{\|\mathcal{T}f\|^2}{\|f\|^2} \\
&= \frac{\langle \mathcal{T}f, f\rangle}{\|f\|^2} \\
&= \frac{\langle \widehat{\mathcal{T}}F, F\rangle}{\|F\|^2} \\
&= \frac{\langle \widehat{\mathcal{B}\mathcal{T}}F, F\rangle}{\|F\|^2} \\
&\leq \frac{\|\widehat{\mathcal{B}\mathcal{T}}F\| \, \|F\|}{\|F\|^2} \\
&= \frac{\|\widehat{\mathcal{B}\mathcal{T}} \sum_{n=0}^{\infty} a_n \varphi_n\|}{\|F\|}
\end{aligned}$$

$$
\begin{aligned}
&= \frac{\| \sum_{n=0}^{\infty} a_n \lambda_n \varphi_n \|}{\|F\|} \\
&\leq \lambda_0 \frac{\| \sum_{n=0}^{\infty} a_n \varphi_n \|}{\|F\|} \\
&= \lambda_0. \qquad (2.133)
\end{aligned}
$$

To achieve the above bound with equality we can now pick a bandlimited function F to be the solution corresponding to λ_0 of the eigenvalue equation

$$\widehat{\mathcal{B}}\widehat{\mathcal{T}}F = \alpha^2 F, \qquad (2.134)$$

which corresponds to (2.55).

2.2 Determine $\lim_{c_0 \to 0} \lambda_n(c_0)$ for all n. (Hint: recall the energy interpretation of λ_n).

2.3 Determine the relation between ΩT in Slepian's concentration problem and the geometry of the prolate spheroidal coordinates in which the Helmholtz equation can be solved. What does the limit $T \to \infty$ in Slepian's case correspond to in the context of the time-independent solution of the wave equation in prolate spheroidal coordinates?

2.4 Verify that (2.113) leads to (2.66), and use it to derive the Fourier transforms (2.69) and (2.70).

2.5 Verify the orthonormality relations (2.71), (2.72) and (2.74), (2.75).

2.6 Rewrite the uncertainty principle for arbitrarily measurable sets (2.42) in the case that the signal is subject to the energy constraint

$$\int_{-\infty}^{\infty} f^2(t) \leq E, \qquad (2.135)$$

and provide the corresponding proof by modifying the proof in Appendix A.2.

2.7 Show that the set of functions $\{1, x, x^2, \ldots\}$ is not pairwise orthogonal. Is it possible to provide a representation of $f \in \mathcal{B}_\Omega$ using such a set?

2.8 Show that a signal with Gaussian instantaneous power achieves the highest simultaneous time–frequency concentration permitted by Heisenberg's uncertainty principle,

$$\sigma_t^2 \sigma_\omega^2 = \frac{1}{4}. \qquad (2.136)$$

Solution

Let $f(t)$ be the unit-energy signal with Gaussian instantaneous power,

$$f^2(t) = \frac{1}{\sqrt{2\pi}\,\sigma} \exp[-t^2/(2\sigma^2)]. \qquad (2.137)$$

The Fourier transform of the Gaussian in (2.137) can be computed to be

$$\mathcal{F}f^2 = \exp(-\omega^2\sigma^2/2). \qquad (2.138)$$

The Fourier transform of the square root of (2.137) can be computed using (2.138) to be

$$\mathcal{F}f = \frac{\sqrt{2\pi}\sqrt{2}\sigma}{\sqrt{\sqrt{2\pi}\sigma}} \exp(-\sigma^2\omega^2). \tag{2.139}$$

From this, it follows that

$$\begin{aligned}(\mathcal{F}f)^2 &= \sqrt{2\pi}2\sigma \exp(-2\sigma^2\omega^2) \\ &= 2\pi \frac{2\sigma}{\sqrt{2\pi}} \exp\left\{-\frac{\omega^2}{2[1/(2\sigma)^2]}\right\}.\end{aligned} \tag{2.140}$$

We then have

$$\sigma_\omega^2 = \frac{1}{2\pi}\int_{-\infty}^{\infty} \omega^2(\mathcal{F}f)^2(\omega)d\omega = \frac{1}{4\sigma^2}. \tag{2.141}$$

Since

$$\sigma_t^2 = \int_{-\infty}^{\infty} t^2 f^2(t)dt = \sigma^2, \tag{2.142}$$

the result now follows.

2.9 Show that a signal with Gaussian instantaneous power achieves the highest entropic time–frequency concentration

$$h(f^2) + h(|F|^2/(2\pi)) = \log(\pi e). \tag{2.143}$$

Solution

Using (2.141) and, since in the case of the Gaussian given by (2.137) we have

$$h(f^2) = \frac{1}{2}\log(2\pi e\sigma^2), \tag{2.144}$$

it follows that

$$h(f^2) + h(|F|^2/(2\pi)) = \frac{1}{2}\log(2\pi e\sigma^2) + \frac{1}{2}\log\left(\frac{2\pi e}{4\sigma^2}\right) = \log(\pi e). \tag{2.145}$$

2.10 Provide an elementary proof of the impossibility of time–frequency limiting.

Solution

Consider a square-integrable, timelimited signal such that $f(t) = 0$ for $|t| > T/2$, and whose Fourier transform is $F(\omega)$. Let $T_0 > T$, and consider the periodic signal

$$g(t) = \sum_{n=-\infty}^{\infty} f(t + nT_0). \tag{2.146}$$

From (2.8) it follows that the Fourier series coefficients of $g(t)$ are

$$c_n = F(n\omega_0)/T_0, \tag{2.147}$$

where $\omega_0 = 2\pi/T_0$. Assume by contradiction that $F(\omega)$ is also bandlimited, namely $F(\omega) = 0$ for $|\omega| > \Omega$. It follows that $c_n = 0$ for $|n| > \Omega/\omega_0$, and $g(t)$ is a trigonometric polynomial

$$P(z) = \sum_{n=-M}^{M} c_n z^n, \tag{2.148}$$

where $z = \exp(j\omega_0 t)$ and $M = \lfloor \Omega/\omega_0 \rfloor$. Since $g(t) = 0$ in the interval $[T/2, T_0 - T/2]$, then $P(z)$ must have infinitely many roots on the unit circle, namely it must be zero at every point between $\exp(j\omega_0 T/2)$ and $\exp[j\omega_0(T_0 - T/2)] = \exp(-j\omega_0 T/2)$, which is impossible.

3 Functional Approximation

Functional analysis of infinite-dimensional spaces is never fully convincing; you don't get a feeling of having done an honest day's work.[1]

3.1 Signals and Functional Spaces

The set of signals for the communication engineer corresponds to an infinite-dimensional functional space for the mathematician. This can be viewed as a vector space on which norm (i.e., length), inner product (i.e., angle), and limits can be defined. The engineering problem of determining the *effective dimension* of the signals' space then falls in the mathematical framework of approximation theory that is concerned with finding a sequence of functions with desirable properties whose linear combination best approximates a given limit function.

Approximation is a well-studied problem in analysis. In terms of abstract Hilbert spaces, the problem is to determine what functions can asymptotically generate a given Hilbert space in the sense that the closure of their span (i.e., finite linear combinations and limits of those) is the whole space. Such a set of functions is called a basis, and its cardinality, taken in a suitable limiting sense, is the Hilbert dimension of the space.

Various differential equations arising in physics have orthogonal solutions that can be interpreted as bases of Hilbert spaces. One example is the solution of the wave equation leading to the prolate spheroidal wave functions examined in the previous chapter. Another notable example arises in quantum mechanics in the context of the Schrödinger differential equation.

In this chapter, we describe the connection between physical properties, such as the energy concentration of a wave function, and the mathematics of Hilbert spaces, showing that Slepian's concentration problem is a special case of the eigenvalue problem arising from the spectral decomposition of a self-adjoint operator on a Hilbert space. It turns out that this decomposition provides the *optimal* approximation for any function in the space. The effective dimension, or degrees of freedom, of the space is then defined as the cardinality of such an optimal representation. The notion of cardinality here is intended in a limiting sense, made precise with the notion of the

[1] G.-C. Rota (1985). Mathematics, philosophy and artificial intelligence. A dialogue with Gian-Carlo Rota and David Sharp. *Los Alamos Science*, 12, pp. 94–104. Reprinted with permission.

Kolmogorov N-width. The effective dimension of the signals' space is then given by the minimum number of orthogonal eigenfunctions that are required to approximate any signal in the space up to arbitrary precision.

These mathematical results tell us all about N-widths, in the sense that they give an explicit calculation in terms of subspaces and operators. However, they also tell us very little, since in order to compute them we have to determine the eigenfunctions and eigenvalues of these operators. Only when this is possible, as in the case of N-widths of bandlimited signals observed over an interval, is the dimension of the space determined completely and the approximating eigenfunctions shown to be the orthogonal prolate spheroidal ones.

3.2 Kolmogorov *N*-Width

Let $\mathscr{X}$ be a normed linear space and $\mathscr{X}_N$ any N-dimensional subspace of $\mathscr{X}$. For any $f \in \mathscr{X}$, the *distance* of $\mathscr{X}_N$ from f is defined as

$$\inf\{\|f-g\| : g \in \mathscr{X}_N\}. \tag{3.1}$$

If instead of a single element $f \in \mathscr{X}$ we are given a subset $\mathscr{A} \subseteq \mathscr{X}$, a measure of an N-dimensional approximation of $\mathscr{A}$ is given by the extent to which the "worst element" of $\mathscr{A}$ can be approximated from $\mathscr{X}_N$. This *deviation* is defined as

$$D_{\mathscr{X}_N}(\mathscr{A}) = \sup_{f\in\mathscr{A}} \inf_{g\in\mathscr{X}_N} \|f-g\|. \tag{3.2}$$

One might also ask how well it is possible to approximate $\mathscr{A}$ by N-dimensional subspaces of $\mathscr{X}$. The idea of allowing $\mathscr{X}_N$ to vary within $\mathscr{X}$ was first proposed by Kolmogorov (1936), and the level of approximation goes under the name of the *Kolmogorov N-width* of $\mathscr{A}$ in $\mathscr{X}$, defined as

$$d_N(\mathscr{A}, \mathscr{X}) = \inf_{\mathscr{X}_N \subseteq \mathscr{X}} D_{\mathscr{X}_N}(\mathscr{A}). \tag{3.3}$$

When the space $\mathscr{X}$ is clear from the context we use the shorthand notation $d_N(\mathscr{A})$. Since $d_N(\mathscr{A})$ measures the extent to which $\mathscr{A}$ can be approximated by N-dimensional subspaces of $\mathscr{X}$, it is in a certain sense a measure of the "massiveness" or "thickness" of $\mathscr{A}$.

Consider now the set of bandlimited signals that are in $\mathscr{E}(\epsilon_T)$ as defined by the pair of constraints (2.43) and (2.44), and assume that $\epsilon_T \leq \sqrt{1-\lambda_N}$. We equip these signals with the $L^2(-\infty,\infty)$ norm

$$\|f(t)\|^2 = \int_{-\infty}^{\infty} f^2(t)dt. \tag{3.4}$$

In this case, we have the following result:

$$d_N(\mathscr{E}(\epsilon_T)) = \frac{\epsilon_T}{\sqrt{1-\lambda_N}}, \tag{3.5}$$

and the optimal spanning eigenfunctions are the prolate spheroidal ones in the set $\mathscr{Y}_N = \{\psi_0(t), \psi_1(t), \ldots, \psi_{N-1}(t)\}$, namely

$$D_{\mathscr{Y}_N}(\mathscr{E}(\epsilon_T)) = d_N(\mathscr{E}(\epsilon_T)). \tag{3.6}$$

From this result we can immediately recover (2.129). Letting $N = N_0 + o(1)$ as $N_0 \to \infty$, and using (2.131), we have

$$d_N^2(\mathscr{E}(\epsilon_T)) = \frac{\epsilon_T^2}{1 - \lambda_{N_0}} \to 2\epsilon_T^2. \tag{3.7}$$

Similarly, letting $N = N_0 + O(\log N_0)$ as $N_0 \to \infty$, for any $0 < \epsilon < 1$ and T large enough we have

$$d_N^2(\mathscr{E}(\epsilon_T)) = \frac{\epsilon_T^2}{1 - \lambda_N} \leq \frac{\epsilon_T^2}{1 - \epsilon}, \tag{3.8}$$

which recovers (2.130) since ϵ is arbitrarily small.

A similar result holds by considering the $L^2[-T/2, T/2]$ norm

$$\|f\|^2 = \int_{-T/2}^{T/2} f^2(t)dt. \tag{3.9}$$

In this case, the signal is observed over a finite time interval and the accuracy of the approximation is evaluated over the observation interval rather than over the whole real axis. We have

$$d_N(\mathscr{E}(\epsilon_T)) = \frac{\epsilon_T \sqrt{\lambda_N}}{\sqrt{1 - \lambda_N}}. \tag{3.10}$$

Compared to (3.3), since $\lambda_N \to 0$ the accuracy of the fit in this case is higher, as expected. Letting $N = N_0 + o(1)$ as $N_0 \to \infty$, and using (2.131), we have

$$d_N^2(\mathscr{E}(\epsilon_T)) = \frac{\epsilon_T^2 \lambda_{N_0}}{1 - \lambda_{N_0}} \to \epsilon_T^2, \tag{3.11}$$

showing that in this case using N_0 eigenfunctions we achieve a better approximation compared to (3.7). Similarly, letting $N = N_0 + O(\log N_0)$ as $N_0 \to \infty$, for any $0 < \epsilon < 1$ and T large enough we have

$$d_N^2(\mathscr{E}(\epsilon_T)) = \frac{\epsilon_T^2 \lambda_N}{1 - \lambda_N} \leq \frac{\epsilon}{1 - \epsilon}\epsilon_T^2, \tag{3.12}$$

showing that an additional logarithmic number of eigenfunctions can drive the N-width as close to zero as desired, since ϵ is arbitrarily small. This should be compared with (3.8), where using the same number of eigenfunctions the N-width is made only as close to ϵ_T as possible.

We also have the following additional results for bandlimited signals subject to the single constraint (2.43). In the $L^2[-T/2, T/2]$ norm

$$d_N(\mathscr{B}_\Omega) = \sqrt{\lambda_N}, \tag{3.13}$$

and in the $L^2(-\infty, \infty)$ norm

$$d_N(\mathscr{B}_\Omega) = 1. \tag{3.14}$$

A derivation of (3.5), (3.10), (3.13), and (3.14) is shown in the next section and provides the link between the behavior of the eigenvalues in Slepian's concentration problem and the approximation error bounds given in the previous chapter.

3.3 Degrees of Freedom of Bandlimited Signals

We apply the Kolmogorov N-width to determine the number of degrees of freedom of bandlimited signals with respect to a given norm. In the context of physics, the number of degrees of freedom is defined as:

> The minimum number of parameters sufficient to describe a system within a given precision.

In an analogy to the physical definition, we consider a subset $\mathscr{A}$ of a normed linear space $\mathscr{X}$ of signals of norm at most one, and define the number of degrees of freedom at level ϵ as

$$N_\epsilon(\mathscr{A}) = \min\{N : d_N^2(\mathscr{A}) \leq \epsilon\}. \tag{3.15}$$

In this way, the number of degrees of freedom corresponds to the dimension of the minimal subspace representing the elements of $\mathscr{A}$ within ϵ accuracy.

Letting $\mathscr{B}_\Omega$ be the set of all square-integrable, bandlimited signals subject to the energy constraint (2.43) and equipped with the $L^2[-T/2,T/2]$ norm, using (3.13) and (2.132), we have

$$\begin{aligned} N_\epsilon(\mathscr{B}_\Omega) &= \min\{N : d_N^2(\mathscr{B}_\Omega) \leq \epsilon\} \\ &= \min\{N : \lambda_N \leq \epsilon\} \\ &= N_0 + O(\log N_0), \quad \text{as } N_0 \to \infty. \end{aligned} \tag{3.16}$$

This basic result follows directly from the computation of the N-width of bandlimited signals and from the behavior of the eigenvalues in Slepian's concentration problem. It shows that any bandlimited signal can be described over a finite observation interval and up to arbitrary precision by a number of coefficients only slightly larger than N_0. The dependence on ϵ appears hidden as a pre-constant of the second-order term $O(\log N_0)$ in the number of degrees of freedom, and we have

$$\lim_{N_0 \to \infty} \frac{N_\epsilon(\mathscr{B}_\Omega)}{N_0} = 1. \tag{3.17}$$

In addition, by (2.132) the pre-constant of $O(\log N_0)$ grows only as $\log(1/\epsilon)$ as $\epsilon \to 0$. A completely analogous result holds if the signals are subject to the constraint (1.1) by defining the number of degrees of freedom as the minimum index N such that the largest energy of the error becomes an arbitrarily small fraction of the largest energy of the signal, namely

$$N_\epsilon(\mathscr{B}_\Omega) = \min\{N : d_N^2(\mathscr{B}_\Omega)/E \leq \epsilon\}. \tag{3.18}$$

Table 3.1 Degrees of freedom of bandlimited signals.

$L^2[-T/2, T/2]$	$L^2(-\infty, \infty)$
$N_\epsilon(\mathscr{B}_\Omega) = N_0 + O(\log N_0)$	$N_\epsilon(\mathscr{B}_\Omega) = \infty$
$N_{\epsilon_T}(\mathscr{E}(\epsilon_T)) = N_0 + o(1)$	$N_{\epsilon_T/(1-\epsilon)}(\mathscr{E}(\epsilon_T)) = N_0 + O(\log N_0)$

On the other hand, when signals in $\mathscr{B}_\Omega$ are observed over the whole real line, the appropriate norm used to compute the number of degrees of freedom is the $L^2(-\infty, \infty)$ one, and by virtue of (3.14) the number of degrees of freedom is undetermined. In this case, since it is not possible to approximate all signals in the space using any finite number of coefficients, we posit that $N_\epsilon(\mathscr{B}_\Omega) = \infty$.

We now consider the set of bandlimited signals $\mathscr{E}(\epsilon_T)$ that are subject to the pair of constraints (2.43) and (2.44). In this case, in the $L^2[-T/2, T/2]$ norm, using (3.11) we have

$$N_{\epsilon_T}(\mathscr{E}(\epsilon_T)) = N_0 + o(1), \quad \text{as } N_0 \to \infty. \tag{3.19}$$

In the $L^2(-\infty, \infty)$ norm, by (3.8) we have

$$N_{\epsilon_T/(1-\epsilon)}(\mathscr{E}(\epsilon_T)) = N_0 + O(\log N_0), \quad \text{as } N_0 \to \infty. \tag{3.20}$$

These results are summarized in Table 3.1. By (3.19), a number N_0 of coefficients can represent bandlimited signals over a finite observation interval, up to a degree of approximation ϵ_T given by the unconcentrated part of the signal. On the other hand, to obtain a representation with a degree of approximation arbitrarily close to ϵ_T over the entire real line, by (3.20) we need to add an extra logarithmic number of terms. By (3.16), this same number of extra terms is needed to represent any bandlimited signal up to arbitrary accuracy over a finite observation interval, while it is impossible to obtain such an approximation over the entire real line.

In all cases where an approximation is feasible, the dependence of the degrees of freedom on the level of approximation ϵ is weak. By (2.132), this number grows at most as $\log(1/\epsilon)$ as $\epsilon \to 0$, and it affects only a second-order term as $N_0 \to \infty$. This makes the *effective dimensionality* of the signals' space asymptotically independent of the accuracy of the approximation. This is very desirable, as it makes the degrees of freedom a principal feature of our mathematical model of the real world of transmitted signals, one that is practically insensitive to small changes of a secondary feature of the model, such as the accuracy of the measurement with which the signals are detected, and is a direct consequence of the phase transition of the eigenvalues expressed by (2.131). A physical regularity constraint of smoothness of the transmitted signal, expressed by their bandlimitation property, allows us to define the degrees of freedom as an intrinsic feature of the model.

3.3.1 Computation of the N-Widths

We now show the validity of the results in (3.13) and (3.14), as well as in (3.5) and (3.10). The derivations follow the same outline, with some minor variations. We proceed in two steps. First, we determine an upper bound on $d_N(\mathscr{B}_\Omega)$ by computing $D_{\mathscr{X}_N}(\mathscr{B}_\Omega)$ for a certain choice of the approximating subspace $\mathscr{X}_N$. In a second step, we show that the obtained quantity is a lower bound as well. We start by describing the proof of (3.13), then we show the variation needed to obtain (3.5) and (3.10).

Recall, as discussed in Section 2.5, that the $\{\psi_n(t)\}$ are the solutions of the Fredholm integral equation

$$\int_{-T/2}^{T/2} \frac{\sin\Omega(t-t')}{\pi(t-t')}\psi_n(t')dt' = \lambda_n\psi_n(t), \quad t\in[-T/2,T/2], \tag{3.21}$$

and satisfy the following properties:

$$\psi_n(t) \in \mathscr{B}_\Omega, \tag{3.22}$$

$$\int_{-T/2}^{T/2} \psi_n(t)\psi_m(t)dt = \begin{cases} \lambda_n & \text{if } n=m, \\ 0 & \text{otherwise.} \end{cases} \tag{3.23}$$

When these functions are extended over the whole real line, we also have

$$\int_{-\infty}^{\infty} \psi_n(t)\psi_m(t)dt = \begin{cases} 1 & \text{if } n=m, \\ 0 & \text{otherwise.} \end{cases} \tag{3.24}$$

Let $f(t) \in \mathscr{B}_\Omega$, then by the completeness of the $\{\psi_n\}$ we have

$$f(t) = \sum_{n=0}^{\infty} a_n\psi_n(t). \tag{3.25}$$

Consider the N-dimensional approximation

$$f_N(t) = \sum_{n=0}^{N-1} a_n\psi_n(t). \tag{3.26}$$

By (2.127), we have, in the $L^2[-T/2,T/2]$ norm,

$$\|f-f_N\|^2 = \inf_{g\in\mathscr{X}_N} \|f-g\|^2 = \sum_{n=N}^{\infty} a_n^2\lambda_n. \tag{3.27}$$

From the monotonicity of the $\{\lambda_n\}$, we have

$$\begin{aligned} \inf_{g\in\mathscr{X}_N} \|f-g\|^2 &= \sum_{n=N}^{\infty} a_n^2\lambda_n \\ &\leq \lambda_N \sum_{n=N}^{\infty} a_n^2. \end{aligned} \tag{3.28}$$

On the other hand, by Parseval's theorem, we have

$$\int_{-\infty}^{\infty} f^2(t)dt = \sum_{n=0}^{\infty} a_n^2 \leq 1. \tag{3.29}$$

Combining (3.28) and (3.29), it then follows that

$$d_N(\mathscr{B}_\Omega) \leq \sqrt{\lambda_N}. \tag{3.30}$$

Next, we show that the upper bound is achievable. Consider the $N+1$-dimensional ball U_{N+1} in the $L^2[-T/2,T/2]$ norm, defined by

$$g(t)=\sum_{n=0}^{N} a_n\psi_n(t), \quad \|g\| \le \sqrt{\lambda_N}. \tag{3.31}$$

We have (see Problem 3.7) that

$$d_N(U_{N+1})=\sqrt{\lambda_N}. \tag{3.32}$$

Since, by (3.22), the $\{\psi_n\}$ are bandlimited functions, $U_{N+1}\in\mathscr{B}_\Omega$. We only have to verify that the energy over $(-\infty,\infty)$ is at most one. By (3.23) and (3.31), we have

$$\begin{aligned}\|g\|^2 &= \sum_{n=0}^{N} a_n^2\|\psi_n\|^2 \\ &= \sum_{n=0}^{N} a_n^2\lambda_n \le \lambda_N.\end{aligned} \tag{3.33}$$

From the monotonicity of the $\{\lambda_n\}$ and (3.33), we finally have

$$\begin{aligned}\int_{-\infty}^{\infty} g^2(t)dt &= \sum_{n=0}^{N} a_n^2 \\ &\le \sum_{n=0}^{N} a_n^2\frac{\lambda_n}{\lambda_N} \le 1,\end{aligned} \tag{3.34}$$

and the proof of (3.13) is complete. As for (3.14), this follows by the same steps and using (3.24) instead of (3.23) when computing the $L^2(-\infty,\infty)$ norm.

We now turn to the variation needed to obtain (3.10). In this case, the proof of the upper bound proceeds in the same way until (3.28). Then, the following different upper bound is used:

$$\begin{aligned}\lambda_N\sum_{n=N}^{\infty} a_n^2 &\le \lambda_N\sum_{n=0}^{\infty} a_n^2\frac{1-\lambda_n}{1-\lambda_N} \\ &\le \frac{\epsilon_T^2}{1-\lambda_N}\lambda_N,\end{aligned} \tag{3.35}$$

where the first inequality follows from the monotonicity of the $\{\lambda_n\}$. The second inequality follows by recalling that $f\in\mathscr{E}(\epsilon_T)$ and, assuming the validity of term by term integration of the series below, formally performing the computation:

$$\begin{aligned}\epsilon_T^2 &\ge \int_{-\infty}^{\infty}(f(t)-\mathcal{T}f(t))^2dt \\ &= \int_{-\infty}^{\infty}\left(\sum_{n=0}^{\infty} a_n\psi_n(t)-a_n\mathcal{T}\psi_n(t)\right)^2 dt\end{aligned}$$

$$= \int_{-\infty}^{\infty} \left(\sum_{n=0}^{\infty} a_n \psi_n(t) - a_n \sqrt{\lambda_n} \xi_n(t) \right)^2 dt$$

$$= \sum_{n=0}^{\infty} a_n^2 (1 - 2\lambda_n + \lambda_n)$$

$$= \sum_{n=0}^{\infty} a_n^2 (1 - \lambda_n), \tag{3.36}$$

where $\mathcal{T}$ is the projection operator (2.87) onto the subspace of timelimited signals and ξ_n is defined in (2.70). A completely rigorous derivation is possible, but requires a more detailed argument.

To show that the upper bound is achievable, one needs to consider the $N+1$-dimensional ball U_{N+1} in the $L^2[-T/2, T/2]$ norm, defined by

$$g(t) = \sum_{n=0}^{N} a_n g_n(t), \quad \|g\| \leq \frac{\epsilon_T \sqrt{\lambda_N}}{\sqrt{1-\lambda_N}}, \tag{3.37}$$

with

$$g_n(t) = \frac{\epsilon_T}{\sqrt{1-\lambda_n}} \psi_n(t), \tag{3.38}$$

and assume that $\epsilon_T / \sqrt{1-\lambda_N} \leq 1$. With the same argument as in Problem 3.7, we have

$$d_N(U_{N+1}) = \frac{\epsilon_T \sqrt{\lambda_N}}{\sqrt{1-\lambda_N}}. \tag{3.39}$$

We now need to check that the energy of g is at most one, and that the amount of energy outside $[-T/2, T/2]$ is at most ϵ_T^2. We have

$$\|g\|^2 = \sum_{n=0}^{N} a_n^2 \|g_n\|^2$$

$$= \sum_{n=0}^{N} a_n^2 \frac{\epsilon_T^2 \lambda_n}{1-\lambda_n}$$

$$\leq \frac{\epsilon_T^2 \lambda_N}{1-\lambda_N}, \tag{3.40}$$

where the inequality follows from (3.37). From the monotonicity of the $\{\lambda_n\}$, we have that, for all $n = 0, 1, \ldots, N$,

$$\frac{\epsilon_T^2 \lambda_n}{1-\lambda_n} \geq \frac{\epsilon_T^2 \lambda_N}{1-\lambda_N}, \tag{3.41}$$

which, combined with (3.40), shows that

$$\sum_{n=0}^{N} a_n^2 \leq 1. \tag{3.42}$$

Using (3.42), it follows that

$$\begin{aligned}\int_{-\infty}^{\infty} g^2(t)dt &= \sum_{n=1}^{N} a_n^2 \frac{\epsilon_T^2}{1-\lambda_n} \\ &\le \sum_{n=1}^{N} a_n^2 \frac{\epsilon_T^2}{1-\lambda_N} \\ &\le \frac{\epsilon_T^2}{1-\lambda_N},\end{aligned} \tag{3.43}$$

which is at most one. Combining (3.43) and (3.40), we also have

$$\int_{-\infty}^{\infty} (g(t) - \mathcal{T}g(t))^2 dt \le \epsilon_T^2. \tag{3.44}$$

The proof of (3.5) proceeds with computations analogous to those above, using the $L^2(-\infty,\infty)$ norm, and it is left as an exercise.

3.4 Hilbert–Schmidt Integral Operators

The results on the N-width and degrees of freedom of bandlimited signals can be viewed in a more general setting by considering the behavior of the singular values of Hilbert–Schmidt integral operators, which are a class of operators that were extensively studied by the great German mathematician David Hilbert and his student Erhard Schmidt. The optimal orthogonal representation of any signal in $\mathscr{B}(\Omega)$ is obtained by solving Slepian's concentration problem (3.21) and then constructing the finite-dimensional approximation (3.26) in terms of prolate spheroidal wave functions. When the accuracy of the approximation is evaluated over the observation interval $[-T/2, T/2]$, by (3.13) it turns out that this is at most $\sqrt{\lambda_N}$.

We now view bandlimited signals as the image of square-integrable signals through a bandlimiting operator, show that their optimal representation corresponds to the Hilbert–Schmidt decomposition of this image space, and that the accuracy of the approximation corresponds to the square root of the Nth eigenvalue of the associated self-adjoint operator.

Consider the Hilbert spaces $\mathscr{X} = L^2(a,b)$, $\mathscr{Y} = L^2(c,d)$. A *Hilbert–Schmidt kernel* $K(x,y)$ on $(a,b) \times (c,d)$ is defined by the property

$$\int_c^d \int_a^b |K(x,y)|^2 dx dy < \infty. \tag{3.45}$$

This naturally induces an operator $\mathcal{K} : \mathscr{Y} \to \mathscr{X}$ such that, for any $g \in \mathscr{Y}$,

$$(\mathcal{K}g)(x) = \int_c^d K(x,y) g(y) dy. \tag{3.46}$$

Similarly, consider another operator $\mathcal{K}' : \mathcal{X} \to \mathcal{Y}$ induced by the Hilbert–Schmidt kernel $K'(x,y)$ on $(a,b) \times (c,d)$ such that, for any $h \in \mathcal{X}$,

$$(\mathcal{K}'h)(y) = \int_a^b K'(x,y)h(x)dx. \tag{3.47}$$

The operator $\mathcal{K}'$ is called an *adjoint operator* of $\mathcal{K}$ if and only if, for any $g \in \mathcal{Y}$ and $h \in \mathcal{X}$,

$$\int_a^b (\mathcal{K}g)(x)\, h^*(x)dx = \int_c^d g(y)\, (\mathcal{K}'h)^*(y)dy. \tag{3.48}$$

A *self-adjoint operator* is its own adjoint. Consider the operator $\mathcal{K}'\mathcal{K}$, with symmetric Hermitian kernel

$$K'K(x,y) = (K'K)^*(y,x) = \int_a^b K(z,x)K^*(z,y)dz \tag{3.49}$$

of Hilbert–Schmidt type

$$\int_c^d \int_c^d |K'K(x,y)|^2 dxdy < \infty. \tag{3.50}$$

This operator is self-adjoint since, for all $g,h \in \mathcal{Y}$,

$$\int_c^d \mathcal{K}'\mathcal{K}g(x)h^*(x)dx = \int_c^d g(x)(\mathcal{K}'\mathcal{K}h)^*(x)dx. \tag{3.51}$$

It is also non-negative; namely, for all $g \in \mathcal{Y}$,

$$\int_c^d \mathcal{K}'\mathcal{K}g(x)g^*(x)dx \geq 0. \tag{3.52}$$

A basic property of Hilbert–Schmidt operators is that they are *compact*, meaning that for any bounded sequence $\{x_n\}$ in $\mathcal{X}$, there exists a bounded subsequence of $\{\mathcal{K}x_n\}$ that converges in $\mathcal{Y}$. The *spectral theorem* states that compact, self-adjoint operators admit either a finite or a countably infinite orthonormal basis of eigenvectors $\{\varphi_n\}$, and that the corresponding eigenvalues $\{\lambda_n\}$ are real and such that $\lambda_n \to 0$. In the case that the operator is non-negative, the eigenvalues are also non-negative. The eigenvalues and eigenvectors are solutions of the equation

$$\mathcal{K}'\mathcal{K}\varphi_n = \lambda_n \varphi_n. \tag{3.53}$$

The singular values of $\mathcal{K}$ are defined as the square roots of the eigenvalues $\lambda_0 \geq \lambda_1 \geq \lambda_2 \geq \cdots$. Letting $\psi_n = \mathcal{K}\varphi_n$, $n = 1,2,\ldots$, it can be easily verified that the function ψ_n is an eigenfunction, with corresponding eigenvalue λ_n, of the operator $\mathcal{K}\mathcal{K}'$, namely

$$\mathcal{K}\mathcal{K}'\psi_n = \lambda_n \psi_n. \tag{3.54}$$

Consider now the subset $\mathcal{A} \subset \mathcal{X}$ that is the image of the unit-norm elements of $\mathcal{Y}$ through the Hilbert–Schmidt operator induced by the kernel K, namely

$$\mathcal{A} = \left\{ f(x) : f(x) = (\mathcal{K}g)(x) = \int_c^d K(x,y)g(y)dy,\ \|g\| \leq 1 \right\}. \tag{3.55}$$

Another basic property of Hilbert–Schmidt operators is that the N-width of the set of signals in (3.55) is given by their Nth singular value, when these are arranged in non-increasing order, namely:

> The N-width of the image of the unit-norm elements of $\mathscr{Y}$ through the operator $\mathcal{K}$ is
>
> $$d_N(\mathscr{A}) = \sqrt{\lambda_N}, \tag{3.56}$$
>
> and the eigenfunctions in the set $\{\psi_0, \psi_1, \ldots, \psi_{N-1}\}$ span an optimal approximating N-dimensional subspace $\mathscr{X}_N$.

It follows that the singular values of a Hilbert–Schmidt operator determine the level of approximation for the optimal decomposition of the elements of the image space of the operator into basis functions. This decomposition is useful when one wants to study the asymptotic dimensionality of a Hilbert–Schmidt operator. The solution to Slepian's concentration problem corresponds to the decomposition of one such operator, and leads to the exact determination of the asymptotic dimensionality of the associated operator.

The basic result (3.56) follows directly from a duality result through the Schwarz inequality, and from the variational characterization of the eigenvalues of compact self-adjoint operators on Hilbert spaces.

By the definition of N-width, we have

$$d_N(\mathscr{A}) = \inf_{\mathscr{X}_n} \sup_{\|g\|\leq 1} \inf_{h\in\mathscr{X}_n} \|\mathcal{K}g - h\|. \tag{3.57}$$

The duality result referred to above is

$$\inf_{h\in\mathscr{X}_n} \|\mathcal{K}g - h\| = \sup_{z\perp\mathscr{X}_n, \|z\|=1} \langle \mathcal{K}g, z\rangle, \tag{3.58}$$

where by $z \perp \mathscr{X}_n$ we mean that z is orthogonal to every element of $\mathscr{X}_n$, and $\langle u, v\rangle$ indicates the inner product in $L^2(c,d)$:

$$\langle u, v\rangle = \int_c^d u(x)v^*(x)dx. \tag{3.59}$$

Letting $\mathcal{K}g = f$, we write

$$f = f_\perp + f_-, \tag{3.60}$$

where $f_- \in \mathscr{X}_n$ and $f_\perp \perp \mathscr{X}_n$, so that the distance of f from $\mathscr{X}_n$ is

$$\inf_{h\in\mathscr{X}_n} \|f - h\| = \|f_\perp\|, \tag{3.61}$$

and to establish the duality result (3.58) we need to prove that

$$\|f_\perp\| = \sup_{z\perp\mathscr{X}_n, \|z\|=1} \langle f, z\rangle. \tag{3.62}$$

By again using the orthogonal decomposition (3.60) and the Schwarz inequality, we have that, for all $z \perp X_n$,

$$\begin{aligned} \langle f, z\rangle &= \langle f_\perp, z\rangle \\ &\leq \|f_\perp\| \, \|z\|, \end{aligned} \tag{3.63}$$

where the equality holds for $z = f_\perp/\|f_\perp\|$, establishing (3.62) and hence (3.58).

Substituting (3.58) into (3.57), by the definition of an adjoint operator and using once again the Schwarz inequality in the case of equality, it follows that

$$\begin{aligned} d_N(\mathscr{A}) &= \inf_{\mathscr{X}_n} \sup_{\|g\|\leq 1} \sup_{z\perp \mathscr{X}_n, \|z\|=1} \langle \mathcal{K}g, z\rangle \\ &= \inf_{\mathscr{X}_n} \sup_{z\perp \mathscr{X}_n, \|z\|=1} \sup_{\|g\|\leq 1} \langle g, \mathcal{K}'z\rangle \\ &= \inf_{\mathscr{X}_n} \sup_{z\perp \mathscr{X}_n, \|z\|=1} \|\mathcal{K}'z\| \\ &= \left(\inf_{\mathscr{X}_n} \sup_{z\perp \mathscr{X}_n, \|z\|=1} \langle \mathcal{K}\mathcal{K}'z, z\rangle \right)^{1/2} \\ &= \sqrt{\lambda_N}, \end{aligned} \tag{3.64}$$

where the last equality follows from the min-max principle, see Appendix A.4.1, and it is easy to see that the infimum and supremum in (3.64) are obtained by letting $\mathscr{X}_n =$ span $\{\psi_0, \psi_n, \ldots, \psi_{n-1}\}$ and $z = \psi_n$.

The derivation above did not explicitly calculate the optimal approximation for the set $\mathscr{A}$. In Section 3.3.1, we gave an explicit computation of the approximation when $\mathscr{A}$ is the space of bandlimited signals observed over an interval. In this case, it turns out that the prolate spheroidal functions can optimally approximate bandlimited signals up to arbitrary precision, and that timelimiting and bandlimiting operators are the corresponding adjoint operators associated with this approximation.

3.4.1 Timelimiting and Bandlimiting Operators

A special case of Hilbert–Schmidt integral operators is that of timelimiting and bandlimiting operators with complex exponential kernels that can be viewed as adjoint operators acting on L^2 space. In this case, the decomposition can be computed explicitly, and it turns out that the prolate spheroidal functions can optimally approximate bandlimited signals up to arbitrary precision.

First, we check that the timelimiting operator $\mathcal{K}'$ with complex exponential kernel is indeed the adjoint of the bandlimiting operator $\mathcal{K}$ with corresponding conjugate kernel. For any real $h \in L^2[-T/2, T/2]$, we let

$$(\mathcal{K}'h)(\omega) = \frac{1}{\sqrt{2\pi}} \int_{-T/2}^{T/2} h(t) \exp(-j\omega t) dt \tag{3.65}$$

and

$$(\mathcal{K}'h)^*(\omega) = \frac{1}{\sqrt{2\pi}} \int_{-T/2}^{T/2} h(t) \exp(j\omega t) dt. \tag{3.66}$$

For any $g \in L^2[-\Omega, \Omega]$, we let

$$(\mathcal{K}g)(t) = \frac{1}{\sqrt{2\pi}} \int_{-\Omega}^{\Omega} g(\omega) \exp(j\omega t) d\omega. \tag{3.67}$$

We can now verify (3.48) by writing

$$\int_{-T/2}^{T/2} \int_{-\Omega}^{\Omega} g(\omega) \exp(j\omega t) h(t) d\omega dt = \int_{-\Omega}^{\Omega} \int_{-T/2}^{T/2} h(t) \exp(j\omega t) g(\omega) dt d\omega. \tag{3.68}$$

In the case of prolate spheroidal functions, using (2.67)–(2.70), we have

$$\begin{aligned}(\mathcal{K}'\psi_n)(\omega) &= \frac{1}{\sqrt{2\pi}} \int_{-T/2}^{T/2} \psi_n(t) \exp(-j\omega t) dt \\ &= \frac{1}{\sqrt{2\pi}} \sqrt{\lambda_n} \Xi_n(\omega),\end{aligned} \tag{3.69}$$

$$\begin{aligned}(\mathcal{K}\Xi_n)(t) &= \frac{1}{\sqrt{2\pi}} \int_{-\Omega}^{\Omega} \Xi_n(\omega) \exp(j\omega t) d\omega \\ &= \sqrt{2\pi} \sqrt{\lambda_n} \psi_n(t).\end{aligned} \tag{3.70}$$

Using (3.69) and (3.70), we have that (3.48) can also be verified as follows:

$$\begin{aligned}\frac{1}{\sqrt{2\pi}} \int_{-\Omega}^{\Omega} (\mathcal{K}'\psi_n)^*(\omega) \Xi_m(\omega) d\omega &= \frac{\sqrt{\lambda_n}}{2\pi} \int_{-\Omega}^{\Omega} \Xi_n^*(\omega) \Xi_m(\omega) d\omega \\ &= \begin{cases} \sqrt{\lambda_n} \lambda_n & \text{if } n = m, \\ 0 & \text{otherwise;} \end{cases}\end{aligned} \tag{3.71}$$

$$\begin{aligned}\frac{1}{\sqrt{2\pi}} \int_{-T/2}^{T/2} \mathcal{K}\Xi_n(t) \psi_m(t) dt &= \sqrt{\lambda_n} \int_{-T/2}^{T/2} \psi_n(t) \psi_m(t) dt \\ &= \begin{cases} \sqrt{\lambda_n} \lambda_n & \text{if } n = m, \\ 0 & \text{otherwise.} \end{cases}\end{aligned} \tag{3.72}$$

Similarly, the eigenvalue equation (3.54) follows from (3.69) and (3.70), writing

$$\begin{aligned}(\mathcal{K}\mathcal{K}'\psi_n)(t) &= \sqrt{\lambda_n} \sqrt{\lambda_n} \psi_n(t) \\ &= \lambda_n \psi_n(t).\end{aligned} \tag{3.73}$$

We can finally recover (3.21) by rewriting the above eigenvalue equation as

$$\begin{aligned}(\mathcal{K}\mathcal{K}'\psi_n)(t) &= \frac{1}{2\pi} \int_{-\Omega}^{\Omega} \int_{-T/2}^{T/2} \psi_n(t') \exp(-j\omega t') dt' \exp(j\omega t) d\omega \\ &= \int_{-T/2}^{T/2} \psi_n(t') \int_{\Omega}^{\Omega} \exp(j\omega(t - t')) d\omega dt'\end{aligned}$$

$$= \int_{-T/2}^{T/2} \frac{\sin \Omega(t-t')}{\pi(t-t')} \psi(t')dt'$$
$$= \lambda_n \psi_n(t). \tag{3.74}$$

It follows that the Fredholm integral equation arising from Slepian's concentration problem is a special case of the eigenvalue equation determining the singular values associated with a given Hilbert–Schmidt kernel – that is, the complex exponential. This equation also defines the N-width of the set of signals that are the image of the bandlimiting operator with such a kernel. The equation can be solved explicitly, and the solution is found in terms of the optimal prolate spheroidal basis, which are real singular functions providing the optimal approximation.

3.4.2 Hilbert–Schmidt Decomposition

We now provide a construction of the optimal approximation of every element in the image space of a Hilbert–Schmidt operator in terms of singular functions, and give a geometric interpretation of the Hilbert–Schmidt decomposition. Consider the following approximation problem for a Hilbert–Schmidt kernel $K(x,y)$:

$$e_N(K) = \min \left\{ \int_a^b \int_c^d \left| K(x,y) - \sum_{n=0}^{N-1} u_n(x) v_n(y) \right|^2 dxdy : \right.$$
$$\left. u_n \in L^2(a,b), v_n \in L^2(c,d) \right\}. \tag{3.75}$$

Schmidt (1907) proved that the minimum of (3.75) is obtained by choosing the functions $\{u_n\}$ and $\{v_n\}$ to be the eigenfunctions of the self-adjoint operator $\mathcal{K}'\mathcal{K} : L^2(c,d) \to L^2(c,d)$, with symmetric Hermitian kernel

$$K'K(x,y) = (K'K)^*(y,x) = \int_a^b K(z,x)K^*(z,y)dz \tag{3.76}$$

of Hilbert–Schmidt type

$$\int_c^d \int_c^d |K'K(x,y)|^2 dxdy < \infty. \tag{3.77}$$

This operator is non-negative, compact, and self-adjoint. Call its eigenvalues $\lambda_1 \geq \lambda_2 \geq \cdots > 0$ and associated orthonormal eigenvectors φ_n, $n = 1,2,\ldots$, left singular functions. Let $\psi_n(x) = (\mathcal{K}\varphi_n)(x) = \int_c^d K(x,y)\varphi_n(y)dy$. The orthogonal functions ψ_n, $n = 1,2,\ldots$, are eigenvectors of KK', are called right singular functions, and $\int_a^b \psi(x)\psi^*(x)dx = \lambda_n$. A minimizer of (3.75) is then given by

$$K_N(x,y) = \sum_{n=0}^{N-1} \psi_n(x)\varphi_n^*(y), \tag{3.78}$$

and the minimum value of the error equals the tail sum of the eigenvalues:

$$e_N(K) = \int_a^b \int_c^d |K(x,y) - K_N(x,y))|^2 dxdy$$

$$= \sum_{n=N}^{\infty} \lambda_n. \tag{3.79}$$

It follows that the optimal way to represent the operator induced by $K(x,y)$ is given by the combination (3.78) of its left and right singular functions, and the error of the approximation is given by the sum of the tail of the eigenvalues, or singular values squared, as a function of their indices.

This also implies that an optimal N-dimensional approximation of any function that is the image of the operator induced by the kernel K on (a,b), namely of the elements in the set $\mathscr{A}$ in (3.55), is

$$\begin{aligned} f_N(x) &= \int_c^d K_N(x,y)g(y)dy \\ &= \int_c^d \sum_{n=0}^{N-1} \psi_n(x)\varphi_n^*(y)g(y)dy \\ &= \sum_{n=0}^{N-1} \int_c^d \varphi_n^*(y)g(y)dy\ \psi_n(x) \\ &= \sum_{n=0}^{N-1} \int_c^d \varphi_n^*(y)g(y)dy\ \sqrt{\lambda_n}\xi_n(x), \end{aligned} \tag{3.80}$$

where $\xi_n(x)$ is the normalized version of $\psi_n(x)$, so that both $\xi_n(x)$ and $\varphi_n(y)$ have unit norm over $L^2(a,b)$ and $L^2(c,d)$, respectively. The error of the approximation, over all possible functions f in the space, is in this case given in terms of Kolmogorov's N-width by (3.56).

Recalling the inner product definition (2.80), the geometric interpretation of (3.80) is that of writing the image of g through the operator $\mathcal{K}$ by first projecting g onto the coordinate axes corresponding to the eigenvectors $\{\varphi_n\}$, then scaling these coordinates by the corresponding singular values, and finally obtaining the new representation in terms of these scaled coordinates viewed in the image space through the operator $\mathcal{K}$ of the eigenvectors $\{\varphi_n\}$, namely the space spanned by the eigenvectors $\{\xi_n\}$. A sketch of this geometrical interpretation is depicted in Figure 3.1.

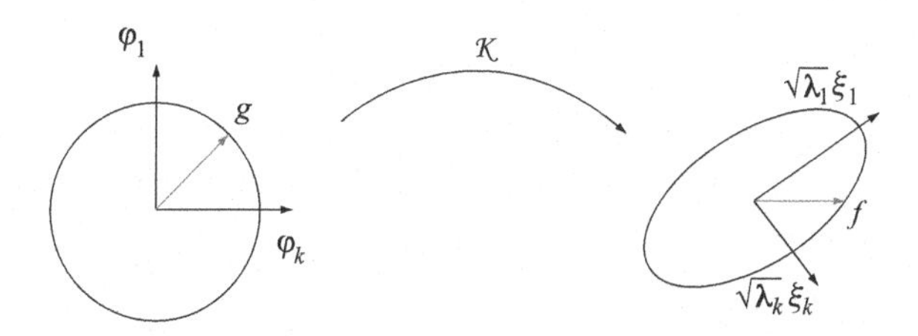

Fig. 3.1 Geometric view of the Hilbert–Schmidt decomposition.

Substituting (3.78) into (3.76), we also have the Hilbert–Schmidt decomposition of the kernel $K'K$,

$$K'K_N(x,y) = \sum_{n=0}^{N-1} \lambda_n \varphi_n(x)\varphi_n^*(y), \tag{3.81}$$

that achieves the minimum approximation error

$$\begin{aligned} e_N(K'K) &= \int_a^b \int_c^d \left| K'K(x,y) - \sum_{n=0}^{N-1} \lambda_n \varphi_n(x)\varphi_n^*(y) \right|^2 dxdy \\ &= \sum_{n=N}^{\infty} \lambda_n^2. \end{aligned} \tag{3.82}$$

The approximation errors in (3.79) and (3.82), as well as the Kolmogorov N-width (3.56), tend to zero as $N \to \infty$, ensuring L^2 convergence of the Hilbert–Schmidt representation. Sometimes, it is convenient to obtain stronger convergence properties. This can be done provided that some regularity assumptions are made besides square-integrability. For example, for any Hilbert–Schmidt, self-adjoint operator that is non-negative and has a continuous kernel, *Mercer's theorem*, stated in Appendix A.4.2, ensures that the representation (3.81) converges absolutely and uniformly (and hence point-wise) as $N \to \infty$.

3.4.3 Singular Value Decomposition

When working on his decomposition, Schmidt was apparently unaware of the parallel work on singular values of finite matrices. His representation is the continuous version of the *diagonalization* of a finite-dimensional matrix operator. Any (real or complex) matrix A of size $M \times N$ has the singular value decomposition (SVD)

$$\mathrm{A} = \mathrm{U}\Sigma\mathrm{V}^\dagger = \sum_{n=0}^{R-1} \sigma_n \mathrm{U}[n]\mathrm{V}^\dagger[n], \tag{3.83}$$

where U is an $M \times M$ unitary matrix, U$[n]$ is its nth column, $\mathrm{V}^\dagger$ is an $N \times N$ unitary matrix representing the conjugate transpose of V, and V$[n]$ is its nth column. The columns of V are the eigenvectors of $\mathrm{A}^\dagger\mathrm{A}$, called the right singular vectors. The columns of U are the eigenvectors of $\mathrm{AA}^\dagger$, called left singular vectors. The matrix Σ is an $M \times N$ diagonal matrix of non-negative real numbers, partitioned in the form

$$\Sigma = \begin{pmatrix} \Sigma_R & 0 \\ 0 & 0 \end{pmatrix}, \tag{3.84}$$

with Σ_R an $r \times r$ square diagonal matrix with positive entries

$$\Sigma_R = \mathrm{diag}\,(\sigma_0, \sigma_1, \ldots, \sigma_{R-1}) \tag{3.85}$$

called the singular values of A and corresponding to the square roots of the eigenvalues of $\mathrm{AA}^\dagger$ and $\mathrm{A}^\dagger\mathrm{A}$. These are arranged in decreasing order, so that $\sigma_0 \geq \sigma_1 \geq \cdots \geq$

$\sigma_{R-1} > 0$. Geometrically, the SVD can then be viewed as follows. A is decomposed into three transformations: a rotation $V^\dagger$, a scaling Σ along the coordinate axes $\{\varphi_n\}$, and a second rotation U. The singular values represent the lengths of the axes of the ellipsoid obtained.

Associated with the SVD there is a family of reduced-rank matrices $\{A_K\}$ obtained by keeping only the first $K < R$ terms in the expansion:

$$A_K = \sum_{n=0}^{K-1} \sigma_n U[n] V^\dagger[n]. \tag{3.86}$$

The matrices A and A_K agree in their largest K singular values, but the last $R-K$ singular values of A are replaced by zeros in A_K.

The L^2 matrix norm is defined in terms of the vector norm as

$$\|A\|_{L^2} = \sup_{\|X\|_{L^2} \neq 0} \frac{\|AX\|_{L^2}}{\|X\|_{L^2}} = \sup_{\|X\|_{L^2}=1} \|AX\|_{L^2}, \tag{3.87}$$

where X is an $N \times 1$ column vector and its norm is the square root of the sum of its absolute value elements,

$$\|X\|_{L^2} = \sqrt{|x_1|^2 + |x_2|^2 + \cdots + |x_N|^2}. \tag{3.88}$$

The Frobenius norm – which is not derived from a vector norm – is the square root of the sum of the absolute values squared of all matrix elements,

$$\|A\|_F = \sqrt{\sum_{n=0}^{M-1} \sum_{k=0}^{N-1} |A_{n,k}|^2} = \sqrt{(A^\dagger A)} = \sqrt{\text{tr } (A^* A)}. \tag{3.89}$$

The reduced-rank approximation theorem states that within the set of $M \times N$ matrices of rank $K < R$, the matrix that most closely approximates A in the L^2 or the Frobenius norm is the matrix A_K. Namely, over all $N \times M$ matrices B of rank K, the distance $\|A - B\|$ is minimized when $B = A_K$. Furthermore, the minimized distance is

$$\|A - A_K\|_{L^2} = \sigma_K, \tag{3.90}$$

$$\|A - A_K\|_F = (\sigma_K^2 + \cdots + \sigma_{R-1}^2)^{1/2}. \tag{3.91}$$

This result is an essential tool in signal processing, data compression, statistics, and many areas in physics. In these applications one often works with matrices that are full rank, but their singular values tend to cluster into two groups, those that are large and those that are small, so that one can perform *dimensionality reduction* and define an effective rank of the matrix that is the discrete analog of the Kolmogorow N-width in continuous spaces. The approximation result (3.90) corresponds to (3.56), and (3.91) corresponds to (3.79).

3.5 Extensions

We have shown that the N-width of signals in $\mathscr{B}_\Omega$ can be determined in terms of the asymptotic behavior of the eigenvalues of a certain Fredholm integral equation. This leads to a natural definition of the number of degrees of freedom of bandlimited signals that represents the effective dimensionality of the signals' space. As $T \to \infty$, any bandlimited signal can be represented by specifying only slightly more than N_0 real numbers, corresponding to its coordinates in the space. The folk theorem has been made precise.

In practice, however, it is often the case that signals are only approximately bandlimited, and it is therefore necessary to study the approximation involved in the substitution of signals that are not strictly bandlimited with elements in the set $\mathscr{B}_\Omega$. Another extension regards signals that are not bandlimited to single intervals in frequency, or observed over single intervals in time, but are supported over the union of a finite number of fixed, disjoint intervals. Finally, signals can also be functions of several variables, and be supported over arbitrary measurable sets, where these variables and their transformed pairs can take values. All of the above extensions are relevant for electromagnetic waveforms that are functions of space and time, are only approximately filtered by the propagation process, and that, depending on the application, can span multiple bands in the spatial, temporal, frequency, and wavenumber domains.

In the following, we sketch the theory in the most general form and apply the results in subsequent chapters in the specific context of the propagation of electromagnetic signals.

3.5.1 Approximately Bandlimited Signals

Consider signals in a subset $\mathscr{E} \subset L^2[-T/2, T/2]$ that are not necessarily strictly bandlimited. Their Kolmogorov N-width can be bounded by the sum of the deviation from the set of bandlimited signals plus the N-width of bandlimited signals:

$$d_N(\mathscr{E}) \leq D_{\mathscr{B}_\Omega}(\mathscr{E}) + d_N(\mathscr{B}_\Omega). \tag{3.92}$$

This bound can be evaluated by first projecting the space of signals onto $\mathscr{B}_\Omega$, determining the error associated with this projection, and then adding it to the error associated with the best N-dimensional subspace approximation of $\mathscr{B}_\Omega$.

To derive the result, we consider the deviation $D_{\mathscr{X}_n}(\mathscr{E})$ between $\mathscr{E}$ and an arbitrary N-dimensional subspace

$$\begin{aligned} D_{\mathscr{X}_N}(\mathscr{E}) &= \sup_{f \in \mathscr{E}} \inf_{g \in \mathscr{X}_N} \|f - g\| \\ &= \sup_{f \in \mathscr{E}} D_{\mathscr{X}_N}(f). \end{aligned} \tag{3.93}$$

From this, we can derive the following "triangle" inequality:

$$D_{\mathscr{X}_N}(\mathscr{E}) \leq D_{\mathscr{B}_\Omega}(\mathscr{E}) + D_{\mathscr{X}_N}(\mathscr{B}_\Omega). \tag{3.94}$$

The derivation of the inequality is as follows. For any $h \in \mathscr{B}_\Omega$, we write

$$\begin{aligned}
D_{\mathscr{X}_N}(f) &= \inf_{g \in \mathscr{X}_N} \|f - g\| \\
&\le \inf_{g \in \mathscr{X}_N} (\|f - h\| + \|h - g\|) \\
&= \|f - h\| + \inf_{g \in \mathscr{X}_N} \|h - g\| \\
&= \|f - h\| + D_{\mathscr{X}_N}(h) \\
&\le \|f - h\| + D_{\mathscr{X}_N}(\mathscr{B}_\Omega).
\end{aligned} \tag{3.95}$$

Since h is arbitrary, we have

$$\begin{aligned}
D_{\mathscr{X}_N}(f) &\le \inf_{h \in \mathscr{B}_\Omega} \|f - h\| + D_{\mathscr{X}_N}(\mathscr{B}_\Omega) \\
&= D_{\mathscr{B}_\Omega}(f) + D_{\mathscr{X}_N}(\mathscr{B}_\Omega).
\end{aligned} \tag{3.96}$$

Now, taking the supremum of f on both sides, we have

$$\begin{aligned}
\sup_{f \in \mathscr{E}} D_{\mathscr{X}_N}(f) &\le \sup_{f \in \mathscr{E}} D_{\mathscr{B}_\Omega}(f) + D_{\mathscr{X}_N}(\mathscr{B}_\Omega) \\
&= D_{\mathscr{B}_\Omega}(\mathscr{E}) + D_{\mathscr{X}_N}(\mathscr{B}_\Omega).
\end{aligned} \tag{3.97}$$

The inequality (3.94) then follows from (3.93). This is useful to get an upper bound on the N-width of $\mathscr{E}$ in terms of the N-width of the set of signals $\mathscr{B}_\Omega$. Taking the infimum with respect to $\mathscr{X}_N$ of (3.94), we obtain (3.92).

It follows that in order to bound the degree to which any set of signals can be approximated by finite-dimensional subspaces, it is sufficient to determine their deviation from the set of bandlimited signals and then apply the approximation results for bandlimited signals. In contrast to the first term, the second term in (3.92) depends on Ω. A typical technique is to project the space $\mathscr{E}$ onto $\mathscr{B}_\Omega$ and then determine the error associated with this projection. This is then added to the approximation error obtained by a finite interpolation of a strictly bandlimited signal. We make use of this technique extensively in subsequent chapters to determine the number of degrees of freedom of electromagnetic signals propagating in multiple scattering environments.

3.5.2 Multi-band Signals

We now consider signals that are supported over multiple intervals. These can model, for example, transmission in communication systems where users multiplex messages over disjoint time or frequency intervals of the electromagnetic spectrum, and the receiver observes a multi-band signal composed of the superposition of the different transmitted waveforms in the time or frequency domain. An example is depicted in Figure 3.2, where an ideal real-valued signal

$$f(t) = f_0(t) + f_1(t)\cos(\omega_1 t) + f_2(t)\cos(\omega_2 t) \tag{3.98}$$

is composed of five disjoint frequency bands $\{\Delta_n\}$, multiplexing information from three different users, and the occupied portion of the spectrum is $\Omega_5 = \{\cup_n \Delta_n\}$ for $-2 \le n \le 2$.

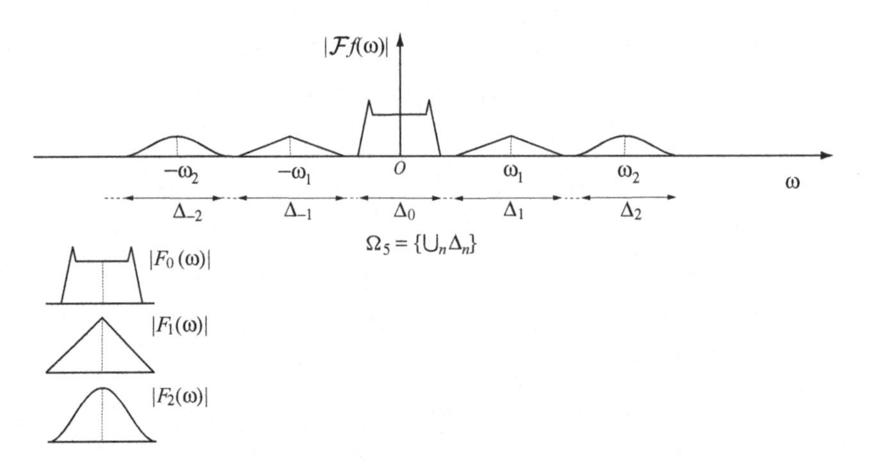

Fig. 3.2 Multi-band signal resulting from transmission from three different users.

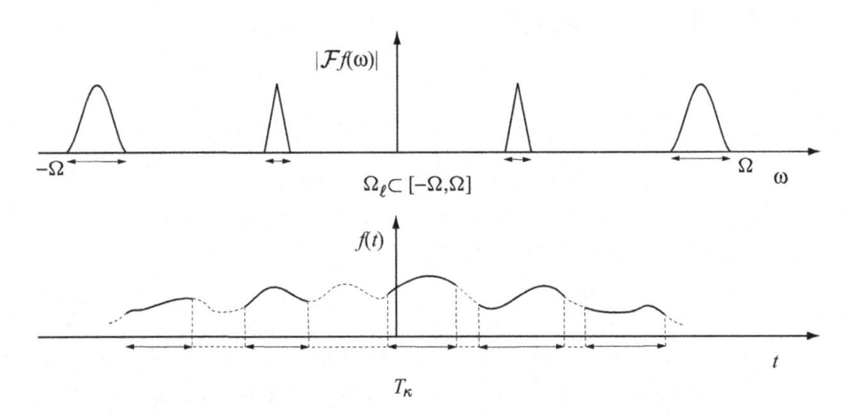

Fig. 3.3 Multi-band signal and its observation set.

In general, the different modulation techniques employed for transmission determine the occupied portion of the spectrum.

To compute the number of degrees of freedom, we let $\Omega_\ell \subset [-\Omega, \Omega]$ be the union of ℓ disjoint, fixed intervals in the frequency domain, over which the signal is transmitted, and T_κ be the union of κ disjoint, fixed intervals in the time domain where the signal is observed – see Figure 3.3. We assume the normalized energy constraint (2.43).

The Fredholm integral equation (2.57) arising from Slepian's concentration problem becomes, in this case,

$$\mathcal{B}_{\Omega_\ell} \mathcal{T}_{T_\kappa} \psi(t) = \lambda \psi(t), \quad t \in T_\kappa, \tag{3.99}$$

where the operators $\mathcal{T}_{T_\kappa}$ and $\mathcal{B}_{\Omega_\ell}$ correspond to timelimiting over the set T_κ, and bandlimiting the signal over the set Ω_ℓ. Using $\mathcal{F}$ to indicate the Fourier transform, and defining the indicator function

$$\mathbb{1}_{T_\kappa}(t) = \begin{cases} 1 & \text{if } t \in T_\kappa, \\ 0 & \text{otherwise,} \end{cases} \tag{3.100}$$

they are defined as

$$\mathcal{T}_{T_\kappa} f = \mathbb{1}_{T_\kappa} f, \tag{3.101}$$

$$\mathcal{B}_{\Omega_\ell} f = \mathcal{F}^{-1} \mathbb{1}_{\Omega_\ell} \mathcal{F} f. \tag{3.102}$$

The operator in (3.99) can be compactly represented as $\mathcal{T}_{T_\kappa} \mathcal{B}_{\Omega_\ell} \mathcal{T}_{T_\kappa}$ and is positive, self-adjoint, compact, and bounded by one. It admits a countable set of eigenvalues, representing the asymptotic dimension of the signals in $\mathscr{B}_{\Omega_\ell}$, while its eigenfunctions correspond to the optimal approximating basis. Considering the norm

$$\|f\|^2 = \int_{T_\kappa} f^2(t) dt, \tag{3.103}$$

we have that the number of eigenvalues in (3.99) that are above level ϵ corresponds to the number of degrees of freedom $N_\epsilon(\mathscr{B}_{\Omega_\ell})$ of multi-band signals of total spectral support Ω_ℓ, observed over the time support T_κ, namely to the dimension of the minimal subspace approximating the elements of $\mathscr{B}_{\Omega_\ell}$ within ϵ accuracy over the set T_κ.

To obtain a concentration behavior, we let the spectral support Ω_ℓ be a fixed set, while the observation set scales as αT_κ, with $\alpha \to \infty$. Due to symmetry, the case in which the observation set is fixed and the spectral support scales as $\alpha \Omega_\ell$, with $\alpha \to \infty$, is completely analogous. For all $0 < \epsilon < 1$, we have

$$\begin{aligned} N_\epsilon(\mathscr{B}_{\Omega_\ell}) &= \frac{m(\Omega_\ell) m(T_\kappa)}{2\pi} \alpha + \frac{\kappa \ell}{\pi^2} \log\left(\frac{1-\epsilon}{\epsilon}\right) \log \alpha + o(\log \alpha) \\ &= \frac{m(\Omega_\ell) m(T_\kappa)}{2\pi} \alpha + O(\log \alpha), \end{aligned} \tag{3.104}$$

where $m(\cdot)$ indicates Lebesgue measure. This result should be compared with (2.132). The latter equation is immediately recovered from (3.104) by letting $\ell = \kappa = 1$, $m(\Omega_1) = 2\Omega$, $\alpha m(T_1) = T$. In the multi-band case, for ℓ and κ greater than one, the time bandwidth product in (2.132) is replaced by the product of the measures of the occupied portions of the spectrum in the time and frequency domains, and we have that:

> The occupied portion of the spectrum in the time and frequency domain determines, up to first order, the scaling of the number of degrees of freedom.

Compared to the single-interval case, the weak dependence of the number of degrees of freedom on the approximation level ϵ still holds, since ϵ appears only in the higher-order terms of the number of degrees of freedom. However, in the case of (3.104) we also have a linear dependence in the second-order term on the product of the number of occupied intervals $\kappa\ell$. When one of the two support sets is scaled by the factor $\alpha \to \infty$, having fixed the number of intervals and the level of approximation, the phase transition is revealed, and we have

$$\lim_{\alpha \to \infty} \frac{N_\epsilon(\mathscr{B}_{\Omega_\ell})}{\alpha} = \frac{m(\Omega_\ell) m(T_\kappa)}{2\pi}, \tag{3.105}$$

which is the analog of (3.17) in the multi-band case. The optimal approximating eigenfunctions $\{\psi_n(t)\}$ that are solutions of (3.99) depend in this case on the multi-band sets Ω_ℓ and T_κ.

We can also ask what it takes to drive the approximation error to zero. For all $\nu > 0$, interpolating a number of eigenfunctions

$$N = (1+\nu)\frac{m(\Omega_\ell)m(T_\kappa)}{2\pi}\alpha, \tag{3.106}$$

it is possible to approximate any multi-band signal with vanishing error as $\alpha \to \infty$. The analogous result for signals supported over single intervals in the time and frequency domains is given in (2.131), namely, interpolating $(1+\nu)\Omega T/\pi$ prolate spheroidal wave functions, the approximation error tends to zero as $N_0 \to \infty$.

3.5.3 Signals of Multiple Variables

The notion of number of degrees of freedom and asymptotic dimensionality of the signals' space can be extended from signals of a single variable to signals of multiple variables. This extension is particularly useful for electromagnetic waveforms that are functions of space and time.

Consider two subsets P and Q of $\mathbb{R}^N$ having finite measure; an arbitrary point $\mathbf{p} \in \mathbb{R}^N$; and a positive scalar $\alpha > 0$. We indicate by αP the set of points of the form $\alpha \mathbf{x}$ with $\mathbf{x} \in P$; we immediately have that

$$m(\alpha P) = \alpha^N m(P), \tag{3.107}$$

where $m(\cdot)$ represents Lebesgue measure in $\mathbb{R}^N$. Similarly, for $\mathbf{x} = (x_1, x_2, \ldots, x_N)$, we have

$$m\{(\alpha x_1, x_2, \ldots, x_n) : \mathbf{x} \in P\} = \alpha m(P). \tag{3.108}$$

For any two points $\mathbf{x} = (x_1, \ldots, x_N)$, $\mathbf{u} = (u_1, \ldots, u_N)$ we let $\mathbf{x} \cdot \mathbf{u} = x_1 u_1 + \cdots + x_N u_N$. We consider square-integrable signals $f \in \mathbb{R}^N$ with Fourier transform

$$\mathcal{F}f = \int_{\mathbb{R}^N} f(\mathbf{x}) e^{-j\mathbf{x}\cdot\mathbf{u}} d\mathbf{x}, \tag{3.109}$$

and subject to the energy constraint

$$\int_{\mathbb{R}^N} f^2(\mathbf{x}) d\mathbf{x} \leq 1. \tag{3.110}$$

Among these, we pick the subsets of signals supported in P and those whose Fourier transform is supported in Q. These will play the role of the sets of timelimited and bandlimited signals considered previously. Accordingly, we let

$$\mathscr{T}_P = \{f : f(\mathbf{x}) = 0,\ \mathbf{x} \notin P\}, \tag{3.111}$$

$$\mathscr{B}_Q = \{f : \mathcal{F}f(\mathbf{u}) = 0,\ \mathbf{u} \notin Q\}. \tag{3.112}$$

Letting the indicator function

$$\mathbb{1}_P(\mathbf{x}) = \begin{cases} 1 & \text{if } \mathbf{x} \in P, \\ 0 & \text{otherwise,} \end{cases} \tag{3.113}$$

we consider the following operators:

$$\mathcal{T}_P f = \mathbb{1}_P(\mathbf{x}) f(\mathbf{x}), \tag{3.114}$$

$$\mathcal{B}_Q f = \mathcal{F}^{-1} \mathbb{1}_Q \mathcal{F} f = \int_{\mathbb{R}^N} h(\mathbf{x} - \mathbf{y}) f(\mathbf{y}) d\mathbf{y}, \tag{3.115}$$

where

$$\mathcal{F} h = \mathbb{1}_Q. \tag{3.116}$$

It should be clear that for signals of a single variable supported over intervals, operators analogous to (3.114) and (3.115) are the timelimiting and bandlimiting operators corresponding to the multiplication by a rect $(\cdot)$ and convolution by a sinc$(\cdot)$ of the time-domain representation of the signal, and the convolution by a sinc$(\cdot)$ and multiplication by a rect $(\cdot)$ of the frequency-domain representation of the signal. We consider here the general case of arbitrarily measurable sets in arbitrary dimensions.

The Fredholm integral equation (2.57) arising from Slepian's concentration problem for signals of a single variable is then generalized as

$$\mathcal{B}_Q \mathcal{T}_P \psi(\mathbf{x}) = \lambda \psi(\mathbf{x}), \quad \mathbf{x} \in P, \tag{3.117}$$

and it can be compactly represented with the operator $\mathcal{T}_P \mathcal{B}_Q \mathcal{T}_P$, which is positive, self-adjoint, compact, and bounded by one. It admits a countable set of eigenvalues, representing the asymptotic dimension of $\mathscr{B}_Q$, while its eigenfunctions correspond to the optimal approximating basis. Considering the norm

$$\|f\| = \left(\int_P f^2(\mathbf{x}) d\mathbf{x} \right)^{1/2}, \tag{3.118}$$

the number of eigenvalues above level ϵ corresponds to the number of degrees of freedom $N_\epsilon(\mathscr{B}_Q)$, namely to the dimension of the minimal subspace approximating the elements of $\mathscr{B}_Q$ within ϵ accuracy over the set P.

In order to compute the number of degrees of freedom, we let Q be a fixed set, while P varies over the family $\alpha P'$, with P' fixed. Due to symmetry, the asymptotic dimension of $\mathscr{T}_P$ can similarly be obtained by considering the operator $\mathcal{B}_Q \mathcal{T}_P \mathcal{B}_Q$ and scaling $Q = \alpha Q'$ while keeping P fixed. The situation is depicted in Figure 3.4 in the case of a space–time field with $N = 4$.

As $\alpha \to \infty$ and one of the two support sets is scaled to infinity, the number of significant eigenvalues of the operator $\mathcal{T}_{\alpha P'} \mathcal{B}_Q \mathcal{T}_{\alpha P'}$ undergoes a phase transition around α^N and the width of the transition is $o(\alpha^N)$, so that the eigenvalues arranged in non-increasing order appear, as a function of their indexes, as a step function when viewed at the scale of α^N. The number of degrees of freedom corresponds to the transition point of this step function, that is of the order of α^N, when the support set is appropriately scaled by "blowing up" all of its coordinates.

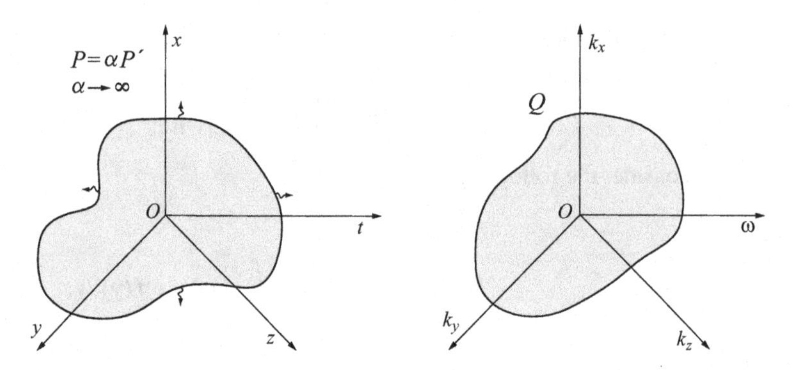

Fig. 3.4 Scaling of the support set P.

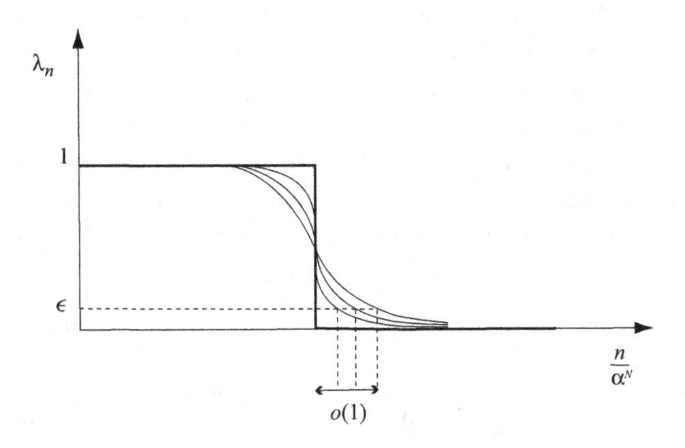

Fig. 3.5 Phase transition of the eigenvalues.

This result is depicted in Figure 3.5 and should be compared to the case of signals of a single variable defined over intervals depicted in Figure 2.13. In the case of signals of multiple variables the width of the transition is characterized as $o(1)$, as $\alpha \to \infty$; in the single-variable case the transition window is more precisely characterized as $O[(\log N_0)/N_0]$. The optimal approximating eigenfunctions are the prolate spheroidal ones in the case of one-dimensional intervals, while in the general case they depend on the sets of concentration P and Q. Despite these differences, the phase transition determining the effective dimensionality of the space also occurs for bandlimited signals of multiple variables defined over arbitrary measurable sets, and is obtained by blowing up either one of the support sets in the original or transformed domain, at rate α.

The precise result is stated as follows. For any arbitrary level of precision $0 < \epsilon < 1$, we have

$$\lim_{\alpha \to \infty} \frac{N_\epsilon(\mathscr{B}_Q)}{\alpha^N} = (2\pi)^{-N} m(P') m(Q). \tag{3.119}$$

As in the single-variable case, the scale α^N of the transition is independent of the accuracy level ϵ of the approximation, and the ϵ dependence appears hidden as a pre-constant of the second-order term of the number of degrees of freedom. This is evident by rewriting (3.119) as

$$N_\epsilon(\mathscr{B}_Q) = (2\pi)^{-N} m(P')m(Q)\alpha^N + o(\alpha^N). \quad (3.120)$$

The pre-constant of the term $o(\alpha^N)$ also contains the dependence on the geometry of the sets of concentration. This is made explicit, in terms of number of intervals, along with the dependence on the approximation level, for the single-variable case in (3.104). The analogous statement for signals of a single variable bandlimited to Ω and observed over an interval of size T is easily recovered by letting $N = 1$, Q be the angular frequency support, P be the time support, $m(P') = 1$, $m(Q) = 2\Omega$, $\alpha = T$, and letting $T \to \infty$, so that the number of degrees of freedom in (3.119) becomes

$$N_\epsilon(\mathscr{B}_\Omega) = \Omega T/\pi + o(T) \quad \text{as } T \to \infty, \quad (3.121)$$

in agreement with (3.16). In the case of (3.16), the transition is more precisely characterized as being of the order of $O(\log T)$, and by (2.132) we also have an explicit characterization of the ϵ dependence that diverges at most as $\log(1/\epsilon)$ as $\epsilon \to 0$, as already noted.

We now summarize the results with the following statement:

> The number of degrees of freedom of signals of multiple variables scales as the product of the measure of the spectral support set times the measure of the observation set, as all coordinates of one of the two sets tend to infinity.

3.5.4 Hybrid Scaling Regimes

So far, spectral concentration has been achieved by scaling all coordinates of one of the two support sets P and Q by the factor α. An analogous result holds by scaling any combination of coordinates of both support sets. For example, for $N = 2$ we can let P and Q be fixed; consider the families

$$P_\alpha = \{(\alpha x_1, x_2) : (x_1, x_2) \in P\}, \quad (3.122)$$

$$Q_\alpha = \{(x_1, \alpha x_2) : (x_1, x_2) \in Q\}, \quad (3.123)$$

and scale the first coordinate of P and the second coordinate of Q by the factor α, while maintaining the other coordinates fixed. For any arbitrary level of precision $\epsilon > 0$, let $N_\epsilon(\alpha)$ be the number of eigenvalues of $\mathcal{T}_{P_\alpha}\mathcal{B}_{Q_\alpha}\mathcal{T}_{P_\alpha}$ not smaller than ϵ; then we have

$$\lim_{\alpha \to \infty} \frac{N_\epsilon(\alpha)}{\alpha^2} = (2\pi)^{-2} m(P)m(Q). \quad (3.124)$$

The coordinates' scaling of the two support sets does not need to be uniform, and we can also use two distinct scaling parameters in place of α. Consider the families

$$P_\tau = \{(\tau x_1, x_2) : (x_1, x_2) \in P\}, \tag{3.125}$$

$$Q_\rho = \{(x_1, \rho x_2) : (x_1, x_2) \in Q\}; \tag{3.126}$$

letting $N_\epsilon(\tau, \rho)$ be the number of eigenvalues of $\mathcal{T}_{P_\tau} \mathcal{B}_{Q_\rho} \mathcal{T}_{P_\tau}$ not smaller than ϵ, we have

$$\lim_{\tau, \rho \to \infty} \frac{N_\epsilon(\tau, \rho)}{\tau \rho} = (2\pi)^{-2} m(P) m(Q). \tag{3.127}$$

These results can be extended to any number of variables, where any combination of coordinates' scaling in the original or transformed domain can be performed. It follows that the number of degrees of freedom of signals of multiple variables scales as the product of the measure of the spectral support set times the measure of the observation set, as any subset of coordinates of one set and the remaining coordinates of the other set tend to infinity.

A more general result holds for any invertible linear transform of the support sets, subject to a limiting condition, that includes the scalings described above as special cases. Let P and Q be measurable sets in $\mathbb{R}^N$. For any real matrix A of size $N \times N$, we indicate by AP the set of points of the form Ax, where x is the column vector composed of the elements of $\mathbf{x} \in P$. We also indicate with |A| the determinant of A, with A^T the transpose of A, and with U the unit ball in $L^2(\mathbb{R}^N)$. We let $\mathrm{A} = \mathrm{A}(\tau)$ and $\mathrm{B} = \mathrm{B}(\rho)$ for real parameters τ and ρ. The case in which either matrix is constant, or depends on multiple parameters, is completely analogous.

For any $0 < \epsilon < 1$, let $N_\epsilon(\mathrm{A}, \mathrm{B})$ be the number of eigenvalues of $\mathcal{T}_{\mathrm{A}P} \mathcal{B}_{\mathrm{B}Q} \mathcal{T}_{\mathrm{A}P}$ not smaller than ϵ. If

$$\lim_{(\tau, \rho) \to \infty} \mathrm{B}^\mathrm{T} \mathrm{A} U = \mathbb{R}^N, \tag{3.128}$$

then we have

$$\lim_{(\tau, \rho) \to \infty} \frac{N_\epsilon(\mathrm{A}, \mathrm{B})}{|\mathrm{A}|\,|\mathrm{B}|} = (2\pi)^{-N} m(P) m(Q). \tag{3.129}$$

Equation (3.119) is a special case where A is a scalar matrix and B is the identity. Equation (3.127) corresponds to the special case

$$\mathrm{A} = \begin{pmatrix} \tau & 0 \\ 0 & 1 \end{pmatrix}, \quad \mathrm{B} = \begin{pmatrix} 1 & 0 \\ 0 & \rho \end{pmatrix}. \tag{3.130}$$

This result is shown in Appendix A.6 and is applied in Chapter 9 to determine the number of degrees of freedom of bandlimited electromagnetic waveforms in time and space.

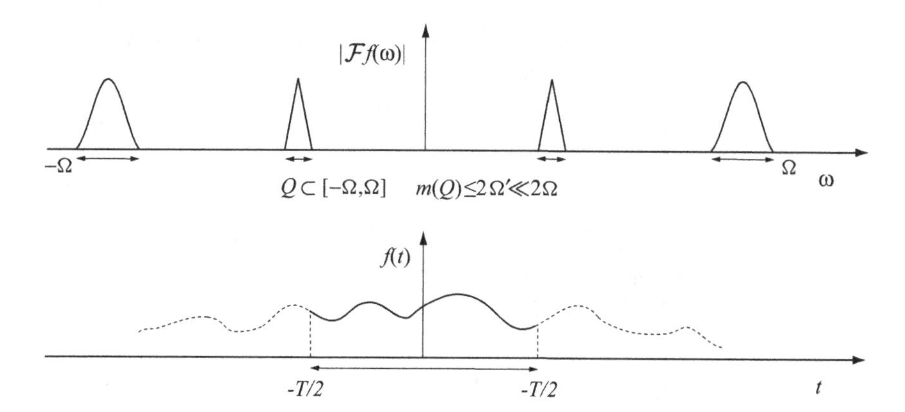

Fig. 3.6 Illustration of a sparse multi-band signal observed over a single time interval.

3.6 Blind Sensing

The results in Sections 3.5.2 and 3.5.3 indicate that it is possible to reconstruct square-integrable signals of spectral support Q from a number of measurements performed along the dimensions of the space spanned by the eigenfunctions of (3.117). The reconstruction procedure amounts to first solving (3.117) to obtain the basis functions $\{\psi_n\}$ and the eigenvalues $\{\lambda_n\}$. From the behavior of the eigenvalues we can then determine the number of functions in the approximation

$$f_N(\mathbf{x}) = \sum_{n=1}^{N} a_n \psi_n(\mathbf{x}) \tag{3.131}$$

that give a satisfactory error

$$e_N = \int_P [f(\mathbf{x}) - f_N(\mathbf{x})]^2 d\mathbf{x} \tag{3.132}$$

as the observation set P is scaled up to infinity. The coefficients $\{a_n\}$ are obtained by integration:

$$a_n = \frac{1}{\lambda_n} \int_P f(\mathbf{x}) \psi_n(\mathbf{x}) d\mathbf{x}. \tag{3.133}$$

Consider now a signal $f(t) \in \mathscr{B}_\Omega$ of a single variable t, subject to (2.43), and such that its Fourier transform

$$F(\omega) = 0, \text{ for } \omega \notin Q, \tag{3.134}$$

where Q is a subset of $[-\Omega, \Omega]$ whose measure is at most $2\Omega' \ll 2\Omega$, and the observation set is $P = [-T/2, T/2]$. Since the portion of the occupied spectrum is much smaller than the total bandwidth, we call this signal "sparse" in the frequency domain – see Figure 3.6.

We define a measurement vector

$$\mathbf{y} = \mathcal{M} f(t), \tag{3.135}$$

where $\mathcal{M}(\cdot)$ is a linear measurement operator that maps multi-band signals into M-dimensional vectors. Since $f(t) \in \mathscr{B}_\Omega$, by (2.131) and (3.13) for any $\nu > 0$ the signal can be reconstructed with vanishing error over the observation interval as $T \to \infty$ using N measurements, where

$$N = (1+\nu)\Omega T/\pi. \tag{3.136}$$

In this case, each component of $\mathbf{y}$ corresponds to the projection of the signal onto a prolate spheroidal basis function that is a solution of

$$\mathcal{B}_\Omega \mathcal{T}_T \psi(t) = \lambda \psi(t), \quad t \in [-T/2, T/2]. \tag{3.137}$$

On the other hand, by (3.106), if we have knowledge of the support set Q, namely of the size and positions of all the sub-bands, then reconstruction with vanishing error is also possible using S measurements, where

$$S = (1+\nu)\Omega' T/\pi. \tag{3.138}$$

In this case, each component of $\mathbf{y}$ corresponds to the projection of the signal onto an eigenfunction of (3.117).

Assume now that we only know that $f(t)$ is a multi-band signal supported over a set of size $2\Omega'$, but we do not know the sizes and positions of its sub-bands. Without further investigation we can only conclude from the results above that a number of measurements M sufficient to perfectly recover $f(t)$ over the observation interval satisfies

$$S \leq M \leq N. \tag{3.139}$$

Determining the minimum value of M within these bounds is called the *blind sensing* problem.

A basic result in blind sensing states that roughly $2S$ measurements are necessary and sufficient to blindly recover any multi-band signal $f(t)$ without any error. It follows that the price to pay for blind reconstruction compared to reconstruction with full knowledge of the support set is a factor of two. More precisely, for any real multi-band signal f, using a number of measurements

$$M = 2S, \tag{3.140}$$

we can construct a signal f_M such that the approximation error

$$\|f - f_M\| = \left(\int_{-T/2}^{T/2} [f(t) - f_M(t)]^2 dt \right)^{1/2} \tag{3.141}$$

tends to zero as $T \to \infty$. On the other hand, there exist signals that cannot be reconstructed without error as $T \to \infty$ using a number of measurements

$$M = 2\Omega' T/\pi + o(T). \tag{3.142}$$

Combining (3.138), (3.140), and (3.142), defining the measurement rate

$$R = \frac{M}{T}, \tag{3.143}$$

and recalling that ν is arbitrary, we have that as $T \to \infty$ the smallest number of measurements per unit time that guarantees a vanishing error must satisfy

$$R > 2\Omega'/\pi. \tag{3.144}$$

> Twice the occupied portion of the frequency bandwidth determines the smallest measurement rate sufficient for blind reconstruction of multi-band signals.

3.6.1 Robustness of Blind Sensing

As elegant as they are, the information-theoretic results described above lack some practical considerations. For example, we may ask whether the approximation obtained is robust to small perturbations of the measurement, and whether this approximation can be obtained in a computationally efficient way.

To start with the first question, we consider the measurement

$$\mathbf{y} = \mathcal{M}f(t) + \mathbf{e}, \tag{3.145}$$

where $\mathbf{e}$ is a measurement error having vector norm

$$\|\mathbf{e}\| = \left(\sum_{n=1}^{M} e_n^2 \right)^{1/2}. \tag{3.146}$$

We would like a small measurement error not to lead to a huge approximation error. For this reason, we require that the norm of the approximation error (3.141) and the norm of the measurement error (3.146) be linearly related. This requirement can be satisfied if the measurement somewhat preserves distances; namely, if, for all f satisfying (3.134), we have

$$1 - \mu \leq \frac{\|\mathcal{M}f\|}{\|f\|} \leq 1 + \mu, \tag{3.147}$$

where $\mu < 1$ is a positive constant. The smallest constant μ for which (3.147) holds is called the *isometry constant* of the measurement operator $\mathcal{M}$.

One can show that there exist operators satisfying (3.147), provided that

$$M = c_\mu S, \tag{3.148}$$

where $c_\mu \geq 2$ increases with decreasing μ. In this case, we can construct a signal $f_M(t)$ such that, as $T \to \infty$,

$$\|f - f_M\| \leq k_\mu \|\mathbf{e}\|. \tag{3.149}$$

It follows that by appropriately choosing the operator $\mathcal{M}$ we can ensure that small perturbations in the measurements lead to correspondingly small values of the approximation error.

Measurement operators with smaller isometry numbers require larger values of c_μ and lead to smaller values of k_μ. Finding computationally efficient reconstruction methods using measurement maps satisfying (3.147), and having reasonably small values for both constants c_μ and k_μ, is an area of research. The smallest possible value $k_\mu = 1$ in (3.149) is easily achieved by using a number of measurements $M = (1+\nu)\Omega T/\pi$, where ν is an arbitrarily small constant. In this case, using the noisy measurements

$$b_n = \frac{1}{\lambda_n}\int_{-T/2}^{T/2} f(t)\psi_n(t)dt + e_n, \tag{3.150}$$

we can construct

$$f_N(t) = \sum_{n=1}^{N} b_n \psi_n(t), \tag{3.151}$$

where the $\{\psi_n\}$ are the eigenfunctions of (3.137). It follows that

$$\begin{aligned} f_N(t) &= \sum_{n=1}^{N} \frac{1}{\lambda_n}\int_{-T/2}^{T/2} f(t)\psi_n(t)dt\, \psi_n(t) + \sum_{n=1}^{N} e_n\psi_n(t) \\ &= \sum_{n=1}^{N} a_n\psi_n(t) + \sum_{n=1}^{N} e_n\psi_n(t). \end{aligned} \tag{3.152}$$

Using the orthogonality of the $\{\psi_n\}$, and $\lambda_n \le 1$ for all n, it follows that

$$\begin{aligned} \|f - f_N\| &\le \left\| \sum_{n=1}^{N} e_n\psi_n(t) \right\| + \left\| f(t) - \sum_{n=1}^{N} a_n\psi_n(t) \right\| \\ &\le \left(\sum_{n=1}^{N} \lambda_n^2 e_n^2 \right)^{1/2} + \left\| f(t) - \sum_{n=1}^{N} a_n\psi_n(t) \right\| \\ &\le \|\mathbf{e}\| + \left\| f(t) - \sum_{n=1}^{N} a_n\psi_n(t) \right\|. \end{aligned} \tag{3.153}$$

For $M = (1+\nu)\Omega T/\pi$, letting $T \to \infty$, the second term in (3.153) tends to zero.

Comparing this result with (3.149), we realize that using this number of measurements may be overkill to have a reasonable approximation error. In practice, for sufficiently sparse signals blind sensing can achieve an approximation error very close to $\|\mathbf{e}\|$ using a considerably smaller number of measurements. If $\Omega' \ll \Omega$, then larger values of c_μ in (3.148) may still ensure a small number of measurements while leading to smaller values of k_μ.

3.6.2 Fractal Dimension

The problem of signal reconstruction is related to that of determining the effective dimensionality, or degrees of freedom, of the signals' space. For bandlimited signals, the effective number of dimensions can be identified with the Nyquist number $N_0 = \Omega T/\pi$.

For multi-band signals for which the location and widths of all the sub-bands is known *a priori*, it can be identified with "the sparsity number" $S_0 = \Omega' T/\pi$. On the other hand, without any *a priori* knowledge, we need to account for the additional degrees of freedom of allocating the sub-bands in the frequency domain, and the effective dimensionality increases to $2S_0$.

To make these considerations precise, we consider an information-theoretic quantity that measures the effective dimensionality of a set in metric space, namely its fractal dimension, which corresponds to the rate of growth of the ϵ-entropy of successively finer discretizations of the space, and represents the "degree of fractality" of the set. This is also known as the Minkowksi–Bouligand dimension, and we examine it in detail in Chapter 12.

Consider multi-band signals of energy at most one that are in $\mathscr{B}_\Omega$ and whose spectral support is of measure at most $2\Omega'$. Since these signals are bandlimited, they can be approximated, with vanishing error as $T \to \infty$, by a set $\mathscr{X}$ of vectors of size $N = (1+\nu)\Omega T/\pi$. Each vector contains the coefficients of the optimal interpolation basis for bandlimited signals.

The fractal dimension of $\mathscr{X}$ is defined as

$$\dim_{\mathrm{F}}(\mathscr{X}) = \lim_{\epsilon \to 0} \frac{H_\epsilon(\mathscr{X})}{-\log \epsilon}, \tag{3.154}$$

where H_ϵ is the Kolmogorov ϵ-entropy introduced in Chapter 1.

This definition is related to the number of measurements required for reconstruction. By considering the Minkowski sum

$$\mathscr{X} \oplus \mathscr{X} = \{\mathbf{x}_1 + \mathbf{x}_2 : \mathbf{x}_1, \mathbf{x}_2 \in \mathscr{X}\}, \tag{3.155}$$

we have the following basic result: with a number of measurements greater than

$$M = \dim_{\mathrm{F}}(\mathscr{X} \oplus \mathscr{X}), \tag{3.156}$$

it is possible to blindly reconstruct any multi-band signal with vanishing error as $T \to \infty$. On the other hand, there exist signals that cannot be blindly reconstructed with vanishing error as $T \to \infty$ using a smaller number of measurements.

An intuitive interpretation of these results is as follows. The set of all multi-band signals is the union of infinitely many subsets, each corresponding to the multi-band signals of a given sub-band allocation. The Minkowski sum in (3.155) takes into account the additional degrees of freedom of allocating the sub-bands in the frequency domain. Within any subset, any multi-band signal is specified by essentially $\dim_{\mathrm{F}}(\mathscr{X})$ coordinates, but when considering the union of all subsets, it is specified by essentially $\dim_{\mathrm{F}}(\mathscr{X} \oplus \mathscr{X})$ coordinates. For multi-band signals of a given sub-band allocation we have $\dim_{\mathrm{F}}(\mathscr{X} \oplus \mathscr{X}) = \dim_{\mathrm{F}}(\mathscr{X})$, while for multi-band signals of arbitrary sub-band allocation we have $\dim_{\mathrm{F}}(\mathscr{X} \oplus \mathscr{X}) = 2\dim_{\mathrm{F}}(\mathscr{X})$. It then follows that, in general, the relevant information-theoretic quantity that characterizes the possibility of reconstruction is the fractal dimension of the dilation set in (3.155).

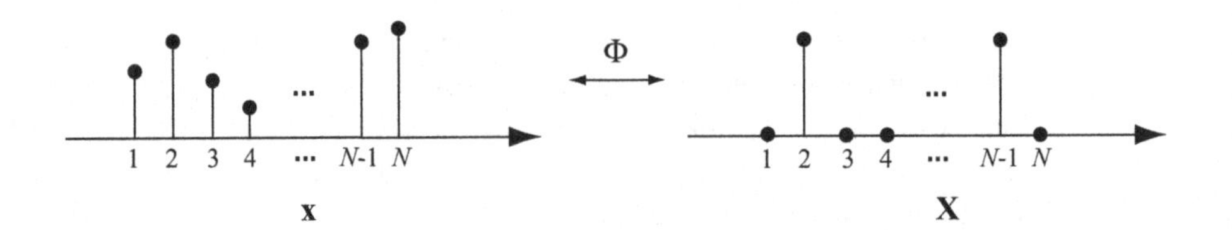

Fig. 3.7 A discrete vector with a sparse representation.

3.7 Compressed Sensing

The blind sensing problem is closely related to the one of *compressed sensing*, which attempts to reconstruct a sparse N-dimensional vector from a small number of measurements. In this case, we consider a vector $\mathbf{x}$ such that

$$\mathbf{x} = \Phi\mathbf{X}, \tag{3.157}$$

where Φ is an $N \times N$ orthogonal matrix that plays a role similar to an inverse Fourier transform in the continuous setting, and $\mathbf{X}$ is an N-dimensional vector with $S < N$ non-zero elements. If $S \ll N$, we say that $\mathbf{X}$ is a sparse representation of $\mathbf{x}$ – see Figure 3.7.

We define a measurement vector

$$\mathbf{y} = \mathrm{A}\mathbf{x}, \tag{3.158}$$

where A is an $M \times N$ matrix, and M is the number of measurements.

Cleary, $\mathbf{x}$ can be recovered from N measurements by observing all the elements of $\mathbf{x}$. In this case, the $N \times N$ measurement matrix A is diagonal. If we know the position of the non-zero elements of $\mathbf{X}$, then S measurements are also enough to perfectly reconstruct $\mathbf{x}$. In this case, each measurement extracts the nth coefficient of $\mathbf{X}$ from $\Phi^{-1}\mathbf{x}$, and the signal is recovered by performing a final multiplication by Φ. However, assuming we only know that $\mathbf{x}$ has a sparse representation, but we do not know the positions of the non-zero elements of $\mathbf{X}$, without further investigation we can only conclude that a number of measurements M sufficient for reconstruction satisfies $S \leq M \leq N$.

> Determining the minimum value of M within these bounds in this discrete setting is called the *compressed sensing* problem.

A little thought reveals that any sparse vector can be blindly recovered from $M = 2S$ measurements, by choosing a measurement matrix A such that any $2S$ columns of A are linearly independent. If we assume by contradiction that this is not the case, then there are two distinct vectors $\mathbf{x} \neq \mathbf{x}'$ with at most S non-zero elements such that $\mathrm{A}\mathbf{x} = \mathrm{A}\mathbf{x}'$. This implies that $\mathrm{A}(\mathbf{x} - \mathbf{x}') = 0$. However, $(\mathbf{x} - \mathbf{x}')$ has at most $2S$ non-zero elements, so there must be a linear dependence between $2S$ columns of A, which is impossible.

A reconstruction procedure suggested from the above argument is the following: find the sparsest solution $\mathbf{x}^*$ compatible with the measurements (3.158); namely, let

$$\mathbf{x}^* = \operatorname{argmin} \left\{ \sum_{n=1}^{N} \mathbb{1}_{\{x_n \neq 0\}} : \mathbf{A}\mathbf{x}^* = \mathbf{y} \right\}. \tag{3.159}$$

Unfortunately, solving this optimization problem is computationally prohibitive.

A more attractive alternative is to consider a measurement matrix whose elements are the complex exponentials

$$\mathrm{A}_{n,k} = \exp(-j2\pi nk/N), \quad n \in \{1,2,\ldots,2S\},\ k \in \{1,2,\ldots,N\}, \tag{3.160}$$

so that the $2S$ measurements correspond to $2S$ consecutive discrete Fourier coefficients of $\mathbf{x}$:

$$y_n = \sum_{k=1}^{N} x_k \exp(-k2\pi nk/N), \quad n \in \{1,2,\ldots,2S\}. \tag{3.161}$$

The objective is now to recover the whole vector $\mathbf{x}$ from these coefficients. A procedure called Reed–Solomon decoding allows us to transform the problem into that of solving an $S \times S$ Toeplitz system, and then taking an inverse discrete Fourier transform.

It can be shown that $2S$ measurements are also necessary for reconstruction – see Problem 3.12. It follows that the same phase transition behavior occurring in the continuous setting for blind sensing also occurs in the discrete setting for compressed sensing.

> In both the continuous and discrete settings, the number of linear measurements necessary and sufficient for reconstruction is equal to twice the sparsity level of the signal.

The main differences between the two settings are as follows. The compressed sensing formulation assumes knowledge of the matrix Φ, corresponding to the basis where the discrete signal is sparse. In the case of blind sensing, it is only assumed that the signal does not occupy the whole frequency spectrum, but the discrete basis set required for the optimal representation is unknown *a priori*. In addition, in blind sensing the reconstruction error tends to zero as $T \to \infty$, while in compressed sensing perfect reconstruction occurs for all N.

3.7.1 Robustness of Compressed Sensing

While computationally attractive and using the smallest possible number of measurements, the reconstruction procedure based on Reed–Solomon decoding is not entirely clear of obstacles either. It turns out that it is terribly ill-conditioned and very sensitive to perturbations in the measurements: small measurement errors lead to huge reconstruction errors.

Assuming a measurement model

$$\mathbf{y} = \mathrm{A}\mathbf{x} + \mathbf{e}, \tag{3.162}$$

we would like the norm of the reconstruction error to be linearly related to the norm of the measurement error. Such a robust reconstruction can be performed if, for all vectors $\mathbf{x}$ having at most S non-zero elements, the measurement matrix A satisfies

$$1 - \mu \leq \frac{\|\mathrm{A}\mathbf{x}\|}{\|x\|} \leq 1 + \mu, \tag{3.163}$$

where $\mu < 1$ is a positive constant. The smallest constant μ for which (3.163) holds is the isometry constant of the measurement matrix A.

One can show that there exist matrices satisfying (3.163), provided that

$$M = c_\mu S \log(N/S), \tag{3.164}$$

where c_μ increases with decreasing μ. In this case, we can construct $\mathbf{x}^*$ by solving

$$\mathbf{x}^* = \operatorname{argmin} \left\{ \sum_{n=1}^{N} |x_n| : \mathrm{A}\mathbf{x}^* = \mathbf{y} \right\}, \tag{3.165}$$

and the reconstructed vector satisfies

$$\|\mathbf{x} - \mathbf{x}^*\| \leq k_\mu \|\mathbf{e}\|. \tag{3.166}$$

Since (3.165) is a convex optimization problem, it can be efficiently solved by linear programming methods. The same procedure can also be applied to reconstruct vectors $\mathbf{x}$ that are not necessarily sparse, namely they can have more than S non-zero elements, using only S measurements. In this case, we have

$$\|\mathbf{x} - \mathbf{x}^*\| \leq k_\mu \|\mathbf{e}\| + k' S^{-1/2} \sum_{n=1}^{N} |x_n - x_n^{(S)}|, \tag{3.167}$$

where $\mathbf{x}^{(S)}$ is the vector $\mathbf{x}$ with all but the largest S components set to zero. If $\mathbf{x}$ has only S non-zero components then $\mathbf{x} = \mathbf{x}^{(S)}$, and (3.167) reduces to (3.166). In general, the error is the sum of two contributions. The first is proportional to the measurement error. The second contribution corresponds to the error one would obtain in the absence of measurement error and by knowing the location of the S largest values of $\mathbf{x}$ and measuring those directly. In this case a converse result also holds, namely if there exists a matrix A and a reconstruction algorithm such that

$$\|\mathbf{x} - \mathbf{x}^*\| \leq k' S^{-1/2} \sum_{n=1}^{N} |x_n - x_n^{(S)}|, \tag{3.168}$$

then the number of measurements must satisfy

$$M \geq c S \log(N/S), \tag{3.169}$$

where c depends on k' but does not depend on the isometry number of A.

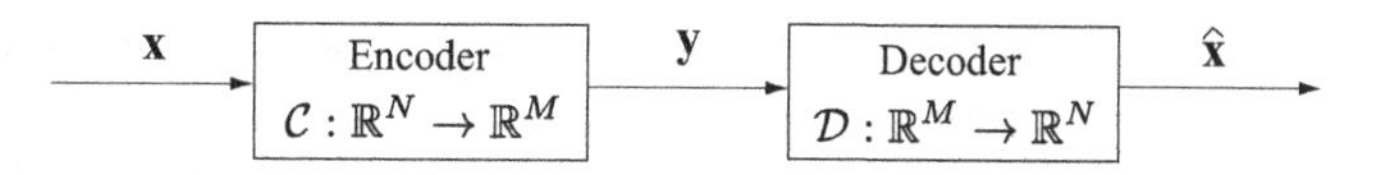

Fig. 3.8 Source coding view of compressed sensing.

3.7.2 Probabilistic Reconstruction

The problem of compressed sensing can also be formulated in a probabilistic setting. In this case, the discrete signal to be recovered is modeled as a stochastic process and the objective is to reconstruct the signal with arbitrarily small probability of error, given a sufficiently long observation sequence. Viewing the measurement operator as an encoder and the reconstruction operator as a decoder acting on a sequence of independent, identically distributed (i.i.d.) real-valued random variables, in the parlance of communication theory this setup corresponds to lossless source coding of analog memoryless sources when the encoding operation is multiplication by a real-valued matrix. This coding setup is depicted in Figure 3.8.

A real vector $\mathbf{x} \in \mathbb{R}^N$ is mapped into $\mathbf{y} \in \mathbb{R}^M$ by a linear encoder (compressor) $\mathcal{C}: \mathbb{R}^N \to \mathbb{R}^M$ corresponding to the given measurement matrix. The decoder (decompressor) $\mathcal{D}: \mathbb{R}^M \to \mathbb{R}^N$ corresponds to the reconstruction operator. The decoder receives the measured signal $\mathbf{y}$ and outputs the reconstructed signal $\hat{\mathbf{x}}$. The compression rate is given by the number of measurements divided by the signals' dimension,

$$R = \frac{M}{N}. \tag{3.170}$$

Modeling $\mathbf{x}$ as a random vector composed of N independent random variables distributed as X, we define $R^*(\mathsf{X}, \epsilon)$ as the smallest number of measurements per signal dimension $R > 0$ such that there exist a sequence of encoders $\mathcal{C}^{(N)}: \mathbb{R}^N \to \mathbb{R}^{\lfloor RN \rfloor}$ and decoders $\mathcal{D}^{(N)}: \mathbb{R}^{\lfloor RN \rfloor} \to \mathbb{R}^N$ that for sufficiently large N make the probability of error arbitrarily small, namely

$$\mathbb{P}(\mathcal{D}^{(N)}\mathcal{C}^{(N)}(\mathsf{X}_1, \mathsf{X}_2, \ldots, \mathsf{X}_N) \neq (\mathsf{X}_1, \mathsf{X}_2, \ldots, \mathsf{X}_N)) \leq \epsilon. \tag{3.171}$$

Compared to the deterministic compressed sensing setup, where reconstruction is required for all possible source signals, here the performance is measured on a probabilistic basis by considering long block lengths and averaging with respect to the distribution of the source signal. The model can be easily extended to include a measurement error, and imposing different constraints on the design of the encoder and decoder.

To capture the notion of sparsity in a probabilistic setting, we consider the following mixture distribution for the source signal:

$$p_{\mathsf{X}}(x) = (1 - \gamma)\delta(x) + \gamma p'(x), \tag{3.172}$$

where $\delta(\cdot)$ is Dirac's distribution, $0 \leq \gamma \leq 1$, and p' is an absolutely continuous probability measure. By the law of large numbers, the weight γ in (3.172) represents, for large values of N, the level of sparsity of the signal in terms of the fraction of its non-zero elements. Given this source model, a basic result for probabilistic reconstruction is that the threshold R^* for the smallest measurement rate that guarantees reconstruction with arbitrarily small probability of error depends only on the sparsity level and is independent of p', and we have

$$R^*(\mathsf{X},\epsilon) = \gamma. \tag{3.173}$$

It follows that, in a probabilistic setting the number of linear measurements necessary and sufficient for reconstruction with negligible probability of error is equal to the sparsity level of the signal and is independent of the prior distribution of its non-zero elements.

This result should be compared with the analogous one in the deterministic setting, where a number of measurements equal to twice the sparsity level is required to perfectly reconstruct any signal in the space. The weaker requirement of probabilistic reconstruction compared to reconstruction for all instances of the problem yields an improvement in the number of measurements of a factor of two.

3.7.3 Information Dimension

The operational definition of $R^*(\mathsf{X},\epsilon)$ can be given an equivalent information-theoretic formulation in terms of "information dimension," introduced by Rényi (1959). For any real random variable X, consider its quantized version X^ϵ obtained from the discrete probability measure induced by partitioning the real line into intervals of size ϵ and assigning to the quantized variable the probability of lying in each interval, as described in Section 1.4.4.

The information dimension of X is then defined as

$$\dim_{\mathrm{I}}(\mathsf{X}) = \lim_{\epsilon \to 0} \frac{H_{\mathsf{X}^\epsilon}}{-\log \epsilon}, \tag{3.174}$$

where H is the Shannon entropy introduced in Chapter 1.

In the case that the limit in (3.174) does not exist, then lower and upper information dimensions are defined by taking lim inf and lim sup, respectively.

The definition immediately extends to random vectors and should be compared to its deterministic counterpart (3.154). For a vector of N independent random variables distributed as X, we have

$$\dim_{\mathrm{I}}(\mathsf{X}_1, \mathsf{X}_2, \ldots, \mathsf{X}_N) = N \dim_{\mathrm{I}}(\mathsf{X}). \tag{3.175}$$

Assuming the limit exists, if $H(\lfloor \mathsf{X} \rfloor)$ is finite we have

$$0 \leq \dim_{\mathrm{I}}(\mathsf{X}) \leq 1, \tag{3.176}$$

where the value zero is achieved if the variable is discrete, and the value one is achieved if the variable is continuous – see Problem 3.13. For a mixture distribution such as (3.172), we also have

$$\dim_{\mathrm{I}}(\mathsf{X}) = \gamma. \tag{3.177}$$

It follows by combining (3.173), (3.175), and (3.177) that the number of linear measurements necessary and sufficient for reconstruction of sparse random vectors composed of N independent components having the mixture distribution (3.172) and such that $H(\lfloor \mathsf{X} \rfloor) < \infty$ scales with the number of dimensions:

$$\dim_{\mathrm{I}}(\mathsf{X}_1, \mathsf{X}_2, \ldots, \mathsf{X}_N) = \gamma N. \tag{3.178}$$

On the other hand, the number of linear measurements necessary and sufficient for the reconstruction of random vectors composed of N independent components having an absolutely continuous distribution and such that $H(\lfloor \mathsf{X} \rfloor) < \infty$ is

$$\dim_{\mathrm{I}}(\mathsf{X}_1, \mathsf{X}_2, \ldots, \mathsf{X}_N) = N. \tag{3.179}$$

By comparing (3.178) and (3.179), it follows that while the geometrical and information-theoretical concepts of dimension coincide for absolutely continuous probability distributions, the information-theoretic dimension shrinks for sparse signals. Mathematically, this is due to the presence of the discrete measure $\delta(\cdot)$, which forces a fraction of elements of the vector to zero. The practical implication is that this allows reconstruction of the signal using a smaller number of measurements.

Similar considerations apply to the continuous case for the fractal dimension introduced in Section 3.6.2. In Chapter 12 we show that for any $A \subset \mathbb{R}^N$ such that $m(A) < \infty$, we have

$$\dim_{\mathrm{F}}(A) = N. \tag{3.180}$$

On the other hand, for sets in an infinite-dimensional space such as multi-band signals, the effective dimensionality can be computed using an approximation argument, and this leads to the number of measurements (3.156).

3.8 Summary and Further Reading

Slepian's problem has been placed in a more general setting considering N-widths in approximation theory. We have limited the treatment to N-widths in the sense of Kolmogorov (1936), but alternative definitions also exist. The main reference for this area of analysis is the book by Pinkus (1985). The concept of N-width naturally leads to the definition of degrees of freedom of signals, which again we have taken from an analogy with physics. An historical perspective on the singular value decomposition can be found in Stewart (1993). Classic reference texts for functional representations are by Hilbert and Courant (1953), Kolmogorov and Formin (1954), and Riesz and Nagy (1955). More recent treatments, among others, are by Reed and Simon (1980) and Conway (1990).

We have pointed out that the concentration problem is a special case of a self-adjoint Hilbert–Schmidt kernel acting on the space spanned by the prolate spheroidal functions. Other bounded operators can also be studied in the same framework, and their information content is limited by an analogous analysis of their singular values. This is further explored in subsequent chapters in the context of electromagnetic propagation and scattering.

For the computation of the N-widths we followed Jagerman (1969) and Pinkus (1985). The triangle inequality trick of Section 3.5.1 is taken from Bucci and Franceschetti (1989). Results for multi-band signals are by Landau and Widom (1980). Extensions to higher dimensions are by Landau (1975). The hybrid scaling results in higher dimensions are by Franceschetti (2015).

The problem of optimally reconstructing multi-band signals without knowledge of their support sets in the time and frequency domains is a topic of research called blind sensing, and related to the field of compressed sensing. The main results for blind sensing that we presented are by Lim and Franceschetti (2017a). Earlier results that require knowledge of the number of sub-bands and their widths were given by Feng and Bresler (1996a, 1996b), Venkataramani and Bresler (1998), Mishali and Eldar (2009), Izu and Lakey (2009), and Davenport and Wakin (2012). The Minkowski–Bouligand dimension is a way of determining the fractal dimension of a set in a metric space and is related to the number of degrees of freedom for reconstruction of multi-band signals. It is named after the German mathematician Hermann Minkowski and the French mathematician Georges Bouligand – see Falconer (1990).

There is a huge literature on compressed sensing that includes both theoretical and practical developments, and we have given only a sample of some of the key results in the area. The topic was ignited by the papers by Candés, Romberg, and Tao (2006) and Donoho (2006). Some key results are also summarized in a note by Candés (2008). The reviews by Candés (2006) and Candés and Wakin (2008) provide excellent introductions. The monograph by Foucart and Rauhut (2013) gives a rigorous and complete description of the main theoretical aspects. Wu and Verdú (2010, 2012) studied the problem from an information-theoretic perspective and related the possibility of reconstruction to the information dimension, introduced by Rényi (1959). Additional information-theoretic results are given by Donoho, Javanmard, and Montanari (2014). A first operational characterization of the Rényi information dimension appeared in Kawabata and Dembo (1994).

3.9 Test Your Understanding

Problems

3.1 Prove (3.5) following the steps given in Section 3.3.1.

3.2 Verify Eq. (3.54).

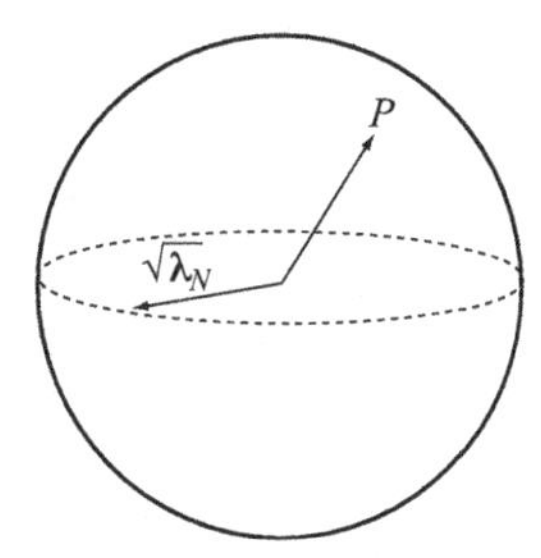

Fig. 3.9 Approximating a sphere with a two-dimensional plane.

3.3 "A Slepian's theorem." Using the results on the Kolmogorov N-width and on the number of degrees of freedom, prove the following theorem that appears in Slepian (1976):

Consider the set of bandlimited signals $\mathscr{E}(\epsilon_T)$ and the $L_2(-\infty,\infty)$ norm. For any $\epsilon > \epsilon_T$, we have

$$\lim_{T\to\infty} \frac{N_\epsilon(\mathscr{E}(\epsilon_T))}{T} = \frac{\Omega}{\pi}. \tag{3.181}$$

3.4 Provide the analogous statement of the theorem in Problem 3.3 for the set of signals $\mathscr{E}(\epsilon_\Omega)$ timelimited to $[-T/2,T/2]$ whose fraction of energy outside the interval $[-\Omega,\Omega]$ is at most ϵ_Ω^2.

3.5 Compute the number of degrees of freedom $N_{\sqrt{2}\epsilon_T}(\mathscr{E}(\epsilon_T))$ in the $L_2(-\infty,\infty)$ norm, as $T\to\infty$.

3.6 Compute the number of degrees of freedom $N_{\epsilon\epsilon_T}(\mathscr{E}(\epsilon_T))$ in the $L_2[-T/2,T/2]$ norm, as $T\to\infty$.

3.7 Show that $d_N(U_{N+1}) = \sqrt{\lambda_N}$, where U_{N+1} is defined in (3.31).

Solution

This can be proven to an arbitrarily high level of rigor. We first sketch the proof using some simple geometric considerations. It is useful to visualize the problem in three dimensions, i.e., let $N+1=3$. In this case, we wish to approximate points in a sphere of radius $\sqrt{\lambda}$ by points in an intersecting two-dimensional plane. According to the definition (3.3), we wish to choose a two-dimensional plane for which the "worst" point in the sphere can be "best" approximated by a point on the plane. The best such choice is given by any plane passing through the center of the sphere, for which the "worst element" (farthest point) on the sphere is within a distance of $\sqrt{\lambda_N}$ from the point on the plane corresponding to the center of the sphere – see Figure 3.9.

To make the above sketch more precise, consider an N-dimensional subspace $\mathscr{X}_N$ and pick a point $y \notin \mathscr{X}_N$. Let x be a point of $\mathscr{X}_N$ that minimizes the distance to y and let the

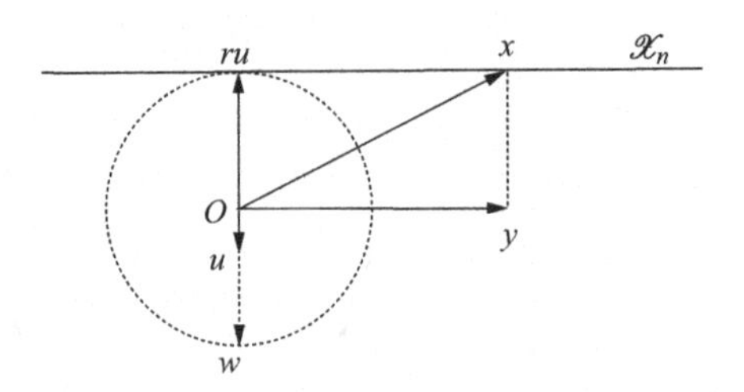

Fig. 3.10 Construction of the best approximation.

unit norm vector

$$u = \frac{y-x}{\|y-x\|}. \tag{3.182}$$

Clearly, there exists a scalar $r \in \mathbb{R}$ such that $ru \in \mathscr{X}_N$. Let

$$w = \begin{cases} -\sqrt{\lambda_N}u & \text{if } r \geq 0, \\ \sqrt{\lambda_N}u & \text{if } r < 0. \end{cases} \tag{3.183}$$

Now observe that the given choice of w maximizes the distance of any point $z \in U_{N+1}$ from $\mathscr{X}_N$, namely

$$\sup_{z \in U_{N+1}} \inf_{a \in \mathscr{X}_N} \|z-a\| = \inf_{a \in \mathscr{X}_N} \|w-a\|, \tag{3.184}$$

and this maximal distance can be computed as follows:

$$\begin{aligned} \inf_{a \in \mathscr{X}_N} \|w-a\| &= \|w-ru\| \\ &= \sqrt{\lambda_N} + |r|. \end{aligned} \tag{3.185}$$

The above expression is then minimized by letting y lie on $\mathscr{X}_N$. In this case $r = 0$, the subspace $\mathscr{X}_N$ passes through the center of the ball of radius $\sqrt{\lambda_N}$, and we have

$$\begin{aligned} d_N(U_{N+1}) &= \inf_{\mathscr{X}_N} \sup_{z \in U_{N+1}} \inf_{a \in \mathscr{X}_N} \|z-a\| \\ &= \|w\| \\ &= \sqrt{\lambda_N}; \end{aligned} \tag{3.186}$$

see Figure 3.10.

3.8 Show that for $\epsilon_T > \sqrt{1-\lambda_N}$, we have $d_N(\mathscr{E}(\epsilon_T)) = 1$ using the $L^2(-\infty,\infty)$ norm, and $d_N(\mathscr{E}(\epsilon_T)) = \sqrt{\lambda_N}$ using the $L^2[-T/2,T/2]$ norm.

Solution

We give only a sketch of the proof. A signal $f \in \mathscr{E}(\epsilon_T)$ that can have an energy spill outside $[-T/2,T/2]$ greater than $1-\lambda_N$ must lie outside the span of the most concentrated basis functions $\{\psi_0, \psi_1, \ldots, \psi_{N-1}\}$, since a linear combination of those can only generate signals with an energy spill at most equal to $1-\lambda_N$. It follows that f can have some energy along a direction orthogonal to all of these basis functions. Letting all of its energy lie along this orthogonal direction leads to the desired result on the N-width.

3.9 Show how the N-widths of bandlimited signals change if instead of a unit energy constraint we assume that

$$\int_{-\infty}^{\infty} f^2(t)dt \leq E. \tag{3.187}$$

Does the number of degrees of freedom change as well?

3.10 Consider the Hilbert–Schmidt operator $\mathcal{K}'\mathcal{K} : L^2(-\Omega,\Omega) \to L^2(-\Omega,\Omega)$ with kernel

$$K'K(\omega,\omega') = \int_{-T/2}^{T/2} K(t,\omega)K^*(t,\omega')dt. \tag{3.188}$$

Verify that it is self-adjoint on $\mathscr{B}_\Omega$.

Solution

For all $g \in \mathscr{B}_\Omega$,

$$\mathcal{K}'\mathcal{K}g(\omega) = \int_{-\Omega}^{\Omega}\int_{-T/2}^{T/2} K(t,\omega)K^*(t,\omega')g(\omega')dt\, d\omega'. \tag{3.189}$$

On the other hand, for all $h \in \mathscr{B}_\Omega$,

$$(\mathcal{K}'\mathcal{K}h)^*(\omega) = \int_{-\Omega}^{\Omega}\int_{-T/2}^{T/2} K^*(t,\omega)K(t,\omega')h^*(\omega')dt\, d\omega'. \tag{3.190}$$

It then easily follows, using the compact notation defined in Section 2.5.3, that

$$\langle \mathcal{K}'\mathcal{K}g(\omega), h(\omega)\rangle = \langle g(\omega), \mathcal{K}'\mathcal{K}h(\omega)\rangle. \tag{3.191}$$

3.11 Show that (3.79), (3.82), and (3.56) tend to zero as $N \to \infty$, ensuring L^2 convergence of the Hilbert–Schmidt representation.

Solution

The vanishing behavior of (3.56) is clear by compactness. For (3.79), we have that the energy of the error is

$$\begin{aligned}
e_N(K) &= \int_a^b\int_c^d |K(x,y) - K_N(x,y))|^2 dxdy \\
&= \int_a^b\int_c^d \left| K(x,y) - \sum_{N=0}^{N-1} \psi_n(x)\varphi_n^*(y)\right|^2 dxdy \\
&= \int_a^b\int_c^d |K(x,y)|^2 dxdy - 2\sum_{N=0}^{N-1}\lambda_n + \sum_{N=0}^{N-1}\lambda_n
\end{aligned}$$

$$= \int_a^b \int_c^d |K(x,y)|^2 dx dy - \sum_{n=0}^{N-1} \lambda_n. \tag{3.192}$$

Since the energy of the error is positive for all N, and $K(x,y)$ is square-integrable, it follows that

$$\sum_{n=0}^{\infty} \lambda_n < \infty, \tag{3.193}$$

and

$$\lim_{N\to\infty} \sum_{n=N}^{\infty} \lambda_n = 0. \tag{3.194}$$

A similar computation takes care of the convergence of (3.82).

3.12 Show that in the context of compressed sensing, $2S$ measurements are necessary for reconstruction.

Solution

For reconstruction to be possible the measurement matrix A should map distinct elements of its domain to distinct elements of its codomain. It follows that if $\mathrm{A}(\mathbf{x}_1 - \mathbf{x}_2) = 0$ then $\mathbf{x}_1 = \mathbf{x}_2$ for all sparse vectors $\mathbf{x}_1$ and $\mathbf{x}_2$. Consider the set $\mathscr{X}$ of all sparse vectors in $\mathbb{R}^N$, and the set

$$\mathscr{Y} = \mathscr{X} \oplus \mathscr{X} = \{\mathbf{x}_1 + \mathbf{x}_2 : \mathbf{x}_1, \mathbf{x}_2 \in \mathscr{X}\} = \{\mathbf{x}_1 - \mathbf{x}_2 : \mathbf{x}_1, \mathbf{x}_2 \in \mathscr{X}\}. \tag{3.195}$$

Letting $\mathbf{y} = \mathbf{x}_1 - \mathbf{x}_2$, we have that if $\mathrm{A}\mathbf{y} = 0$ then $\mathbf{y} = 0$ for all $\mathbf{y} \in \mathscr{Y}$, or, equivalently, non-zero elements of $\mathscr{Y}$ cannot be in the null space of A. The set $\mathscr{Y}$ contains all the vectors in $\mathbb{R}^N$ that have a sparse representation with at most $2S$ non-zero elements. We now argue that if the number of rows of A is smaller than $2S$ then there exists a non-zero element of $\mathscr{Y}$ in the null space of A, which is a contradiction. To see why this is the case, consider the subset $\mathscr{Z} \subset \mathscr{Y}$ of all vectors $\mathbf{z}$ that have a sparse representation where all but the first $2S$ coefficients are zero. It is easy to see that $\mathscr{Z}$ is a subspace of $\mathbb{R}^N$ of dimension $2S$. The null space condition is now equivalent to having the dimension of the projection $\mathrm{A}\mathbf{z}$ be at least $2S$. This cannot occur if the number of rows of A (corresponding to the number of measurements) is less than $2S$.

3.13 Show that the information dimension of a discrete random variable of finite entropy is zero, and of a continuous random variable of finite differential entropy is one.

Solution

For a discrete random variable, the numerator of (3.174) coincides with its entropy and the denominator diverges, so that the ratio tends to zero. For a continuous random variable, as $\epsilon \to 0$ by (1.50) the numerator of (3.174) diverges as $h_{\mathsf{X}} - \log \epsilon$ and the denominator also diverges as $-\log \epsilon$, so that the ratio tends to one.

4 Electromagnetic Propagation

"O tell me, when along the line
From my full heart the message flows,
What currents are induced in thine?
One click from thee will end my woes."
Through many an Ohm the Weber flew,
And clicked the answer back to me, —
"I am thy Farad, staunch and true,
Charged to a Volt with love for thee."[1]

4.1 Maxwell's Equations

In this chapter we specify the signals used to transfer information as being of electromagnetic type. We provide the background on electromagnetic waves that is needed to apply the theory of orthogonal signal decomposition described in the previous chapters, and characterize the information content of electromagnetic space–time fields.

The consideration at the basis of what is discussed here is that electric charges and electric currents act as sources for electric and magnetic fields. An accurate model of this phenomenon is provided by Maxwell's equations. Presented by James Clerk Maxwell to the Royal Society of London on December 8, 1864, these equations mark one of the greatest advancements in science. Quantum theorist Richard Feynman went as far as saying that their discovery should be regarded as the most significant event in the history of the nineteenth century. They predict the existence of electromagnetic waves at all frequencies, traveling at the speed of light. They were first verified experimentally by Heinrich Hertz in 1887, eight years after Maxwell's death, with the detection of electromagnetic radiation at microwave frequencies; and were perhaps best exploited commercially by the entrepreneur Guglielmo Marconi, founder in 1897 of the Wireless Telegraph & Signal Company, which remained active until 2006, when it was eventually acquired by the Swedish firm Ericsson.

[1] James Clerk Maxwell, "Valentine by a telegraph clerk ♂ to a telegraph clerk ♀" (circa 1860). Reprinted in J. Warburg, ed. (1958) *The Industrial Muse: The Industrial Revolution in English Poetry*, Oxford University Press.

Table 4.1 Field constituents.

Symbol	Name	Units
$\mathbf{e}$	Electric field	$\mathrm{V\,m^{-1}}$
$\mathbf{h}$	Magnetic field	$\mathrm{A\,m^{-1}}$
$\mathbf{d}$	Electric induction	$\mathrm{C\,m^{-2}}$
$\mathbf{b}$	Magnetic induction	$\mathrm{W\,m^{-2} = T}$
$\mathbf{i}$	Electric current density	$\mathrm{A\,m^{-2}}$
ρ	Electric charge density	$\mathrm{C\,m^{-3}}$

In modern form, due to Oliver Heaviside, the equations are:

$$\begin{cases} \nabla \times \mathbf{e} = -\dfrac{\partial \mathbf{b}}{\partial t}, & (4.1\text{a}) \\ \nabla \times \mathbf{h} = \dfrac{\partial \mathbf{d}}{\partial t} + \mathbf{i}, & (4.1\text{b}) \\ \nabla \cdot \mathbf{d} = \rho, & (4.1\text{c}) \\ \nabla \cdot \mathbf{b} = 0, & (4.1\text{d}) \end{cases}$$

where all the fields involved are space ($\mathbf{r}$) and time (t) dependent. The field constituents, together with their measurement units, are summarized in Table 4.1. We can easily derive the relation between the current and charge densities by substituting (4.1c) into the divergence of (4.1b), yielding

$$\begin{aligned} \nabla \cdot (\nabla \times \mathbf{h}) &= \frac{\partial}{\partial t}(\nabla \cdot \mathbf{d}) + \nabla \cdot \mathbf{i} \\ &= \frac{\partial \rho}{\partial t} + \nabla \cdot \mathbf{i} = 0. \end{aligned} \tag{4.2}$$

The electric current and charge densities may include both induced $(\mathbf{i}_\mathrm{d}, \rho_\mathrm{d})$ and impressed $(\mathbf{i}_\mathrm{s}, \rho_\mathrm{s})$ components, the latter being identified as sources of the field and excited by appropriate devices. Accordingly, we have

$$\mathbf{i} = \mathbf{i}_\mathrm{d} + \mathbf{i}_\mathrm{s}, \tag{4.3}$$

$$\rho = \rho_\mathrm{d} + \rho_\mathrm{s}, \tag{4.4}$$

where, by definition,

$$\nabla \cdot \mathbf{i}_\mathrm{d} + \frac{\partial \rho_\mathrm{d}}{\partial t} = 0, \tag{4.5}$$

$$\nabla \cdot \mathbf{i}_\mathrm{s} + \frac{\partial \rho_\mathrm{s}}{\partial t} = 0. \tag{4.6}$$

The asymmetry in the equations (4.1) is due to the presence of electric charges and currents only, not magnetic ones. Since there are no magnetic charges, magnetic induction field lines neither begin nor end but make loops or extend to infinity and back, and the total magnetic flux through any Gaussian surface is zero. In other words, the magnetic induction is a *solenoidal* vector field. The asymmetry, however, is only an apparent one. The equations can be extended in the natural way allowing for a density of magnetic charge $\rho^{(m)}$ and a magnetic current density $\mathbf{i}^{(m)}$:

$$\begin{cases} \nabla \times \mathbf{e} = -\dfrac{\partial \mathbf{b}}{\partial t} - \mathbf{i}^{(m)}, & (4.7a) \\ \nabla \times \mathbf{h} = \dfrac{\partial \mathbf{d}}{\partial t} + \mathbf{i}, & (4.7b) \\ \nabla \cdot \mathbf{d} = \rho, & (4.7c) \\ \nabla \cdot \mathbf{b} = \rho^{(m)}. & (4.7d) \end{cases}$$

Dirac's (1931) work on relativistic quantum electrodynamics showed that the hypothetical existence of magnetic monopoles in Maxwell's equations, leading to symmetric equations under the interchange of the electric and magnetic fields, would explain the quantization of electric charges that is observed in practice.[2] Another symmetric representation is obtained when all electric charges and electric currents are zero. This happens outside the region where sources are located, in materials where induced currents and charges are not generated.

4.2 Propagation Media

We classify propagation media based on the properties of the electric and magnetic induction. We focus on materials where the electric induction depends only on the electric field and the magnetic induction only on the magnetic field. This is a common case in practice, except for chiral media that include some classes of sugar solutions, amino acids, and DNA. A first distinction is between media that are *dispersive* and those that are *non-dispersive*. For spatially non-dispersive media the induction at any point in space depends only on the corresponding field value at the same location in space. Similarly, for time non-dispersive media the induction at any point in time depends only on the corresponding field value at that point in time. For dispersive media that are *linear* and *space–time invariant* the relations between induction and field are expressed in terms of convolutions. For example, for spatially non-dispersive, time-dispersive, linear, time-invariant media, we have

$$\mathbf{d}(\mathbf{r},t) = \int_{-\infty}^{\infty} f_{\mathrm{e}}(\mathbf{r},t-\tau)\mathbf{e}(\mathbf{r},\tau)d\tau, \tag{4.8}$$

$$\mathbf{b}(\mathbf{r},t) = \int_{-\infty}^{\infty} f_{\mathrm{h}}(\mathbf{r},t-\tau)\mathbf{h}(\mathbf{r},\tau)d\tau, \tag{4.9}$$

where the functions $f_{\mathrm{e}}(\cdot)$ and $f_{\mathrm{h}}(\cdot)$ are the electric and magnetic impulse responses of the medium, i.e., the induction generated by Dirac-shaped unit fields.

[2] In modern physics, grand unified theories and superstring theories allow for magnetic monopoles. The detection of magnetic monopoles is an open problem in experimental physics and is considered by some as the safest bet that one can make about physics not yet seen.

Anisotropic media do not behave the same way in all directions, and have at least one impulse response that is a dyad. On the other hand, for *isotropic* media the impulse responses in (4.8) and (4.9) are scalar functions. Causality requires that induction cannot appear at a time preceding its source application. It follows that $f_i(\mathbf{r},t) = 0$ for $t < 0, i \in \{\mathrm{e},\mathrm{h}\}$, and (4.8) and (4.9) become

$$\mathbf{d}(\mathbf{r},t) = \int_{-\infty}^{t} f_{\mathrm{e}}(\mathbf{r},t-\tau)\mathbf{e}(\mathbf{r},\tau)d\tau, \tag{4.10}$$

$$\mathbf{b}(\mathbf{r},t) = \int_{-\infty}^{t} f_{\mathrm{h}}(\mathbf{r},t-\tau)\mathbf{h}(\mathbf{r},\tau)d\tau. \tag{4.11}$$

This shows that the induction at time t and point $\mathbf{r}$ depends upon the field existing at the same point for all times preceding and including time t. The medium possesses a sort of "hereditary'" property that enforces its effect on the induction itself. This explains the meaning of the term "time dispersive."

If all impulse responses do not depend on $\mathbf{r}$, the medium is called *homogeneous*. For homogeneous, isotropic, (space and time) non-dispersive media, we have

$$f_{\mathrm{e}}(t) = \epsilon\delta(t), \tag{4.12}$$

$$f_{\mathrm{h}}(t) = \mu\delta(t), \tag{4.13}$$

and substituting into (4.10) and (4.11), we have

$$\mathbf{d}(\mathbf{r},t) = \epsilon\,\mathbf{e}(\mathbf{r},t), \tag{4.14}$$

$$\mathbf{b}(\mathbf{r},t) = \mu\,\mathbf{h}(\mathbf{r},t), \tag{4.15}$$

where ϵ [F m^{-1}] and μ [H m^{-1}] are the *permittivity* and *permeability* of the medium. It is common to normalize these to the permittivity and permeability of the vacuum, hence $\epsilon = \epsilon_{\mathrm{r}}\epsilon_0$, $\mu = \mu_{\mathrm{r}}\mu_0$, where the normalization constants are $\epsilon_0 = 8.84 \times 10^{-12}$ and $\mu_0 = 1.256 \times 10^{-6}$.

The relation between the electric field and the induced current density for linear, time-invariant media is typically of non-dispersive type,

$$\mathbf{i}_{\mathrm{d}}(\mathbf{r},t) = \sigma\mathbf{e}(\mathbf{r},t), \tag{4.16}$$

where σ [S m^{-1}] is the conductivity of the medium.

It follows from (4.14), (4.15), and (4.16) that for homogeneous, isotropic, non-dispersive media, the first two Maxwell equations (4.1a) and (4.1b) simplify to

$$\begin{cases} \nabla \times \mathbf{e} = -\mu \dfrac{\partial \mathbf{h}}{\partial t}, & \text{(4.17a)} \\ \nabla \times \mathbf{h} = \epsilon\dfrac{\partial \mathbf{e}}{\partial t} + \sigma\mathbf{e} + \mathbf{i}_{\mathrm{s}}, & \text{(4.17b)} \end{cases}$$

where the impressed current density $\mathbf{i}_{\mathrm{s}}$, i.e., the source of the field, has been explicitly indicated.

Table 4.2 Examples of homogeneous isotropic media.

	Non-dispersive	Time dispersive
$\sigma = 0$	Vacuum	Pure water
$\sigma \neq 0$	Fog, rain, moisture	Ocean

Table 4.3 Examples of inhomogeneous media.

	Isotropic	Anisotropic
$\sigma \neq 0$	Stratified air	Stratified ionosphere

The simplest homogeneous, isotropic, non-dispersive medium is the vacuum, where we also have $\sigma = 0$. A good approximation to the vacuum is clean air, provided that the rate of change of the propagating fields in time is not too large (up to the GHz range). In the presence of fog, rain, clouds, moisture, etc., the air can still be considered homogeneous, isotropic, and non-dispersive, but with $\sigma \neq 0$ and, as we shall see below, the field undergoes an exponential attenuation as it propagates, due to absorption.

Pure (distilled) water is a homogeneous, isotropic, time-dispersive medium with $\sigma = 0$. Water is composed of polar molecules that are forced to be oriented in the direction of the propagating electric field and contribute to the electric induction. This process requires time, resulting in the dispersive effect and hereditary property modeled by (4.10). Hence, dispersion in the water is due to its electric properties, while its magnetic response is the same as that of vacuum, i.e., $f_{\rm h}(t) = \mu_0 \delta(t)$ and $\mathbf{b} = \mu_0 \mathbf{h}$. When water is not pure and free electrons are present, due for example to dissolved salts and minerals, it can still be considered homogeneous, isotropic, and time dispersive, but with $\sigma \neq 0$.

The upper part of the atmosphere, where free electrons are present due to ionization from the Sun, is called the ionosphere. This is an inhomogeneous medium, composed of homogeneous layers, each anisotropic, time dispersive, and with $\sigma \neq 0$. The anisotropic effect is due to the free electrons, pushed by the propagating electric field, interacting with the earth's static magnetic field, so that the induced current is not necessarily aligned with the electric field and σ in (4.16) becomes a dyad.

The stratified layers of air immediately above sea level contain a vapor density profile and constitute another example of an inhomogeneous medium. The change of the vapor content provides a corresponding local change in the electromagnetic properties of the material. Examples of typical propagation media are given in Tables 4.2 and 4.3.

4.2.1 Perfectly Conductive Media

A perfect electric conductor is a homogeneous, isotropic, non-dispersive medium with $\sigma \to \infty$. It shields itself from an elecromagnetic field through the induction of a surface charge and a surface current density. Inside the conductor, propagation is inhibited: the

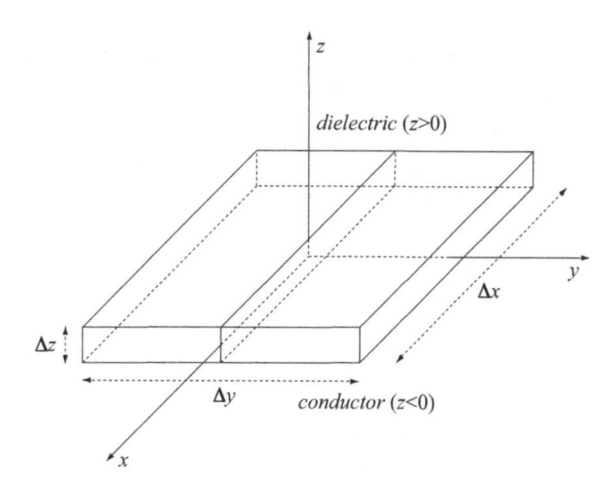

Fig. 4.1 The boundary of a conductive medium.

electric field is zero and there can only be a *static* magnetic field. At the boundary, the normal component of the electric field is discontinuous, being zero inside and non-zero outside the conductor. The tangential component of the electric field is continuous and equal to zero on both internal and external sides of the conductor. The time-varying normal component of the magnetic field is also continuous and equal to zero on both sides of the conductor, while the time-varying tangential component of the magnetic field is discontinuous, being zero inside and non-zero outside the conductor.

All of the above properties follow directly from Maxwell's equations. The electric field inside the conductor is zero, otherwise by (4.16) an infinite current density would be induced by any finite field. This implies by (4.14) that the electric induction is also zero. On the other hand, by (4.1a) and by (4.15) it follows that only a *static* magnetic field and induction can be present inside the conductor.

At the boundary of the conductor a field discontinuity arises if an electromagnetic field is present outside it. Due to the outside field, a surface charge density $\tilde{\rho}_{\rm d}$ [$\mathrm{C\,m^{-2}}$] and a corresponding surface current density $\tilde{\mathbf{i}}_{\rm d}$ [$\mathrm{A\,m^{-1}}$] are induced on the surface boundary of the conductive medium represented by the plane $z = 0$ in Figure 4.1. Accordingly, we have

$$\rho_{\rm d} = \tilde{\rho}_{\rm d}\,\delta(z)\ [\mathrm{C\,m^{-3}}], \tag{4.18}$$

$$\mathbf{i}_{\rm d} = \tilde{\mathbf{i}}_{\rm d}\,\delta(z)\ [\mathrm{A\,m^{-3}}]. \tag{4.19}$$

Integrating Maxwell's equation (4.1c) over the small volume $\Delta V = \Delta x \Delta y \Delta z$ placed at the boundary of the conductor, as depicted in Figure 4.1, and applying Gauss' theorem (B.34), we have

$$\begin{aligned} &\lim_{\Delta z \to 0} \iiint_{\Delta V} \nabla \cdot \mathbf{d}\, dV = (\mathbf{d} \cdot \bar{\mathbf{z}})\, \Delta x \Delta y \\ &= \lim_{\Delta z \to 0} \iiint_{\Delta V} \tilde{\rho}_d \delta(z)\, dV = \tilde{\rho}_d \Delta x \Delta y, \end{aligned} \tag{4.20}$$

where the first equality follows from $\mathbf{d}$ being different from zero only outside the conductor, so that the outgoing flux is only computed for $z > 0$, and from the width Δz tending to zero, so that the outgoing flux is only through the upper surface $\Delta x \Delta y$. It then follows that the component of the electric induction normal to the surface boundary of the conductor is different from zero outside the conductor, due to the presence of $\tilde{\rho}_{\mathrm{d}} \neq 0$. It also follows from (4.14) that the normal component of the electric field is different from zero outside the conductor. The normal components of the electric field and induction are discontinuous at the boundary, being zero on one side and different from zero on the other side of the boundary.

We can also integrate (4.1d) over the volume ΔV, and by a similar argument obtain

$$(\mathbf{b} \cdot \bar{\mathbf{z}}) \Delta x \Delta y = 0, \tag{4.21}$$

from which it follows that the time-varying component of the magnetic induction normal to the surface boundary of the conductor is continuous and equal to zero on both sides of the boundary. From (4.15), it follows that the same holds for the time-varying normal component of the magnetic field.

Next, we focus on the components of the fields tangential to the boundary. Assuming that the tangential component of the electric field is directed along $\bar{\mathbf{x}}$, we start computing the flux of (4.1a) over the cross section $\Delta x \Delta z$ along the $\bar{\mathbf{y}}$ direction. By Stokes' theorem (B.35) we have

$$\begin{aligned}
&\lim_{\Delta z \to 0} \iint_{\Delta S} \nabla \times \mathbf{e}\, dS = \mathbf{e} \times \bar{\mathbf{z}}\, \Delta x \\
&= \lim_{\Delta z \to 0} \iint_{\Delta S} -\frac{\partial \mathbf{b}}{dt}\, dS = \lim_{\Delta z \to 0} -\frac{\partial b_y}{dt} \Delta x \Delta z \\
&= 0,
\end{aligned} \tag{4.22}$$

where the first equality follows from $\mathbf{e}$ being different from zero only outside the conductor, so that the contour integral along the perimeter of ΔS is only computed for $z > 0$, and from the width Δz tending to zero, so that the contour integral reduces to only a line integral along the segment Δx. It then follows that the component of the electric field tangential to the boundary of the conductor is continuous and equal to zero on both sides of the boundary. Proceeding in the same way for (4.1b) and assuming that the tangential component of the magnetic field is along $\bar{\mathbf{x}}$, we obtain

$$\mathbf{h} \times \bar{\mathbf{z}}\, \Delta x = \tilde{\mathbf{i}}_{\mathrm{d}} \Delta x, \tag{4.23}$$

showing that the tangential component of the magnetic field is different from zero on the outside boundary of the conductor due to the presence of $\tilde{\mathbf{i}}_{\mathrm{d}} \neq 0$. It follows that:

> The induced current on the surface of the conductor is proportional to the tangential component of the magnetic field there.

4.2.2 Dielectric Media

Dielectric media have $\sigma < \infty$ and in contrast to perfect conductors they permit propagation. Propagation is lossless for $\sigma = 0$ (perfect dielectric), lossy for $\sigma > 0$, and completely inhibited for $\sigma \to \infty$ (perfect conductors). The losses are due to small amounts of induced currents and charges inside the dielectric that have an *absorption effect* on the propagation process. As we shall see below, in the homogeneous, isotropic, non-dispersive case, the propagation loss due to absorption is of exponential type and depends on the ratio σ/ϵ, which dictates the rate of the exponential attenuation with the propagation distance.

To determine the induced current inside the dielectric, it is useful to consider the Fourier transform of (4.1b),

$$\nabla \times \mathbf{H}(\mathbf{r},\omega) = j\omega\mathbf{D} + \mathbf{I}, \tag{4.24}$$

where, assuming the dielectric to be time non-dispersive, we have

$$\mathbf{D}(\mathbf{r},\omega) = \epsilon\mathbf{E}(\mathbf{r},\omega) \tag{4.25}$$

and

$$\mathbf{I}(\mathbf{r},\omega) = \sigma\mathbf{E}(\mathbf{r},\omega). \tag{4.26}$$

It then follows that

$$\nabla \times \mathbf{H} = j\omega\epsilon\mathbf{E} + \sigma\mathbf{E}, \tag{4.27}$$

which can be rewritten in two different forms:

$$\nabla \times \mathbf{H} = j\omega\left(\epsilon + \frac{\sigma}{j\omega}\right)\mathbf{E}, \tag{4.28}$$

$$\nabla \times \mathbf{H} = (\sigma + j\omega\epsilon)\mathbf{E}. \tag{4.29}$$

The first case models propagation in a dielectric medium of (frequency-) dispersive dielectric constant $\hat{\epsilon} = \epsilon + \sigma/(j\omega)$. The second case models propagation in a conductive medium of (frequency-) dispersive conductivity $\hat{\sigma} = \sigma + j\omega\epsilon$. This latter case can be also written as

$$\nabla \times \mathbf{H} = j\omega\epsilon_0\mathbf{E} + (\sigma + j\omega(\epsilon - \epsilon_0))\mathbf{E}, \tag{4.30}$$

where the induced current term is given by

$$\mathbf{I}_{\mathrm{d}} = (\sigma + j\omega(\epsilon - \epsilon_0))\mathbf{E}. \tag{4.31}$$

It follows that:

> The induced current inside the dielectric is linearly related to the electric field there.

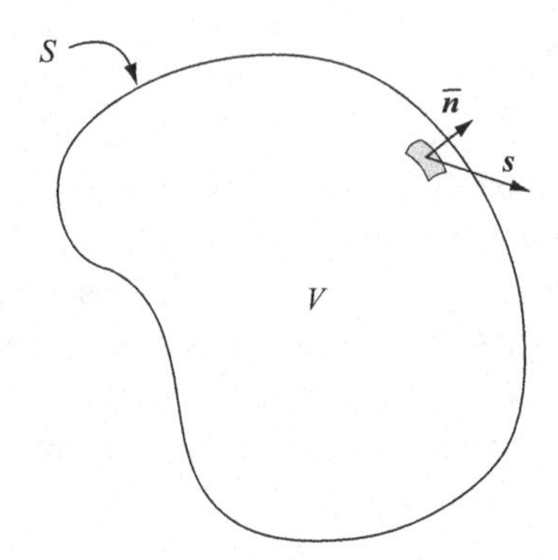

Fig. 4.2 The geometry corresponding to the Poynting–Umov theorem.

4.3 Conservation of Power

The (instantaneous, local) power surface density associated with the electromagnetic field is given by the Poynting–Umov vector, named after John Henry Poynting and Nikolay Alekseevich Umov,

$$\mathbf{s}(\mathbf{r},t) = \mathbf{e}(\mathbf{r},t) \times \mathbf{h}(\mathbf{r},t) \ [\mathrm{W\,m^{-2}}]. \tag{4.32}$$

Computing the divergence of $\mathbf{s}$,

$$\nabla \cdot \mathbf{s} = \nabla \cdot (\mathbf{e} \times \mathbf{h}) = \mathbf{h} \cdot (\nabla \times \mathbf{e}) - \mathbf{e} \cdot (\nabla \times \mathbf{h}), \tag{4.33}$$

and using (4.17), we obtain for homogeneous, isotropic, non-dispersive media,

$$\nabla \cdot \mathbf{s} + \frac{\partial w}{\partial t} + \sigma \mathbf{e}^2 = -\mathbf{e} \cdot \mathbf{i}_{\mathrm{s}}, \tag{4.34}$$

where

$$w = \frac{1}{2}\epsilon\, \mathbf{e}^2 + \frac{1}{2}\mu \mathbf{h}^2 \ [\mathrm{J\,m^{-3}}]. \tag{4.35}$$

Integration over a volume V closed by a bounding surface S, as depicted in Figure 4.2, leads to

$$\oint\!\!\oint_S \mathbf{s} \cdot \bar{\mathbf{n}}\; dS + \frac{\partial W}{\partial t} + P_{\mathrm{d}} = P_{\mathrm{s}}, \tag{4.36}$$

where

$$P_{\mathrm{s}}(t) = -\iiint_V \mathbf{e} \cdot \mathbf{i}_{\mathrm{s}}\, dV, \tag{4.37}$$

$$P_{\mathrm{d}}(t) = \iiint_V \sigma \mathbf{e}^2 dV, \tag{4.38}$$

and

$$W(t) = \iiint_V w\, dV. \tag{4.39}$$

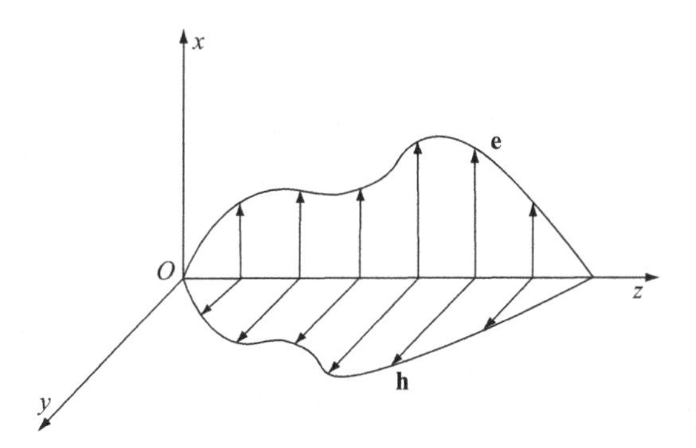

Fig. 4.3 Plane waves.

Relation (4.36), known as the *Poynting–Umov theorem*, is a conservation equation if we read $P_s(t)$ as the instantaneous power delivered to the field by the sources inside V, $P_d(t)$ as the instantaneous power dissipated for joule effect inside the volume, $W(t)$ as the electromagnetic energy stored inside the volume at time t, and $\partial W/\partial t$ as the corresponding instantaneous power. Hence, the power delivered by the sources, minus the power converted into heat inside the volume, minus the outgoing flux of the Poynting vector outside the volume corresponds to the power stored inside the volume. For lossless media $\sigma = 0$, so that by (4.38) there is no dissipated power and by (4.16) there are no induced currents.

4.4 Plane Wave Propagation

We now study the canonical case of propagation far away from the sources and in a one-dimensional space interval. In this interval propagation is described by the solution of Maxwell's equations in the absence of sources, enforcing a boundary condition corresponding to the value of the electric field at the beginning of the space interval and a radiation condition along the positive z axis.

Consider an unbounded, homogeneous, isotropic, non-dispersive medium described by the parameters $\epsilon = \epsilon_0\epsilon_r$, $\mu = \mu_0\mu_r$, σ, and fields $\mathbf{e}(z,t)$, $\mathbf{h}(z,t)$ that depend only on one spatial coordinate z. In addition, we assume that the electric field is polarized along the x axis, namely $\mathbf{e}(z,t) = e(z,t)\bar{\mathbf{x}}$. Let $\mathbf{h} = h_x\bar{\mathbf{x}} + h_y\bar{\mathbf{y}} + h_z\bar{\mathbf{z}}$. By (4.17a), we have

$$\frac{\partial e}{\partial t}\bar{\mathbf{y}} = -\mu\left(\frac{\partial h_x}{\partial t}\bar{\mathbf{x}} + \frac{\partial h_y}{\partial t}\bar{\mathbf{y}} + \frac{\partial h_z}{\partial t}\bar{\mathbf{z}}\right), \tag{4.40}$$

showing that the magnetic field components h_x and h_z are time invariant. Considering propagation along z and disregarding the static components of the magnetic field by letting $h_x = h_z = 0$, we have that the magnetic and electric fields are mutually orthogonal, and orthogonal to the direction of propagation z – see Figure 4.3. Letting $h_y = h$,

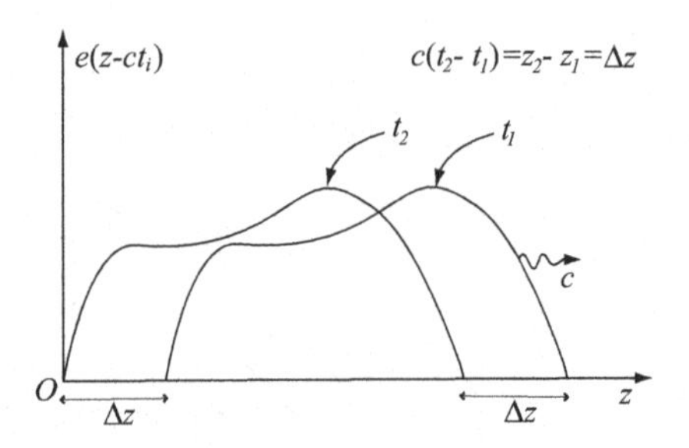

Fig. 4.4 Propagation in the lossless case.

expansion in component form of (4.17b) leads to

$$-\frac{\partial h}{\partial z} = \epsilon \frac{\partial e}{\partial t} + \sigma e, \tag{4.41}$$

while (4.17a) becomes

$$\frac{\partial e}{\partial z} = -\mu \frac{\partial h}{\partial t}. \tag{4.42}$$

Taking the derivative with respect to z of (4.42) and substituting into (4.41), we obtain

$$\frac{\partial^2 e}{\partial z^2} - \frac{1}{c^2}\frac{\partial^2 e}{\partial t^2} - \frac{1}{c^2 \tau}\frac{\partial e}{\partial t} = 0, \tag{4.43}$$

where $c = 1/\sqrt{\epsilon\mu}$ is the propagation velocity [m s^{-1}] of the waveform in the medium and $\tau = \epsilon/\sigma$ [s] is called the *relaxation time* of the medium.

4.4.1 Lossless Case

In the lossless case, $\sigma = 0$, $\tau \to \infty$, and (4.43) reduces to

$$\frac{\partial^2 e}{\partial z^2} - \frac{1}{c^2}\frac{\partial^2 e}{\partial t^2} = 0, \tag{4.44}$$

showing that any waveform with functional dependence of the type $e(z \pm ct)$ is consistent with (4.44), because $\partial e/\partial z = -c\partial e/\partial t$. In the following, we consider the solution $e(z-ct)$, namely propagation along the positive z axis. Accordingly, the electric field propagates along the positive direction of z, without any deformation, with velocity c – see Figure 4.4.

The associated magnetic field is obtained from either (4.41) or (4.42) by substituting $\partial/\partial z = -c\partial/\partial t$, so that

$$\zeta h = e, \tag{4.45}$$

where $\zeta = \sqrt{\mu/\epsilon}$ [ohm] is the *intrinsic impedance* (in this case a resistance) of the space. In free space, $c = 3 \times 10^8$ m s^{-1} and $\zeta = 377$ [ohm].

Finally, the (local, instantaneous) surface density power

$$\mathbf{s} = \mathbf{e} \times \mathbf{h} = \frac{1}{\zeta} e^2(z-ct)\bar{\mathbf{z}} = \zeta h^2(z-ct)\bar{\mathbf{z}} \tag{4.46}$$

shows that the instantaneous power transported by the field through a unitary surface orthogonal to the z axis is proportional to the square of the field and propagates along the positive direction of the z axis with velocity c and without any attenuation.

4.4.2 Lossy Case

In the lossy case, we have to account for the presence of the last term in (4.43). Assuming that the solution is of the type

$$e(z,t) = C(z,t)e(z - ct), \tag{4.47}$$

we consider the case of small losses, so that $C(z,t)$ is a slowly varying function of z and t, and its second-order derivatives can be neglected. Substituting into (4.43) and taking into account that $\partial e/\partial z = -\partial e/\partial(ct)$, we get

$$\frac{\partial e}{\partial z}\left(2\frac{\partial C}{\partial z} + \frac{1}{c\tau}C\right) - \frac{1}{c^2}\frac{\partial C}{\partial t}\left(2\frac{\partial e}{\partial t} + \frac{e}{\tau}\right) = 0. \tag{4.48}$$

A possible solution to (4.48) is

$$2\frac{\partial C}{\partial z} + \frac{1}{c\tau}C = 0, \quad \frac{\partial C}{\partial t} = 0. \tag{4.49}$$

By solving for $C(z)$ and substituting into (4.47), we obtain that the plane wave solution for the electric field in a slightly lossy medium takes the form

$$e(z,t) = e(z - ct)\exp\left(-\frac{z}{2c\tau}\right). \tag{4.50}$$

The magnetic field $h(z,t)$ is computed by substituting the solution (4.50) for $e(z,t)$ into (4.42), obtaining

$$\frac{\partial e(z-ct)}{\partial t}\exp\left(-\frac{z}{2c\tau}\right) + \frac{1}{2\tau}e(z-ct)\exp\left(-\frac{z}{2c\tau}\right) = \zeta\frac{\partial h(z,t)}{\partial t}. \tag{4.51}$$

If the second term on the left-hand side can be neglected, using (4.50) we obtain

$$\zeta h(z,t) = e(z,t). \tag{4.52}$$

It follows that the electric and magnetic fields propagate along the positive direction of z with velocity c, and their waveforms are not deformed but only exponentially scaled – see Figure 4.5.

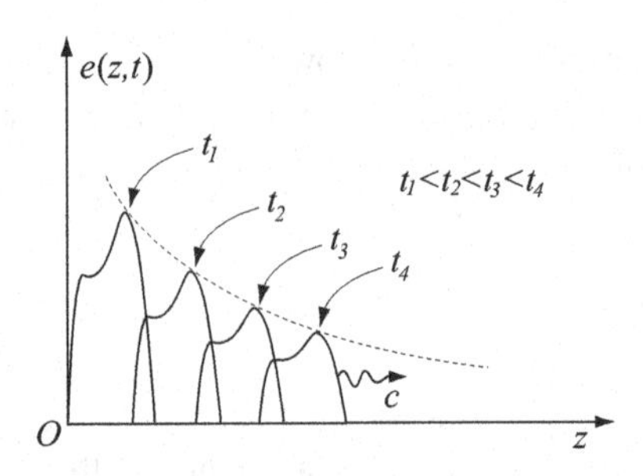

Fig. 4.5 Propagation in a medium with absorption. The electromagnetic wave undergoes exponential attenuation with the propagation distance.

The approximation made to obtain (4.52) from (4.51) is possible if

$$\left|\frac{1}{2\tau}e(z-ct)\right| \ll \left|\frac{\partial e(z-ct)}{\partial t}\right|, \tag{4.53}$$

which holds for fast-varying waveforms inside media with large relaxation time. It can also be easily shown that if this condition is satisfied, then the approximation made to obtain (4.48) from (4.43), namely that the second derivatives of $C(z,t)$ can be neglected, also holds. An integral condition can be used for practical purposes by defining the effective time rate of change of the waveform as

$$\frac{1}{T'} = \sqrt{\frac{\int_{-\infty}^{\infty}(\partial e/\partial t)^2 dt}{\int_{-\infty}^{\infty} e^2 dt}}, \tag{4.54}$$

and imposing the condition

$$T' \ll 2\tau. \tag{4.55}$$

Finally, the (local, instantaneous) surface density power transported by the field through a unitary surface orthogonal to the z axis,

$$\mathbf{s} = \mathbf{e}\times\mathbf{h} = \frac{e^2(z-ct)}{\zeta}\exp\left(-\frac{z}{c\tau}\right)\bar{\mathbf{z}} = \zeta h^2(z-ct)\exp\left(-\frac{z}{c\tau}\right)\bar{\mathbf{z}}, \tag{4.56}$$

exponentially attenuates as the wave propagates in the medium, due to its ohmic losses.

4.4.3 Boundary Effects

We now consider plane wave propagation across two homogeneous lossless half-spaces separated by the plane $z = 0$. A plane wave from the left half-space $z \le 0$ hits the right half-space $z \ge 0$ with an incident angle θ_1. We let the incident electric field be polarized along $\bar{\mathbf{y}}$,

$$\mathbf{e}(s-c_1t) = e(s-c_1t)\,\bar{\mathbf{y}}, \tag{4.57}$$

and propagate at velocity

$$c_1 = \frac{1}{\sqrt{\epsilon_1\mu_1}} = \frac{c}{n_1}, \tag{4.58}$$

where

$$n_1 = \sqrt{\frac{\epsilon_1 \mu_1}{\epsilon_0 \mu_0}} \tag{4.59}$$

is the normalized refractive index of the medium, and s is the coordinate along the direction of propagation. The magnetic field is orthogonal to the electric one and to the direction of propagation.

Considering two constant wavefronts orthogonal to the direction of propagation, the distance between them is

$$ds = dx \sin\theta_1 = dz \cos\theta_1, \tag{4.60}$$

from which it follows that the velocities of propagation along the x and z axes are

$$c_x = \frac{dx}{dt} = \frac{c_1}{\sin\theta_1}, \tag{4.61}$$

$$c_z = \frac{dz}{dt} = \frac{c_1}{\cos\theta_1}. \tag{4.62}$$

Due to the discontinuity, reflected and transmitted waves are formed in the two half-spaces, moving in reflected and transmitted directions at angles θ_1' and θ_2 respectively. The values of c_x for the reflected and transmitted waves must be coincident with the corresponding value for the incident field in order to be able to enforce continuity conditions of the tangential component of the field at the boundary between the two media. It follows that

$$\frac{c_1}{\sin\theta_1} = \frac{c_1}{\sin\theta_1'}, \tag{4.63}$$

$$\frac{c_1}{\sin\theta_1} = \frac{c_2}{\sin\theta_2}, \tag{4.64}$$

where c_2 is the velocity of propagation of the transmitted field. We then obtain Snell's law, according to which the incident angle equals the reflection angle,

$$\theta_1 = \theta_1', \tag{4.65}$$

and the ratio of the sines of the angles of incidence and transmission equals the reciprocal of the ratio of the indices of refraction,

$$\frac{\sin\theta_1}{\sin\theta_2} = \frac{n_2}{n_1}. \tag{4.66}$$

4.4.4 Evanescent Waves

As the wave passes the border between media, depending upon the refractive indices of the two media, it will approach either the normal direction to the separating boundary, or the tangent direction to the separating boundary – see Figure 4.6. When $n_2 < n_1$, and the angle of incidence is large enough, Snell's law (4.66) seems to require that the sine of the angle of the transmitted wave is greater than one. This is, of course, impossible, and the largest critical angle occurs at $\theta_2 = \pi/2$, so that the transmitted wave travels along the separating surface in the y direction and its intensity is constant along the wavefront. For larger values of $\sin\theta_1$, the transmitted wave continues to travel in the y direction along the separating surface, but becomes *evanescent*, exponentially attenuating along the z coordinate – see Figure 4.7 and Problem 4.8.

4.5 The Wave Equation for the Potentials

We assume propagation in a homogeneous, isotropic, non-dispersive, lossless medium, so that $\mathbf{d} = \epsilon\mathbf{e}$, $\mathbf{b} = \mu\mathbf{h}$, the permittivity ϵ and permeability μ are scalar constants, and $\sigma = 0$. Since the magnetic induction $\mathbf{b}$ is solenoidal, it can be represented as

$$\mathbf{b} = \nabla \times \mathbf{a}, \tag{4.67}$$

where $\mathbf{a}$ [W m^{-1}] is the *vector potential*.

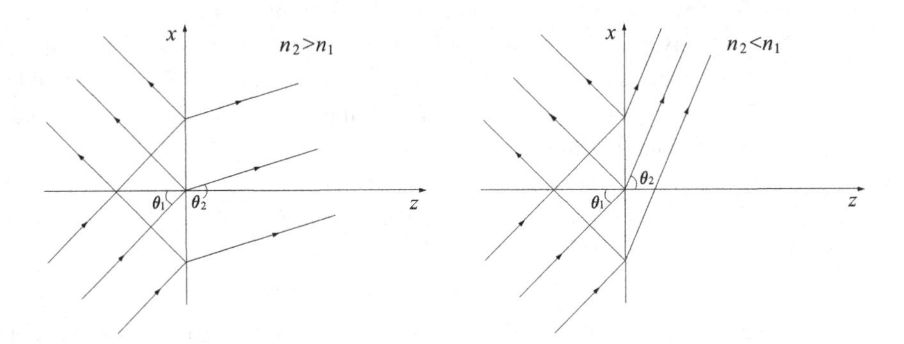

Fig. 4.6 Snell's law.

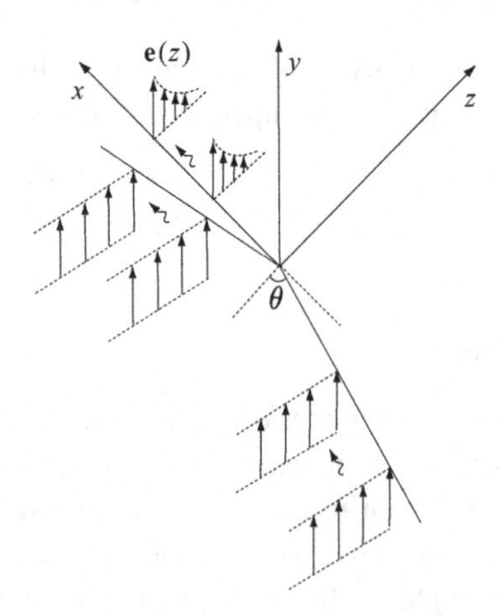

Fig. 4.7 Evanescent waves.

Substitution into (4.1a) leads to

$$\nabla \times \left(\mathbf{e} + \frac{\partial \mathbf{a}}{\partial t} \right) = 0, \tag{4.68}$$

which defines an irrotational vector that can be represented as

$$\mathbf{e} + \frac{\partial \mathbf{a}}{\partial t} = -\nabla \phi, \tag{4.69}$$

where $\phi(r,t)$ [V] is the *scalar potential.*

It follows that the electromagnetic fields, $\mathbf{e}$ and $\mathbf{h} = \mathbf{b}/\mu$, can be represented in terms of the vector and scalar potentials via (4.67) and (4.69), respectively. Accordingly, we proceed to the derivation of the defining equations for $\mathbf{a}$ and ϕ as functions of the impressed sources $\mathbf{i}_s(\mathbf{r},t)$ and $\rho_s(\mathbf{r},t)$. Substituting (4.67) and (4.69) into (4.1b), with $\mathbf{i}_d = 0$ by virtue of (4.16) and $\sigma = 0$, leads to

$$\nabla \times \nabla \times \mathbf{a} + \epsilon\mu \frac{\partial^2 \mathbf{a}}{\partial t^2} + \epsilon\mu \nabla \frac{\partial \phi}{\partial t} = \mu \mathbf{i}_s, \tag{4.70}$$

which can be rewritten in more convenient form by using the vectorial relation $\nabla^2 \mathbf{a} = \nabla\nabla \cdot \mathbf{a} - \nabla \times \nabla \times \mathbf{a}$ and recalling that $\epsilon\mu = 1/c^2$, yielding

$$\nabla^2 \mathbf{a} - \frac{1}{c^2} \frac{\partial^2 \mathbf{a}}{\partial t^2} = -\mu \mathbf{i}_s + \nabla \left(\nabla \cdot \mathbf{a} + \frac{1}{c^2} \frac{\partial \phi}{\partial t} \right). \tag{4.71}$$

Similarly, substituting (4.69) into (4.1c), where $\rho_d = 0$ is constant by virtue of (4.5) and can be taken equal to zero, and adding at both sides the term $(\partial^2 \phi / \partial t^2)/c^2$, we get

$$\nabla^2 \phi - \frac{1}{c^2} \frac{\partial^2 \phi}{\partial t^2} = -\frac{\rho_s}{\epsilon} - \frac{\partial}{\partial t} \left(\nabla \cdot \mathbf{a} + \frac{1}{c^2} \frac{\partial \phi}{\partial t} \right). \tag{4.72}$$

The two equations (4.71) and (4.72), defining the vector and scalar potentials, are now coupled by the common term in the brackets, which can be made equal to zero by enforcing the *gauge invariance*

$$\nabla \cdot \mathbf{a} + \frac{1}{c^2} \frac{\partial \phi}{\partial t} = 0. \tag{4.73}$$

To enforce the gauge invariance, notice that by redefining the potentials,

$$\mathbf{a} \to \mathbf{a} + \nabla \psi, \; \phi \to \phi - \frac{\partial \psi}{\partial t}, \tag{4.74}$$

where ψ is an arbitrary differentiable function, and substituting the new potentials (4.74) into (4.67) and (4.69), the magnetic induction $\mathbf{b}$ and the electric field $\mathbf{e}$ do not change. Now, substituting (4.74) into the gauge invariance equation (4.73), we obtain

$$\nabla \cdot \mathbf{a} + \frac{1}{c^2} \frac{\partial \phi}{\partial t} + \nabla^2 \psi - \frac{1}{c^2} \frac{\partial \psi}{\partial^2 t^2} = 0, \tag{4.75}$$

and we can choose the function ψ to satisfy the equation

$$\nabla^2 \psi - \frac{1}{c^2}\frac{\partial \psi}{\partial t^2} = -\left(\nabla \cdot \mathbf{a} + \frac{1}{c^2}\frac{\partial \phi}{\partial t}\right). \tag{4.76}$$

Considering the right-hand side of (4.76) as a source $f(\mathbf{r},t)$, we obtain the *wave equation*

$$\nabla^2 \psi - \frac{1}{c^2}\frac{\partial \psi}{\partial t^2} = f(\mathbf{r},t). \tag{4.77}$$

This need not be solved explicitly; since any solution ψ ensures that the gauge invariance condition is satisfied, the potentials decouple and we get the equivalent wave equation forms

$$\nabla^2 \mathbf{a} - \frac{1}{c^2}\frac{\partial^2 \mathbf{a}}{\partial t^2} = -\mu \mathbf{i}_s, \tag{4.78}$$

$$\nabla^2 \phi - \frac{1}{c^2}\frac{\partial^2 \phi}{\partial t^2} = -\frac{\rho_s}{\epsilon}, \tag{4.79}$$

and the electromagnetic fields do not change.

4.6 Radiation

The electromagnetic potentials introduced in the previous section can be used to study the propagation of the field radiated by given sources in the whole three-dimensional space. As before, we consider a homogeneous, isotropic, non-dispersive, lossless medium.

Let the source be an elementary electric dipole, composed of two opposite charges $\pm q(t)$ located along the z axis and separated by a distance ℓ. The dipole corresponds to a source charge density [C m^{-3}]

$$\begin{aligned}\rho_s(t,\mathbf{r}) &= q(t)\,\delta(x)\,\delta(y)\left[\delta\left(z-\frac{\ell}{2}\right)-\delta\left(z+\frac{\ell}{2}\right)\right]\\ &= -q(t)\,\ell\,\delta(x)\,\delta(y)\,\delta'(z),\end{aligned} \tag{4.80}$$

where the last equality holds in the distributional limit as $\ell \to 0$. A physical model of the above idealized situation is that of a short conductive wire of length ℓ with capacitors at its endpoints, where charges can accumulate. We now wish to determine the corresponding source current density $\mathbf{i}_s = i_s\mathbf{z}$ that flows along the wire. From (4.6), we get the scalar equation

$$\begin{aligned}\nabla \cdot \mathbf{i}_s + \frac{\partial \rho_s}{\partial t} &= \frac{\partial i_s}{\partial z} + \frac{\partial \rho_s}{\partial t}\\ &= \frac{\partial i_s}{\partial z} - \frac{\partial q(t)\ell}{\partial t}\delta(x)\delta(y)\delta'(z)\\ &= 0,\end{aligned} \tag{4.81}$$

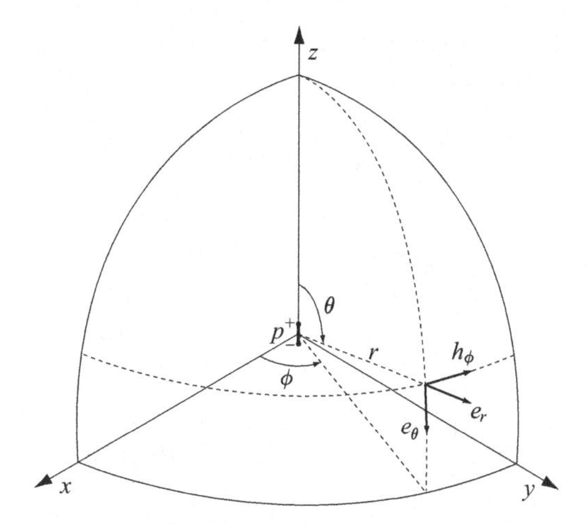

Fig. 4.8 Radiation from an elementary dipole.

and by integrating with respect to z and letting the dipole moment be $p(t) = q(t)\ell$, we obtain the source current density vector

$$\mathbf{i}_s(\mathbf{r},t) = \frac{\partial p(t)}{\partial t}\delta(\mathbf{r})\bar{\mathbf{z}}. \tag{4.82}$$

Substituting the source current density (4.82) into the wave equation (4.78) and solving the differential equation with the radiation condition at infinity leads to the vector potential

$$\mathbf{a}(\mathbf{r},t) = \frac{\mu}{4\pi r}\frac{\partial p(t-r/c)}{\partial t}\bar{\mathbf{z}}; \tag{4.83}$$

using (4.67) we can get the magnetic induction field **b**, and from (4.15) the magnetic field **h**. Then, the electric field **e** can be obtained from either (4.1a) or (4.1b). With reference to a system of spherical coordinates (r,θ,ϕ), the final result is depicted in Figure 4.8 and is given by

$$\begin{cases} e_r(r,t^*) = \dfrac{\zeta}{2\pi}\left(\dfrac{1}{r^2}\dfrac{dp(t^*)}{dt} + \dfrac{c}{r^3}p(t^*)\right)\cos\theta, & (4.84a)\\[2ex] e_\theta(r,t^*) = \dfrac{\zeta}{4\pi}\left(\dfrac{1}{cr}\dfrac{d^2p(t^*)}{dt^2} + \dfrac{1}{r^2}\dfrac{dp(t^*)}{dt} + \dfrac{c}{r^3}p(t^*)\right)\sin\theta, & (4.84b)\\[2ex] h_\phi(r,t^*) = \dfrac{1}{4\pi}\left(\dfrac{1}{cr}\dfrac{d^2p(t^*)}{dt^2} + \dfrac{1}{r^2}\dfrac{dp(t^*)}{dt}\right)\sin\theta, & (4.84c)\end{cases}$$

where $t^* = t - r/c$ is the retarded time, namely the propagation time of the signal from the origin of the coordinate system where the dipole is located to the receiving point at distance r, and $\zeta = \sqrt{\mu/\epsilon}$ is the intrinsic impedance of the space.

The (local, instantaneous) density power is given by

$$\mathbf{s} = \mathbf{e}\times\mathbf{h} = e_\theta h_\phi\bar{\mathbf{r}} - e_r h_\phi\bar{\boldsymbol{\theta}}. \tag{4.85}$$

Performing some calculations, the density power is divided into the sum of irreversible and reversible components,

$$\mathbf{s} = \mathbf{s}_{\text{irr}} + \mathbf{s}_{\text{rev}}, \tag{4.86}$$

where

$$\begin{cases} \mathbf{s}_{\text{irr}}(r,t^*) = \dfrac{\zeta \sin^2\theta}{(4\pi cr)^2}\left(\dfrac{d^2p}{dt^2}\right)^2 \bar{\mathbf{r}}, & (4.87\text{a}) \\ \mathbf{s}_{\text{rev}}(r,t^*) = \dfrac{\zeta \sin\theta}{(4\pi cr)^2}\dfrac{c}{r}\dfrac{d}{dt}\left[\left(\dfrac{dp}{dt}\right)^2(\bar{\mathbf{r}}\sin\theta - \bar{\boldsymbol{\theta}}\cos\theta)\right. \\ \qquad \left. +\dfrac{cp}{r}\left(\dfrac{dp}{dt} + \dfrac{c}{2r}p\right)(\bar{\mathbf{r}}\sin\theta - 2\bar{\boldsymbol{\theta}}\cos\theta)\right]. & (4.87\text{b}) \end{cases}$$

To express the instantaneous radiated power associated with the vector $\mathbf{s}_{\text{irr}}$, it is convenient to provide the relationship between the source current and the dipole moment. Integration of (4.82) over (x,y) provides the source current vector $\mathbf{i}$. With further integration along z, we get

$$\mathbf{i}\ell = \frac{\partial p}{\partial t}\bar{\mathbf{z}}. \tag{4.88}$$

The instantaneous radiated power associated with the vector $\mathbf{s}_{\text{irr}}$ in a solid angle $d\theta d\phi$ is now given by the flux of (4.87a) over the surface $r^2 d\theta d\phi$,

$$\begin{aligned} \mathbf{s}_{\text{irr}} \cdot \bar{\mathbf{r}}\, r^2 d\theta d\phi &= \frac{\zeta}{(4\pi cr)^2}\sin^2\theta\left(\frac{d^2p}{dt^{*2}}\right)^2 r^2 d\theta d\phi \\ &= \frac{\zeta\ell^2}{(4\pi c)^2}\sin^2\theta\left(\frac{\partial i}{\partial t^*}\right)^2 d\theta d\phi, \end{aligned} \tag{4.89}$$

where the second equality follows from (4.88). From (4.89) it follows that there is a power flux along the positive radial direction flowing within any constant solid angle. The total instantaneous power flux radiated in space is given by

$$\int_0^{2\pi}\int_0^{\pi}\frac{\zeta\ell^2}{(4\pi c)^2}\sin^2\theta\left(\frac{\partial i}{\partial t^*}\right)^2 d\theta d\phi = \frac{\zeta\ell^2}{(4c)^2}\left(\frac{\partial i}{\partial t^*}\right)^2; \tag{4.90}$$

this flux is a *constant*, independent of r, for any given retarded time t^*. It follows that there is a continuous transfer of energy between the antenna and infinity. This energy can be used for communication over long distances and, not being recovered from the transmitting antenna, it is irreversibly lost – see Figure 4.9(a).

On the other hand, the power flux associated with the vector $\mathbf{s}_{\text{rev}}$ for any given t^* *decreases* along the radial direction so that there is no power flux at infinity; namely, in any solid angle $d\theta d\phi$ we have

$$\lim_{r\to\infty} \mathbf{s}_{\text{rev}} \cdot \bar{\mathbf{r}}\, r^2 d\theta d\phi = 0. \tag{4.91}$$

In this case, from (4.87b) it follows that the power flux circulates around the dipole and is continuously emitted and reversed into the generator associated with the transmitting

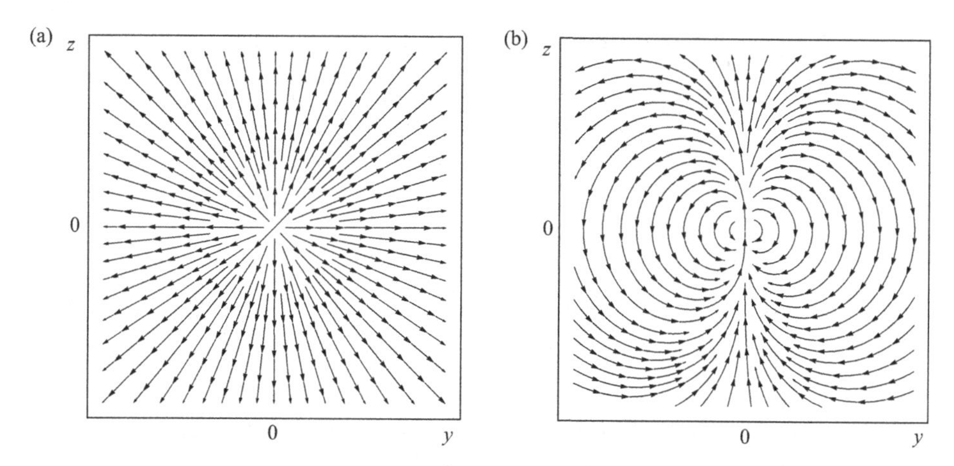

Fig. 4.9 Orientation of the power density vector field in the (y,z) plane: (a) irreversible component, (b) reversible component.

antenna – see Figure 4.9(b). Furthermore, since by (4.87b) $\mathbf{s}_{\rm rev}$ is given in terms of a time derivative, it follows that integration over time at any point in space of the reversible component gives zero total energy density, provided that the source current is of finite duration, namely its extremal points are zero. This implies that this energy is eventually completely recovered by the generator. The electromagnetic signal associated with the reversible component of the power flux is typically not used for communication but finds other applications, like medical ones, wireless power transfer, and near-field microscopy.

4.6.1 The Far-Field Region

From (4.84), and using (4.88), it follows that at large distances from the source, the dominant field components are

$$\begin{cases} e_\theta(r,t^*) = \dfrac{\zeta}{4\pi}\dfrac{1}{cr}\dfrac{d^2p}{dt^2}\sin\theta = \dfrac{\zeta}{4\pi}\dfrac{\ell}{cr}\dfrac{di}{dt}\sin\theta, & \text{(4.92a)} \\ h_\phi(r,t^*) = \dfrac{1}{4\pi}\dfrac{1}{cr}\dfrac{d^2p}{dt^2}\sin\theta = \dfrac{1}{4\pi}\dfrac{\ell}{cr}\dfrac{di}{dt}\sin\theta. & \text{(4.92b)} \end{cases}$$

We want to determine the condition on the distance at which the above approximation holds. First, we define the effective time width of a pulsed waveform of maximum amplitude e_0 as the width of the corresponding rectangular window of amplitude e_0 having the same energy as the waveform, namely

$$T = \frac{\int_{-\infty}^{\infty} e^2 dt}{e_0^2}. \tag{4.93}$$

As a preliminary example, consider now a Gaussian pulse

$$p(t) = p_0 \exp\left(-\frac{t^2}{2T_0^2}\right). \tag{4.94}$$

This has effective time width $T = \sqrt{\pi}\, T_0$, and we have

$$\frac{dp}{dt} = -\frac{\pi t}{T^2} p(t) \tag{4.95}$$

$$\frac{d^2p}{dt^2} = -\frac{\pi}{T^2} p(t) \left(1 - \frac{\pi t^2}{T^2}\right). \tag{4.96}$$

Substituting into (4.84), we note that since the ratios of the $1/r$ and the $1/r^2$, $1/r^3$ fields are of the order of r/cT and $(r/cT)^2$ respectively, the $1/r$ fields are dominant for $r \gg cT$. Accordingly, (4.92) are referred to as *far fields*, in contrast to the other field components scaled by $1/r^2$ and $1/r^3$ that are referred to as *near fields*. Notice also that the far fields are associated with the irreversible component of the power $\mathbf{s}_{\text{irr}}$, while the near fields are associated with the reversible one $\mathbf{s}_{\text{rev}}$.

The above example shows that the far field definition (4.97) for the Gaussian waveform depends on its effective time width. The same result holds referring to the effective time width of general waveforms, and we have that the far-field condition for pulsed waveforms is

$$r \gg cT. \tag{4.97}$$

However, when the waveform is modulated by a sinusoidal carrier of fundamental period $T_c \ll T$, one needs to refer to the effective width of the carrier, given by

$$\int_{-T_c/2}^{T_c/2} \sin^2 \omega t \, dt = \frac{T_c}{2}. \tag{4.98}$$

It follows that the far-field condition for modulated waveforms is

$$r \gg \frac{cT_c}{2} = \frac{\lambda_c}{2}, \tag{4.99}$$

where λ_c is the wavelength of the sinusoidal carrier.

Finally, the (local, instantaneous) density power in the far field is

$$\mathbf{s} = \mathbf{e} \times \mathbf{h} = e_\theta h_\phi \bar{\mathbf{r}} = \frac{1}{\zeta} e_\theta^2 \bar{\mathbf{r}} = \zeta h_\phi^2 \bar{\mathbf{r}} = \frac{1}{(4\pi)^2} \frac{\zeta}{(cr)^2} \sin^2\theta \left(\ell \frac{di}{dt}\right)^2 \bar{\mathbf{r}}, \tag{4.100}$$

which coincides with the irreversible one and decreases as $1/r^2$. In the case of a lossy medium, an additional exponential attenuation factor arises, as in the case of (4.56).

The total instantaneous radiated power in an angle $d\theta d\phi$ is given by

$$dP = \frac{1}{(4\pi)^2} \frac{\zeta}{(cr)^2} \sin^2\theta \left(\ell \frac{di}{dt}\right)^2 r^2 d\theta d\phi, \tag{4.101}$$

which remains constant with the distance from the antenna, whereas the density power decreases as $1/r^2$. The total instantaneous radiated power over a sphere enclosing

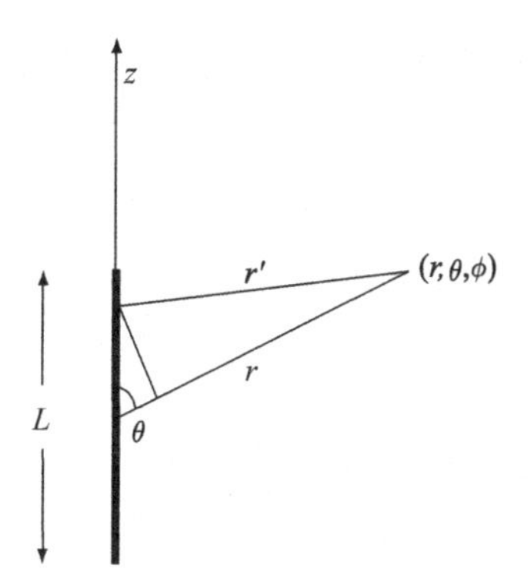

Fig. 4.10 Radiation from a linear antenna.

the transmitter is obtained by integration, yielding

$$P(t^*) = \zeta \frac{\ell^2}{16c^2} \left(\frac{di}{dt} \right)^2, \tag{4.102}$$

which coincides with (4.90).

4.6.2 The Fraunhofer Region

The radiation from a linear wire antenna of length L along z in the far field is immediately given by generalizing (4.92) via superposition of the radiations from elementary dipoles. Referring to Figure 4.10, we have

$$r'^2 = r^2 + z^2 - 2rz\cos\theta. \tag{4.103}$$

By solving for r' and using the Maclaurin expansion of the square root up to terms of order $(z/r)^2$, we get

$$r' \simeq r - z\cos\theta. \tag{4.104}$$

The truncated expansion holds for $z^2 \ll 2r$. This condition leads to the definition of the Fraunhofer condition, named after Bavarian physicist Joseph von Fraunhofer, where geometrically r and r' can be considered parallel. The Fraunhofer condition for antennas of length L is

$$r \gg \frac{L}{2}\sqrt{2}. \tag{4.105}$$

In this case, we have

$$\begin{cases} e_\theta(r,t^*) = \zeta \dfrac{\sin\theta}{4\pi cr} \displaystyle\int_{-L/2}^{L/2} \dfrac{\partial i(z, t^* + (z/c)\cos\theta)}{\partial t} dz, & \text{(4.106a)} \\[2ex] h_\phi(r,t^*) = \dfrac{1}{\zeta} e_\theta(r,t^*). & \text{(4.106b)} \end{cases}$$

Notice that we have defined the far field with respect only to the *time length* of the source, whereas we have defined the Fraunhofer region with respect only to the *spatial*

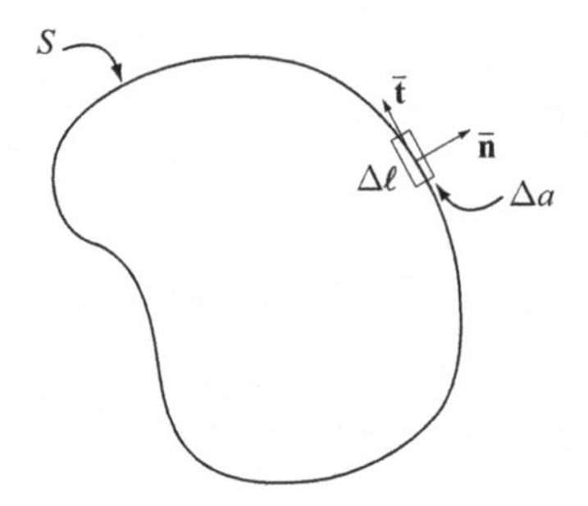

Fig. 4.11 Flux through an elementary surface crossing the boundary S.

length of the source. The radiation from a linear antenna has been obtained assuming both the far-field and Fraunhofer region. Following the same line of reasoning, the radiated power can also be generalized to the case of a linear antenna by simply substituting $\ell di/dt$ in (4.100) with the appropriate integral along the antenna.

4.7 Equivalence and Uniqueness

Consider an electromagnetic field $(\mathbf{e},\mathbf{h})$ generated by sources bounded by a closed surface S, and a new field $(\mathbf{e}',\mathbf{h}')$ coincident with the previous one outside S and equal to zero inside S. We show that this new field can be generated by removing all sources inside S and placing electric and magnetic surface density currents on the boundary S:

$$\begin{cases} \tilde{\mathbf{i}} = \bar{\mathbf{n}} \times \mathbf{h}, & (4.107a) \\ \tilde{\mathbf{i}}^{(m)} = \bar{\mathbf{n}} \times \mathbf{e}. & (4.107b) \end{cases}$$

An immediate consequence of this result is the following:

> Knowledge of the field on the boundary surface S uniquely determines the field outside it.

To derive the result, notice that both the field outside S and the null field inside S are separate solutions of Maxwell's equations and are discontinuous at the boundary S. The new impressed currents ensure that the discontinuity at the boundary S is verified. We show the computation for the electric field only – the magnetic case is completely analogous. We compute the flux of (4.7a) and (4.7b) across an elementary surface ΔS orthogonal to S, of length $\Delta\ell$ and height $\Delta a \to 0$, as depicted in Figure 4.11, and apply Stokes' theorem (B.35). Letting $\bar{\mathbf{t}}'$ be the unit vector tangent to S and normal to ΔS, and $\bar{\mathbf{t}}$ the unit vector tangent to S and such that

$$\bar{\mathbf{t}} = \bar{\mathbf{t}}' \times \bar{\mathbf{n}}, \tag{4.108}$$

we have

$$\lim_{\Delta a \to 0} \iint_{\Delta s} \nabla \times \mathbf{e} \cdot \bar{\mathbf{t}}' ds = \mathbf{e} \cdot \bar{\mathbf{t}} \Delta \ell$$

$$= \lim_{\Delta a \to 0} \left(\iint_{\Delta s} -\frac{\partial \mathbf{b}}{\partial t} \cdot \bar{\mathbf{t}}' ds - \iint_{\Delta s} \mathbf{i}^{(\mathrm{m})} \cdot \bar{\mathbf{t}}' ds \right)$$

$$= -\mathbf{i}^{(\mathrm{m})} \cdot \bar{\mathbf{t}}' \Delta \ell, \tag{4.109}$$

where the first equality follows from Stokes' theorem (B.35) and the field being zero inside S, the second equality follows from Maxwell's equation (4.7a), and the last equality follows from having impressed the surface density current (4.107b). By dividing both sides of (4.109) by $\Delta \ell$, it follows that

$$\mathbf{e} \cdot \bar{\mathbf{t}} = -\mathbf{i}^{(\mathrm{m})} \cdot \bar{\mathbf{t}}'. \tag{4.110}$$

Using (4.108), we also have

$$\mathbf{e} \cdot \bar{\mathbf{t}} = \mathbf{e} \cdot \bar{\mathbf{t}}' \times \bar{\mathbf{n}} = -\mathbf{e} \cdot \bar{\mathbf{n}} \times \bar{\mathbf{t}}' = -\mathbf{e} \times \bar{\mathbf{n}} \cdot \bar{\mathbf{t}}'. \tag{4.111}$$

Substituting (4.111) into (4.110), we obtain (4.107b).

The uniqueness of the solution of Maxwell's equations ensures that the constructed possible field $(\mathbf{e}', \mathbf{h}')$ is the only field associated with the new sources satisfying the given boundary conditions. In practice, the effect of the surface currents (4.107a) and (4.107b) is to provide null boundary conditions on the internal face of S, so that the field in the region inside S is zero, while providing boundary conditions on the external face of S corresponding to the field values generated by the original sources, so that the new field outside S is the same as the old one.

The equivalence result can be regarded as a rigorous statement of the *Huygens' principle*: each point over S becomes a source of radiation for the outside space, while the internal field is zero. This does not imply that the equivalent sources do not individually generate any field inside S, only that their collective result is zero. By symmetry, the same result holds by removing all sources external to S and considering a field that is zero outside S and coincident with $(\mathbf{e}, \mathbf{h})$ inside S.

The equivalence result can greatly simplify certain radiation problems. A complicated boundary value problem, such as a multiple scattering problem, can be reduced to the problem of radiation from a closed surface surrounding the scattering system, provided that one can compute the field on this boundary. We make use of this property in the context of determining the information content of scattered fields in Chapter 8.

4.8 Summary and Further Reading

We have illustrated the main elements of Maxwell's theory of electromagnetic wave propagation. Rather than following the classical approach of starting with steady-state propagation of harmonic waves in the phasor domain, we developed the theory for general waveforms in the time domain. This less standard approach favors the comparison with the theory of signals elaborated in the previous chapters. We believe that starting with the time domain is more natural, and it has the additional advantage of being better suited to the information-theoretic analysis that is the main focus of this book. The theory presented should provide the reader with a bit more than just a flavor of electromagnetics. A different presentation of the material can be found in many classic texts, including the reprint of Maxwell's (1873) original treatise. Among the many modern classics, Jackson (1962), Stratton (1941), Papas (1965), and Van Bladel (1985) are standard references. Another book that favors the time domain approach is Franceschetti (1997).

We have shown that in homogeneous, isotropic, non-dispersive media, plane waves propagate without deformation, and in the presence of small absorption they undergo an exponential attenuation with the propagation distance. In Chapter 9 we show that exponential attenuation due to absorption also occurs in the presence of multiple scattering, where the medium effectively becomes analogous to a homogeneous but dispersive one.

Radiation in free space shows that the elecromagnetic field is generally related to the derivative of the source currents. The far-field condition is defined with respect to the effective time-width of the radiated pulse, or to the wavelength of the sinusoidal carrier in the case of a modulated waveform. The Fraunhofer condition is defined with respect to the length of the transmitting antenna. The power associated with the electromagnetic field is subject to a conservation equation (the Poynting–Umov theorem), and the radiated power is constant in any solid angle centered at the source. The (local, instantaneous) density power is singular at the origin and is divided into a reversible component that does not propagate at infinity, and an irreversible component that propagates in the radial direction. Finally, an electromagnetic field generated by sources bounded by a closed surface and measured in the space external to it can be generated by placing equivalent sources on the boundary surface that depend only on the field there.

4.9 Test Your Understanding

Problems

4.1 Verify that summing (4.87a) and (4.87b) yields (4.86).

4.2 Determine the equivalence result (4.107a) for the magnetic field.

4.3 Write the energy density for the electric and magnetic fields.

4.4 Show that the energy density for a plane wave in free space equals $\epsilon_0 e^2$, where e indicates the electric field.

4.5 The momentum density of the electromagnetic field can be obtained from Maxwell's equations by computing the force density exerted by the sources of the field, and it is given by

$$\mathbf{g} = \mathbf{s}/c^2, \tag{4.112}$$

where $\mathbf{s}$ is the Poynting vector and c is the propagation velocity of the electromagnetic field. Use this formula to show that the magnitude of the momentum density for a plane wave in free space equals $\epsilon_0 e^2/c$.

4.6 The electromagnetic field carries energy and momentum in the direction of the Poynting vector with velocity c. This results in a certain amount of pressure exerted by the field impinging on material objects. Compute the radiation pressure exerted by the electromagnetic field in free space on a totally absorbing screen. (Hint: the rate of change in momentum over time is a force, and pressure is force spread over an area).

4.7 Recall from the discussion in Section 1.5.3 that radiation occurs in discrete quanta of energy, called photons. The radiation pressure computed in Problem 4.6 can be viewed as the aggregate effect of quantum particles hitting a given area. Use the results of Problems 4.4 and 4.5 and Planck's equation of energy (1.85) to compute the momentum of a single photon, then compare the result with the relativistic calculation performed in Section 2.3.2, leading to (2.34). The aggregate effect of these quanta impinging over an area yields the radiation pressure.

4.8 Apply Snell's law with $n_1 > n_2$ and $\sin\theta_1 > n_2/n_1$, and discuss the features of the transmitted wave for a sinusoidal signal.

Solution

For a sinusoidal transmitted waveform we have

$$\cos(\omega t - s/c_2) = \Re \exp[j(\omega t - s\omega/c_2)], \tag{4.113}$$

so that in the following we can refer to the corresponding complex exponential signal. From Snell's law (4.66) we have $\theta_2 = \pi/2$ if $\sin\theta_1 = n_2/n_1$. For larger values of θ_1 we consider the complex extension

$$\theta_2 = \pi/2 - j\theta_2'. \tag{4.114}$$

The phase changes of the signal along x and z are given by

$$\Delta\Phi_x = \omega t - \frac{\omega x \sin\theta_2}{c_2}, \tag{4.115}$$

$$\Delta\Phi_z = \omega t - \frac{\omega z \cos\theta_2}{c_2}, \tag{4.116}$$

where c_2 is the velocity of the transmitted wave. We now have

$$\sin\theta_2 = \sin(\pi/2 - j\theta_2') = \cos(-j\theta_2') = \cosh\theta_2', \tag{4.117}$$

$$\cos\theta_2 = \cos(\pi/2 - j\theta_2') = \sin(-j\theta_2') = -j\sinh\theta_2'. \tag{4.118}$$

It follows that

$$\Delta\Phi_x = \omega t - \frac{\omega x \cosh\theta_2'}{c_2} \tag{4.119}$$

and

$$\Delta\Phi_z = \omega t + j\frac{\omega z \sinh\theta_2'}{c_2}. \tag{4.120}$$

Accordingly, the field along the x direction is given by

$$\Re\exp\left[j\left(\omega t - \frac{\omega x \cosh\theta_2'}{c_2}\right)\right] = \cos\left(\omega t - \frac{\omega x \cosh\theta_2'}{c_2}\right), \tag{4.121}$$

which implies propagation along the x direction. On the other hand, along the z direction the field is given by

$$\Re\exp\left[j\left(\omega t + j\frac{\omega z \sinh\theta_2}{c_2}\right)\right] = \exp\left[\frac{-\omega z \sinh\theta_2'}{c_2}\right]\cos(\omega t), \tag{4.122}$$

which implies that there is no propagation along the z direction but only exponential attenuation along the z direction. This surface waveform is called an evanescent wave.

4.9 Find the relation between the incident and the imaginary part of the angle of transmission for sinusoidal evanescent waves.

Solution

We enforce continuity of the phase change along x at $z = 0$. For the incident and reflected waves, we have

$$\Delta\Phi_x = \omega t - \frac{\omega x \sin\theta_1}{c_1}, \tag{4.123}$$

and using (4.119) for the transmitted wave, we get

$$\frac{\omega x \sin\theta_1}{c_1} = \frac{\omega x \cosh\theta_2'}{c_2}, \tag{4.124}$$

from which it follows that

$$\frac{\sin\theta_1}{\cosh\theta_2'} = \frac{c_1}{c_2} = \frac{n_2}{n_1}. \tag{4.125}$$

The obtained expression (4.125) should be compared with (4.66).

5 Deterministic Representations

I present myself to you in a form suitable to the relationship I wish to achieve with you.[1]

5.1 The Spectral Domains

An information-theoretic analysis of the electromagnetic field requires an appropriate mathematical representation of this physical quantity. The linearity of the Maxwell equations allows us to view the field as the output of a linear system, excited by a source signal. This yields a discrete representation in terms of the Hilbert–Schmidt decomposition that provides the number of degrees of freedom, and thus the number of channels that can be used for communication. This representation occurs in the time–frequency and space–wavenumber domains, in a completely symmetric fashion.

5.1.1 Four Field Representations

Field quantities are functions of space and time, defined in the domain provided by their initial and boundary conditions. A suitable mathematical representation for physically realizable sources and fields is the L^2 space of square-integrable functions. Occasionally, the space can be extended to include Dirac δ-distributions to model, for example, idealized concentrated sources, such as point charges and line currents. All mathematical derivations can be extended in this case, following appropriate limiting arguments. It follows that fields can be represented as superpositions of spectral components via space and time Fourier transforms. For example, the electric field on the whole space–time is represented by superposition of angular frequency components as

$$\mathbf{e}(\mathbf{r},t) = \frac{1}{2\pi}\int_{-\infty}^{\infty} \mathbf{E}(\mathbf{r},\omega)\exp(j\omega t)d\omega, \tag{5.1}$$

and by superposition of wavenumber components as

$$\mathbf{e}(\mathbf{r},t) = \frac{1}{(2\pi)^3}\int_{\mathbb{R}^3} \widehat{\mathbf{e}}(\mathbf{k},t)\exp(j\mathbf{k}\cdot\mathbf{r})d\mathbf{k}, \tag{5.2}$$

[1] L. Pirandello (1917). *The Pleasure of Honesty*, act 1, scene 8. Reprinted in 1936 as *Each in His Own Way and Two Other Plays*, E. P. Dutton.

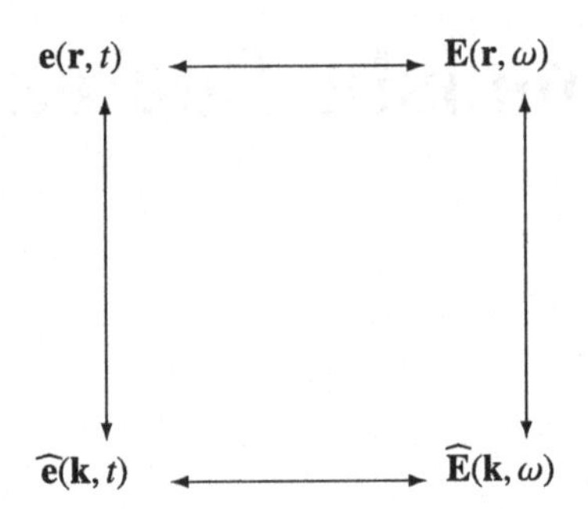

Fig. 5.1 Spectral solution Fourier transform pairs.

where

$$\mathbf{E}(\mathbf{r},\omega) = \int_{-\infty}^{\infty} \mathbf{e}(\mathbf{r},t)\exp(-j\omega t)dt, \tag{5.3}$$

$$\widehat{\mathbf{e}}(\mathbf{k},t) = \int_{\mathbb{R}^3} \mathbf{e}(\mathbf{r},t)\exp(-j\omega\mathbf{k}\cdot\mathbf{r})d\mathbf{r}. \tag{5.4}$$

Accordingly, we define the natural *space–time* $(\mathbf{r},t)$ domain, the partial spectral *space–frequency* $(\mathbf{r},\omega)$ and *wavenumber–time* $(\mathbf{k},t)$ domains, and the full spectral *wavenumber–frequency* $(\mathbf{k},\omega)$ domain. The spectral solutions in these domains are $\mathbf{e}(\mathbf{r},t)$, $\mathbf{E}(\mathbf{r},\omega)$, $\widehat{\mathbf{e}}(\mathbf{k},t)$, and $\widehat{\mathbf{E}}(\mathbf{k},\omega)$. Their Fourier transform pairs are depicted in Figure 5.1.

5.1.2 The Space–Frequency Spectral Domain

Maxwell's equations (4.1) in the space–frequency domain are

$$\begin{cases} \nabla\times\mathbf{E} = -j\omega\mathbf{B}, & (5.5a)\\ \nabla\times\mathbf{H} = j\omega\mathbf{D}+\mathbf{I}, & (5.5b)\\ \nabla\cdot\mathbf{D} = \rho, & (5.5c)\\ \nabla\cdot\mathbf{B} = 0. & (5.5d) \end{cases}$$

By (4.8) and (4.9), we have that for linear, space non-dispersive, time-invariant media,

$$\mathbf{D}(\mathbf{r},\omega) = \epsilon(\mathbf{r},\omega)\mathbf{E}(\mathbf{r},\omega), \tag{5.6}$$

$$\mathbf{B}(\mathbf{r},\omega) = \mu(\mathbf{r},\omega)\mathbf{H}(\mathbf{r},\omega), \tag{5.7}$$

where $\epsilon(\mathbf{r},\omega)$ and $\mu(\mathbf{r},\omega)$ are the Fourier transforms of the electric and magnetic impulse responses of the medium. For isotropic media they are scalars; for anisotropic media they are dyads. For homogeneous media they do not depend on $\mathbf{r}$, and for time non-dispersive media they also do not depend on ω.

5.2 System Representations

The electromagnetic field can be viewed as the output of a linear system excited by a source signal. Viewing the source $x(t)$ as the superposition of Dirac's pulses,

$$x(t) = \int_{-\infty}^{\infty} x(\tau)\delta(t-\tau)d\tau, \tag{5.8}$$

we define the *Green's function*, named after British mathematical physicist George Green, as the electromagnetic response to a unit pulse excitation. This function leads to a number of field representations that are used for the analysis and design of communication systems, and to determine the amount of information carried by the propagating wave.

5.2.1 Linear, Time-Invariant Systems

A linear, time-invariant system transforms an input signal $x(t)$ into an output signal $y(t)$ that is given by the convolution integral

$$y(t) = \int_{-\infty}^{\infty} g(t-\tau)x(\tau)d\tau = \int_{\infty}^{\infty} g(\tau)x(t-\tau)d\tau, \tag{5.9}$$

where $g(t)$ is the response to the Dirac impulse $\delta(t)$, and $g(t)$ is the Green's function of the system,

$$g(t) = \int_{-\infty}^{\infty} g(t-\tau)\delta(\tau)d\tau = \int_{-\infty}^{\infty} g(\tau)\delta(t-\tau)d\tau; \tag{5.10}$$

see Figure 5.2. In the frequency domain, we immediately get, from (5.9),

$$Y(\omega) = G(\omega)X(\omega), \tag{5.11}$$

where $G(\omega)$ is the *spectral Green's function*, and

$$y(t) = \frac{1}{2\pi}\int_{-\infty}^{\infty} G(\omega)X(\omega)\exp(j\omega t)d\omega. \tag{5.12}$$

Figure 5.3 gives the corresponding block diagram.

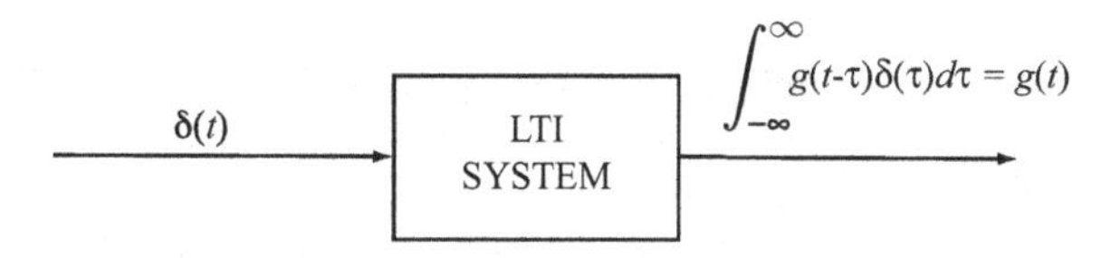

Fig. 5.2 Green's function of a linear, time-invariant (LTI) system.

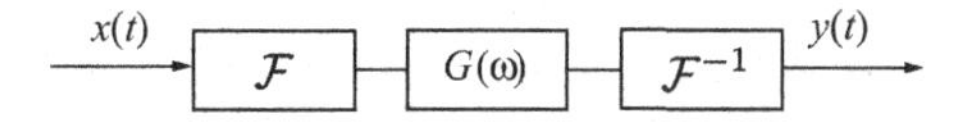

Fig. 5.3 Input–output representation of a linear, time-invariant system.

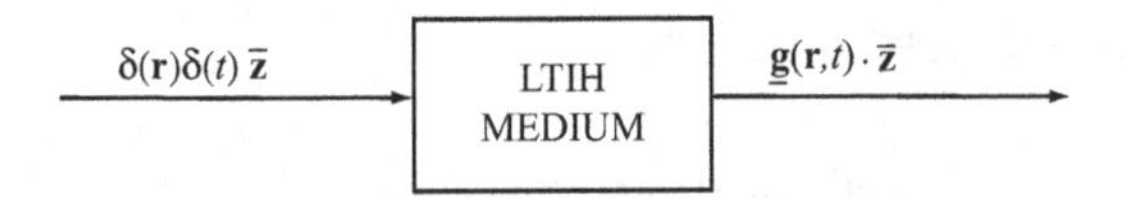

Fig. 5.4 Green's function of a linear, time-invariant, homogeneous (LTIH) medium.

5.2.2 Linear, Time-Invariant, Homogeneous Media

We now place the definitions above in the context of electromagnetic signals. For linear, time-invariant, homogeneous media, the system is the propagation environment, the input is the applied current at the transmitter, the output is the electromagnetic field at the receiver. The associated Green's function is the field radiated by a unit pulse current density vector,

$$\mathbf{i}(\mathbf{r},t) = \delta(\mathbf{r})\delta(t)\bar{\mathbf{z}}. \tag{5.13}$$

As described in Section 4.6, the field is not generally oriented as the source current. It follows that the Green's function is a dyad, $\underline{\mathbf{g}}(\mathbf{r},t)$, referred to as the *dyadic Green's function* – see Appendix B.5. A block diagram representation is given in Figure 5.4.

With reference to the electric field in the natural domain, we have the space–time convolution relation

$$\mathbf{e}(\mathbf{r},t) = \iiint_V \int_{-\infty}^{\infty} \underline{\mathbf{g}}(|\mathbf{r}-\mathbf{r}'|,t-\tau)\cdot\mathbf{i}(\mathbf{r}',\tau)d\mathbf{r}'d\tau, \tag{5.14}$$

where the first integral is extended to the transmitting source volume V. Analogous relations are immediately obtained in the partial spectral domains,

$$\widehat{\mathbf{e}}(\mathbf{k},t) = \int_{-\infty}^{\infty} \underline{\widehat{\mathbf{g}}}(\mathbf{k},t-\tau)\cdot\widehat{\mathbf{i}}(\mathbf{k},\tau)d\tau, \tag{5.15}$$

$$\mathbf{E}(\mathbf{r},\omega) = \iiint_V \underline{\mathbf{G}}(|\mathbf{r}-\mathbf{r}'|,\omega)\cdot\mathbf{I}(\mathbf{r}',\omega)d\mathbf{r}', \tag{5.16}$$

and in the full spectral domain,

$$\widehat{\mathbf{E}}(\mathbf{k},\omega) = \underline{\widehat{\mathbf{G}}}(\mathbf{k},\omega)\cdot\widehat{\mathbf{I}}(\mathbf{k},\omega). \tag{5.17}$$

The geometry of the space convolution is depicted in Figure 5.5.

5.2.3 Green's Function in Free Space for the Potential

A simple example of a linear, time-invariant, homogeneous medium is free space. In this case, the Green's function can be computed directly from Maxwell's equations. To compute the Green's function in free space for the field, we first evaluate the corresponding function for the vector potential. By (4.82) and (4.83), the vector potential has the same orientation as the source current, so that the Green's function for the potential is a monad, and can be obtained by solving the wave equation (4.78) with the radiation condition at infinity and $\mathbf{i}_s$ given by (5.13). This physically corresponds to placing a source dipole moment concentrated at the origin, which appears as a step

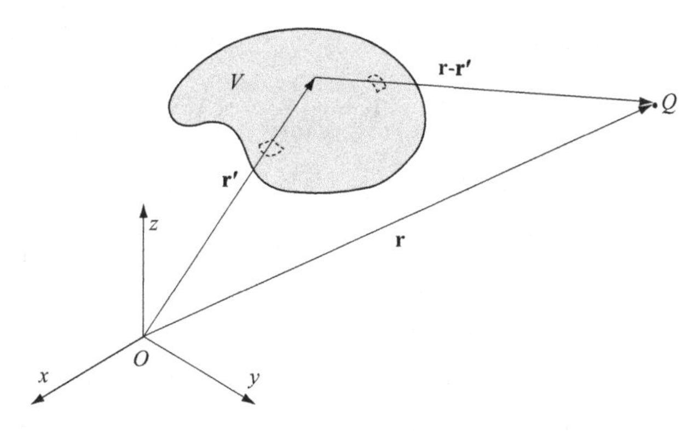

Fig. 5.5 Geometry of the three-dimensional space convolution at point Q for a radiating volume V.

function at $t = 0$ and is oriented along the $\bar{\mathbf{z}}$ direction. By (4.82) and (4.83), we have the four representations of the Green's function in free space for the potential:

$$\begin{cases} g(\mathbf{r},t)\,\bar{\mathbf{z}} = \dfrac{\mu}{4\pi r}\delta(t-r/c)\,\bar{\mathbf{z}}, & (5.18a) \\ G(\mathbf{r},\omega)\,\bar{\mathbf{z}} = \dfrac{\mu}{4\pi r}\exp(-j\omega r/c)\,\bar{\mathbf{z}}, & (5.18b) \\ \widehat{g}(\mathbf{k},t)\,\bar{\mathbf{z}} = \dfrac{\mu}{4\pi t}\exp(-jkct)\,\bar{\mathbf{z}}, & (5.18c) \\ \widehat{G}(\mathbf{k},\omega)\,\bar{\mathbf{z}} = \dfrac{\mu}{4\pi}U(kc+\omega)\,\bar{\mathbf{z}}, & (5.18d) \end{cases}$$

where $U(\cdot)$ is the Heaviside step function. From (5.18a) it follows that the space–time pulse propagates with velocity c. From (5.18b) and (5.18c) it follows that the propagation delay in the natural domain corresponds to a phase shift in frequency or wavenumber in the corresponding partial spectral domains.

5.2.4 Green's Function in Free Space for the Field

The electromagnetic field can be computed from the vector potential following the outline described in Section 4.6. Proceeding in the space–frequency domain, from (4.67) and (5.7) we get

$$\mu\mathbf{H} = \nabla \times \mathbf{A}, \tag{5.19}$$

and from (5.5b) at any point away from the source, and (5.6), we get

$$\nabla \times \mathbf{H} = j\omega\epsilon\,\mathbf{E}. \tag{5.20}$$

From which it follows that

$$\begin{aligned}\mathbf{E} &= \frac{1}{j\omega\mu\epsilon}\nabla\times\nabla\times\mathbf{A}\\ &= \frac{1}{j\omega\mu\epsilon}(\nabla\nabla\cdot\mathbf{A}-\nabla^2\mathbf{A}).\end{aligned} \tag{5.21}$$

Using the homogeneous wave equation resulting from (4.78) with $\mathbf{i}_s = 0$ in the space–frequency domain, and using $c^2 = 1/(\epsilon\mu)$, we finally obtain

$$\mathbf{E}(\mathbf{r},\omega) = -j\omega\mathbf{A}(\mathbf{r},\omega) + \frac{1}{j\omega\epsilon\mu}\nabla(\nabla\cdot\mathbf{A}(\mathbf{r},\omega)), \tag{5.22}$$

from which it follows that the Green's function in free space for the electric field in the space–frequency domain is:

$$\underline{\mathbf{G}}(\mathbf{r},\omega) = -\frac{j\omega\mu}{4\pi}\left[\underline{\mathbf{I}} + \frac{\nabla\nabla}{\beta^2}\right]\frac{\exp(-j\beta r)}{r}, \tag{5.23}$$

where $\beta = \omega\sqrt{\epsilon\mu}$ and the dyadic operator in the square brackets in Cartesian coordinates is represented by the matrix

$$\begin{pmatrix} 1+\frac{1}{\beta^2}\frac{\partial^2}{\partial x^2} & \frac{1}{\beta^2}\frac{\partial^2}{\partial x\partial y} & \frac{1}{\beta^2}\frac{\partial^2}{\partial x\partial z} \\ \frac{1}{\beta^2}\frac{\partial^2}{\partial x\partial y} & 1+\frac{1}{\beta^2}\frac{\partial^2}{\partial y^2} & \frac{1}{\beta^2}\frac{\partial^2}{\partial y\partial z} \\ \frac{1}{\beta^2}\frac{\partial^2}{\partial x\partial z} & \frac{1}{\beta^2}\frac{\partial^2}{\partial y\partial z} & 1+\frac{1}{\beta^2}\frac{\partial^2}{\partial z^2} \end{pmatrix}.$$

Finally, using (5.23) we obtain the convolution equation for the electric field:

$$\begin{aligned}\mathbf{E}(\mathbf{r},\omega) &= \iiint_V \underline{\mathbf{G}}(|\mathbf{r}-\mathbf{r}'|,\omega)\cdot\mathbf{I}(\mathbf{r}',\omega)d\mathbf{r}'\\ &= -\frac{j\omega\mu}{4\pi}\iiint_V\left[\underline{\mathbf{I}}+\frac{\nabla\nabla}{\beta^2}\right]\cdot\frac{\exp(-j\beta|\mathbf{r}-\mathbf{r}'|)}{|\mathbf{r}-\mathbf{r}'|}\mathbf{I}(\mathbf{r}',\omega)d\mathbf{r}'.\end{aligned} \tag{5.24}$$

5.2.5 Green's Function for Cylindrical Propagation

In the special two-dimensional case of cylindrical propagation when fields and sources do not depend on z, Maxwell's equations take a particularly simple form. In the case of an infinitely long wire antenna oriented along $\bar{\mathbf{z}}$, the electric field is also oriented along $\bar{\mathbf{z}}$ and satisfies the wave equation in the space–frequency domain

$$\nabla^2\mathbf{E} + \frac{\omega^2}{c^2}\mathbf{E} = j\omega\mu\mathbf{I}. \tag{5.25}$$

By solving (5.25) with $\mathbf{I} = \delta(\mathbf{r})\bar{\mathbf{z}}$ and the radiation condition at infinity, we obtain the Green's function for the electric field in free space with cylindrical symmetry in the space–frequency domain:

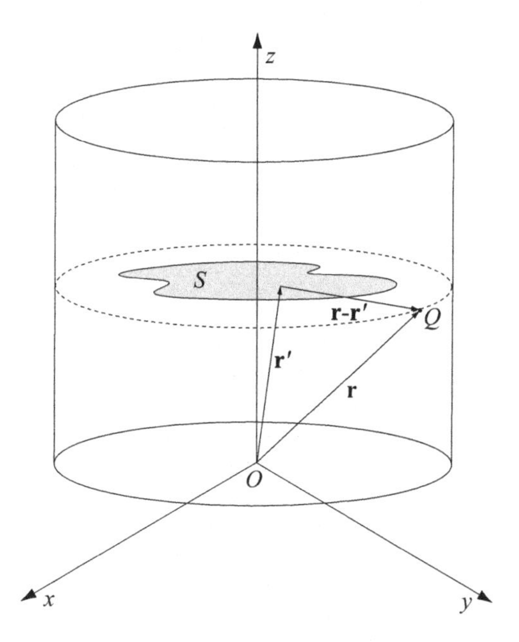

Fig. 5.6 Geometry of the two-dimensional space convolution at point Q for a radiating surface S.

$$G(\mathbf{r},\omega)\,\bar{\mathbf{z}} = -\frac{\omega\mu}{4} H_0^{(2)}(\beta r)\,\bar{\mathbf{z}}, \tag{5.26}$$

where $H_0^{(2)}(\cdot)$ is the Hankel function of the second kind and of order zero – see Appendix E.3.

We then have the convolution integral

$$\begin{aligned}\mathbf{E}(\mathbf{r},\omega) &= \iint_S G(|\mathbf{r}-\mathbf{r}'|,\omega) I(\mathbf{r}',\omega) d\mathbf{r}'\bar{\mathbf{z}} \\ &= -\frac{\omega\mu}{4} \iint_S H_0^{(2)}(\beta|\mathbf{r}-\mathbf{r}'|) I(\mathbf{r}',\omega) d\mathbf{r}'\bar{\mathbf{z}},\end{aligned} \tag{5.27}$$

where the integral is over a surface perpendicular to $\bar{\mathbf{z}}$. The geometry of the two-dimensional space convolution is depicted in Figure 5.6.

From Maxwell's equations, it immediately follows that the magnetic field is oriented along ϕ and is given by

$$\mathbf{H}(\mathbf{r},\omega) = -\frac{1}{j\omega\mu} \nabla \times \mathbf{E}. \tag{5.28}$$

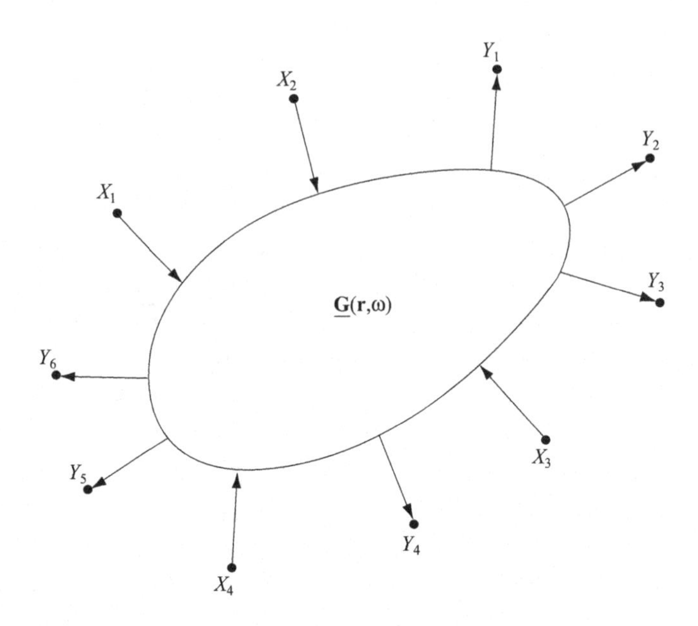

Fig. 5.7 Communication system with four inputs and six outputs.

5.3 Communication Systems: Discrete Radiating Elements

Systems using electromagnetic waves for communication of information range from single a transmitter–receiver pair to multiple antenna systems and spatially distributed networks, where cooperation between the transmitters and the receivers can occur in many different ways. A general model considers a discrete set of $|\mathscr{T}|$ radiating elements and $|\mathscr{R}|$ receiving ones. Every radiating element produces an input signal that is transmitted through the propagation medium. Every receiving element measures the linear superposition of all the received signals. The Green's function can then be used to describe the input–output relationships, treating the environment as a linear filter acting on the transmitted signals. Figure 5.7 gives a representation of this input–output model.

5.3.1 Single Transmitter–Receiver Pair

The free-space electromagnetic channel for a single transmitter–receiver pair is a linear time-invariant system transforming a signal $x(t)$ at the transmitter into the corresponding signal $y(t)$ at the receiver. This system is modeled as having impulse response

$$g(t) = a\delta(t - r/c), \tag{5.29}$$

where r is the distance between transmitter and receiver, c is the velocity of the propagating waveform, and the scaling coefficient a is the signal attenuation along the straight path connecting the two. The corresponding frequency response is

$$G(\omega) = a\exp(-j\omega r/c), \tag{5.30}$$

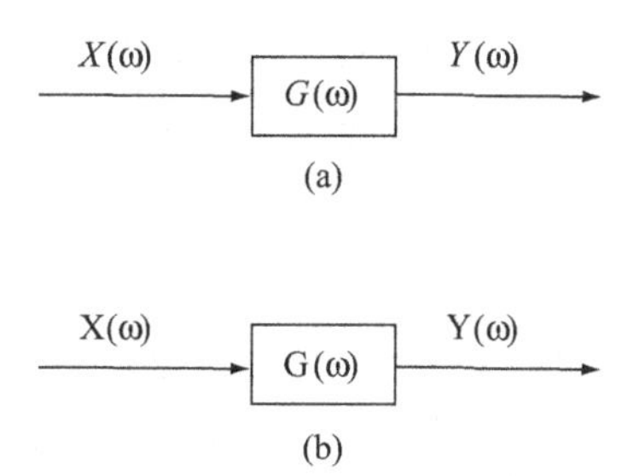

Fig. 5.8 System transfer function in free space: (a) single antenna pair system, (b) multiple antenna system. $G(\omega)$ is a scalar function, $\mathrm{G}(\omega)$ is an $|\mathscr{R}| \times |\mathscr{T}|$ matrix; $X(\omega)$ and $Y(\omega)$ are scalar functions, $\mathrm{X}(\omega)$ is a $|\mathscr{T}| \times 1$ column vector, and $\mathrm{Y}(\omega)$ is an $|\mathscr{R}| \times 1$ column vector.

and the received signal is given by

$$y(t) = a \int_{-\infty}^{\infty} x(\tau)\delta(t - r/c - \tau)d\tau = ax(t - r/c). \tag{5.31}$$

It follows that the output signal is a time-delayed version of the input, scaled by an appropriate multiplicative factor. By (5.31), in the frequency domain the output is a phase-shifted, magnitude-scaled version of the input,

$$Y(\omega) = G(\omega)X(\omega) = a\exp(-j\omega r/c)X(\omega); \tag{5.32}$$

see Figure 5.8(a).

The representation in Figure 5.8 is appropriate to describe the input–output relationship between an applied current at the transmitter and the vector potential at the receiver. By (5.18a), the vector potential has the same orientation as the applied current and is simply a time-shifted, magnitude-scaled version of the current signal. In this case, the scaling factor a is

$$a = \mu/4\pi r. \tag{5.33}$$

The model is also appropriate to describe the electric and magnetic *far fields* generated by elementary dipoles, provided that one considers the derivative of the current as the input signal to the linear system. In this case, according to (4.92), the output fields, polarized orthogonally to the direction of propagation and orthogonally to each other, are a magnitude-scaled, time-delayed version of the derivative of the current. In the near field, however, the response (5.31) applies only to the vector potential, as the fields are related to the sources in a more involved way expressed by (4.84) and the appropriate Green's function of the form given in Section 5.2.4.

The presence of a singular point at $r = 0$ in the scaling coefficient (5.33) is often a source of confusion, since from a system engineering perspective it seems to lead to an unbounded signal gain as the receiver gets closer to the transmitter. This apparent inconsistency is often considered as being a near-field effect not observable in practice. Physical insight, however, should bring clarity to the picture: the Green's function being singular at the origin is consistent with Maxwell's theory and does not violate any physical law. As shown in Section 4.6, although the instantaneous *density* power diverges at the origin, the total instantaneous *radiated* power is a constant, independent of r. Integration over a solid angle makes the singularity disappear.

5.3.2 Multiple Transmitters and Receivers

The system's representation for a single transmitter and receiver can be easily extended to multiple transmitters and receivers. As usual, the simplest case to model is that of propagation in free space. Consider a pair (i,k) of transmitting and receiving antennas separated by a distance $r_{i,k}$. By (5.30), we have

$$G_{i,k}(\omega) = a_{i,k} \exp(-j\omega r_{i,k}/c). \tag{5.34}$$

In the presence of $|\mathscr{T}|$ transmitting and $|\mathscr{R}|$ receiving antennas, and assuming these do not change the free-space characteristics of the propagation environment, the signal at the kth receiving antenna is given by the superposition

$$Y_k(\omega) = \sum_{i \in \mathscr{T}} G_{i,k}(\omega) X_i(\omega). \tag{5.35}$$

This linear relation can be written in convenient matrix form as

$$\mathrm{Y}(\omega) = \mathrm{G}(\omega)\mathrm{X}(\omega), \tag{5.36}$$

where Y is an $|\mathscr{R}| \times 1$ column vector, G is an $|\mathscr{R}| \times |\mathscr{T}|$ matrix whose entries are given by (5.34), and X is a $|\mathscr{T}| \times 1$ column vector. Figure 5.8(b) shows the corresponding block diagram.

Equation (5.35) is the discrete analog of the two-dimensional spatial convolution (5.14), where the integral over the whole transmitting surface is replaced by a discrete sum over the antenna ensemble. For homogeneous media, each element $G_{i,k}$ in the sum (5.35) depends only on the distance between the ith transmitting and the kth receiving antenna, and not on their individual positions, consistent with (5.14) where the Green's function inside the convolution integral depends only on the distance $|\mathbf{r} - \mathbf{r}'|$.

5.3.3 Singular Value Decomposition

The matrix $\mathrm{G}(\omega)$ can be decomposed into singular values, see Section 3.4.3, yielding

$$\begin{aligned} \mathrm{G} &= \mathrm{U}\Sigma\mathrm{V}^\dagger \\ &= \sum_{n=0}^{R-1} \sigma_n \mathrm{U}[n]\mathrm{V}^\dagger[n], \end{aligned} \tag{5.37}$$

where

$$R = \operatorname{rank} \mathrm{G} \le \min\{|\mathscr{T}|, |\mathscr{R}|\}, \tag{5.38}$$

$\sigma_0 \ge \sigma_2 \ge \cdots \ge \sigma_{R-1}$ are the singular values of G, U[n] is the nth column (left singular vector) of the unitary $|\mathscr{R}| \times |\mathscr{R}|$ matrix U, and V[n] is the nth column (right singular vector) of the unitary $|\mathscr{T}| \times |\mathscr{T}|$ matrix V. By (5.36), it then follows that

$$\mathrm{Y} = \mathrm{U}\Sigma\mathrm{V}^\dagger\mathrm{X}. \tag{5.39}$$

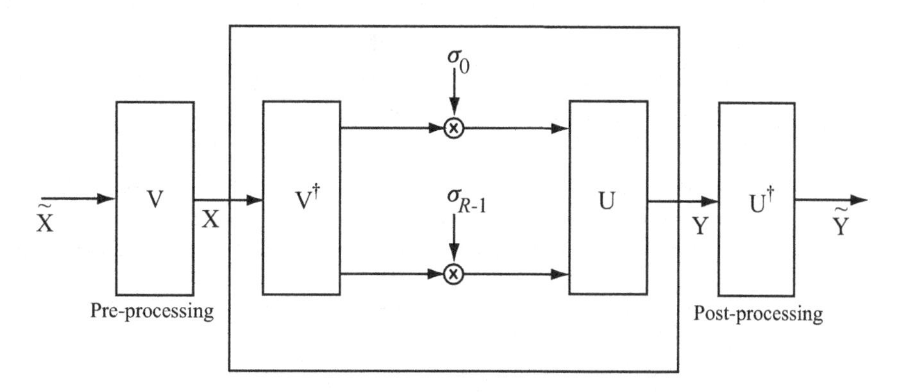

Fig. 5.9 Parallel spatial channels arising from the singular value decomposition.

Letting $\widetilde{X} = V^\dagger X$, $\widetilde{Y} = U^\dagger Y$, and recalling that for any unitary matrix U we have $U^\dagger U = UU^\dagger = I$, pre-multiplication of (5.39) by $U^\dagger$ yields

$$\widetilde{Y} = \Sigma \widetilde{X}. \tag{5.40}$$

Thus, we have an equivalent input–output representation in terms of R parallel spatial channels:

$$\widetilde{Y}_n = \sigma_n \widetilde{X}_n, \quad n = 0, \ldots, R-1; \tag{5.41}$$

see Figure 5.9 for the corresponding block diagram. The obtained representation can be interpreted in terms of coordinate transformations. With appropriate choice of pre-processing and post-processing matrices V and $U^\dagger$, the input is expressed in a coordinate system defined by the columns of V, and the output is expressed in terms of a coordinate system defined by the columns of U. In this new coordinate system the input–output relationship has the very simple form given by (5.41).

5.4 Communication Systems: Arbitrary Radiating Elements

The multiple transmitters and receivers representation in (5.41) is an orthogonal spatial representation for a discrete ensemble of transmitting and receiving antennas obtained starting from (5.35) and decomposing the corresponding linear operator into singular values. If instead we start from the spatial convolution representations (5.14) or (5.27) for a continuous radiating volume or surface, that can include both sources and scatterers, and perform the Hilbert–Schmidt decomposition of the linear operator induced by the Green's function, we obtain an analogous orthogonal representation in continuous space for the field generated by an arbitrary configuration of sources and scatterers. Performing the same decomposition in the time domain leads to the corresponding orthogonal time representation.

Performing the Hilbert–Schmidt decomposition requires a square-integrable kernel. For an arbitrary radiating volume in free space, the Green's function is square-integrable over the whole space outside the volume where the sources are located. On the other hand, it follows from (5.29) and (5.32) that the Green's function is not square-integrable over the whole time axis, since ideal free space provides a constant response at all frequencies. In practice, in any reasonable propagation environment the high-frequency components of the signal are highly attenuated and the environment acts as a linear filter with an effectively finite bandwidth, so that instead of having an impulsive Green's function with a constant spectrum, we can assume a continuous, square-integrable Green's function, and the Hilbert–Schmidt decomposition can be performed.

5.4.1 Hilbert–Schmidt Decomposition

We derive the orthogonal time representation in the case of cylindrical propagation, so that both field and sources are directed along $\bar{\mathbf{z}}$ and we can refer to the scalar component of the field only. We have

$$e(\mathbf{r},t) = \iint_S \int_{-\infty}^{\infty} g(|\mathbf{r}-\mathbf{r}'|, t-\tau) i(\mathbf{r}',\tau) d\tau d\mathbf{r}'. \tag{5.42}$$

Dropping, for convenience of notation, the dependence on the spatial variables, we write the inner integral as the convolution

$$(\mathcal{G}i)(t) = \int_{-\infty}^{\infty} g(t-\tau)\, i(\tau) d\tau. \tag{5.43}$$

The Hilbert–Schmidt representation is obtained by following the same steps that lead to (3.80). We first approximate the kernel in (5.43) by

$$g_N(t-\tau) = \sum_{n=0}^{N-1} \psi_n(t)\varphi_n^*(\tau), \tag{5.44}$$

where $\{\psi_n\}$ and $\{\varphi_n\}$ are the left and right singular functions of the operator $\mathcal{G}$. These can be obtained by the eigendecomposition of the self-adjoint compact operators $\mathcal{G}'\mathcal{G}$ and $\mathcal{G}\mathcal{G}'$. Using angle brackets to denote inner product, and letting $\mathcal{G}'$ be the adjoint operator of $\mathcal{G}$, defined by

$$\langle \mathcal{G}i, e\rangle = \langle i, \mathcal{G}'e\rangle, \tag{5.45}$$

we have

$$\begin{cases} (\mathcal{G}'\mathcal{G}\varphi_n)(\tau) = \lambda_n \varphi_n(\tau), \\ (\mathcal{G}\mathcal{G}'\psi_n)(t) = \lambda_n \psi_n(t). \end{cases} \tag{5.46}$$

It follows that an N-dimensional approximation of (5.43) is

$$\begin{aligned} (\mathcal{G}i)_N(t) &= \int_{-\infty}^{\infty} \sum_{n=0}^{N-1} \psi_n(t) i(\tau) \varphi_n^*(\tau) d\tau \\ &= \sum_{n=0}^{N-1} \int_{-\infty}^{\infty} \varphi_n^*(\tau) i(\tau) d\tau \, \sqrt{\lambda_n} \xi_n(t) \\ &= \sum_{n=0}^{N-1} \sqrt{\lambda_n} \langle i, \varphi_n \rangle \xi_n(t) \end{aligned}$$

$$= \sum_{n=0}^{N-1} a_n \xi_n(t), \tag{5.47}$$

where $\xi_n(t)$ is the normalized version of $\psi_n(t)$, so that the basis functions in the interpolation have unit norm over the finite interval $[-T/2, T/2]$. The representation (5.47) shows that an N-dimensional approximation of the field due to the current $i(t)$ is obtained by a linear combination of the (normalized) left singular functions $\{\xi_n\}$, and that the coefficients $\{a_n\}$ in the interpolation are given by the product of the nth singular value $\sqrt{\lambda_n}$ times the projection of $i(t)$ onto the nth right singular function $\varphi_n(t)$.

As discussed in Chapter 3, the representation (5.47) optimally approximates the image of any function $i(t)$ under $\mathcal{G}$ using N basis functions. The error associated with the approximation over the interval $[-T/2, T/2]$ is $\sqrt{\lambda_N}$, and since $\lambda_N \to 0$, it is possible to optimally approximate the radiated field at any desired level of accuracy ϵ by choosing N large enough. The number of degrees of freedom of the field radiated by an arbitrary configuration of sources and scatterers placed inside the transmitting domain corresponds to the index of the first singular value having magnitude smaller than ϵ. We can then write the infinite series expansion

$$(\mathcal{G}i)(t) = \sum_{n=0}^{\infty} \sqrt{\lambda_n} \langle i, \varphi_n \rangle \xi_n(t), \tag{5.48}$$

where convergence is intended in the energy sense.

For timelimited fields in the set $\mathscr{T}_T$, letting the bandwidth of observation $\Omega \to \infty$, the phase transition of the number of degrees of freedom allows truncation of the series (5.48) to essentially $N_0 = \Omega T/\pi$ terms as $\Omega \to \infty$. It follows that an arbitrary timelimited field can be represented by the superposition of N_0 wide-band pulses $\xi_n(t)$ that concentrate more and more in time as $\Omega \to \infty$. The analogous result holds for bandlimited fields in the set $\mathscr{B}_\Omega$ observed over a time interval of width $T \to \infty$, as they can be represented by the superposition of N_0 bandlimited functions that concentrate more and more in frequency as $T \to \infty$.

A completely analogous picture arises by deriving the orthogonal spatial representation of the system. From the spatial convolution relation (5.27), we obtain the Hilbert–Schmidt representation

$$(\mathcal{G}I)(\mathbf{r}) = \sum_{n=0}^{\infty} \sqrt{\lambda_n} \langle I, \varphi_n \rangle \xi_n(\mathbf{r}), \tag{5.49}$$

and it follows that any field observed in a bounded spatial domain of size S and of wavenumber bandwidth $W \to \infty$ can be represented by $N_0 = WS/\pi$ spatial pulses that concentrate more and more in space as $W \to \infty$. The corresponding physical picture has been anticipated in Section 1.2.3, and is discussed in detail in Chapter 8.

These orthogonal representations essentially diagonalize the channel into parallel subchannels that can be used to multiplex different streams of information. The diagonalization provides orthogonal division of the time–frequency or spatial–wavenumber resource that can then be used in parallel by a given system's implementation. The optimal division corresponds to the system architecture with the highest multiplexing

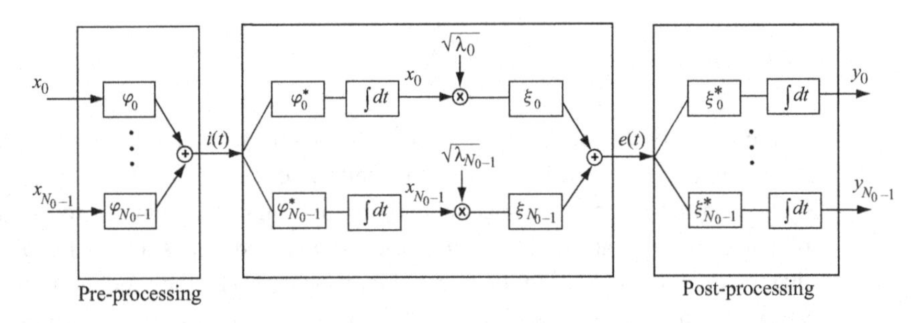

Fig. 5.10 Communication architecture based on orthogonal parallel channels.

capability. The largest number of channels obtained in this way corresponds to the time–frequency $\Omega T/\pi$ or space–wavenumber WS/π degrees of freedom of the propagating field, and provides a fundamental limit on the amount of information that the electromagnetic waveform can carry across the boundary of its radiating volume.

5.4.2 Optimal Communication Architecture

In terms of system representation, letting the input to the system be

$$i(t) = \sum_{n=0}^{N_0-1} x_n \varphi_n(t) \tag{5.50}$$

and the output

$$e(t) = \sum_{n=0}^{N_0-1} y_n \xi_n(t), \tag{5.51}$$

it follows that for each input coefficient

$$x_n = \langle i(t), \varphi_n(t) \rangle \tag{5.52}$$

we have a corresponding output

$$y_n = \langle e(t), \xi_n(t) \rangle. \tag{5.53}$$

By (5.47), we then have the input–output representation in terms of N_0 parallel channels:

$$y_n = \sqrt{\lambda_n} x_n, \quad n = 0, \ldots, N_0 - 1 = \Omega T/\pi. \tag{5.54}$$

This leads to the communication architecture depicted in Figure 5.10, which performs multiplexing of N_0 parallel streams of information through the propagation channel based on the Hilbert–Schmidt representation of the signal. The N_0 input coefficients $\{x_0, x_1, \ldots, x_{N_0-1}\}$, representing parallel communication streams, are recovered at the receiver by integration of the received signal $e(t)$, after multiplication by the appropriate basis functions $\{\xi_n^*(t)\}$. Thus, the architecture uses N_0 real numbers to communicate a signal of essentially N_0 degrees of freedom, and is information-theoretically optimal. Each coefficient of the signal's representation can be interpreted as transporting one degree of freedom of the signal, or as serving as one channel from transmitter to receiver. In this way, a communication rate of one real number per degree of freedom, corresponding to one real number per channel use, is achieved.

The same architecture could also be implemented using a suboptimal orthogonal basis representation. For example, we could use the cardinal series with the sampled field's values as coefficients, or the Fourier series restricted to a given interval of observation. In this case, however, to communicate a signal of N_0 degrees of freedom we would need $N > N_0$ coefficients, which results in a less efficient usage of the physical resource, yielding a communication rate of $N_0/N < 1$ real numbers per degree of freedom.

Finally, we can derive the input–output representation in terms of N_0 spatial channels formally identical to (5.54). Each input coefficient in this case is

$$\widetilde{X}_n = \langle I(\mathbf{r}), \varphi_n(\mathbf{r}) \rangle, \tag{5.55}$$

and the corresponding output is

$$\widetilde{Y}_n = \langle E(\mathbf{r}), \xi_n(\mathbf{r}) \rangle. \tag{5.56}$$

The parallel spatial channels are

$$\widetilde{Y}_n = \sqrt{\lambda_n} \widetilde{X}_n, \quad n = 0, \ldots, N_0 - 1 = WS/\pi. \tag{5.57}$$

This representation should be compared with its discrete analog (5.41), where the rank R plays the role of the number of degrees of freedom N_0.

5.5 Summary and Further Reading

The electromagnetic field is a signal of space and time and it has four spectral representations, linked by Fourier transforms. The medium where propagation occurs can be viewed as a linear system and the (dyadic) Green's function as its impulse response. The field at any point in space radiated by an arbitrary source volume including transmitters and scatterers can be computed via a convolution equation and decomposed using appropriate orthogonal basis sets. We have applied the theory developed in the previous chapters to derive optimal decompositions that yield the number of degrees of freedom of the field in the time–frequency and space–wavenumber domains. These representations apply to waveforms generated by arbitrary sources immersed in arbitrary scattering environments, and they provide a limit on the number of channels that can be used for communication, suggesting an optimal architecture for communication systems.

The material discussed in this chapter draws on standard concepts in electromagnetics, linear systems, and analysis. Some references have been given in Chapters 3 and 4.

5.6 Test Your Understanding

Problems

5.1 Provide a complete derivation of (5.18).

5.2 Check that (5.26) is a solution for (5.25).

5.3 Show that the Green's function for an arbitrary radiating volume in free space is square-integrable over the space outside the volume.

5.4 Show that the Green's function for an arbitrary radiating volume in free space is not square-integrable over time.

5.5 Provide a physical justification of the mathematical results in Problems 5.3 and 5.4.

5.6 Describe the relationship between the Hilbert–Schmidt and the singular value decomposition of the radiated field.

5.7 Explain the physical significance of the Green's function being singular at the origin.

5.8 How many orthogonal parallel channels can a radiating system of radius r and angular frequency bandwidth Ω support?

6 Stochastic Representations

One should always be a little improbable.[1]

6.1 Stochastic Models

Sometimes it is convenient to use stochastic representations of the electromagnetic field in place of deterministic ones to describe average observations in complex environments, and in this case the number of degrees of freedom depends on the parameters of the stochastic process used to represent the field. These should be chosen so that the model is consistent with the physics, and can predict average observations.

While representations in a deterministic setting consider the field radiated by an *arbitrary* environment, in a stochastic setting we consider the field radiated by a *random* environment. In this case, the analog of the number of degrees of freedom is the amount of stochastic diversity of the received waveform. A larger diversity corresponds to more unpredictable waveforms that require, on average, a larger number of coefficients to be represented to a given accuracy.

Both deterministic degrees of freedom and stochastic diversity have applications in communications. The number of degrees of freedom provides an upper bound over all possible environments on the number of channels that can be used to multiplex different streams of information over different dimensions of the signals' space. In the stochastic setting, the amount of diversity provides a limit on the reliability that can be achieved by performing transmissions over multiple realizations of the channel. If the received signal is modeled as a random process in time, frequency, and space, then redundant transmissions over multiple frequency bands, multiple time slots, or multiple antennas can improve the probability that at least one of these transmissions is received successfully. In short, the number of degrees of freedom is used to measure the *rate gain* that can be achieved by performing multiple parallel transmissions over the channel, and the stochastic diversity is used to measure the *reliability gain* that can be achieved by performing repeated transmissions over multiple realizations of the channel.

In a stochastic setting, the analog of the Hilbert–Schmidt representation leading to the number of degrees of freedom is the *Karhunen–Loève* representation leading to the stochastic diversity. This was named after the Finnish and Israeli mathematicians

[1] O. Wilde (1894). Phrases and philosophies for the use of the young. *The Chameleon*, 1(1), p. 2.

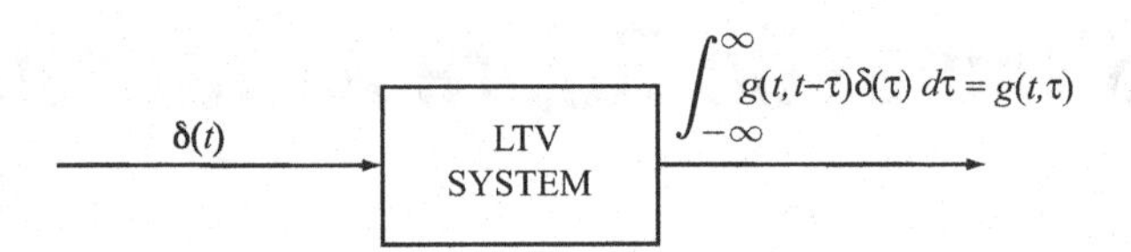

Fig. 6.1 Time-varying Green's function for a linear, time-varying (LTV) system.

Kari Karhunen and Michel Loève, and provides the optimal finite-dimensional approximation of a stochastic process of given autocorrelation function in terms of orthogonal deterministic basis functions and uncorrelated stochastic coefficients.

6.2 Green's Function for a Random Environment

We consider stochastic models of the Green's function that describe, on average, the effect of a random configuration of radiating elements on the received waveform. These stochastic models should be applied with care. In the wrong context, they easily lead to non-physical results. Stochastic assumptions may hold in specific physical situations of interest, for example rich scattering, and may not be applicable in all cases.

While in Chapter 5 we have considered time-invariant radiating systems composed of an arbitrary configuration of fixed sources and scatterers, we now consider systems composed of a stochastic configuration of radiating elements that can also be time varying. We consider a discrete ensemble of transmitters and receivers, representing the inputs and outputs of the communication system, and model the Green's function as randomly varying in frequency, time, and space across different antennas. The frequency and space variations are due to the presence of multiple scattering; the time variations are due to mobility in the environment.

The stochastic convolution integral used to represent the received waveform is intended in the mean square sense – see Appendix D.

6.2.1 Linear, Time-Varying Systems

We first consider the input–output relationship for a linear, time-varying system in a deterministic setting. In the time domain, we have

$$y(t) = \int_{-\infty}^{\infty} g(t,\tau)x(t-\tau)d\tau = \int_{-\infty}^{\infty} g(t,t-\tau)x(\tau)d\tau, \tag{6.1}$$

where $g(t,\tau)$ is the time-varying Green's function – see Figure 6.1. This input–output relationship should be compared with its time-invariant counterpart (5.9). In the time-variant case, the Green's function has four possible representations, linked by Fourier transforms, that are depicted in Figure 6.2, where $\tau \leftrightarrow \omega$ and $t \leftrightarrow \tilde{\omega}$ are Fourier variable pairs. Of particular interest are the time-varying impulse response $g(t,\tau)$ and the time-varying transfer function $G(t,\omega)$. The physical interpretation of $g(t,\tau)$ is the response of the system to an impulse applied at time τ. This response varies with time,

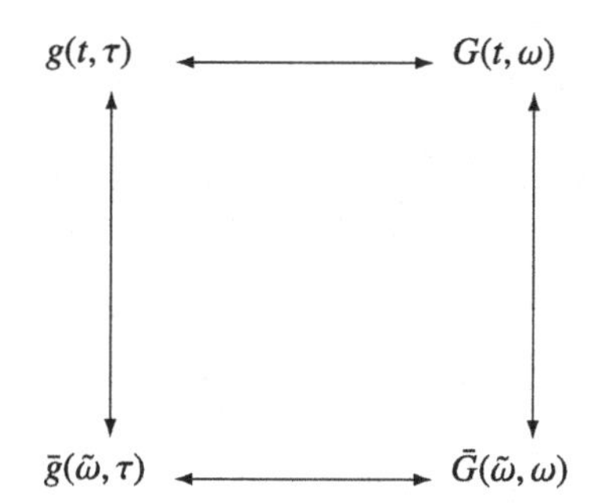

Fig. 6.2 Green's function Fourier transform pairs.

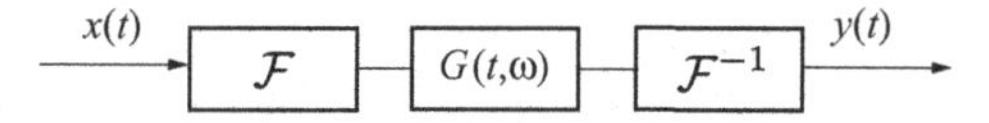

Fig. 6.3 Input–output representation of a linear, time-varying system.

and its Fourier transform with respect to the time variable is $\bar{g}(\tilde{\omega},\tau)$. Analogously, the physical interpretation of $G(t,\omega)$ is the response of the system to a sinusoidal excitation of angular frequency ω. This response varies with time, and its Fourier transform with respect to the time variable is $\bar{G}(\tilde{\omega},\omega)$.

We now derive the relationship between input and output in the frequency domain using the time-varying transfer function

$$G(t,\omega) = \int_{-\infty}^{\infty} g(t,\tau)\exp(-j\omega\tau)d\tau. \tag{6.2}$$

Writing the input as a Fourier integral of its spectrum,

$$x(t-\tau) = \frac{1}{2\pi}\int_{-\infty}^{\infty} X(\omega)\exp(-j\omega\tau)\exp(j\omega t)d\omega, \tag{6.3}$$

and substituting into (6.1), we obtain

$$y(t) = \frac{1}{2\pi}\int_{-\infty}^{\infty} X(\omega)G(t,\omega)\exp(j\omega t)d\omega, \tag{6.4}$$

which shows that the output in the time domain is given by the inverse transform of the product of the time-varying transfer function and the Fourier transform of the input – see Figure 6.3. This input–output relationship should be compared with its time-invariant counterpart (5.12), depicted in Figure 5.3. Finally, by taking the Fourier transform with respect to the time variable, it follows that the relationship between input and output in the frequency domain is given by

$$Y(\tilde{\omega}) = \frac{1}{2\pi}\int_{-\infty}^{\infty} \bar{G}(\tilde{\omega}-\omega,\omega)X(\omega)d\omega, \tag{6.5}$$

which should be compared with its time-invariant counterpart (5.11).

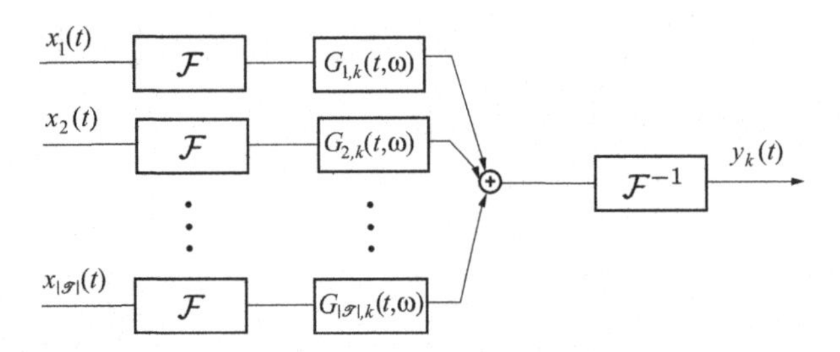

Fig. 6.4 Response at the kth receiving antenna.

6.2.2 Linear, Space–Time-Varying Systems

When communication occurs using multiple antennas at different locations in space, we have to account for the superposition of signals from the different transmitters. For a discrete ensemble of antennas, the signal at the kth receiving antenna is given by the linear superposition

$$Y_k(t,\omega) = \sum_{i\in\mathcal{T}} G_{i,k}(t,\omega)X_i(\omega), \tag{6.6}$$

which is the generalization of (5.35) to a time-varying medium. For homogeneous media, each element $G_{i,k}$ in the sum (6.6) depends only on the distance between the ith transmitting and the kth receiving antenna and not on their individual positions, so that the sum is the discrete equivalent of the continuous spatial convolution (5.14). Figure 6.4 gives the corresponding block diagram representation. The relationship can be put in convenient matrix form as

$$\mathrm{Y}(t,\omega) = \mathrm{G}(t,\omega)\mathrm{X}(\omega), \tag{6.7}$$

where Y is an $|\mathcal{R}| \times 1$ column vector, G is an $|\mathcal{R}| \times |\mathcal{T}|$ matrix, and X is a $|\mathcal{T}| \times 1$ column vector. Each element of the matrix G represents the time-varying transfer function between the ith transmitting and the kth receiving antenna.

Stochastic models consider the entries of G to be realizations of the elements of a random matrix, forming a discrete spatial random process, each being a two-dimensional time–frequency random process, and average responses are considered.

6.3 Multi-path

We begin our stochastic description by considering the response to a given input signal in a time-invariant setting and at a given radiation frequency. The model is later extended to a time-varying setting.

Random multi-path is a simplified model of propagation where the radiating elements represent a random configuration of scattering elements. Like the multi-antenna channel representation, it is based on the linearity of the electromagnetic propagation channel. By (5.35), the signal at each receiving antenna is the superposition of the signals from all transmitting antennas. Each term in this sum is now modeled as a linear superposition of different signals traveling along different multiple scattered paths from one transmitter to one receiver.

A signal along the mth path from the ith transmitter to the kth receiver carries a phase shift $-j\omega\tau_{i,k}^{(m)}$, where $\tau_{i,k}^{(m)}$ is the propagation delay along the path, and a geometric real attenuation coefficient $d_{i,k}^{(m)}$. In addition, there is a complex scattering coefficient $b_{i,k}^{(m)}\exp(j\beta_{i,k}^{(m)})$ that accounts for the absorption and phase shifts occurring at all scattering points. Letting the set of multiple scattered paths be $\mathscr{P}$ and

$$\begin{aligned} d_{i,k}^{(m)}b_{i,k}^{(m)} &= a_{i,k}^{(m)}, \\ \beta_{i,k}^{(m)} - \omega\tau_{i,k}^{(m)} &= \phi_{i,k}^{(m)}, \end{aligned} \tag{6.8}$$

we rewrite (5.35) as

$$\begin{aligned} Y_k(\omega) &= \sum_{i\in\mathscr{T}} X_i(\omega)G_{i,k}(\omega) \\ &= \sum_{i\in\mathscr{T}} X_i(\omega)a_{i,k}\exp(j\phi_{i,k}) \\ &= \sum_{i\in\mathscr{T}} X_i(\omega)\sum_{m\in\mathscr{P}} a_{i,k}^{(m)}\exp(j\phi_{i,k}^{(m)}) \\ &= \sum_{i\in\mathscr{T}} X_i(\omega)\sum_{m\in\mathscr{P}} \Re(a_{i,k}^{(m)}\exp(j\phi_{i,k}^{(m)})) + j\Im(a_{i,k}^{(m)}\exp(j\phi_{i,k}^{(m)})). \end{aligned} \tag{6.9}$$

We now model the attenuations and the phase shifts as independent random variables and assume that the phase of the resulting random vector in the second sum of (6.9) is uniformly distributed in $[0, 2\pi]$. This latter assumption is reasonable if the phase of each component in the sum is random and uniformly distributed so that the resulting sum vector does not have a preferred phase. In the case that a direct field path is present, the assumption is not valid and a modification of the model is required that leads to a more general form of the resulting distribution – see Problem 6.1.

In the presence of a large (ideally infinite) number of paths, by the central limit theorem the sums over all paths of the real and imaginary parts in (6.9) approach two Gaussian random variables $\Re\mathsf{G}_{i,k}$ and $\Im\mathsf{G}_{i,k}$ of zero mean and variance $\sigma_{i,k}^2$. Considering the resulting complex random variable

$$\mathsf{G}_{i,k} = \Re\mathsf{G}_{i,k} + j\Im\mathsf{G}_{i,k} = \mathsf{A}_{i,k}\exp(j\Phi_{i,k}), \tag{6.10}$$

we have

$$\begin{aligned} \mathbb{E}(\Re\mathsf{G}_{i,k}\Im\mathsf{G}_{i,k}) &= \mathbb{E}(\mathsf{A}_{i,k}^2\sin\Phi_{i,k}\cos\Phi_{i,k}) \\ &= \mathbb{E}(\mathsf{A}_{i,k}^2)\,\mathbb{E}(\sin\Phi_{i,k}\cos\Phi_{i,k}) \\ &= 0, \end{aligned} \tag{6.11}$$

where the second equality follows from independence, and the last equality follows from the uniform phase assumption. Since the real and imaginary parts are uncorrelated and Gaussian, they are also independent, and their joint distribution over the complex plane factorizes into the product form

$$f_{\Re \mathsf{G}_{i,k}, \Im \mathsf{G}_{i,k}}(a_{i,k}) = \frac{1}{2\pi \sigma_{i,k}^2} \exp(-a_{i,k}^2/(2\sigma_{i,k}^2)), \tag{6.12}$$

which depends only on the amplitude $a_{i,k}$ and not on the phase $\phi_{i,k}$ of the complex vector. It follows that each transmitted signal in the frequency domain is multiplied by a complex variable distributed as a circularly symmetric Gaussian, and is characterized by a single parameter $\sigma_{i,k}$.

Noting the differential relationships

$$f_{\Re \mathsf{G}_{i,k}, \Im \mathsf{G}_{i,k}} d\Re \mathsf{G}_{i,k} d\Im \mathsf{G}_{i,k} = f_{\mathsf{A}_{ik}, \Phi_{i,k}} d\mathsf{A}_{i,k} d\Phi_{i,k}, \tag{6.13}$$

$$d\Re \mathsf{G}_{i,k} d\Im \mathsf{G}_{i,k} = \mathsf{A}_{i,k} d\mathsf{A}_{i,k} d\Phi_{i,k}, \tag{6.14}$$

$$f_{\mathsf{A}_{i,k}}(a_{i,k}) = \int_0^{2\pi} f_{\mathsf{A}_{i,k}, \Phi_{i,k}}(a_{i,k}, \phi_{i,k}) d\phi_{i,k}, \tag{6.15}$$

we obtain the Rayleigh distribution for the amplitude

$$f_{\mathsf{A}_{i,k}}(a_{i,k}) = \frac{a_{i,k}}{\sigma_{i,k}^2} \exp(-a_{i,k}^2/(2\sigma_{i,k}^2)), \tag{6.16}$$

which is also characterized by the single parameter $\sigma_{i,k}$. Fixing this parameter corresponds to fixing the average attenuation between transmitter i and receiver k, namely

$$\mathbb{E}(\mathsf{A}_{i,k}) = \sqrt{\frac{\pi}{2}} \sigma_{i,k}. \tag{6.17}$$

From (6.16), it immediately follows that $\mathsf{A}_{i,k}^2$ has the exponential distribution

$$f_{\mathsf{A}_{i,k}^2}(a_{i,k}) = \frac{1}{2\sigma_{i,k}^2} \exp(-a_{i,k}/2\sigma_{i,k}^2), \tag{6.18}$$

and the choice of $\sigma_{i,k}$ also controls the average attenuation power

$$\mathbb{E}(\mathsf{A}_{i,k}^2) = 2\sigma_{i,k}^2. \tag{6.19}$$

Since the variance

$$\mathbb{E}(\mathsf{A}_{i,k}^2) - \mathbb{E}^2(\mathsf{A}_{i,k}) = 4\sigma_{i,k}^4, \tag{6.20}$$

it follows that the standard deviation of the power equals its average, indicating that fluctuations occur at the order of the entire power. This relation can be further extended to higher moments:

$$\mathbb{E}(\mathsf{A}_{i,k}^n) = n! \, \mathbb{E}^n(\mathsf{A}_{i,k}). \tag{6.21}$$

We now summarize. The random multi-path model leads to a stochastic Green's function distributed as a zero-mean complex Gaussian random variable in the frequency domain, with Rayleigh-distributed magnitude representing the field's attenuation,

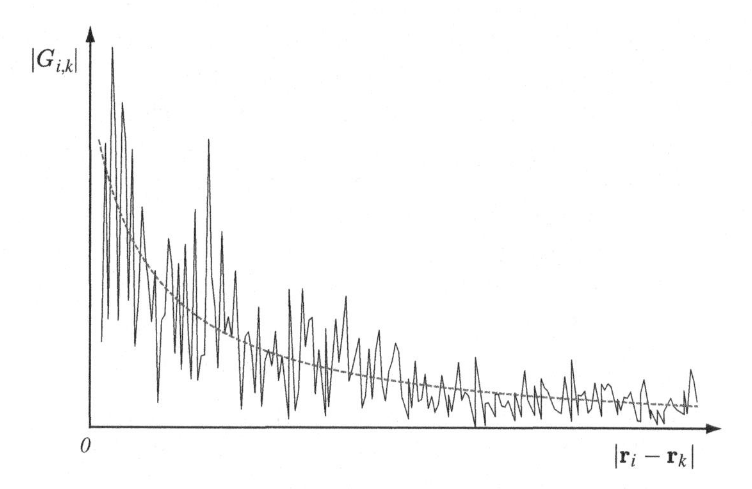

Fig. 6.5 Realization of Rayleigh fading for the random multi-path model.

uniform phase representing the field's phase shift, and exponentially distributed squared magnitude, representing power attenuation.

The parameter $\sigma_{i,k}$, corresponding to the average attenuation between transmitter i and receiver k, is usually modeled as a decreasing function of the distance between transmitter and receiver, called the average *path loss*. Models used in the literature include exponential and inverse power-law path loss functions. The average path loss in the case of a uniform distribution of point scatterers can also be computed analytically using multiple scattering theory in random media – see Chapter 10. This theory predicts an exponential decay with distance, due to absorption, and a singular point at zero distance from the transmitter. The exponential attenuation is consistent with the case of propagation without scattering in a continuum absorbing medium examined in Section 4.4.2. The singular behavior at the origin is consistent with free-space propagation near the transmitting antenna, and it disappears when integration is performed to compute the radiated power. A plot of a typical realization of the magnitude of the Green's function for the random multi-path model with a $\sigma_{i,k}$ that decays with the distance between transmitter and receiver $|\mathbf{r}_i - \mathbf{r}_k|$, is depicted in Figure 6.5. Fluctuations around the average value indicated with a dashed line are called *random fading*.

6.3.1 Frequency-Varying Green's Function: Coherence Bandwidth

The notation used to derive the probabilistic model for the Green's function in multiple scattering environments concealed the fact that the random variables $\mathsf{G}_{i,k}$ depend on ω. By (6.8), each random path carries a random phase shift that depends on ω through a random propagation delay. Furthermore, scattering by different objects along the path introduces additional phase shifts and attenuations that also depend on frequency. For narrow-band signals the Green's function can be considered flat for all frequencies within the band of the signal, so that the response can be described by a single random variable that in the random multi-path model has a Rayleigh magnitude with average decreasing with distance, and uniform phase. However, for wide-band signals this is not the case, and the statistics of the response can change within the frequency band of the signal. In this case, different frequency components of the transmitted signal are affected in different ways by the propagation environment, and the signal tends to lose coherence due to propagation.

The corresponding stochastic model for this situation is obtained by letting the Green's function be a zero-mean random process $\mathsf{G}_{i,k}(\omega)$, whose frequency autocorrelation

$$s_{\mathsf{G}_{i,k}}(\omega,\omega+\omega') = \mathbb{E}(\mathsf{G}_{i,k}(\omega)\mathsf{G}^*_{i,k}(\omega+\omega')) \tag{6.22}$$

vanishes for $\omega' \to \infty$. The critical value for the frequency difference ω' at which the correlation practically disappears, or decreases to a specified level, is called the *coherence bandwidth* of the channel.

The coherence bandwidth, measuring how rapidly the channel changes with frequency, is inversely related to the time broadening of the signal, or delay spread, due to multi-path propagation. We first give an informal picture. Consider two paths carrying phase shifts $\omega\tau_1(t)$ and $\omega\tau_2(t)$ respectively. The phase difference per unit frequency is

$$\frac{d\phi}{d\omega} = \tau_1(t) - \tau_2(t), \tag{6.23}$$

so that a phase variation of 2π occurs for

$$d\omega = \frac{2\pi}{\tau_1(t) - \tau_2(t)}. \tag{6.24}$$

It follows that, neglecting contributions due to different scattering coefficients along the paths, significant variations in the received signal occur for frequency intervals of the order of the inverse of the time delay between paths. Since the coherence bandwidth measures the frequency interval along which channel variations are negligible, its inverse is also an approximate measure of the time delay spread among different received paths. This reasoning can be made more rigorous by taking the inverse Fourier transform of (6.22) with respect to the increment frequency ω',

$$\check{s}_{\mathsf{G}_{i,k}}(\omega,\bar{t}) = \frac{1}{2\pi}\int_{-\infty}^{\infty} s_{\mathsf{G}_{i,k}}(\omega,\omega+\omega')\exp(j\omega'\bar{t})d\omega'. \tag{6.25}$$

We then have

$$s_{\mathsf{G}_{i,k}}(\omega,\omega+\omega') = \int_{-\infty}^{\infty} \check{s}_{\mathsf{G}_{i,k}}(\omega,\bar{t})\exp(-j\omega'\bar{t})d\bar{t}, \tag{6.26}$$

$$\begin{aligned} s_{\mathsf{G}_{i,k}}(\omega,\omega) &= \int_{-\infty}^{\infty} \check{s}_{\mathsf{G}_{i,k}}(\omega,\bar{t})d\bar{t} \\ &= \sigma^2_{i,k}(\omega), \end{aligned} \tag{6.27}$$

where the last equality follows directly from (6.22). The quantity $\check{s}_{\mathsf{G}_{i,k}}(\omega,\bar{t})$ is called the *power delay profile* and can be identified with the distribution of the average power as a function of the delay due to propagation. Its integral along the time axis is the total average power response of the stochastic channel at angular frequency ω. The range of values for which the power delay profile is non-zero is the *delay spread* of the propagation channel. The broader the delay spread, the narrower the coherence bandwidth representing the support of its transform.

If the autocorrelation (6.22) depends only on the increment ω' and not on ω then the channel is *frequency wide-sense stationary*. In this case, although the Green's function varies randomly in frequency, its first- and second-order statistics remain constant. This is immediately evident from (6.22), having

$$\begin{aligned} s_{\mathsf{G}_{i,k}}(\omega,\omega) &= \mathbb{E}(\mathsf{G}_{i,k}(\omega)\mathsf{G}^*_{i,k}(\omega)) \\ &= \sigma^2_{i,k}(\omega) \\ &= \sigma^2_{i,k}, \end{aligned} \tag{6.28}$$

and from the zero-mean assumption. The physical significance of the frequency wide-sense stationary model is that the power delay profile is the Dirac impulse

$$\check{s}_{\mathsf{G}_{i,k}}(\bar{t}) = \sigma^2_{i,k}\delta(\bar{t}), \tag{6.29}$$

corresponding to the average power output of the channel that is concentrated around $t = \bar{t}$, so that the correlation between signals along different paths is essentially zero if the corresponding delay between them is larger than a tiny interval. For this reason, a frequency wide-sense stationary channel model is also called an *uncorrelated scattering* channel. Notice that this assumption is required for the convergence to a Gaussian distribution described in the previous section, where signals along different paths were assumed to be independent.

6.3.2 Time-Varying Green's Function: Coherence Time

We now extend the treatment of multiple-path models to media that are linear, but may be time varying. In this case, the impulse response $g_{i,k}(t,\tau)$ at time t at the receiver k due to an impulsive input at time $t = \tau$ at the transmitter i is a real function of both t and τ. Only when the medium is time invariant does g become a function only of the time difference $t - \tau$. We have, in general,

$$y_k(t) = \int_{-\infty}^{\infty} x_i(t-\tau)g_{i,k}(t,\tau)d\tau. \tag{6.30}$$

Writing the input as a Fourier integral of its spectrum,

$$x_i(t-\tau) = \frac{1}{2\pi}\int_{-\infty}^{\infty} X_i(\omega)\exp(-j\omega\tau)\exp(j\omega t)d\omega, \tag{6.31}$$

and substituting into (6.30), we obtain

$$y_k(t) = \frac{1}{2\pi}\int_{-\infty}^{\infty} X_i(\omega)G_{i,k}(t,\omega)\exp(j\omega t)d\omega, \tag{6.32}$$

where

$$G_{i,k}(t,\omega) = \int_{-\infty}^{\infty} g_{i,k}(t,\tau)\exp(-j\omega\tau)d\tau. \tag{6.33}$$

The physical significance of $G_{i,k}(t,\omega)/(2\pi)$ is that it is the output signal at time t when the input is a time-harmonic signal of angular frequency ω. This follows immediately by letting

$$x_i(t) = \cos(\omega_0 t) = \Re(\exp(-j\omega_0 t)) \tag{6.34}$$

and substituting into (6.32):

$$\begin{aligned} y_k(t) &= \Re\left(\frac{1}{2\pi}\int_{-\infty}^{\infty}\delta(\omega-\omega_0)G_{i,k}(t,\omega)\exp(j\omega t)d\omega\right) \\ &= \frac{1}{2\pi}\Re(G_{i,k}(\omega_0,t)\exp(j\omega_0 t)) \\ &= \frac{1}{2\pi}\Re(G_{i,k}(\omega_0,t)\exp(-j\omega_0 t)). \end{aligned} \tag{6.35}$$

We can then model $G_{i,k}(t,\omega)$ as a zero-mean two-dimensional stochastic process $\mathsf{G}_{i,k}(t,\omega)$ representing the time–frequency varying transfer function of the medium. We have seen that frequency variations are related to multi-path delays and characterized by the coherence bandwidth and time delay spread. Time variations are instead related to changes in the channel due to mobility and are characterized by the coherence time and Doppler frequency spread. The treatment, in this case, is the exact dual of the previous section.

To appreciate this duality, we define the *mutual coherence function* between the output fields due to the time-harmonic inputs at two different frequencies ω and $\omega+\omega'$ as

$$s_{\mathsf{G}_{i,k}}(\omega,\omega+\omega',t,t+t') = \mathbb{E}(\mathsf{G}(\omega,t)\mathsf{G}^*(\omega+\omega',t+t')). \tag{6.36}$$

By the same reasoning as applied for (6.25)–(6.27), the inverse Fourier transform with respect to the frequency increment ω' of (6.36),

$$\check{s}_{\mathsf{G}_{i,k}}(\omega,\bar{t},t,t+t') = \frac{1}{2\pi}\int_{-\infty}^{\infty} s_{\mathsf{G}_{i,k}}(\omega,\omega+\omega',t,t+t')\exp(j\omega'\bar{t})d\omega', \tag{6.37}$$

provides for $t'=0$ the power delay profile at time t as a function of the delay $\bar{t}$. Similarly, the Fourier transform with respect to the time increment t' of (6.36),

$$\hat{s}_{\mathsf{G}_{i,k}}(\omega,\omega+\omega',t,\bar{\omega}) = \int_{-\infty}^{\infty} s_{\mathsf{G}_{i,k}}(\omega,\omega+\omega',t,t+t')\exp(-j\bar{\omega}t')dt', \tag{6.38}$$

provides for $\omega'=0$ the *Doppler power profile* at frequency ω as a function of the Doppler frequency $\bar{\omega}$. The frequency interval in which the Doppler power profile is non-zero indicates the amount of frequency spread of the signal due to propagation in a time-varying medium. The inverse of the Doppler frequency spread is an indication of the time difference at which the time autocorrelation of the signal practically disappears, or decreases to a specified level, and is defined to be the *coherence time*.

A simple heuristic explanation of the Doppler spread and coherence time due to mobility can also be given in this case. Consider one multiple scattered path carrying a delay $\tau(t)$ that changes over time due to the relative motions in the channel. The corresponding phase shift is $\omega\tau(t)$. The phase shift per unit time is

$$\frac{d\phi}{dt} = \omega \frac{d\tau(t)}{dt}, \tag{6.39}$$

so that a phase variation of 2π occurs for

$$\frac{d\tau(t)}{dt} = \frac{2\pi}{\omega}. \tag{6.40}$$

Letting v be the velocity at which the path length changes over time, this phase variation occurs in time $2\pi c/(\omega v)$, where c is the speed of propagation. It follows that significant variations in the phase of the received signal occur for time variations of the order of the inverse of $\omega v/c$, which is the Doppler shift experienced by the signal along the path. The inverse of the overall Doppler spread gives an indication of the coherence time of the channel.

6.3.3 Mutual Coherence Function

We now summarize. According to our stochastic model, when a sinusoidal wave of angular frequency ω propagates in a time-varying medium, the output wave at the receiver randomly fluctuates over time. The correlation of the received wave at two different times t and $t+t'$ decreases as the separation t' increases. The time difference at which the correlation practically disappears, or decreases to a specified level, is the coherence time. It is an indication of how a wave at angular frequency ω is correlated in time. The Fourier transform of the time correlation with respect to t' is the Doppler power profile, whose support indicates the Doppler spread of the signal due to propagation. Its inverse roughly corresponds to the coherence time of the channel. These effects are due to the time-varying nature of the medium due to mobility.

On the other hand, we can compare waves sent at different frequencies, reporting the outputs on the same time axis, and observe the corresponding random fluctuations in frequency. The correlation in frequency decreases as the separation ω' increases. The frequency difference at which the correlation practically disappears, or decreases to a specified level, is the coherence bandwidth. It is a measure of how the waves at two different frequencies are correlated at the same time. The inverse Fourier transform of the frequency correlation with respect to ω' is the power delay profile, whose support indicates the delay spread of the signal due to propagation. Its inverse roughly corresponds to the coherence bandwidth of the channel. These effects are due to multi-path propagation only and are not due to mobility.

These concepts can be illustrated by sketching the magnitude of the mutual coherence function (6.36) as a function of the difference frequency ω' and the difference time t' at a given center time t and a given center frequency ω – see Figure 6.6. When the shapes indicated in the plot do not depend on the center time t, but only on t', the channel is called time wide-sense stationary. If they do not depend on the center frequency ω, but only on ω', the channel is called frequency wide-sense stationary, or uncorrelated scattering. These two cases are time–frequency duals of each other. If

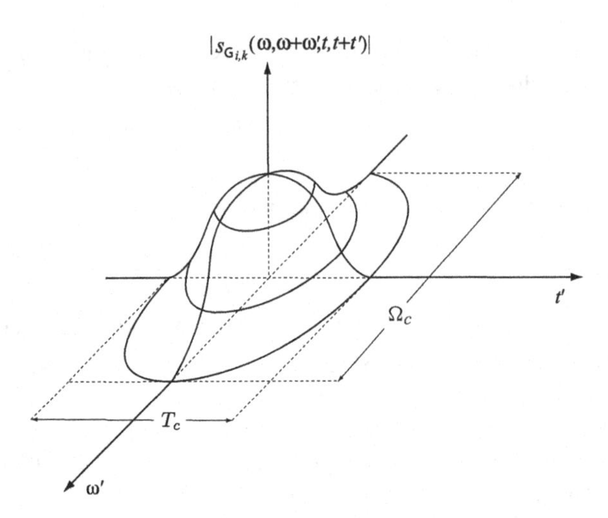

Fig. 6.6 Mutual coherence function, showing the coherence bandwidth Ω_c and the coherence time T_c.

the mutual coherence function is independent of both ω and t, the channel is called wide-sense uncorrelated scattering.

A waveform of bandwidth $\Delta\omega$ smaller than the coherence bandwidth Ω_c of the channel does not undergo significant distortion in frequency. In this case, the response of the channel is said to be frequency flat, and the different frequency components of the signal are treated coherently – see Figure 6.7(a). In this case, the power delay profile appears as a short impulse, and the received waveform has essentially the same shape as the transmitted one. A waveform of bandwidth $\Delta\omega$ larger than Ω_c undergoes significant distortion in frequency due to multi-path delays, and exhibits a lack of correlation among different frequency components at the receiver. The response of the channel is said to be selective in frequency – see Figure 6.7(b). The net effect in the time domain is that the power delay profile appears broader, and the received waveform does not resemble the same shape as the transmitted one, due to the large amount of interference of the delayed returns.

Similarly, a waveform transmitted over a time interval smaller than the coherence time T_c of the channel does not undergo significant distortion in time, and the response is said to be time flat. A signal transmitted over a time interval larger than T_c undergoes a significant distortion in time due to mobility in the environment, and exhibits a lack of correlation among different time components at the receiver. The net effect in the frequency domain is that the Doppler power profile appears broader in frequency and the response of the channel is said to be selective in time.

The different cases are summarized in Figure 6.8, where (a) represents a channel that is essentially frequency flat but time selective, (b) represents a channel that is essentially time flat but frequency selective, and (c) represents a channel that is both time and frequency selective.

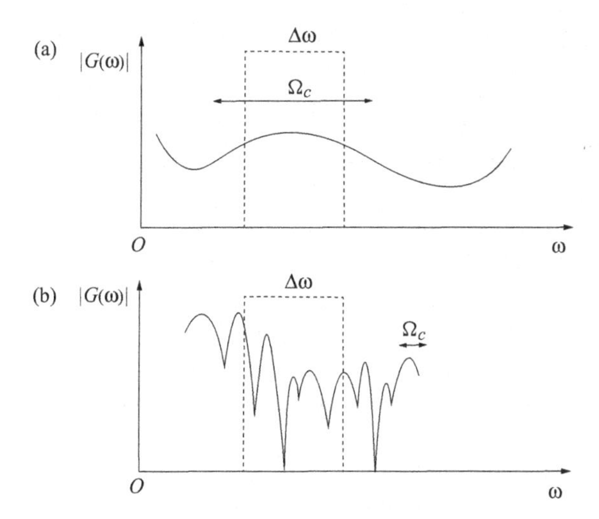

Fig. 6.7 (a) The transmitted signal's bandwidth $\Delta\omega < \Omega_c$, and the signal remains coherent. (b) The transmitted signal's bandwidth $\Delta\omega > \Omega_c$, and the signal is subject to distortion.

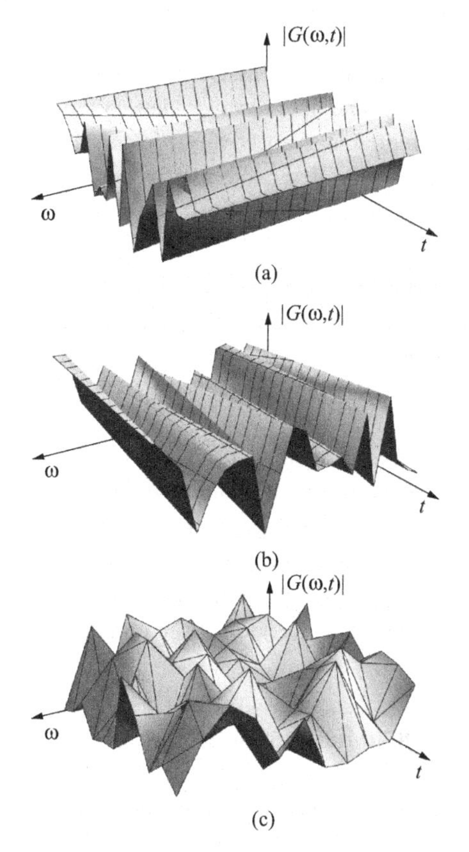

Fig. 6.8 (a) Time-selective Green's function. (b) Frequency-selective Green's function. (c) Time- and frequency-selective Green's function.

6.3.4 Spatially Varying Green's Function: Coherence Distance

We now consider the spatial correlation between received signals. In the presence of multiple scattering, one may expect that the correlation between signals at any two receiving points decreases as the distance between the transmitting antennas increases. Similarly, one may expect the correlation to decrease as the distance between the receiving antennas increases. The idea is that in both cases signals travel along increasingly different multiple paths to the receiver. The minimum distance at which these correlations practically disappear is called the *coherence distance* of the channel.

The corresponding mathematical model considers the elements $G_{i,k}(t,\omega)$ to be realizations of the elements of a random matrix G, each entry $\mathsf{G}_{i,k}(t,\omega)$ being a two-dimensional time–frequency random process as described in the previous section. In the simplest case, these random processes are assumed to be i.i.d. and Gaussian. This corresponds to having a separation distance between the antennas larger than the coherence distance, so that the correlation between any two antenna pairs is zero. On the other hand, in the case that the antennas are in close proximity to each other, we can have a non-zero spatial correlation between channel responses at (i,k) and (i',k'):

$$s_{\mathsf{G}}(i,k;i'k') = \mathbb{E}(\mathsf{G}_{i,k}\mathsf{G}^*_{i',k'}). \tag{6.41}$$

If the spatial correlation is assumed to be a local phenomenon occurring separately at the transmitter and the receiver, letting Θ_{T} and Θ_{R} be deterministic matrices whose entries represent, respectively, the correlation between the transmit and receive antennas, and G_w an i.i.d. random matrix, we have

$$\mathsf{G} = \Theta_{\mathrm{R}}^{1/2}\mathsf{G}_w\Theta_{\mathrm{T}}^{1/2}. \tag{6.42}$$

This means that the correlation between two transmit antennas is the same regardless of the receive antenna at which the observation is made, and vice versa. A slightly more general model is

$$\mathsf{G} = \Theta_{\mathrm{R}}^{1/2}\mathsf{G}_{\mathrm{R}}\mathrm{T}^{1/2}\mathsf{G}_{\mathrm{T}}\Theta_{\mathrm{T}}^{1/2}, \tag{6.43}$$

where G_{R} and G_{T} are both i.i.d. Gaussian matrices, and $\mathrm{T}^{1/2}$ describes the transfer matrix between the transmitter and receiver's environments. The rank of this transfer matrix can be as small as one, in the case, for example, of a single-mode waveguide between transmitter and receiver. In this case, the matrix G is forced to have rank one as well, even if $\Theta_{\mathrm{R}}^{1/2}$, $\Theta_{\mathrm{T}}^{1/2}$ have full rank. This is called the *keyhole* effect and can arise in indoor propagation in corridors, or in outdoor propagation along city streets.

6.4 Karhunen–Loève Representation

The Karhunen–Loève decomposition of the stochastic Green's function is an orthogonal representation of the stochastic process that is convergent with respect to the mean square norm, and provides the minimum number of orthogonal basis functions that can approximate the process in the mean square sense up to arbitrary accuracy.

The representation is based on the spectral decomposition of the autocorrelation function of the process, and it leads to the notion of *stochastic diversity* available for communication.

We first consider the Karhunen–Loève decomposition of the time-varying stochastic process $\mathsf{G}(t)$, then extend it in the natural way to describe the stochastic process $\mathsf{G}(\omega,t)$, which includes random variations in frequency, and finally to describe the stochastic process $\mathsf{G}_{i,k}(\omega,t)$, which includes random variations across different antennas in space.

There is an analogy between the results described here, the singular values representation described in Section 5.3.3, and the time–frequency and space–wavenumber Hilbert–Schmidt representations described in Section 5.4.1. In the deterministic setting we refer to radiation from an arbitrary environment, and the number of degrees of freedom indicates the amount of information in terms of the minimum number of coordinates required to specify the process up to a given level of approximation. In the stochastic case we refer to radiation from a random environment that induces random variations of the field in time, frequency, and space, and the stochastic diversity represents the amount of information, in terms of the minimum number of uncorrelated coordinates required to specify the process up to a given level of approximation, in a certain average sense.

While in a deterministic setting the number of degrees of freedom depends only on the bandwidth and on the size of the interval where the waveform is observed, in the stochastic setting the diversity depends on the form of the autocorrelation function of the process. This should not be surprising: the random environment shapes the response by imposing a certain autocorrelation function, which influences the amount of stochastic variation of the received signal.

6.4.1 Time-Varying Green's Function

We consider a stochastic process $\mathsf{G}(t)$ that is defined over a time interval $[-T/2,T/2]$ and subject to the squared norm constraint

$$\| \mathsf{G}(t) \|^2 = \langle \mathsf{G}(t)\,\mathsf{G}^*(t) \rangle = \mathbb{E}[\mathsf{G}(t)\,\mathsf{G}^*(t)] < \infty. \tag{6.44}$$

This constraint is analogous to imposing square-integrability of the process with respect to its probability measure, namely

$$\int_{\mathscr{S}} |\mathsf{G}(t,z)|^2 d\,\mathbb{P}(z) < \infty, \tag{6.45}$$

where $\mathscr{S}$ is the sample space of the stochastic process

$$\mathsf{G} : [-T/2,T/2] \times \mathscr{S} \to \mathbb{R}, \tag{6.46}$$

equipped by an appropriate σ-algebra over $\mathscr{S}$.

The stochastic process also has zero mean and is mean-square continuous, namely

$$\lim_{\Delta \to 0} \mathbb{E}[(\mathsf{G}(t+\Delta) - \mathsf{G}(t))^2] = 0. \tag{6.47}$$

This immediately implies that the autocorrelation function

$$s_{\mathsf{G}}(t,t') = \mathbb{E}[\mathsf{G}(t)\,\mathsf{G}^*(t')] \tag{6.48}$$

is also continuous.

We consider the integral operator whose kernel is the autocorrelation function

$$(\mathcal{S}f)(t) = \int_{-T/2}^{T/2} s_{\mathsf{G}}(t,t')f(t')dt'. \tag{6.49}$$

This is of Hilbert–Schmidt type since, by continuity,

$$\int_{-T/2}^{T/2}\int_{-T/2}^{T/2} |s_{\mathsf{G}}(t,t')|^2 dt dt' < \infty. \tag{6.50}$$

The Hilbert–Schmidt operator in (6.49) is non-negative, and self-adjoint; it follows that it has a complete set of orthonormal eigenvectors $\{\psi_n\}$, and non-negative, real eigenvalues $\{\lambda_n\}$ that are the solutions of the equation

$$\int_{-T/2}^{T/2} s_{\mathsf{G}}(t,t')\psi_n(t')dt' = \lambda_n \psi_n(t). \tag{6.51}$$

By the results in Section 3.4.2, we have the Hilbert–Schmidt decomposition of the symmetric kernel

$$s_{\mathsf{G}}(t,t') = \sum_{n=0}^{\infty} \lambda_n \psi_n^*(t)\psi_n(t'). \tag{6.52}$$

It then follows that

$$s_{\mathsf{G}}(t,t) = \sum_{n=0}^{\infty} \lambda_n |\psi_n(t)|^2, \tag{6.53}$$

where the equalities in (6.52) and (6.53) hold in the $L^2[-T/2,T/2]$ sense, and also absolutely and uniformly by virtue of Mercer's theorem in Appendix A.4.2. From (6.53), it follows that

$$\int_{-T/2}^{T/2} s_{\mathsf{G}}(t,t)dt = \sum_{n=0}^{\infty} \lambda_n. \tag{6.54}$$

The Karhunen–Loève representation of the stochastic Green's function is obtained by using the deterministic eigenfunctions in (6.51) as an orthonormal basis representation for the process, and we have

$$\mathsf{G}(t) = \sum_{n=0}^{\infty} \mathsf{A}_n \psi_n(t), \tag{6.55}$$

where the coefficients of the expansion are zero-mean random variables, are uncorrelated, and have variances

$$\mathbb{E}(\mathsf{A}_n \mathsf{A}_n^*) = \lambda_n. \tag{6.56}$$

The stochastic convergence of (6.55) is intended in the mean square sense:

$$\lim_{N\to\infty} \mathbb{E}\left(\left|\mathsf{G}(t) - \sum_{n=0}^{N-1} \mathsf{A}_n \psi_n(t)\right|^2\right) = 0. \tag{6.57}$$

The existence of the representation and the convergence result follow from the Hilbert–Schmidt decomposition of the autocorrelation function. Consider the mean squared error

$$e_N(t) = \mathbb{E}\left(\left|\mathsf{G}(t) - \sum_{n=0}^{N-1} \mathsf{A}_n \psi_n(t)\right|^2\right). \tag{6.58}$$

By expanding the square inside the expectation, using (6.56), and since the coefficients $\{\mathsf{A}_\mathsf{n}\}$ are uncorrelated, we have

$$\begin{aligned} e_N(t) &= \mathbb{E}(|\mathsf{G}(t)|^2) + \mathbb{E}\left(\sum_{n=0}^{N-1}\sum_{k=0}^{N-1} \mathsf{A}_n \mathsf{A}_k^* \psi_n(t)\psi_k^*(t)\right) - 2\,\mathbb{E}\left(\mathsf{G}(t)\sum_{n=0}^{N-1}\mathsf{A}_n\psi_n(t)\right) \\ &= s_\mathsf{G}(t,t) + \sum_{n=0}^{N-1}\lambda_n|\psi_n(t)|^2 - 2\,\mathbb{E}\left(\mathsf{G}(t)\sum_{n=0}^{N-1}\mathsf{A}_n\psi_n(t)\right). \end{aligned} \tag{6.59}$$

Substituting

$$\mathsf{A}_\mathsf{n} = \int_{-T/2}^{T/2} \mathsf{G}(t')\psi_n^*(t')dt' \tag{6.60}$$

into (6.59), we have

$$\begin{aligned} e_N(t) &= s_\mathsf{G}(t,t) + \sum_{n=0}^{N-1}\lambda_n|\psi_n(t)|^2 - 2\,\mathbb{E}\left(\sum_{n=0}^{N-1}\mathsf{G}(t)\psi_n(t)\int_{-T/2}^{T/2}\mathsf{G}(t')\psi_n^*(t')dt'\right) \\ &= s_\mathsf{G}(t,t) + \sum_{n=0}^{N-1}\lambda_n|\psi_n(t)|^2 - 2\sum_{n=0}^{N-1}\int_{-T/2}^{T/2}\mathbb{E}[\mathsf{G}(t)\mathsf{G}(t')]\psi_n^*(t')dt'\,\psi_n(t) \\ &= s_\mathsf{G}(t,t) + \sum_{n=0}^{N-1}\lambda_n|\psi_n(t)|^2 - 2\sum_{n=0}^{N-1}\int_{-T/2}^{T/2} s_\mathsf{G}(t,t')\psi_n^*(t')dt'\,\psi_n(t). \end{aligned} \tag{6.61}$$

Substituting (6.51) into (6.61), it follows that

$$e_N(t) = s_\mathsf{G}(t,t) - \sum_{n=0}^{N-1}\lambda_n|\psi_n(t)|^2, \tag{6.62}$$

which tends to zero as $N \to \infty$ for all $t \in [-T/2, T/2]$, since (6.53) converges point-wise in this interval.

One can think of the Karhunen–Loève representation adapting to the process through the choice of the deterministic basis functions based on the autocorrelation of the process, in order to produce the best possible basis for its representation. On the other hand, the joint probability law of the expansion coefficients remains unknown, in the absence of information other than the second-order properties of the process. If the Green's function is distributed as a zero-mean complex Gaussian in the frequency domain, then the coefficients in the expansion are independent and Gaussian, and the series expansion is almost sure convergent.

6.4.2 Optimality of the Karhunen–Loève Representation

The Karhunen–Loève representation is optimal, among all orthonormal basis decompositions of a stochastic process, in a precise energy sense. This property makes it ideal to define the stochastic diversity of a family of processes of a given autocorrelation function in the same way that we defined the number of degrees of freedom of a family of functions that are the image of a Hilbert–Schmidt operator.

Arranging the eigenvalues in decreasing order, by combining (6.53) and (6.62) it follows that the average energy of the error is given by

$$\begin{aligned}\int_{-T/2}^{T/2} e_N(t)dt &= \int_{-T/2}^{T/2} \mathbb{E}\left(\left|\mathsf{G}(t) - \sum_{n=0}^{N-1} \mathsf{A}_n \psi_n(t)\right|^2\right) dt \\ &= \sum_{n=N}^{\infty} \lambda_n. \end{aligned} \tag{6.63}$$

Using a Lagrangian formulation, we show that this is the minimum energy of the error among all orthonormal basis representations of the stochastic process $\mathsf{G}(t)$.

By the orthonormality of the $\{\psi_n\}$ and the linearity of expectation, we have

$$\begin{aligned}\mathbb{E}\left(\int_{-T/2}^{T/2}\left|\mathsf{G}(t) - \sum_{n=0}^{N-1} \mathsf{A}_n \psi_n(t)\right|^2 dt\right) &= \mathbb{E}\left(\int_{-T/2}^{T/2} |\mathsf{G}(t)|^2 dt - \sum_{n=0}^{N-1} |\mathsf{A}_n|^2\right) \\ &= \int_{-T/2}^{T/2} \mathbb{E}\,|\mathsf{G}(t)|^2 dt - \sum_{n=0}^{N-1} \mathbb{E}\,|\mathsf{A}_n|^2. \end{aligned} \tag{6.64}$$

Since the first term is independent of the choice of the $\{\psi_n\}$, the energy of the error is minimized when $\sum_{n=0}^{N-1} \mathbb{E}\,|\mathsf{A}_n|^2$ is maximized. We need to maximize

$$\begin{aligned}\sum_{n=0}^{N-1} \mathbb{E}(|\mathsf{A}_n|^2) &= \sum_{n=0}^{N-1} \mathbb{E}\left(\int_{-T/2}^{T/2} \mathsf{G}(t)\psi_n(t)dt \int_{-T/2}^{T/2} \mathsf{G}^*(t')\psi_n^*(t')dt'\right) \\ &= \sum_{n=0}^{N-1} \int_{-T/2}^{T/2}\int_{-T/2}^{T/2} \mathbb{E}\left(\mathsf{G}(t)\mathsf{G}^*(t')\right) \psi_n(t)\psi_n^*(t')dt'dt \\ &= \sum_{n=0}^{N-1} \int_{-T/2}^{T/2}\int_{-T/2}^{T/2} s_{\mathsf{G}}(t,t')\psi_n(t)\psi_n^*(t')dt'dt, \end{aligned} \tag{6.65}$$

with the orthonormality constraint

$$\int_{-T/2}^{T/2} \psi_n(t)\psi_n^*(t)dt = \begin{cases} 1 & \text{if } n = m, \\ 0 & \text{if } n \neq m. \end{cases} \tag{6.66}$$

We aim to maximize the Lagrangian

$$\Lambda(\{\psi_n\}) = \sum_{n=0}^{N-1} \int_{-T/2}^{T/2} \int_{-T/2}^{T/2} s_{\mathsf{G}}(t,t')\psi_n(t)\psi_n^*(t')dt'dt$$
$$-\lambda_n \left(\int_{-T/2}^{T/2} \psi_n(t)\psi_n^*(t)dt - 1 \right). \tag{6.67}$$

Differentiating with respect to ψ_i^* and setting the derivative to zero, we get

$$\frac{\partial \Lambda(\{\psi_n\})}{\partial \psi_i(t)} = \int_{-T/2}^{T/2} \left(\int_{-T/2}^{T/2} s_{\mathsf{G}}(t,t')\psi_i(t')dt' - \lambda_i \psi_i(t) \right) dt = 0, \tag{6.68}$$

which is satisfied when the integrand is zero, namely when the $\{\psi_i\}$ are solutions of the integral equation (6.51).

6.4.3 Stochastic Diversity

By analogy with the deterministic setting, we define the stochastic diversity of a set of stochastic processes as:

> The minimum number of uncorrelated parameters sufficient for the average description of any process in the set within a given precision.

To give a corresponding mathematical definition, we consider the set of processes of given autocorrelation function

$$\mathscr{S}_s = \left\{ \mathsf{G}(t) : \mathbb{E}[\mathsf{G}(t)\mathsf{G}^*(t')] = s_{\mathsf{G}}(t,t') \right\}, \tag{6.69}$$

whose average energy over the observation interval is

$$E_T = \int_{-T/2}^{T/2} s_{\mathsf{G}}(t,t)dt = \int_{-T/2}^{T/2} \mathbb{E}(|\mathsf{G}(t)|^2)dt. \tag{6.70}$$

We let the minimum energy error associated with the optimal N-dimensional Karhunen–Loève approximation be

$$E_N(\mathscr{S}_s) = \int_{-T/2}^{T/2} e_N(t)dt$$

$$= \mathbb{E}\left(\int_{-T/2}^{T/2} \left| \mathsf{G}(t) - \sum_{n=0}^{N-1} \mathsf{A}_n \psi_n(t) \right|^2 \right) dt$$

$$= \sum_{n=N}^{\infty} \lambda_n, \tag{6.71}$$

and define the stochastic diversity of $\mathscr{S}_s$ as

$$N_\epsilon(\mathscr{S}_s) = \min\{N : E_N(\mathscr{S}_s)/E_T \le \epsilon\}. \tag{6.72}$$

This definition should be compared with (3.18). In the deterministic setting, the number of degrees of freedom corresponds to the dimension of the minimal subspace representing any signal within ϵ accuracy. For bandlimited signals, the number of degrees of freedom depends only on the bandwidth Ω and on the duration of the observation interval T. In the stochastic setting, the diversity of stochastic processes of a given autocorrelation function corresponds to the dimension of the minimal subspace representing on average any process in the set within ϵ accuracy. The stochastic diversity depends on the duration of the observation interval T, but also on the form of the autocorrelation function. The comparison between the deterministic and the stochastic cases is summarized in Table 6.1.

We now consider the error associated with the N-dimensional approximation. In the deterministic setting, by (3.56) the energy of the error is given by the Nth eigenvalue of the self-adjoint operator arising from Slepian's concentration problem. The phase transition of the eigenvalues (2.131) ensures that the number of degrees of freedom of bandlimited signals is essentially given by the time–bandwidth product $N_0 = \Omega T/\pi$. In the stochastic case, by (6.71) the energy of the error is given by the tail sum of the eigenvalues of the self-adjoint operator whose kernel is the autocorrelation function. By (6.51), the tail decay of the eigenvalues as a function of their indexes depends on the particular form of the autocorrelation function. Highly concentrated autocorrelation functions correspond to highly varying processes that require a large number of terms to be represented at a given level of accuracy. On the other hand, processes with spread out correlation functions have somewhat "less wild" stochastic variations and require a smaller number of terms to be represented, on average, at the same level of accuracy. Problems 6.3, 6.4, and 6.5 provide examples of this general behavior.

Table 6.1 Degrees of freedom and stochastic diversity.

	Deterministic	Stochastic
Bandwidth/autocorrelation	Ω	s_{G}
Observation interval	$[-T/2, T/2]$	$[-T/2, T/2]$
Fredholm equation	$\mathcal{BT}\psi(t) = \lambda\psi(t)$	$\mathcal{S}\psi(t) = \lambda\psi(t)$
Degrees of freedom	$N_\epsilon(\mathscr{B}_\Omega) = \min\{N : \lambda_N \le \epsilon\}$	$N_\epsilon(\mathscr{S}_s) = \min\left\{N : (\sum_{n=N}^{\infty} \lambda_n)/E_T \le \epsilon\right\}$
Effective dimensionality	$N_0 = \Omega T/\pi$	Depends on s_{G}

We also have the following general bound on the error for a wide-sense stationary process of autocorrelation $s_{\mathsf{G}}(t,t') = s_{\mathsf{G}}(\tau)$ and power spectral density $S_{\mathsf{G}}(\omega)$. Letting

$$N = \frac{T}{2\pi} \int_{S_{\mathsf{G}}(\omega)\in(a,b)} d\omega, \tag{6.73}$$

we have that, as $T \to \infty$, the error

$$\sum_{n=N}^{\infty} \lambda_n \leq \frac{T}{2\pi} \left(\int_{-\infty}^{\infty} S_{\mathsf{G}}(\omega)\, d\omega - a \int_{S_{\mathsf{G}}(\omega)\in(a,b)} d\omega \right) + o(T). \tag{6.74}$$

This bound is a consequence of the Kac–Murdock–Szegö (1953) theorem – see Appendix A.7.

6.4.4 Constant Power Spectral Density

We now compute the stochastic diversity in the special case of a wide-sense stationary Green's function whose autocorrelation has the form

$$s_{\mathsf{G}}(t,t') = s_{\mathsf{G}}(\tau) = \frac{\Omega}{\pi}\operatorname{sinc}(\Omega\tau), \tag{6.75}$$

where $t' = t - \tau$. By taking the Fourier transform of (6.75), the power spectral density is

$$S_{\mathsf{G}}(\omega) = \begin{cases} 1 & \text{for } \omega \in [-\Omega, \Omega], \\ 0 & \text{otherwise.} \end{cases} \tag{6.76}$$

In this case, an impulsive correlation in the time domain corresponds to a spread out rectangular window for the power spectral density in the frequency domain. We then expect that larger values of Ω, roughly corresponding to a larger bandwidth in the deterministic setting, yield a larger stochastic diversity.

To evaluate the stochastic diversity, we first note that (6.51) now corresponds to

$$\mathcal{S}\psi(t) = \mathcal{B}\mathcal{T}\psi(t) = \lambda\psi(t), \tag{6.77}$$

where $\mathcal{T}$ and $\mathcal{B}$ are the timelimiting and bandlimiting operators defined in (2.87) and (2.88). It follows that the eigenvalues and eigenfunctions in the stochastic and the deterministic cases coincide, as they are both solutions of Slepian's concentration problem.

We then note that from (6.76) we have the variance constraint

$$\mathbb{E}(|\mathsf{G}(t)|^2) = \frac{1}{2\pi} \int_{-\infty}^{\infty} S_{\mathsf{G}}(\omega) d\omega = \frac{\Omega}{\pi}, \tag{6.78}$$

from which it follows that the average energy over the observation interval is

$$\begin{aligned} E_T &= \int_{-T/2}^{T/2} s_{\mathsf{G}}(t,t) dt \\ &= \int_{-T/2}^{T/2} \mathbb{E}(|\mathsf{G}(t)|^2) dt \end{aligned}$$

$$= \frac{\Omega T}{\pi}$$
$$= N_0. \tag{6.79}$$

Finally, substituting (6.71) and (6.79) into (6.72), we have

$$N_\epsilon(\mathscr{S}_s) = \min\left\{N : \frac{1}{N_0}\sum_{n=N}^{\infty}\lambda_n \le \epsilon\right\}. \tag{6.80}$$

From the behavior of the eigenvalues in (2.131), it follows that, as $N_0 \to \infty$,

$$N_\epsilon(\mathscr{S}_s) = (1-\epsilon)N_0. \tag{6.81}$$

This result is illustrated in Figure 6.9. The phase transition of the eigenvalues occurs in a window of size $2\epsilon N_0$, and there are $(1-\epsilon)N_0$ eigenvalues having value one. As $N_0 \to \infty$ the phase transition of the eigenvalues tends to a step function, and the shaded area in the figure that represents the tail sum of the eigenvalues normalized to N_0 tends to ϵ. It follows that by interpolating $(1-\epsilon)N_0$ eigenfunctions we capture most of the energy of the process. A more rigorous derivation is given in Appendix A.7.2.

6.4.5 Frequency-Varying Green's Function

A completely analogous treatment can be given if we assume that the process $\mathsf{G}(\omega)$ varies randomly in frequency rather than in time, and is observed over the interval $[-\Omega, \Omega]$. In this case, we have

$$\mathsf{G}(\omega) = \sum_{n=0}^{\infty}\mathsf{A}_n\psi_n(\omega), \tag{6.82}$$

where the deterministic basis functions $\{\psi_n\}$ are the solutions of the eigenvalue equation

$$\int_{-\Omega}^{\Omega} s_{\mathsf{G}}(\omega,\omega')\psi_n(\omega')d\omega' = \lambda_n\psi_n(\omega), \tag{6.83}$$

where

$$s_{\mathsf{G}}(\omega,\omega') = \mathbb{E}[\mathsf{G}(\omega)\mathsf{G}^*(\omega')], \tag{6.84}$$

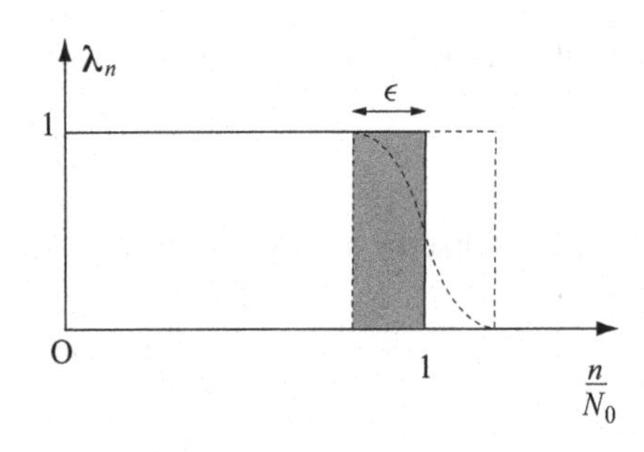

Fig. 6.9 Evaluation of the tail sum of the eigenvalues.

and the random coefficients $\{\mathsf{A}_n\}$ are zero-mean random variables, uncorrelated, and with variances given by the corresponding eigenvalues $\{\lambda_n\}$.

6.4.6 Spatially Varying Green's Function

The number of uncorrelated spatial channels that are available through propagation depends on the spatial correlation structure of the matrix G, and can be obtained from the spatial Karhunen–Loève decomposition that is the discrete analog of the one presented in Section 6.4.2. The Karhunen–Loève image $\tilde{\mathsf{G}}$ of G is the $|\mathscr{R}| \times |\mathscr{T}|$ random matrix with uncorrelated entries

$$\tilde{\mathsf{G}}_{i,k} = \sum_{m \in \mathscr{R}} \sum_{n \in \mathscr{T}} \mathsf{A}_{m,n}\, \psi^*_{i,k}(m,n), \tag{6.85}$$

where the expansion kernel $\{\psi_{i,k}(m,n)\}$ is a set of complete orthonormal basis functions formed by the eigenfunctions of the spatial autocorrelation of G, namely, for all $k \in \mathscr{R}$, $i \in \mathscr{T}$,

$$\sum_{m' \in \mathscr{R}} \sum_{n' \in \mathscr{T}} s_{\mathsf{G}}(m,n;m',n')\psi_{i,k}(m',n') = \lambda_{i,k}\psi_{i,k}(m,n). \tag{6.86}$$

The variances of the entries of the Karhunen–Loève image are given by

$$\mathbb{E}(\tilde{\mathsf{G}}_{i,k}\tilde{\mathsf{G}}^*_{i,k}) = \mathbb{E}(\mathsf{A}_{i,k}\mathsf{A}^*_{i,k}) = \lambda_{i,k}. \tag{6.87}$$

In practice, after estimating the spatial autocorrelation of the process, one can derive through (6.86) and (6.87) the elements of the image $\tilde{\mathsf{G}}$ where the channel response focuses most of the signal's energy. These are the *principal spectral components* of G corresponding to the largest eigenvalues of its autocorrelation function.

Now, if the expansion kernel can be factorized as

$$\psi_{i,k}(m,n) = u_i(m)v_k(n), \tag{6.88}$$

then we also have the decomposition

$$\mathsf{G} = \mathsf{U}\tilde{\mathsf{G}}\mathsf{V}^\dagger, \tag{6.89}$$

where U and V are unitary and deterministic, with $\mathrm{U}_{i,m} = u_i(m)$ and $\mathrm{V}_{n,k} = v^*_k(n)$.

Matrices whose autocorrelations are separable into two marginal correlations that are functions of (m,n) and (m',n') are automatically factorable, and $\lambda_{i,k} = \lambda_i\lambda_k$, where λ_i and λ_k are the ith and kth eigenvalues of the two marginal correlations whose product equals S_{G}.

From (6.89), we can then obtain an approximation of G retaining only the entries in the image corresponding to the largest N principal components. Letting $\mathscr{P}$ be the index pair set of these principal components, we let

$$\mathsf{G}_{N(i,k)} = \sum_{(m,n) \in \mathscr{P}} \mathrm{U}_{i,m}\tilde{\mathsf{G}}_{m,n}\mathrm{V}_{n,k}. \tag{6.90}$$

Of course, we expect that the smaller the ignored eigenvalues, the more accurate the approximation. Indeed, by the uncorrelation of the elements of $\tilde{\mathsf{G}}$, and since U and V are unitary, we have the approximation error in terms of Frobenius norm

$$\begin{aligned} e_N &= \mathbb{E}(\|\mathsf{G} - \mathsf{G}_N\|_\mathrm{F})^2 \\ &= \mathbb{E}\left(\sum_{(m,n)\notin \mathscr{P}} |\tilde{\mathsf{G}}_{m,n}|^2 \right) \\ &= \sum_{(m,n)\notin \mathscr{P}} \lambda_{m,n}. \end{aligned} \tag{6.91}$$

As before, the error is a minimum among all orthonormal basis decompositions of the random matrix G. The error decreases monotonically with the number of terms retained in the expansion, at a rate that depends on the tail behavior of the eigenvalues $\{\lambda_n\}$, and in turn on the autocorrelation function of the spatial process. The more correlated the process, the smaller the number of terms needed to achieve a small error. On the other hand, if the process is weakly correlated, then a larger number of terms is needed.

6.5 Summary and Further Reading

Probabilistic models of multiple scattering attempt an average description of the field in multiple-scattering environments and require a stochastic extension of the propagation model. One of the first stochastic characterizations of the linear, time-varying transfer function for wireless communication appears in Bello (1963). Tse and Viswanath (2005) give a communication-theoretic perspective on different stochastic models used in the literature, and discuss the range of applicability and the modeling philosophy in many practical cases. As the authors point out, these models are generally far less accurate than the stochastic models of additive noise; they are important, however, in guiding the design of real systems. A similar communication-theoretic perspective is provided in Goldsmith (2005).

The classic book by Ishimaru (1978) provides a rigorous treatment of multiple-scattering theory in random media, giving additional physical insight into the different stochastic models. We discuss this theory in Chapter 10. There, we also describe a stochastic model of multiple scattering based on random walks in continuum space that leads to closed-form expressions for the path loss, power-delay profile, and coherence bandwidth of the channel.

For any probabilistic model, the Karhunen–Loève decomposition provides the number of uncorrelated coefficients that describes the evolution of the stochastic process up to arbitrary accuracy. This number depends on the propagation environment through the form of the autocorrelation function of the process, and is the stochastic counterpart of the number of degrees of freedom in the deterministic setting. A proof of the optimality of this decomposition with respect to the mean square error appears in Brown (1960).

By analogy with the number of degrees of freedom in the deterministic case, we have defined the stochastic diversity in terms of the minimum number of coefficients needed to achieve a given mean square truncation error. This led to a parallel between the Hilbert–Schmidt decomposition and the Karhunen–Loève one in determining the optimal functional representation of the process. Another possibility is to minimize the expected number of coefficients required to obtain a given truncation error. In this case, however, the decomposition depends in a detailed manner on the probability distribution of the process and is not determined just by its correlation function, except in the Gaussian case – see Algazi and Sakrison (1969).

In the spatial domain, modeling degrees of freedom and stochastic diversity in the discrete setting of multiple antenna systems is a problem closely related to the study of the spectral properties of random matrices. Several monographs exist on the topic, including Tao (2012), and Tulino and Verdú (2004), the latter focusing on applications of random matrix theory to wireless communication.

When multiple antennas are deployed in a distributed setting to form a network, the typical approach is to consider random *geometric* matrices. In this case, a random spatial distribution for the locations of the transmitters and receivers is assumed. The randomness in the resulting channel's matrix is induced by the underlying random spatial distribution of the transmitters and receivers. This approach is part of a larger research area that models wireless networks using random graphs, percolation theory, and stochastic geometry, and is described in the books by Franceschetti and Meester (2007) and Haenggi (2013), as well as in the tutorial paper by Haenggi et al. (2009).

6.6 Test Your Understanding

Problems

6.1 Consider a random multi-path model in the presence of a deterministic line-of-sight component $A_0 \exp(j\phi_0)$, and determine the resulting distribution for $G_{i,k}(\omega)$.

Solution

We choose the phase reference so that $\phi_0 = 0$ and drop the i,k from the notation for convenience. We then consider the complex random variable

$$\Re(\mathsf{G} - A_0) + j\Im\mathsf{G}. \tag{6.92}$$

Using the central limit theorem, the real and imaginary parts have Gaussian distributions, and by the same reasoning as in Section 6.3, we have that the joint distribution

$$f_{\Re\mathsf{G},\Im\mathsf{G}} = \frac{1}{2\pi\sigma^2} \exp\left(-\frac{(\Re g - A_0)^2 + \Im g^2}{2\sigma^2}\right). \tag{6.93}$$

The probability density for the amplitude A can then be computed to be

$$
\begin{aligned}
f_{\mathsf{A}}(a) &= \int_0^{2\pi} f_{\Re \mathsf{G}, \Im \mathsf{G}}(\Re g, \Im g) a \, d\phi \\
&= \frac{a}{\sigma^2} \exp\left(-\frac{a^2 + A_0^2}{2\sigma^2}\right) I_0\left(\frac{A_0 a}{\sigma^2}\right),
\end{aligned} \tag{6.94}
$$

where $I_0(\cdot)$ denotes the zeroth-order modified Bessel function of the first kind. This is the Rice distribution. The factor

$$
K = \frac{A_0^2}{2\sigma^2} \tag{6.95}
$$

denotes the ratio between the power of the line-of-sight path component and the power of the Rayleigh component. As $K \to 0$, since $I_0(0) = 0$, the Rice distribution approaches a Rayleigh as the line-of-sight component becomes negligible. On the other hand, as $K \to \infty$, the Rice distribution tends to a Gaussian.

6.2 Consider the case in which the stochastic Green's function is white Gaussian, namely

$$
S_{\mathsf{G}}(\omega, \tilde{\omega}) = N\delta(\omega - \tilde{\omega}). \tag{6.96}
$$

Show that the Karhunen–Loève expansion consists of a sequence of i.i.d. Gaussian random variables of variance equal to N.

6.3 Widom (1964) provides several results that are useful in relating the form of the autocorrelation and its concentration properties to the behavior of the eigenvalues in (6.51). Consider the stochastic process $\mathsf{G}(t)$ defined in $[-1, 1]$ with the exponential correlation model

$$
s_{\mathsf{G}}(t, t') = k(\tau) = \frac{1}{2} \exp(-|\tau|/\ell), \tag{6.97}
$$

where $\tau = t - t'$. Use Theorem 1 in Widom (1964) to determine the decay of the eigenvalues of the Fredholm integral equation,

$$
\int_{-1}^{1} k(t - t')\psi(t')dt' = \lambda\psi(t). \tag{6.98}
$$

6.4 Consider the stochastic process $\mathsf{G}(t)$ defined in $[-1, 1]$ with the Gaussian correlation model

$$
s_{\mathsf{G}}(t, t') = k(\tau) = \frac{1}{2} \exp(-\tau^2/\ell^2), \tag{6.99}
$$

where $\tau = t - t'$. Use Theorem 3 in Widom (1964) to determine the decay of the eigenvalues of the Fredholm integral equation,

$$
\int_{-1}^{1} k(t - t')\psi(t')dt' = \lambda\psi(t), \tag{6.100}
$$

and compare the results with the ones in Problem 6.3. Which of the two models suggests a larger number of degrees of freedom?

6.5 Fix the value of ℓ such that the quantity

$$c = \frac{\int_0^\infty \tau k(\tau) d\tau}{\int_0^\infty k(\tau) d\tau} \tag{6.101}$$

is the same for the two models in Problems 6.3 and 6.4. This quantity is a measure of the average spread of the correlation function. Investigate numerically the decay of the eigenvalues and the number of degrees of freedom of the two families of processes. How does this change as c decreases?

7 Communication Technologies

Technology and the machine resurrected San Francisco while Pompeii still slept in her ashes.[1]

7.1 Applications

The theory developed in the previous chapters provides an information-theoretic description of electromagnetic waves in terms of their effective dimensionality. Natural applications of this theory are in communication and sensing systems. In communication systems, waves are used to carry information from transmitters to receivers. In remote sensing systems, they are used to extract information from the environment, probing the surrounding space. When viewed in appropriate asymptotic regimes, the amount of information that waves can carry is limited, and their effective dimensionality, expressed in terms of the number of degrees of freedom, depends on the Green's function that describes the input–output relationship of the system. Only a limited number of orthogonal communication channels can be established between transmitters and receivers, or a limited number of coordinates need to be specified to identify any sensed object using electromagnetic radiation. This limitation is a basic result imposed by the physics of propagation, and mathematically described by the theory of spectral concentration of bandlimited functions.

In recent decades, many technologies have been developed to best exploit the available number of degrees of freedom of electromagnetic waveforms. In this chapter, we focus on some of the communication aspects, showing how different technologies are subject to information-theoretic limits.

7.2 Propagation Effects

We start by considering two ideal canonical cases. In the first case, communication occurs by transmitting a sequence of closely spaced short pulses in a fixed time interval of size T, spread over a large frequency bandwidth Ω. In this case, the total number of degrees of freedom of the signal grows as $\Omega T/\pi$ as $\Omega \to \infty$, and increasingly shorter pulses can be more closely packed together – see Figure 7.1(a). In the second case,

[1] Silas Bent (1930). *Machine Made Man*, Farrar and Rinehart, p. 326.

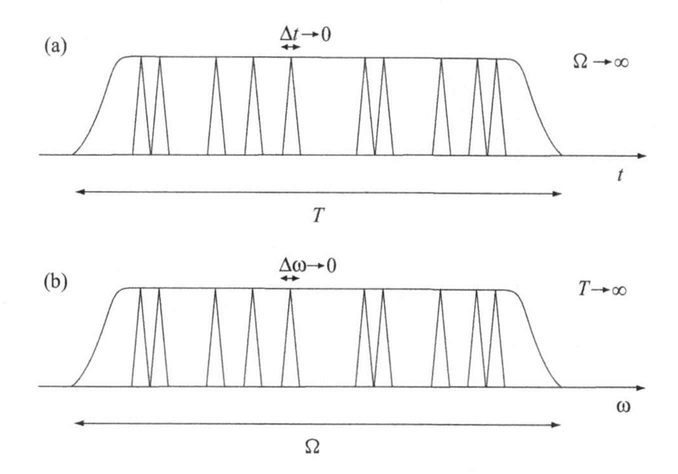

Fig. 7.1 (a) Wide-band transmission. (b) Narrow-band transmission.

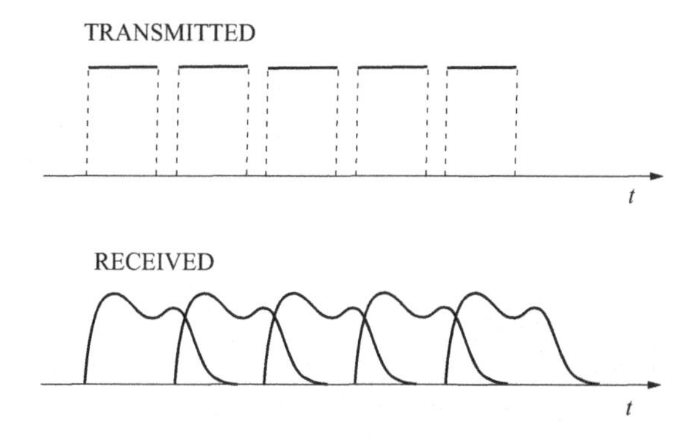

Fig. 7.2 Time broadening and distortion causes inter-symbol interference in wide-band transmission. The time pulses are broadened at the receiver and partially overlap.

communication occurs by transmitting a sequence of narrow frequency pulses in a fixed bandwidth Ω, spread over a long time interval of size T. In this case, the total number of degrees of freedom of the signal grows as $\Omega T/\pi$ as $T \to \infty$ – see Figure 7.1(b). These two basic strategies are employed by communication systems, depending on whether their operation uses *wide-band* or *narrow-band* transmission.

When propagation occurs in a multiple scattering environment, these strategies are affected by environmental limitations. Because of multiple scattering, multiple delayed copies of each transmitted pulse reach the receiver along different paths carrying different phase shifts, and this results in time broadening and distortion of the received waveform. It follows that communication cannot be performed using a train of arbitrarily concentrated time pulses narrowly separated in time, as their broadened versions overlap at the receiver causing *inter-symbol interference* – see Figure 7.2. This effect can also be explained in the frequency domain by considering the environment as a frequency filter. Each transmitted pulse occupies a large frequency band and is

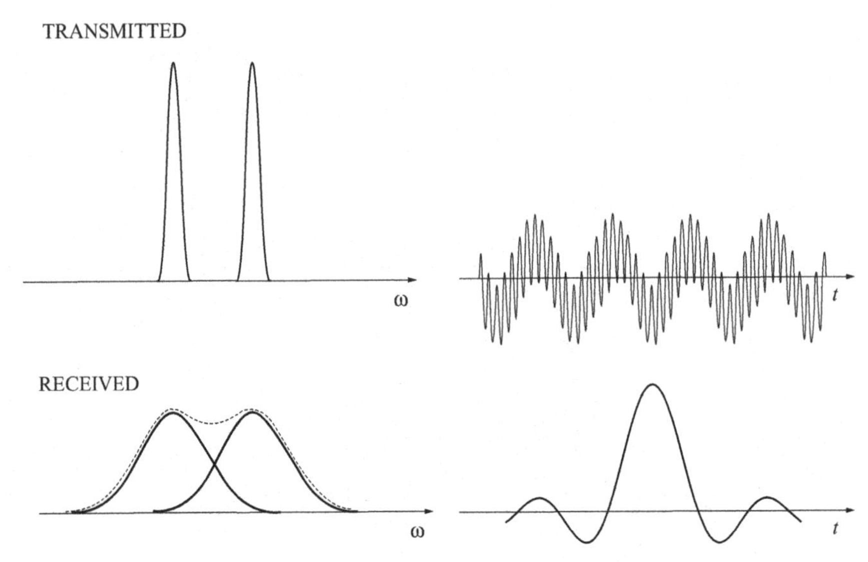

Fig. 7.3 Frequency broadening and distortion causes frequency interference in narrow-band transmission. The frequency pulses are broadened at the receiver and partially overlap.

distorted by the filter, due to multi-path interference. As discussed in Chapter 6 and indicated in Figure 6.7(b), when the bandwidth of a transmitted pulse exceeds the coherence bandwidth of the channel, some frequency components fall into regions of deep fade of the filter's transfer function, and cannot be excited at the receiver. This limits the number degrees of freedom of wide-band transmission, thus limiting the multiplexing capability of any system exploiting this technology. On the other hand, multiple scattering increases the amount of stochastic frequency diversity, as the coherence band decreases and the signal appears to be less correlated over frequency.

Another limitation to the number of degrees of freedom arises for narrow-band communication when the environment is time varying due to mobility of transmitters, receivers, or scatterers. In this case, due to the frequency broadening of the received waveform, communication cannot be performed using highly concentrated narrow-band frequency pulses of arbitrary long time, as the environment acts as a time filtering operator for any given frequency component. The situation is the dual of the time broadening described above – see Figure 7.3. When multiple narrow-band signals are transmitted over the same time interval, the corresponding frequency pulses are broadened by propagation effects and may overlap at the receiver, causing interference in the frequency domain. On the other hand, mobility increases the amount of stochastic time diversity, as the coherence time decreases and the signal appears to be less correlated over time.

Table 7.1 summarizes the different environmental effects. The physics that forms the basis of these effects is described in more detail in the next chapters. Here we concentrate on their influence on system design principles.

Table 7.1 Environmental effects.

	Limits	Increases
Scattering	Wide-band transmission	Frequency diversity
Mobility	Narrow-band transmission	Time diversity

Physical limitations also occur in space. When communication occurs in an arbitrary scattering environment using spatially distributed antennas, the physical size of the system composed by the antennas and the scatterers limits the space–wavenumber degrees of freedom. This effect is studied in detail in Chapter 8.

7.2.1 Multiplexing

Multiplexing different streams of information from different users can be implemented in frequency, time, or space, by dividing the frequency band into small narrow-band portions, the time axis into different time slots, or by sharing the spatial bandwidth among multiple transmitting and receiving antennas. The first case exploits narrow-band communication by dividing the frequency band into small portions allocated to the different users. Since narrow-band signals have longer time durations, they are less affected by time spread due to multiple scattering as the overlap between successive signals is typically a small fraction of the signal's length. In the language of communications, inter-symbol interference is typically negligible. On the other hand, they are more sensitive to frequency spread due to mobility, which can generate significant interference between the narrow-band carriers. The second case divides time into small portions; it is less affected by frequency spread, but is more sensitive to time spread. The third case divides space, and is limited by the spatial spread of the signal.

The three cases correspond to the three different operating regimes depicted in Table 7.2. Frequency multiplexing is exploited in the large time-length regime, when orthogonal frequency signals are multiplexed, each occupying a narrow frequency band. As more and more users transmit using shorter and shorter frequency pulses, their signals are spread over a longer time interval. Time multiplexing is exploited in the wide-band regime, when orthogonal time signals are multiplexed, each occupying a short time interval. As more and more users transmit using shorter and shorter time pulses, their signals are spread over a longer frequency interval. Letting W be the bandwidth in the wavenumber domain, spatial multiplexing is exploited in the large wavenumber band regime, when orthogonal signals are multiplexed in the wavenumber domain, each concentrated in a small spatial interval.

All multiplexing strategies are based on the idea of diagonalizing the channel into parallel time–frequency or space–wavenumber subchannels using appropriate basis functions. As discussed in Chapter 5, the optimal strategy yielding the largest number of parallel channels would require solving an eigenvalue problem for the Green's function at hand. This is not possible in practice, due to incomplete knowledge of the complex scattering environment and its variability over time and space. For this reason, ad hoc strategies have been developed based on different choices of the basis functions used to

Table 7.2 Multiplexing limitations in different regimes.

Regime	Multiplexing	Limitation
$T \to \infty$	Frequency	Frequency spread
$\Omega \to \infty$	Time	Time spread
$W \to \infty$	Space	Spatial spread

perform the decomposition. Each strategy is affected by different trade-offs, and their performance depends on the features of the environment where propagation occurs.

7.2.2 Diversity

The trade-off between diversity and multiplexing is one of the most important concepts in modern communication systems. The degrees of freedom of electromagnetic signals can be exploited for communication by either multiplexing different streams of information from different users, or by improving the reliability of transmission and mitigating the random effects of the channel variations. In the first case, independent signals are sent through the channel to maximize the total information flow. In the second case, correlated signals are sent to achieve reliability by exploiting the stochastic diversity available in the channel. The idea to exploit diversity is that rather than making the success of a transmission entirely dependent on a single realization of the channel, it is possible to hedge the transmission's success across multiple realizations in order to decrease the probability of communication failure due to the signal experiencing a deep fade. A simple example of this is repetition. Consider propagation in a time-varying multiple scattering environment. The electromagnetic signal is affected by random fading occurring in time, frequency, and space. By transmitting the same signal over multiple frequency bands, multiple time slots, or multiple antennas, and if the realizations of the channel are independent over frequency, time, or space, then the probability that at least one transmission does not occur in a deep fade can be made close to one. Of course, this comes at the expense of consuming degrees of freedom, reducing the multiplexing capability of the system. To minimize this consumption, one may note that requiring multiple independently faded copies of the same signal is clearly an overly stringent requirement to achieve reliability. Enter coding: instead of repeating the same signal at different time, frequency, or space blocks, one can encode the signal using an error-correcting code and transmit it over multiple realizations of the channel. In this case, correlated signals are transmitted through the channel providing the desired reliability against fading without sacrificing excessive amounts of resources, as in the case of repetition.

The frequency spread that limits the performance of frequency division technologies introduces time diversity, while the time spread that limits the multiplexing performance of time division technologies introduces frequency diversity. It follows that frequency multiplexing transmission occurring over multiple carriers can exploit time diversity, improving the reliability of transmission through coding over time. Time multiplexing transmission occurring over multiple time slots can exploit frequency diversity, improving reliability of transmission through coding over frequency. Similarly, spatial diversity, occurring when multiple transmitting and receiving antennas are sufficiently spaced apart in a scattering environment, can also be exploited for reliability by coding over space. As a simple example, if two receive antennas are sufficiently spaced apart, the same signal is received over independently faded paths and this squares the probability of experiencing a deep fade.

7.3 Overview of Current Technologies

Communication systems aim to achieve the maximum multiplexing of information, while guaranteeing reliability of transmission. In the following, we restrict our discussion to the fundamental principles of some of the current communication technologies. The interested reader may refer to the wide range of existing literature for a more in-depth technological perspective.

7.3.1 OFDM

The orthogonal frequency division multiple access technology (OFDM) used for broadband internet service and cellular systems in the United States is based on the superposition of many narrow-band signals tightly packed in frequency – see Figure 7.4. These signals partially overlap in the frequency domain, and to control the interference between them the amount of overlap is chosen so that the peak of one signal occurs at the null of the other ones – from which comes the name orthogonal frequency division. The different frequencies are then assigned to different users, which can access the channel simultaneously. User mobility means that OFDM is sensitive to frequency spread, as this can cause loss of orthogonality between signals and interference between the different narrow-band signals. On the other hand, mobility introduces time diversity, and each narrow-band signal experiencing independent time variations allows users to perform coding across time blocks to improve reliability of transmission.

7.3.2 MC-CDMA

In the presence of frequency diversity due to multiple scattering, coding in OFDM systems can also occur over frequency by assigning multiple frequencies to each user. This is an example of the trade-off between multiplexing and reliability, as by coding over different frequencies each user consumes degrees of freedom to provide reliability of transmission rather than using them for multiplexing signals from different users. An extreme case of coding over frequencies is multiple carrier code division multiple access (MC-CDMA), in which every user has access to all frequency channels, encodes its signal across all frequencies to keep it distinct from the other users, and then spreads it over time by transmitting it over the multiple narrow-band channels.

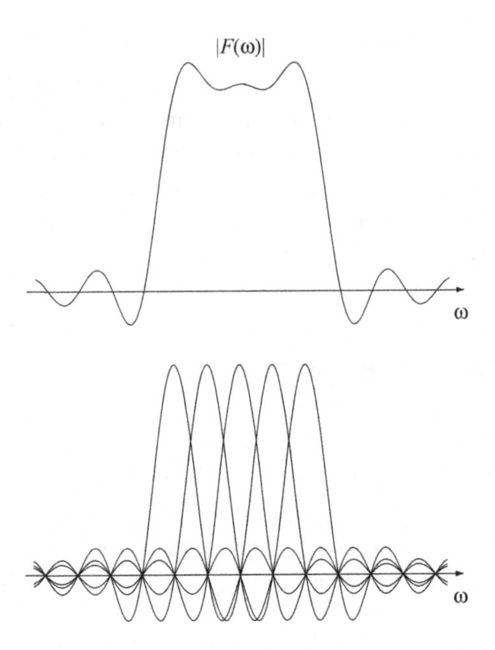

Fig. 7.4 Representation of the OFDM spectrum as a sum of orthogonal narrow-band components.

Users in MC-CDMA systems modulate individual codes to carry their signals. The coding is chosen so that the resulting signals are well separated; ideally, the correlation between the encoded signals is zero. Discrimination among the received signals can then be made by correlating the received signal with the locally generated code of the desired user. If the signal matches the desired user's code then the correlation is high and the system can extract that signal. Otherwise, the correlation is close to zero and the system can reject that signal.

7.3.3 GSM

The global system for mobile communication (GSM), widely used for cellular systems around the world and especially in Europe, is based on orthogonal time division multiple access technology (TDMA), which divides the spectrum into different portions and allocates each portion to different users at different times. Thus, users transmit waveforms occupying the same portion of the spectrum in orthogonal time slots. This technology is sensitive to time spread due to multiple scattering, which can cause excessive overlap and interference between signals in time. On the other hand, multiple scattering introduces frequency diversity, and users can hop between different spectrum allocations to improve reliability of transmission.

7.3.4 DS-CDMA

In the presence of time diversity, coding in TDMA systems can also occur over time by assigning multiple time slots to each user. This is an example of the trade-off between

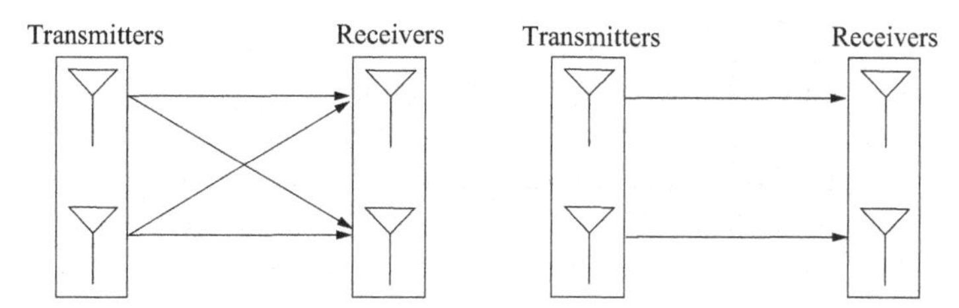

Fig. 7.5 Representation of 2×2 MIMO. The four channels can be diagonalized into two parallel channels providing multiplexing gain, or they can be used for redundant transmissions providing reliability gain.

multiplexing and reliability, as by coding over different time slots each user consumes degrees of freedom to provide reliability of transmission rather than using them for multiplexing signals from different users. An extreme case is direct sequence code division multiple access (DS-CDMA). This is the time dual of MC-CDMA, and is based on spread spectrum transmission. Every user has access to the channel at all times, encodes its signal over time to keep it distinct from the other users, and then spreads it over frequency by transmitting it over the entire spectrum. Reliability of transmission over the frequency-varying channel is obtained by choosing codes that, when correlated with a delayed version of the transmitted signal, also result in a value close to zero. Because the delayed versions of the transmitted codes will have poor correlation with the original code, they will appear as another user, which is ignored at the receiver. In other words, the correlation properties of the codes are such that as long as the multiple scattering introduces a large enough delay among the received signals, these appear uncorrelated with the intended signal, and they are thus ignored.

7.3.5 MIMO

When communication occurs using multiple antennas, the available spatial degrees of freedom can be exploited either by multiplexing signals over independent spatial channels or by improving the reliability of transmission by sending redundant signals through coding to mitigate random spatial channel variations. Figure 7.5 provides a schematic representation of a 2×2 multiple-input multiple-output (MIMO) system. The four channels between transmitter and receiver can be diagonalized into two parallel channels supporting independent streams of information, as described in Section 5.3.2. This requires pre-processing and post-processing operations as depicted in Figure 5.9. Each transmitter antenna sends a linear combination of all input signals through the channel, and these are jointly decoded at the receivers. Alternatively, the four channels can be used for redundant signaling from each transmitting antenna to both receiving antennas. If the channel experiences independent realizations across space, then appropriate coding schemes can be designed to maximize the probability of successful communication despite the random channel realizations.

As for time–frequency transmission technologies, for MIMO there is also an inherent trade-off between multiplexing and diversity. To increase multiplexing, independent signals must be sent over parallel channels. To increase reliability, redundancy needs to be introduced through coding to mitigate the channel variations so that correlated signals are sent over independent realizations of the channel. In the simplest case of repetition coding, the same signal is sent over multiple realizations so that the multiplexing gain is zero, and channel diversity is fully exploited for reliability. With more sophisticated coding schemes, it is possible to trade-off between the two.

7.4 Principles of Operation

The leitmotiv of all digital communication technologies is that they rely on an orthogonal representation of the channel using appropriate basis functions, and then use this representation to provide multiplexing and reliability. Due to multiple scattering and incomplete knowledge of the Green's function, an optimal decomposition providing the largest possible number of parallel channels cannot be performed, and ad hoc basis functions are used in practice that work well under certain assumptions on the channel. Nevertheless, the basic principle remains the same, and the degrees of freedom view developed in the previous chapters can be used to provide upper bounds on the multiplexing capability that can be achieved in practice.

A block diagram for the operation of time–frequency decomposition technologies, such as OFDM, TDMA, and CDMA, is depicted in Figure 7.6. One for the operation of space–wavenumber decomposition technologies, such as MIMO, is shown in Figure 7.7. In these diagrams, the basis functions used for the representation of the input and output signals are generally not the optimal ones arising from the Hilbert–Schmidt decomposition of the Green's function. In the case of the Hilbert–Schmidt representation, the effect of the channel would simply be that of multiplying each coefficient x_n by the singular value $\sqrt{\lambda_n}$ of the channel's decomposition, as indicated in Figure 5.10, and in its discrete equivalent for a spatial channel – Figure 5.9. When a different choice of basis functions is performed, the effect on the channel and the multiplexing capability of the system depend on how well the given representation matches the channel's conditions.

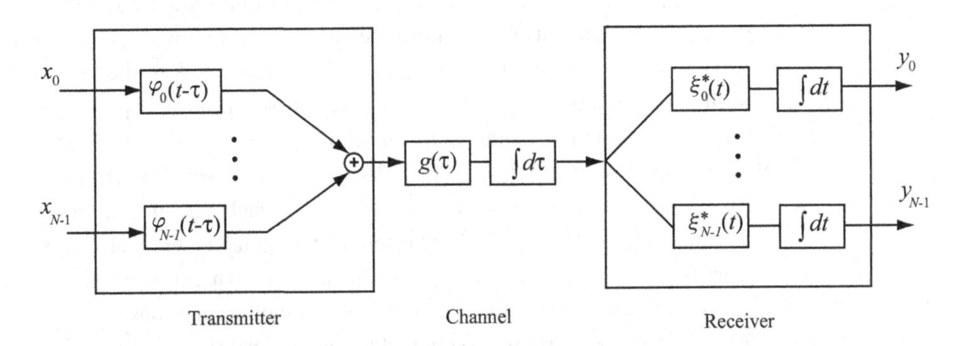

Fig. 7.6 Block diagram of continuous time orthogonal time–frequency decomposition technology.

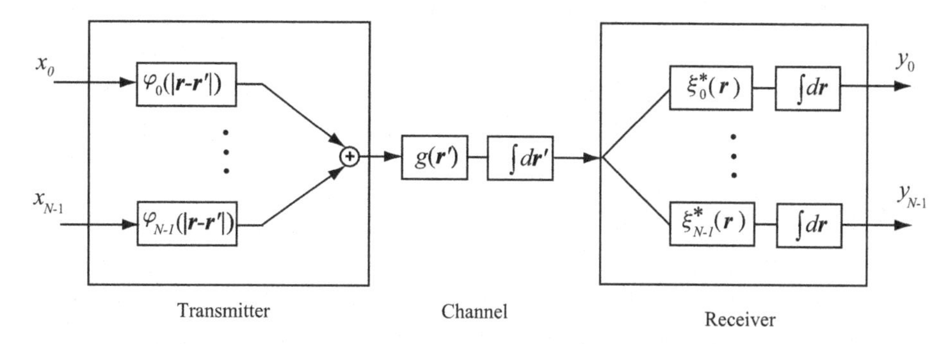

Fig. 7.7 Block diagram of continuous space orthogonal space–wavenumber decomposition technology.

The basis functions for the OFDM technology correspond to N narrow-band orthogonal signals, each carrying a symbol x_m from the mth user. The basis functions for the TDMA technology correspond to N non-overlapping time pulses representing the time allocation slots for the different users, each carrying a symbol x_m from the mth user. The basis functions for the DS-CDMA technology correspond to N wide-band orthogonal signals representing code sequences spread over the entire spectrum, each carrying a symbol x_m for mth user. Finally, in the case of MC-CDMA, the N narrow-band basis functions of the OFDM signal are employed by each user by first encoding the signal in frequency using a wide-band code, and then sending it over the N narrow-band subcarriers. For space–wavenumber technologies, the basis functions for the MIMO system correspond to the columns of the pre-processing and post-processing matrices performing modulation of the signal in space.

These orthogonal representations effectively divide the time, frequency, or space spectrum among the different users, and also allow for coding across the spectrum. To clarify the basic principles and the fundamental limitations of their operation, we consider three examples in more detail. The first example is OFDM. In this case, users transmit over different portions of the frequency spectrum, sharing the time–frequency degrees of freedom provided by the electromagnetic channel. The second example is CDMA. In this case, users transmit over the whole spectrum, but again the effective bandwidth available to each user is only a portion of the entire spectrum. Due to coding, only a portion of the transmitted bits are effectively used as data by each user, so that the total spectral resource is again shared among the different users. The multiplexing capability of the system in this case depends on the number of orthogonal codes that can fit in the given spectrum and it is again limited by the time–frequency degrees of freedom provided by the physical channel. Finally, the same basic principles are applied to MIMO, where rather than the time–frequency degrees of freedom, the resource divided among the users is represented by the space–wavenumber degrees of freedom.

7.4.1 Orthogonal Spectrum Division

Consider an OFDM system with N subcarriers, an (approximate) angular frequency band Ω and an (approximate) signal duration T. The first part of the signal, of duration

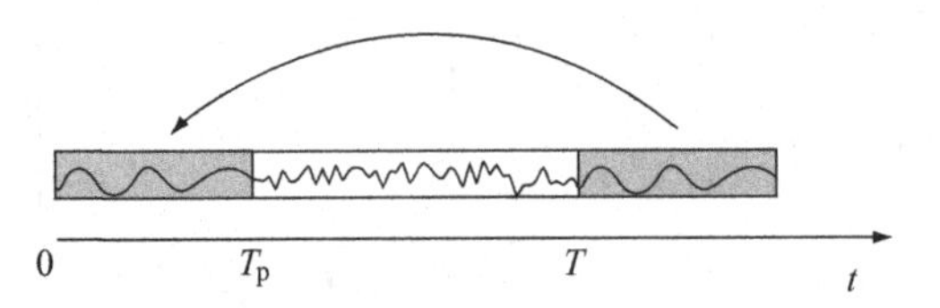

Fig. 7.8 Cyclic prefix for an OFDM signal.

T_p, is the *cyclic prefix* that is a copy of the last part of the signal that is prepended to the transmitted one – see Figure 7.8. The signal transmitted in one transmission period of length T is

$$x(t) = \sum_{m=0}^{N-1} x_m \varphi_m(t), \tag{7.1}$$

where $\{x_m\}$ are points from the signal's constellation and $\{\varphi_m\}$ are functions orthonormal in $[T_p, T]$:

$$\varphi_m(t) = \begin{cases} \sqrt{\dfrac{\Omega}{\pi N}} \exp\left[jm\dfrac{\Omega}{\pi N}(t - T_p)\right] & \text{if } t \in [0, T], \\ 0 & \text{if } t \notin [0, T], \end{cases} \tag{7.2}$$

where

$$T = \frac{\pi N}{\Omega} + T_p. \tag{7.3}$$

In a multi-user system, the interpretation is that the OFDM signal uses N subcarriers, each carrying a symbol x_m from the mth user in a time period of length T. Note that for $T \in [0, T_p]$, we have $\varphi_m(t) = \varphi_m(t + \pi N/\Omega)$.

We now make two assumptions about the channel. First, we assume that the channel is time invariant for the duration of the transmission period. This corresponds to a block fading model where the duration of each block is synchronized with the duration of the transmitted signal. Second, we assume that the length of the cyclic prefix is larger than the support of the impulse response of the channel, so that subsequent signals that are sent through the channel and are affected by time spread due to multi-path, significantly overlap only within the cyclic prefix, which can be ignored at the receiver. With these assumptions, we can write the signal at the receiver as

$$\begin{aligned} y(t) &= \int_0^{T_p} x(t-\tau) g(\tau) d\tau \\ &= \int_0^{T_p} \sum_{m=0}^{N-1} x_m \varphi_m(t-\tau) g(\tau) d\tau. \end{aligned} \tag{7.4}$$

For all $m \in [0, N-1]$ and $t \in [T_p, T]$, we have

$$\int_0^{T_p} \varphi_m(t-\tau) g(\tau) d\tau = \int_0^{T_p} \sqrt{\frac{\Omega}{\pi N}} \exp\left[jm\frac{\Omega}{\pi N}(t - \tau - T_p)\right] g(\tau) d\tau$$

$$= \varphi_m(t) \int_0^{T_p} g(\tau) \exp\left[-jm\frac{\Omega}{\pi N}\tau\right] d\tau$$

$$= G\left(\frac{m\Omega}{\pi N}\right) \varphi_m(t), \tag{7.5}$$

where $G(m\Omega/(\pi N))$ is the sampled frequency response of the channel at the mth subchannel frequency $\omega = m\Omega/(\pi N)$. By combining (7.4) and (7.5) and letting $G_m = G(m\Omega/(\pi N))$, it follows that

$$y(t) = \sum_{m=0}^{N-1} G_m x_m \varphi_m(t), \tag{7.6}$$

and the received signal can be viewed as the superposition of different complex exponentials $\{\varphi_m(t)\}$, spaced in frequency by $\Omega/(N\pi)$, and each weighted by the sampled frequency response.

The nth receiver can now recover the nth input coefficient of the signal, ignoring the cyclic prefix, by computing

$$y_n = \int_{T_p}^{T} y(t)\varphi_n^*(t)dt$$

$$= \int_{T_p}^{T} \int_0^{T_p} \sum_{m=0}^{N-1} x_m \varphi_m(t-\tau) g(\tau) d\tau \, \varphi_n^*(t) dt$$

$$= \sum_{m=0}^{N-1} G_m x_m \int_{T_p}^{T} \varphi_m(t)\varphi_n^*(t)dt$$

$$= G_n x_n. \tag{7.7}$$

The input–output representation in terms of N parallel frequency channels is then

$$y_n = G_n x_n, \quad n = 0, \ldots, N-1, \tag{7.8}$$

and should be compared with the result for the Hilbert–Schmidt orthogonal representation (5.54). By chosing an orthogonal representation where the basis functions are the complex exponentials, the effect of the channel is to multiply each coefficient of the representation by a sampled value of the Green's function in the frequency domain, rather than multiplying it by a singular value of the Green's operator. With the assumptions made on the channel's response, the output at each subcarrier is centered at the same frequency as the input, so that there is no inter-carrier interference in the frequency domain. In addition, the problem of interference between successive signals in the time domain is solved using the cyclic prefix. Thanks to the prefix, the technology is not affected by time spread, as long as the length of the prefix is large enough.

On the other hand, since the cyclic prefix is ignored at reception, this causes the channel to be idle for some time. From (7.3), we have that the number of orthogonal channels that are used is

$$N = \frac{\Omega(T - T_p)}{\pi}, \tag{7.9}$$

causing a loss in degrees of freedom of $\Omega T_{\mathrm{p}}/\pi$. In order to minimize this overhead, one would like to have

$$N \gg \frac{\Omega T_{\mathrm{p}}}{\pi}. \tag{7.10}$$

Letting T_{c} be the coherence time of the channel, the transmission time T over which the channel can be considered constant is necessarily

$$T \ll T_{\mathrm{c}}. \tag{7.11}$$

Letting T_{s} be the time spread of the channel, it must also be that the length of the cyclic prefix is

$$T_{\mathrm{p}} \gg T_{\mathrm{s}}. \tag{7.12}$$

It then follows, by combining (7.9), (7.10), (7.11), and (7.12), that the OFDM technology performs well when

$$\frac{\Omega T_{\mathrm{s}}}{\pi} \ll N \ll \frac{\Omega T_{\mathrm{c}}}{\pi}, \tag{7.13}$$

namely when the coherence time of the channel is much larger than the delay spread, so that the number of subcarriers can be chosen large enough to compensate for the degrees of freedom loss due to the cyclic prefix, but still smaller than the degrees of freedom offered by the channel during the time it can be considered constant. To better understand this latter limitation, recall that as N grows larger and more and more pulses are tightly packed in frequency, the transmission time of each signal over each subcarrier also grows. During this long transmission period the channel may not be considered constant anymore, and this leads to the limitation on the right-hand side of (7.13). The analogous frequency interpretation is that the frequency spread does not allow the transmission of frequency pulses arbitrarily close to each other.

7.4.2 Orthogonal Code Division

Consider now a DS-CDMA system with N codes, an (approximate) angular frequency band Ω, and a signal of (approximate) duration T. The principle of operation represented by the OFDM block diagram in Figure 7.6 still holds. A signal $x(t)$ is constructed and sent through the channel over an interval of time T:

$$x(t) = \sum_{m=0}^{N-1} x_m \varphi_m(t). \tag{7.14}$$

However, in this case the basis functions $\{\varphi_m\}$, rather than being narrow-band carriers centered across the spectrum as depicted in Figure 7.4, are fast-varying orthonormal code sequences, as depicted in Figure 7.9, each occupying the whole bandwidth Ω. The signal $\varphi_m(t)$ is used to transmit a single coefficient x_m over time T using κ short pulses each of time length T_0, called "chips" in CDMA jargon. The spreading factor κ is called the *processing gain* of the system. As κ grows larger, and more and more chips are closely packed together in time, the bandwidth Ω of the transmitted code sequence

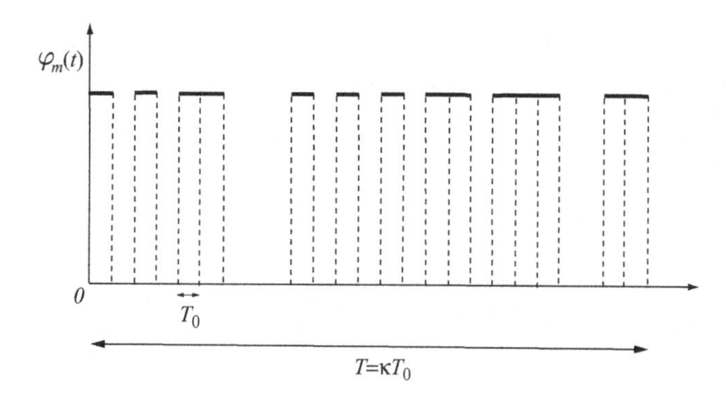

Fig. 7.9 Direct sequence CDMA spreading waveform.

becomes of the order of $1/T_0$, and the time–bandwidth product of $\varphi_m(t)$ becomes of the order of κ. It follows that

$$\kappa = \frac{T}{T_0} \approx \frac{\Omega T}{\pi} \tag{7.15}$$

roughly represents the number of degrees of freedom of the electromagnetic signal in terms of the total time–frequency resource available. For any given length of transmission T, the number of degrees of freedom is increased by transmitting chips of shorter duration T_0, which corresponds to increasing the bandwidth Ω of the transmitted signal. On the other hand, for any given bandwidth Ω, the number of degrees of freedom is increased by transmitting a larger number of chips, increasing the length of transmission T. As in the OFDM case, both of these strategies are limited by the properties of the channel: the time spread does not allow T_0 to be arbitrarily small, and also the channel cannot be considered time invariant for an arbitrarily long duration interval T.

If orthogonality is preserved when the resulting signal is passed through the channel, the transmission coefficients can be recovered at the receiver in the usual way by correlation with the appropriate coding sequence. In free space, the channel only provides delay and attenuation, so that if proper synchronization can be maintained between transmitter and receiver, orthogonality is preserved. However, in the presence of multi-path, multiple copies of the transmitted codeword may reach the receiver. In this case, appropriate codes are required that are not only pair-wise orthogonal, to discriminate among users, but that when correlated with their delayed version received through multi-path propagation also result in a value close to zero. In the case that pseudo-random spreading sequences are used, this amounts to using sequences whose autocorrelation at the receiver is almost impulsive, so that they appear almost as white noise and can be discarded.

Assuming a multi-path model where on each path the received signal is a delayed and attenuated version of the transmitted signal, we have

$$y(t) = \sum_{m \in \mathscr{P}} a_m x(t - \tau_m), \tag{7.16}$$

where $\mathscr{P}$ is the set of multiple scattered paths, a_m is the attenuation along the mth path, and τ_m is the delay along the mth path. The corresponding Green's function is

$$g(t) = \sum_{m \in \mathscr{P}} a_m \delta(t - \tau_m). \tag{7.17}$$

We assume that the coefficients $\{a_m\}$ do not vary during each transmitted block of κ chips. This corresponds to having a channel that is invariant for the duration of each transmitted signal, so that $T \ll T_c$ and, by virtue of (7.15),

$$\kappa \ll \frac{\Omega T_c}{\pi}. \tag{7.18}$$

We also assume that the Green's function (7.17) falls off rapidly, so that the overlap between subsequent signals affected by time spread due to multi-path is negligible. In this way, we can consider the response to each transmitted signal $x(t)$ in isolation. This corresponds to having $T \gg T_s$ and, by virtue of (7.15),

$$\kappa \gg \frac{\Omega T_s}{\pi}. \tag{7.19}$$

Notice that the constraints (7.18) and (7.19) are equivalent to (7.13), and we are again in the case of an *underspread* channel, where the coherence time is significantly larger than the delay spread. The number of different frequencies N used to transmit the OFDM signal over the frequency band Ω corresponds to the number of different chips κ used to transmit the CDMA signal over time T, so that κ plays the role of the multiplexing capability of the system, and in (7.14) we let $N < \kappa$ to ensure the orthogonality of all the transmitted coding sequences.

With these assumptions, we can write the signal at the receiver as simply the convolution

$$y(t) = \int_{-\infty}^{\infty} x(\tau) g(t - \tau) d\tau. \tag{7.20}$$

We also have the cardinal series expansion for the bandlimited signal $x(t)$:

$$x(t) = \sum_{m=-\infty}^{\infty} x\left(\frac{m\pi}{\Omega}\right) \operatorname{sinc}(\Omega t - m\pi). \tag{7.21}$$

By substituting (7.21) into (7.20) and using the Fourier transform pair

$$\operatorname{sinc}(\Omega t) \longleftrightarrow \frac{\pi}{\Omega} \operatorname{rect}\left(\frac{\omega}{2\Omega}\right), \tag{7.22}$$

we have

$$\begin{aligned} y(t) &= \frac{1}{2\pi} \sum_{m=-\infty}^{\infty} \int_{-\Omega}^{\Omega} G(\omega) x\left(\frac{m\pi}{\Omega}\right) \frac{\pi}{\Omega} \operatorname{rect}\left(\frac{\omega}{2\Omega}\right) \exp\left[j\omega\left(t - \frac{m\pi}{\Omega}\right)\right] d\omega \\ &= \frac{\pi}{\Omega} \sum_{m=-\infty}^{\infty} g\left(t - \frac{m\pi}{\Omega}\right) x\left(\frac{m\pi}{\Omega}\right) \\ &= \frac{\pi}{\Omega} \sum_{m=-\infty}^{\infty} x\left(t - \frac{m\pi}{\Omega}\right) g\left(\frac{m\pi}{\Omega}\right). \end{aligned} \tag{7.23}$$

Letting

$$g_m = (\pi/\Omega)g(m\pi/\Omega), \tag{7.24}$$

we have

$$y(t) = \sum_{m=-\infty}^{\infty} x(t - m\pi/\Omega)g_m. \tag{7.25}$$

It follows that the received signal can be seen as the discrete superposition of different time components, spaced in time by π/Ω, and each weighted by the sampled Green's function. This should be compared with (7.6), which provides the analogous frequency domain interpretation for the OFDM case, where the signal is seen as the superposition of different frequency components, spaced in frequency by $\Omega/(N\pi)$.

By retaining only the indexes corresponding to the most significant terms in the series, the corresponding Green's function can be written as

$$g(t) = \sum_{m \in \mathscr{M}} g_m \delta(t - m\pi/\Omega), \tag{7.26}$$

and, comparing it with the multi-path model in (7.17), we deduce that a signal of bandwidth Ω achieves a multi-path delay resolution of π/Ω. We can then model the received signal with a tapped finite delay line with tap spacing π/Ω and tap coefficients g_m – see Figure 7.10. The number of taps $|\mathscr{M}|$ is in practice bounded by $T_s\Omega/\pi \ll \kappa$. The corresponding physical picture is given in Figure 7.11.

If the coding sequences remain orthogonal when subject to multi-path delay, then the nth transmitted coefficient can be recovered by correlating the received signal with the nth spreading waveform, and we have

$$\begin{aligned} y_n &= \int_0^T y(t)\varphi_n(t)dt \\ &= \int_0^T \sum_{m \in \mathscr{M}} x(t - m\pi/\Omega)g_m\varphi_n(t)dt \\ &= \sum_{k=0}^{N-1} \sum_{m \in \mathscr{M}} x_k g_m \int_0^T \varphi_n(t)\varphi_k(t - m\pi/\Omega)dt \\ &= x_n \sum_{m \in \mathscr{M}} g_m, \end{aligned} \tag{7.27}$$

and the input–output representation in terms of N parallel channels is

$$y_n = x_n \sum_{m \in \mathscr{M}} g_m, \quad n = 0, \ldots, N-1. \tag{7.28}$$

Comparing (7.28) and (7.8), it follows that the channel's weight corresponding to one value of the frequency response in the case of OFDM now corresponds to the sum of different samples of the Green's function in time representing the different multi-path components.

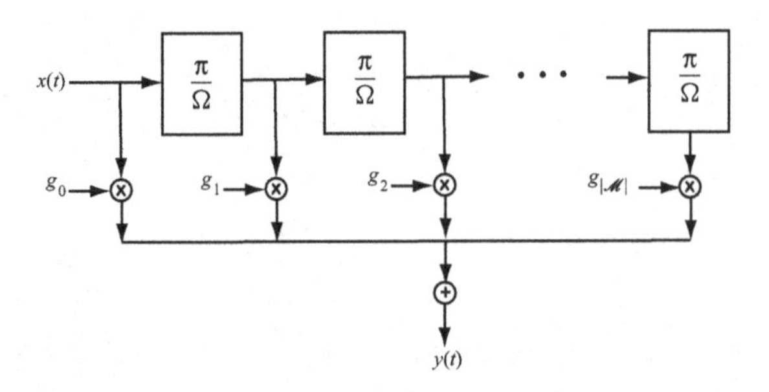

Fig. 7.10 Tapped delay line model of the CDMA channel.

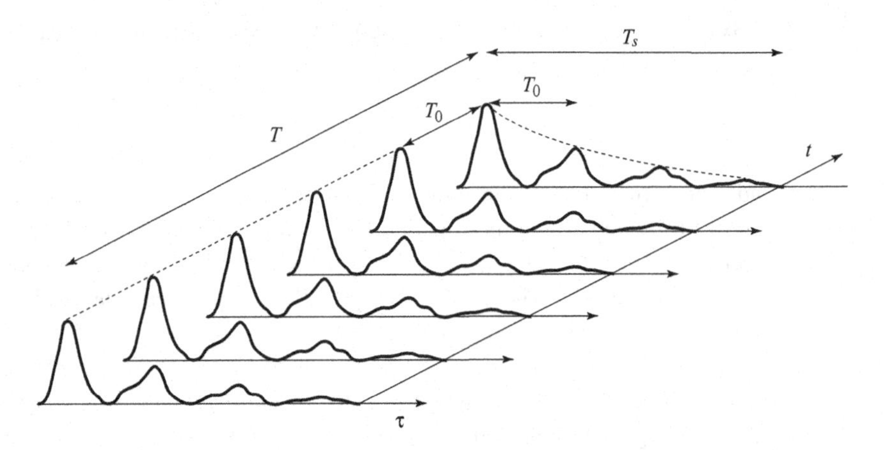

Fig. 7.11 Sequence of chips and corresponding delayed multi-path returns.

7.4.3 Exploiting Diversity

In CDMA, reliability of transmission can be improved by exploiting the frequency diversity provided by the multiple delayed paths. The output can be correlated with different delayed versions of the candidate transmitted sequence φ_n spaced by multiples of π/Ω, which is of the order of the chip time T_0. The result of each correlation provides an estimate for the coefficient x_n; it is weighted by the corresponding channel coefficient, and then summed with all the others. In this way, branches with strong correlations are further amplified, while weak correlations are weighted less – see Figure 7.12. This improves the accuracy of the detection of the desired coefficient x_n, as shown in Problem 7.6.

This type of processing is called a *rake receiver*, and it improves reliability by extracting frequency diversity in an attempt to collect the signal's energy from all the delayed signal paths that fall within the span of the receiver's delay line at resolution π/Ω. With a bit of imagination, its action is somewhat analogous to that of a garden rake, hence its name. Rake processing is effective because transmission of

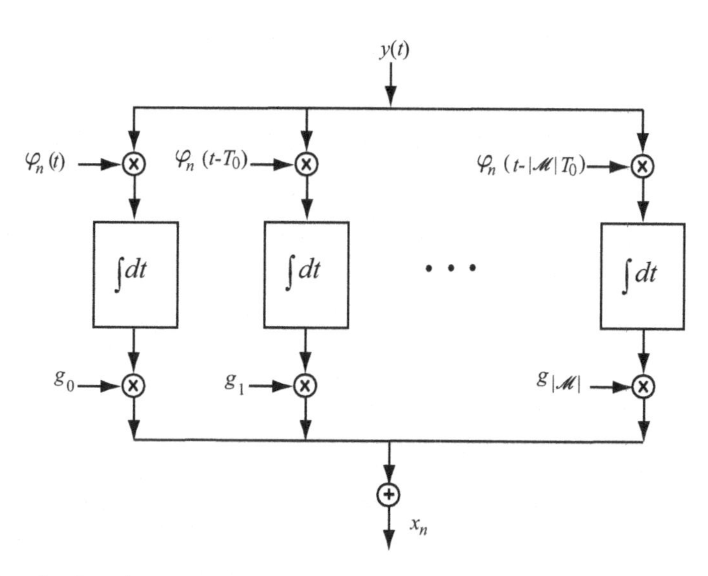

Fig. 7.12 Rake receiver for detection of the mth coefficient of the signal.

each coefficient x_m occurs over a wide-band Ω, allowing for multi-path time resolution of the order of the chip time T_0.

In the case of OFDM, transmission of each coefficient occurs over a narrow frequency band, resulting in a very coarse multi-path time resolution, and a rake receiver is not effective. In this case, frequency diversity for reliability can be exploited by assigning multiple frequencies to each user, which can then perform coding over frequency. Of course, this comes at the expense of consuming degrees of freedom, as correlated coded signals would be transmitted over the channel rather than independent signals from different users.

A similar trade-off between frequency diversity and multiplexing also occurs for CDMA. Achieving higher reliability in the rake receiver requires using a larger number of taps, corresponding to different resolvable paths. This requires having a higher delay spread in the channel, which, by virtue of (7.19), requires a larger number of degrees of freedom.

7.4.4 Orthogonal Spatial Division

Orthogonal frequency division and orthogonal code division technologies share the time–frequency degrees of freedom among the different users of the system. In contrast, technologies using multiple antennas are designed to share the space–wavenumber degrees of freedom. In the time–frequency case, provided that the number of users is smaller than the available time–frequency resource $\Omega T/\pi$, we can allocate an orthogonal channel to each user and transmit over small, non-overlapping time or frequency intervals. In the space–wavenumber case, we can allocate an orthogonal channel to each user, provided that the number of users is smaller than the number of

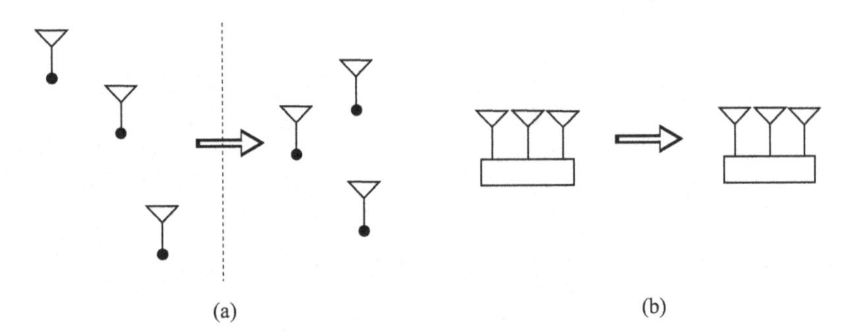

Fig. 7.13 (a) Distributed MIMO. (b) Point-to-point MIMO.

space–wavenumber degrees of freedom WS/π, where W is the wavenumber bandwidth and S is the spatial extension of the domain where the signal is observed. This product corresponds to the number of singular values that are significantly greater than zero in the Hilbert–Schmidt decomposition of the signal in space. In a discrete setting, it corresponds to the rank of the channel's matrix described in Section 5.3.2.

In a MIMO system, the optimal resource allocation is performed by transforming the channel into orthogonal parallel spatial channels using the SVD decomposition. The space–wavenumber degrees of freedom are then divided among the users, much like in the time–frequency case. Establishing these communication channels requires cooperation at the transmitting and receiving antennas. As shown in Figure 5.9, each signal coefficient that is sent through the diagonal channel is a linear combination of all the inputs through the matrix V. Similarly, the received signal coefficients are obtained via a linear combination of all the outputs through the matrix $\mathrm{U}^{\dagger}$. These operations are somewhat easy to perform with little overhead if all the transmitting and receiving antennas are co-located in a single physical device and they can share their input and output signals, but they can be more difficult in the case of spatially distributed systems, where a single transmitter or receiver does not have access to all the signals – see Figure 7.13. This makes the exploitation of the spatial resource more difficult than in the time–frequency case, where each user can transmit and receive signals independently of the others.

A little reflection reveals the core of this problem. Recall that in a spatial setting the coefficients $\{x_n\}$ in Figure 7.7 are multiplied by spatially varying basis functions $\{\varphi_n\}$ and then sent through the channel via a convolution operation that occurs over space. It follows that the effect of each coefficient is present in the field emitted by all the sources. Similarly, the output coefficients $\{y_n\}$ are obtained through a projection operation that requires knowledge of the field at all receiving points. In a distributed system, acquiring this knowledge requires cooperation between the spatially distributed users and must occur for every MIMO transmission. In contrast, each input coefficient in Figure 7.6 only affects the transmitted signal over time, and each output coefficient is obtained through a projection that only requires knowledge of the signal over time. In this case, each user located at a given point in space can independently construct and receive its own signals.

As in the time–frequency case, the number of space–wavenumber degrees of freedom is limited by the physics of propagation. As we shall see in Chapter 8, the wavenumber bandwidth of an arbitrary scattering system depends on the characteristic size of the system, and on the frequency of transmission. This provides an upper bound on the amount of spatial multiplexing that can be achieved using arbitrary technologies and in arbitrary scattering environments. The frequency dependence also shows that the number of space–wavenumber and time–frequency degrees of freedom are related to each other, and the total number of degrees of freedom can be computed by taking into account such dependence. Before examining these physical limitations in detail, we give an overview of some of the proposed techniques to exploit the spatial resource in a distributed network setting.

7.5 Network Strategies

While technologies that exploit the time–frequency resource are well established and widely used in modern communication systems, determining the best way to exploit the spatial resource in a distributed communication system with minimum cooperation overhead is an open problem, and several proposals have been made in the literature. These studies fall into the realm of *network information theory*, which is a research topic that has gained notable momentum in recent years and inspired several innovative engineering designs. Their theoretical performance cannot exceed MIMO systems with co-located transmitters and receivers, which allows full cooperation to encode and decode messages. In this case, for each time–frequency degree of freedom we can have a number of parallel channels at most equal to the number of space–wavenumber degrees of freedom. The proposed strategies attempt to come as close as possible to this limit, to achieve the full multiplexing gain available in the given propagation environment. Which strategy will have the most practical impact, however, has yet to be determined.

7.5.1 Multi-hop

In a spatially distributed system of N users sharing a fixed time–frequency resource without any form of spatial multiplexing, a fraction $1/N$ of the total resource is available to each user. It follows that as $N \to \infty$, the time–frequency degrees of freedom are saturated, and the performance degrades considerably as the number of users increases. This is the case, for example, for the OFDM and CDMA technologies discussed above. The objective of distributed network strategies is to improve this situation by exploiting the additional spatial resource.

A first strategy that exploits spatial reuse is *multi-hop relay*. In this case, routing paths are established between sources and destinations, and point-to-point transmissions are performed at each hop along the paths, treating interference from other nodes as noise. This scheme can be analyzed under a variety of models for the propagation channel and, in a geometric setting in which nodes are scattered in a region of the plane of size proportional to N, sources and destinations are randomly selected, and signals are

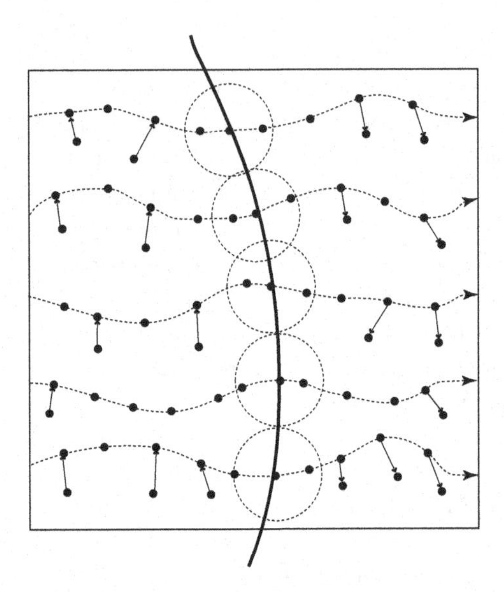

Fig. 7.14 Illustration of the multi-hop strategy exploiting parallel communications across the cut.

strongly attenuated over distances so that the interference from distant users can be neglected, it improves utilization of the resource by a factor $\sqrt{N}$. In other words, a fraction $1/\sqrt{N}$ of the resource is now available to each user, and performance degrades more gracefully compared to the case where the spatial resource is unused.

The result can be explained as follows. Due to the random assignment of sources and destinations, roughly half of the nodes located on one side of the network wish to communicate with the other half located on the other side. The transmission rate that each source can sustain to its intended destination depends on the number of routing paths that can be established simultaneously across the cut that separates these two regions – see Figure 7.14. We can construct these paths by letting each node select, as the next relay, a nearby neighbor located at most a constant distance from it. In this way, each hop transmission generates a small interference footprint around it. If routing paths crossing the cut are sufficiently spaced from each other, so that the corresponding interference footprints do not overlap too much, then the total amount of interference at each hop can be controlled, and the routes can be performed simultaneously, at constant rate. Since the number of nodes accessing these paths is of the order of N, and the number of paths crossing the cut and spaced by at least a constant distance is proportional to the length of the cut, i.e., it is of the order of $\sqrt{N}$, the result follows.

This strategy requires the number of space–wavenumber degrees of freedom to be at least proportional to $\sqrt{N}$. If this is not the case, then the performance is limited by the rank of the channel's matrix between sources and destinations. For example, a keyhole effect, such as the one described in Section 6.3.4, may allow only one interference-free transmission per unit time across the cut.

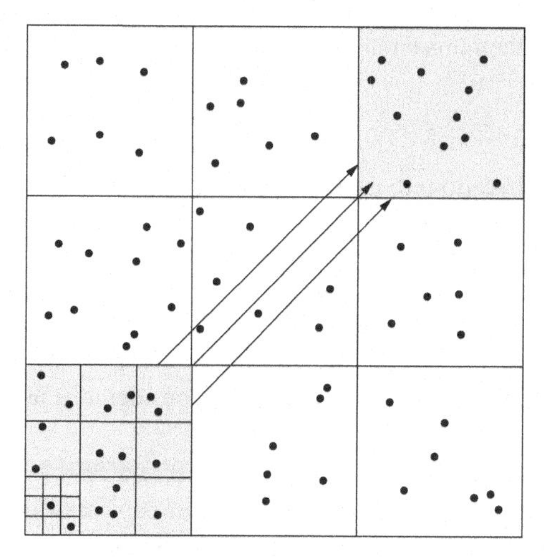

Fig. 7.15 Illustration of the hierarchical cooperation strategy exploiting distributed MIMO transmissions.

7.5.2 Hierarchical Cooperation

A second strategy that exploits the spatial resource is *hierarchical cooperation* using distributed MIMO transmissions between clusters of nodes. This strategy further improves the utilization of the spatial resource, and can make an almost constant fraction of the resource available to each user, under some assumptions on the channel model.

The network area is recursively divided into clusters, and within each cluster nodes form a MIMO system performing joint encoding and communicating with the nodes in another cluster that perform joint decoding. Within each cluster, information is then redistributed recursively by iterating the above scheme. This situation is depicted in Figure 7.15. Each node distributes a constant number of messages among the nodes in its cluster, one for each node, and the nodes within each cluster form a distributed MIMO transmission to send these messages in parallel to the cluster where the destination node lies. Each node in the destination cluster sends its received message to the intended destination within their cluster, which can finally decode all the messages. In this way, the first and last steps of the scheme, corresponding to message distribution and message collection within each cluster, involve only local communication and can be performed in parallel among different clusters, while MIMO transmissions occur sequentially among different clusters. The problem of message distribution and collection within each cluster is treated as a scaled version of the original problem, and can be performed recursively using the same strategy as above, by dividing the cluster into subclusters and letting nodes cooperate to perform MIMO transmissions between subclusters.

A recursive computation shows that if the number of space–wavenumber degrees of freedom is at least of the order of $N^{\ell/(\ell+1)}$, and signals are weakly attenuated over distances, then this strategy allocates

$$\frac{N^{\ell/(\ell+1)}}{N} = N^{-1/(\ell+1)} \tag{7.29}$$

degrees of freedom to each node, where ℓ is the number of recursive clustering layers. For $\ell = 1$, we have

$$\frac{N^{1/2}}{N} = \frac{1}{\sqrt{N}}, \tag{7.30}$$

and the performance of the scheme is the same as in the multi-hop case. For large values of ℓ the performance improves considerably, as each node is allocated an almost constant number of degrees of freedom.

On the other hand, if the number of space–wavenumber degrees of freedom is less than $N^{\ell/(\ell+1)}$, then we are in a limited spatial degree of freedom regime, where performance is limited by the rank of the channel's matrix between all sources and destinations, which depends on environmental constraints.

7.5.3 Interference Alignment

A third strategy that exploits the spatial resource is *interference alignment*. Here, the main idea is to design the transmitted signals so that at each receiver the interference represented by unwanted signals occupies only a small portion of the signals' space. This is achieved by pre-coding over multiple time or frequency realizations of the channel. Each receiver can then recover the desired signal by projecting the received linear combination onto the subspace orthogonal to the interference subspace. The concept is visually illustrated in Figure 7.16, where signals from three users are encoded over three dimensions. Each receiver observes a linear combination of two interfering signals that lie in a two-dimensional subspace, and a third signal that is not fully contained in this subspace. The dimension orthogonal to the interfering subspace can then be used to communicate over an interference-free channel.

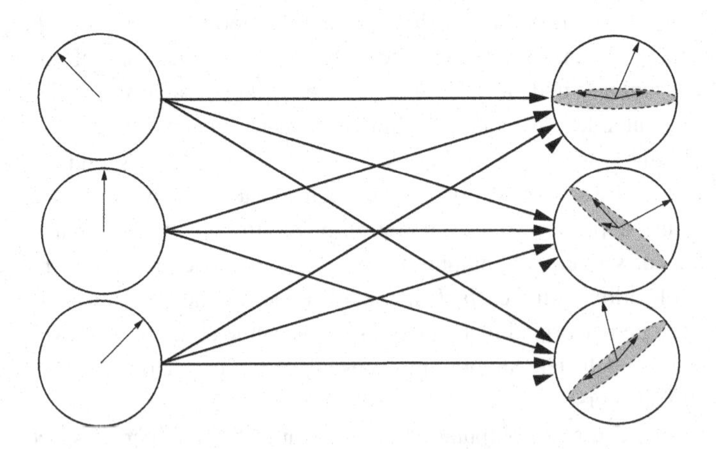

Fig. 7.16 Illustration of interference alignment with three users signaling over three dimensions.

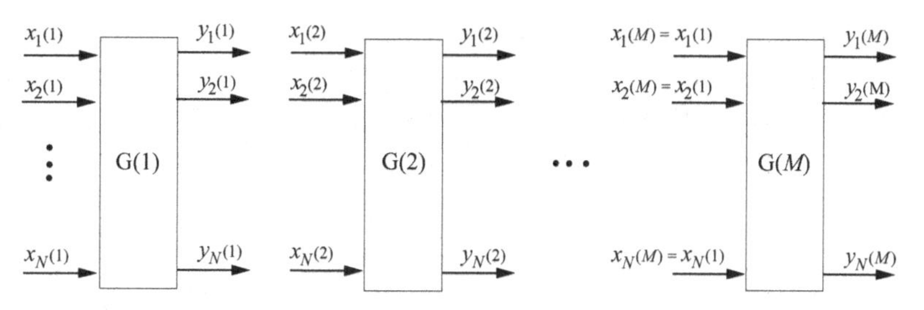

Fig. 7.17 Interference alignment block representation.

The most important feature of an interference alignment scheme is its ability to construct signals which, for a given Green's function, successfully align at the receivers. In a model where the Green's function evolves randomly in time or frequency, and can provide $M \gg N$ independent realizations, it is possible to do so by constructing signals that span multiple realizations and collectively achieve a multiplexing gain of $N/2$, where N is the number of transmitter–receiver pairs. Namely, the N receivers can successfully recover $NM/2$ distinct messages sent over M realizations, yielding a number of messages per channel use of

$$\frac{NM}{2M} = \frac{N}{2}. \tag{7.31}$$

As an example to illustrate a possible construction, consider the repetition scheme depicted in Figure 7.17. In the first realization of the channel, when the channel's matrix is G(1), users send signals $\{x_1(1), x_2(1), \ldots, x_N(1)\}$. Then, they keep sending different signals in subsequent realizations until they repeat in the Mth realization the same signals they sent in the first realization.

Assuming the channel's matrix in the Mth realization is the complement of the one in the first realization,

$$\mathrm{G}(M) = \begin{bmatrix} G_{1,1}(1) & -G_{1,2}(1) & \cdots & -G_{1,N}(1) \\ -G_{2,1}(1) & G_{2,2}(1) & \cdots & -G_{2,N}(1) \\ \vdots & \vdots & \ddots & \vdots \\ -G_{N,1}(1) & -G_{N,2}(1) & \cdots & G_{N,N}(1) \end{bmatrix}, \tag{7.32}$$

then in the Mth realization each receiver has access to

$$y_n(M) = \mathrm{G}_{n,n}(1)x_n(1) - \sum_{k \neq n} \mathrm{G}_{n,k}(1)x_k(1), \tag{7.33}$$

while in the first realization each receiver had access to

$$y_n(1) = \mathrm{G}_{n,n}(1)x_n(1) + \sum_{k \neq n} \mathrm{G}_{n,k}(1)x_k(1). \tag{7.34}$$

Adding the two signals gives

$$y_n(1) + y_n(M) = 2\mathrm{G}_{n,n}(1)x_n. \tag{7.35}$$

It follows that with the assumption that in the Mth realization we have the complementary matrix of the first realization, with two channel uses each user can send one signal over an interference-free channel. This channel has diagonal form

$$\Lambda = \begin{bmatrix} 2G_{1,1}(1) & 0 & \cdots & 0 \\ 0 & 2G_{2,2}(1) & \cdots & 0 \\ \vdots & \vdots & \ddots & \vdots \\ 0 & 0 & \cdots & 2G_{N,N}(1) \end{bmatrix}, \tag{7.36}$$

and provides a multiplexing gain of $N/2$ at the expense of letting the channel evolve over a block of M independent realizations.

The key for this scheme to be successful is the appearance of a complementary matrix every M realizations. Clearly, for a real channel any given complementary matrix has zero probability of occurrence, since a specific set of real numbers can never occur in a random realization. However, in the presence of signal quantization it is enough for the matrix to be only fairly close to its complementary value to obtain the desired result. An ergodic argument then shows that almost all quantized channel states are matched with complementary states when M is large enough, i.e., the fraction of unmatched states approaches zero as $M \to \infty$. This means that at the expense of letting the channel evolve a large number of times, we can in principle match every realization with its complementary one and achieve the desired multiplexing gain.

To obtain the promised gain, interference alignment requires the number of space–wavenumber degrees of freedom to be at least of the order of N; in other words, the rank of the channel's matrix should be large enough to provide concurrent transmissions in each block of Figure 7.17, so that each user can be allocated $1/2$ degrees of freedom at each usage of the channel. If this is not the case, we are in a regime where the number of spatial degrees of freedom is limited by environmental constraints. In practice, these constraints can severely limit the performance of interference alignment as well as hierarchical cooperation schemes.

7.5.4 A Layered View

Interference alignment may be perceived as a scheme by which "everybody gets half of the cake," where the cake here represents the fixed time–frequency resource available to all users. This is only an apparent paradox: the cake has different "layers" corresponding to the space–wavenumber degrees of freedom. This additional spatial resource must be proportional to the number of users in the network and ensures that each user gets half of each layer, rather than of the whole cake, while the other half is lost due to interference.

Another key assumption regards the construction of signals over a large number of independent realizations of the channel. A detailed ergodic argument shows that for alignment to occur with high probability, the logarithm of the number of independent realizations must be of the order of N^2. This means that the amount of stochastic diversity must grow exponentially with the number of users. It follows that interference alignment guarantees that everyone gets half the cake, provided that the cake is layered *and* it can be made arbitrarily large.

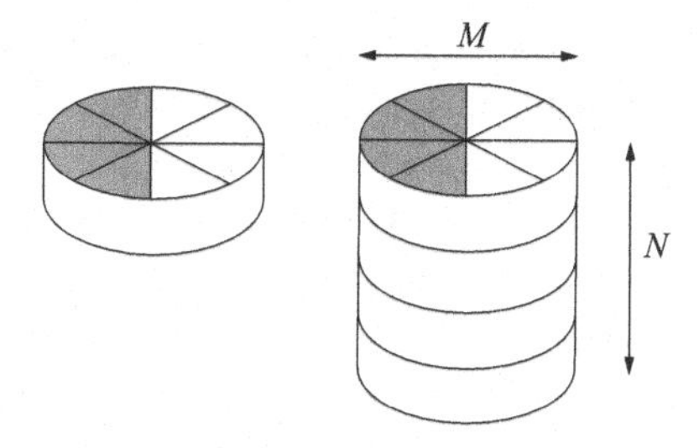

Fig. 7.18 Cake-cutting interpretation of interference alignment.

Table 7.3 Degrees of freedom of different interference management schemes. MIMO: co-located multiple antennas; IA: interference alignment; HC: hierarchical cooperation; MH: multi-hop relay.

	MIMO	IA	HC	MH
Degrees of freedom	N	$N/2$	$N^{\ell/(\ell+1)}$	$N^{1/2}$
Degrees of freedom per user	1	1/2	$N^{-1/(\ell+1)}$	$N^{-1/2}$

This interpretation is illustrated in Figure 7.18, where M corresponds to the amount of stochastic diversity in the time–frequency domain, and N corresponds to the number of space–wavenumber degrees of freedom, representative of the number of spatial dimensions available in the channel.

7.5.5 Degrees of Freedom

The results for interference networks are summarized in Table 7.3. All the schemes are limited by the performance of MIMO with co-located multiple antennas, which allows full cooperation among the transmitters and among the receivers. When the rank of the Green's function is not a bottleneck for the number of degrees of freedom, MIMO provides a multiplexing gain proportional to the number of antennas in the system. Interference alignment comes close to this bound and is order optimal, sacrificing half of the degrees of freedom to align the interference onto low-dimensional subspaces. This is the price to pay in this case for the distributed operation of the network. Hierarchical cooperation is almost order optimal, sacrificing an arbitrarily small power of the number of degrees of freedom to perform cooperation. Multi-hop relay is perhaps the simplest strategy, but needs to sacrifice the square root of the number of degrees of freedom.

All the schemes assume that the rank of the channel matrix is sufficiently large and that it is not a bottleneck for their performance. In practice, several limitations may arise due to specific propagation conditions. Due to propagation and absorption in the environment, signals may be highly attenuated over some channels that become useless for communication. In this case, the effective rank of the channel matrix decreases and becomes the performance bottleneck imposed by physical constraints. In specific propagation environments characterized by high signal attenuation, short-range communication performed via multi-hop is the order-optimal operation strategy. An analysis of the available spatial degrees of freedom in arbitrary environments is provided in Chapter 8.

Additional practical limitations arise from the amount of diversity available, knowledge of the Green's function at all nodes, and the protocol overhead required to set up the network cooperation schemes. The impact of such limitations on the development of future communication technologies exploiting spatial resource in a distributed setting is a current topic of research in wireless communications.

7.6 Summary and Further Reading

We have provided an overview of current communication technologies based on the theory of orthogonal signal representations developed in the previous chapters. These technologies divide the available time–frequency or space–wavenumber resource into parallel channels. The number of degrees of freedom of the electromagnetic field poses a limit on the number of orthogonal channels that can be obtained using these representations. The amount of stochastic diversity poses a limit on the level of reliability that can be obtained with multiple transmissions over these channels.

While an optimal decomposition based on the Hilbert–Schmidt theory is difficult to obtain because it depends on the specific form of the Green's function, practical solutions use basis functions that work well in certain propagation conditions, and this choice forms the basis of the different technologies employed. Technologies exploiting the time–frequency resource, as well as point-to-point MIMO systems, are discussed extensively in wireless communication textbooks, such as Biglieri (2005), Tse and Visvanath (2005), and Goldsmith (2005). Viterbi (1995) is a standard reference for CDMA technology. There are many texts that focus on digital communications, not limited to wireless, each with its own style. Proakis and Salehi (2007), Gallager (2008), and Lapidoth (2009) are some popular examples. The text by El Gamal and Kim (2011) focuses on the information-theoretic aspects of communication in a networked setting.

The optimal operation of interference networks is a topic of research. The information-theoretic performance of multi-hop networks was originally studied by Gupta and Kumar (2000). Subsequent works include Xie and Kumar (2004), Lévêque and Telatar (2005), Franceschetti (2007a), and Franceschetti *et al.* (2007). The general conclusion that can be drawn from these works is that this strategy is order optimal for networks where signals are strongly attenuated over distances.

Hierarchical cooperation was introduced by Özgür, Lévêque, and Tse (2007), and provides an improvement upon multi-hop operation for low signal attenuation channels supporting a large number of space–wavenumber degrees of freedom. Physical limitations to the number of degrees of freedom have been identified by Franceschetti, Migliore, and Minero (2009), and further discussed by Franceschetti *et al.* (2011). The identification of different operating regimes where the network is either degree-of-freedom limited, or amenable to the spatial multiplexing gain offered by hierarchical cooperation, is discussed by Özgür, Lévêque, and Tse (2013) and Lee and Chung (2012). Ghaderi, Xie, and Shen (2009) and Hong and Caire (2015) provide refined analyses of the performance of hierarchical cooperation in the light of these operating regimes.

The idea of interference alignment first appeared in works on coding theory, but the main references in the context of wireless networks are Cadambe and Jafar (2008) and the monograph by Jafar (2011). Practical engineering challenges for the realization of interference alignment are discussed by El Ayach, Peters, and Heath (2013). The interference alignment scheme based on complementary channels is due to Nazer *et al.* (2012). The impact of limited diversity on the performance of interference alignment was investigated by Li and Özgür (2016).

Many works on interference management techniques for next-generation wireless networks are discussed in communication and information theory journals and conferences worldwide.

7.7 Test Your Understanding

Problems

7.1 Digital communication technologies rely on orthogonal signal representations using appropriate basis functions. Under what conditions on the propagation channel do the prolate spheroidal functions provide the optimal representation?

7.2 Verify that the functions $\{\varphi_m(t)\}$ in (7.2) are orthonormal in $[T_\mathrm{p}, T]$.

7.3 Explain how the environmental effects described in Table 7.1 are related to the coherence bandwidth, coherence time, and coherence distance described in Chapter 5.

7.4 Consider a propagation model where there are two significant multi-path components separated by Δt seconds. For an orthogonal code division communication system, what is the minimum bandwidth required to distinguish them?

7.5 Consider an orthogonal spectrum division communication system in which the received signal is the superposition of three frequency components separated by an interval of frequency $\Delta\omega$. Assuming the cyclic prefix occupies negligible time, what is the minimum transmission time required to resolve them? If we account for the length of the cyclic prefix T_p, what is the degree of freedom loss compared to the previous case?

7.6 Explain why the processing of the rake receiver depicted in Figure 7.12 improves the detection capability of the symbol x_n from the nth user. Discuss the quality of the solution.

Solution

From (7.26) and (7.14), we have

$$y(t) = \sum_m g_m x(t - m\pi/\Omega)$$

$$= \sum_m g_m \sum_{i=1}^{N-1} x_i \varphi_i(t - m\pi/\Omega). \tag{7.37}$$

Assuming low autocorrelation of the signal $\varphi_n(t)$ and low cross-correlations between $\varphi_n(t)$ and all the signals $\{\varphi_m(t)\}$, the operation on the mth branch of the rake receiver corresponds to despreading the mth term of the received signal, while the remaining terms contribute to a small error ϵ_m, so that the output to the mth branch is

$$y^{(m)} = g_m x_n + \epsilon_m. \tag{7.38}$$

By repeating over all branches, weighting the result on each branch by a coefficient w_m, and then summing, we get

$$y_n = \sum_m w_m y^{(m)} = x_n \sum_m w_m g_m + w_m \epsilon_m. \tag{7.39}$$

The additive noise terms can be modeled as independent, zero-mean, random variables with variance

$$\mathbb{E}(\epsilon_m^2) = \sigma^2. \tag{7.40}$$

This corresponds to assuming that the noise variance of the despreading error is constant across different branches. It follows that the total noise variance is

$$\mathbb{E}\left[\left(\sum_m w_m \epsilon_m\right)^2\right] = \sigma^2 \sum_m w_m^2. \tag{7.41}$$

On the other hand, by the Cauchy–Schwarz inequality, the square of the signal component in (7.39) is

$$x_n^2 \left(\sum_m w_m g_m\right)^2 \leq x_n^2 \sum_m w_m^2 \sum_m g_m^2. \tag{7.42}$$

The ratio of the signal to the average noise power is therefore

$$\text{SNR} \leq x_n^2 \frac{\sum_m g_m^2}{\sigma^2}, \tag{7.43}$$

where the equality is achieved by choosing $w_m = g_m$ in the Cauchy–Schwarz inequality.

The SNR on any single branch is

$$\text{SNR}_m = \frac{x_n^2 g_m^2}{\sigma^2}; \tag{7.44}$$

it follows that choosing the weights corresponding to the channel's coefficients, the SNR in the rake receiver is improved compared to that of a single branch since

$$\sum_m g_m^2 \geq g_m^2. \tag{7.45}$$

This SNR maximization improves the estimation capability of the system. Dividing both sides of (7.39) by $\sum_m w_m g_m$, we construct the estimate

$$\tilde{x}_n = \frac{y_n}{\sum_m w_m g_m} = x_n + \frac{\sum_m w_m \epsilon_m}{\sum_m w_m g_m}, \tag{7.46}$$

whose mean squared error is

$$\mathbb{E}[(\tilde{x}_n - x_n)^2] = \frac{\mathbb{E}[(\sum_m w_m \epsilon_m)^2]}{(\sum_m w_m g_m)^2}. \tag{7.47}$$

By (7.41) and the Cauchy–Schwarz inequality, we get

$$\mathbb{E}[(\tilde{x}_n - x_n)^2] \geq \frac{\sigma^2}{\sum_m g_m^2}, \tag{7.48}$$

where the equality is achieved by choosing $w_m = g_m$. We conclude that this choice leads to the best linear estimate in terms of mean squared error.

8 The Space–Wavenumber Domain

Space: the final frontier.[1]

8.1 Spatial Configurations

In this chapter we consider the information associated with the different spatial configurations of a waveform. We consider propagation of sinusoidal waves in arbitrary multiple scattering environments, and we compute the spatial bandwidth of the field measured in the space. By "spatial bandwidth" we mean the measure of the support set of the signal in the wavenumber domain – a more appropriate term would be "wavenumber bandwidth." A minor disquieting fact of life is that the name spatial bandwidth is standard, and we shall use the two interchangeably. We obtain a simple formula for the number of spatial degrees of freedom of the received signal, showing that this number is limited by the wavelength-normalized size of the cut through which the information must flow.

A heuristic principle of communication is that in the presence of multiple scattering the number of possible spatial configurations of the field, and thus the amount of information it can carry over space, is increased. Loosely speaking, the superposition in space of many waveforms from many different multiple scattered paths "creates" bandwidth and provides independent parallel channels in the spatial–wavenumber domain. In the jargon of communication theory, the term "rich scattering" is used to denote an environment capable of providing an unlimited number of parallel spatial channels between transmitters and receivers. The intuition used to explain this phenomenon is that if the received field is the superposition of many waveforms coming from many different multiple scattered paths to the receivers, and each path can carry an independent stream of information, then many communications can occur in parallel by using a pair of antennas for communication along each scattered path. If the environment provides a sufficiently large number of independent paths, then this spatial multiplexing capability can grow proportionally to the number of antennas.

For any arbitrary scattering environment, however, the wavenumber bandwidth is not an unlimited resource. A physical limitation is given by the size of the minimum

[1] Opening line of Captain James T. Kirk's monologue in *Star Trek* (1966).

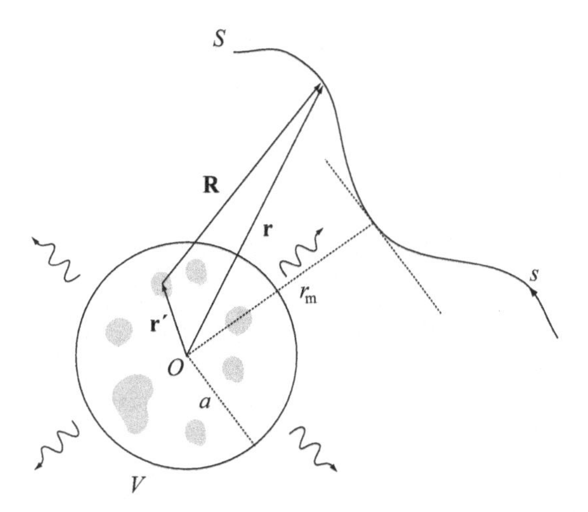

Fig. 8.1 Geometry of the radiation model.

cut-set boundary through which the information must flow as the field propagates from transmitters to receivers. We can view each parallel channel as occupying a certain amount of spatial resource on the cut, proportional to the wavelength of transmission, and if these occupation regions overlap, then the corresponding information streams cannot be resolved at the receivers.

The results described in this chapter can be applied in the context of communication systems, using spatially distributed transmitting antennas, and in the context of remote sensing, providing a limit on the spatial resolution of the obtained image. In the first case, they limit the spatial dimensionality of the signals that can be used for communication across the cut. In the second case, they limit the dimensionality of the image constructed by illuminating an arbitrary scattering system using electromagnetic waveforms and measuring the scattered response at the receivers.

8.2 Radiation Model

We consider a spherical coordinate system $\mathbf{r} = (r,\theta,\phi)$ and a spherical volume V of radius a with center at the origin of the space. We let r_{m} be the minimum distance between the center of V and an indefinite analytic curve S, all external to V. We define a curvilinear coordinate $s = s(\mathbf{r})$ along S, normalized to r_{m}. This geometry is depicted in Figure 8.1.

A density $\mathbf{I}(\mathbf{r}')$ [A m^{-2}], $\mathbf{r}' \in V$, of sinusoidal currents of angular frequency ω is distributed over the radiating elements contained inside V. This may include currents impressed by the sources, or induced due to multiple scattering.

We refer to the field measured over the curve S as $\mathbf{E}(s)$. By (5.16), we have

$$\mathbf{E}(s) = \int_V \underline{\mathbf{G}}(|\mathbf{r}-\mathbf{r}'|)\cdot\mathbf{I}(\mathbf{r}')d\mathbf{r}', \tag{8.1}$$

and we consider the $L^2(-\infty,\infty)$ norm

$$\begin{aligned}\|\mathbf{E}(s)\|^2 &= \int_{-\infty}^{\infty}\left|\int_V \underline{\mathbf{G}}(|\mathbf{r}-\mathbf{r}'|)\cdot\mathbf{I}(\mathbf{r}')d\mathbf{r}'\right|^2 ds \\ &\leq \sup_{\mathbf{r}'\in V}\left(\int_{-\infty}^{\infty}|\underline{\mathbf{G}}(s,\mathbf{r}')|^2 ds\right)\cdot\left(\int_V |\mathbf{I}(\mathbf{r}')|d\mathbf{r}'\right)^2,\end{aligned} \tag{8.2}$$

where the last step follows from Young's inequality – see Appendix A.4.3.

8.3 The Field's Functional Space

We wish to work with the functional space of square-integrable functions so that we can use the analysis tools developed in the previous chapters to determine the spatial degrees of freedom of the scattered field. This requires (8.2) to be finite. We assume that currents inside V are uniformly bounded, so that the second integral in (8.2) is clearly bounded. This assumption also implies that currents are square-integrable functions. On the other hand, the first integral in (8.2) is bounded provided that $|\mathbf{r}-\mathbf{r}'| > 0$, namely the observation curve is external to V. This is immediately evident by examining the form of the Green's function. Letting $R = |\mathbf{r}-\mathbf{r}'| = R(s,\mathbf{r}')$, from (5.23) we have

$$\begin{aligned}\underline{\mathbf{G}}(s,\mathbf{r}') &= -\frac{j\omega\mu}{4\pi}\left[\underline{\mathbf{I}}+\frac{\underline{\nabla\nabla}}{\beta^2}\right]\cdot\frac{\exp(-j\beta R)}{R} \\ &= -\frac{j\omega\mu}{4\pi}\left[\frac{\exp(-j\beta R)}{R}\underline{\mathbf{I}}+\frac{\underline{\nabla\nabla}}{\beta^2}\left(\frac{\exp(-j\beta R)}{R}\right)\right] \\ &= -\exp(-j\beta R)\frac{j\omega\mu}{4\pi R}\left[\underline{\mathbf{I}}+\frac{R}{\beta^2}\exp(j\beta R)\underline{\nabla\nabla}\left(\frac{\exp(-j\beta R)}{R}\right)\right] \\ &= -\exp(-j\beta R)\frac{j\omega\mu}{4\pi R}\underline{\mathbf{N}},\end{aligned} \tag{8.3}$$

where the dyad $\underline{\mathbf{I}}$ is the identity and the dyad $\underline{\nabla\nabla}$ is represented by the matrix

$$\begin{pmatrix} \dfrac{\partial^2}{\partial x^2} & \dfrac{\partial^2}{\partial x\partial y} & \dfrac{\partial^2}{\partial x\partial z} \\ \dfrac{\partial^2}{\partial x\partial y} & \dfrac{\partial^2}{\partial y^2} & \dfrac{\partial^2}{\partial y\partial z} \\ \dfrac{\partial^2}{\partial x\partial z} & \dfrac{\partial^2}{\partial y\partial z} & \dfrac{\partial^2}{\partial z^2} \end{pmatrix}.$$

If $R > 0$, then (8.3) is square-integrable. The expression (8.3) also shows that the field generated by a current flowing through an elementary volume at $\mathbf{r}'$ undergoes a phase shift $\exp(-j\beta R)$, a $1/R$ geometric attenuation, and a dyadic transformation $\underline{\mathbf{N}}$ that affects the radiated field's amplitude and orientation. When $R\to\infty$, (8.3) attains a very simple form, representing the far field radiated from an elementary volume, and given by the Fourier transform of (4.92a) – see Problem 8.1. The total field is given by the superposition of such elementary contributions.

For convenience, we assume that the field is normalized to its maximum norm (8.2), so that the set of scattered fields is contained in the unit sphere of the corresponding functional space. We also extract from the field the propagation factor $\exp(-j\beta r)$, where $\beta = \omega\sqrt{\epsilon\mu} = \omega/c$ is the propagation constant, and refer to the reduced field

$$\mathbf{E}(s) \to \exp(j\beta r)\mathbf{E}(s). \tag{8.4}$$

Letting the reduced Green's function

$$\underline{\mathbf{G}}(s,\mathbf{r}') \to \exp(j\beta r)\underline{\mathbf{G}}(s,\mathbf{r}'), \tag{8.5}$$

the reduced field is then given by

$$\begin{aligned}\mathbf{E}(s) &= -\exp[j\beta(r-R)]\frac{j\omega\mu}{4\pi R}\int_V \underline{\mathbf{N}}\cdot\mathbf{I}(\mathbf{r}')d\mathbf{r}' \\ &= \int_V \underline{\mathbf{G}}(s,\mathbf{r}')\cdot\mathbf{I}(\mathbf{r}')d\mathbf{r}',\end{aligned} \tag{8.6}$$

where the Green's operator acts on the appropriate functional space,

$$\mathcal{G} : L^2(V) \to L^2(S). \tag{8.7}$$

8.4 Spatial Bandwidth

To determine the spatial bandwidth of the radiated field, we apply a rectangular filter of bandwidth $w > 0$ in the wavenumber domain, and evaluate the error associated with the filtered field representation. The study of this error as a function of w reveals the existence of a critical value of the spatial bandwidth beyond which the error quickly drops to zero. The critical bandwidth can then be related to the geometry of the scattering system and used for the computation of the number of spatial degrees of freedom.

The filtered field is given by the convolution in the spatial domain

$$\begin{aligned}\mathbf{E}_w(s) &= \frac{1}{\pi}\frac{\sin(ws)}{s} * \mathbf{E}(s) \\ &= \frac{1}{\pi}\frac{\sin(ws)}{s} * \int_V \underline{\mathbf{G}}(s,\mathbf{r}')\cdot\mathbf{I}(\mathbf{r}')d\mathbf{r}' \\ &= \frac{1}{\pi}\int_{-\infty}^{\infty}\frac{\sin[w(s-s')]}{s-s'}\int_V \underline{\mathbf{G}}(s',\mathbf{r}')\cdot\mathbf{I}(\mathbf{r}')d\mathbf{r}'ds' \\ &= \int_V \underline{\mathbf{G}}_w(s,\mathbf{r}')\cdot\mathbf{I}(\mathbf{r}')d\mathbf{r}',\end{aligned} \tag{8.8}$$

where

$$\underline{\mathbf{G}}_w(s,\mathbf{r}') = \frac{1}{\pi}\int_{-\infty}^{\infty}\frac{\sin[w(s-s')]}{s-s'}\underline{\mathbf{G}}(s',\mathbf{r}')ds'$$

$$= \frac{1}{2\pi j} \int_{-\infty}^{\infty} \left\{ \frac{\exp[jw(s'-s)]}{s'-s} \underline{\mathbf{G}}(s',\mathbf{r}') - \frac{\exp[-jw(s'-s)]}{s'-s} \underline{\mathbf{G}}(s',\mathbf{r}') \right\} ds'. \tag{8.9}$$

We now define

$$\Delta\underline{\mathbf{G}}(s,\mathbf{r}') = \underline{\mathbf{G}}_w(s,\mathbf{r}') - \underline{\mathbf{G}}, \tag{8.10}$$

and

$$\Delta\mathbf{E}(s) = \mathbf{E}_w(s) - \mathbf{E}(s) = \int_V \Delta\underline{\mathbf{G}}(s,\mathbf{r}')\mathbf{I}(\mathbf{r}')d\mathbf{r}'. \tag{8.11}$$

The error obtained using bandlimited fields in the set $\mathscr{B}_w$ to represent the radiated field is then given in terms of the error obtained when the Green's function is substituted by its filtered version,

$$\begin{aligned} \|\Delta\mathbf{E}(s)\| &= \left(\int_{-\infty}^{\infty} |\Delta\mathbf{E}|^2 ds \right)^{1/2} \\ &= \left(\int_{-\infty}^{\infty} \left| \int_V \Delta\underline{\mathbf{G}}(s,\mathbf{r}')\mathbf{I}(\mathbf{r}')d\mathbf{r}' \right|^2 ds \right)^{1/2} \\ &\le \sup_{\mathbf{r}' \in V} \left(\int_{-\infty}^{\infty} |\Delta\underline{\mathbf{G}}(s,\mathbf{r}')|^2 ds \right)^{1/2} \cdot \int_V |\mathbf{I}(\mathbf{r}')| d\mathbf{r}', \end{aligned} \tag{8.12}$$

where, as before, the last step follows from Young's inequality. This bandlimitation error corresponds to the deviation $D_{\mathscr{B}_w}(\mathscr{E})$ of the space of bandlimited fields $\mathscr{B}_w$ from the whole space of radiated fields $\mathscr{E}$. Since the currents are uniformly bounded, the second integral in (8.12) contributes only a constant factor to the error, and the behavior of $D_{\mathscr{B}_w}(\mathscr{E})$ is essentially dictated by the first term. Letting $\beta = 2\pi/\lambda$, where λ is the wavelength of radiation, asymptotic evaluation as $\beta a \to \infty$ reveals that the error has a phase transition, tending to zero for values of w slightly larger than βa, and to its maximum value for values of w slightly smaller than βa. The transition is illustrated in Figure 8.2, and leads to the notion of spatial bandwidth of an arbitrary radiating system of radius a given by $\beta a + o(\beta a)$. In the following, we provide the derivation of this basic result and then use it to compute the number of degrees of freedom of the radiated field.

8.4.1 Bandlimitation Error

In order to bound the error in (8.12), we need to evaluate $\Delta\underline{\mathbf{G}}(s,\mathbf{r}')$. According to (8.10), this amounts to performing the integration in (8.9). The integral is over a real line and has one singularity at $s' = s$. It is convenient to extend it to the complex plane and perform contour integration of the two factors inside the integral separately. Integrating

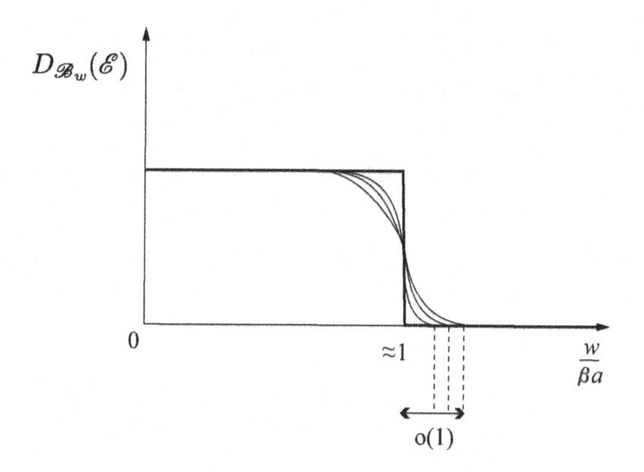

Fig. 8.2 Phase transition of the deviation from the space of bandlimited functions.

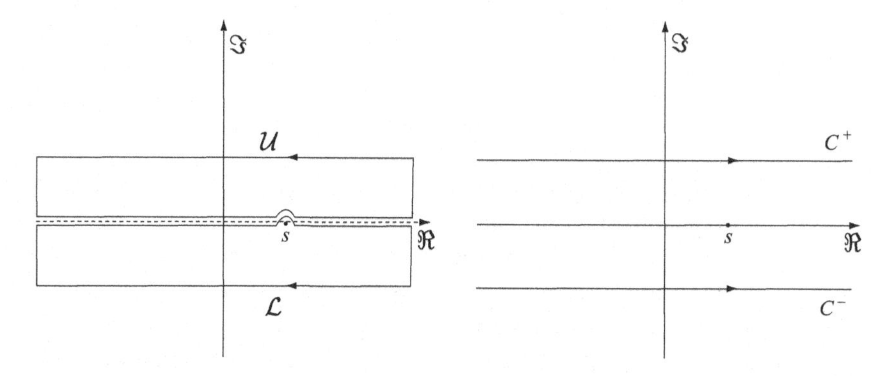

Fig. 8.3 Contour integration in the complex plane.

the first factor along the upper closed contour $\mathcal{U}$ that does not contain the singularity, as depicted in Figure 8.3, we have, by Cauchy's theorem,

$$\frac{1}{2\pi j}\oint_{\mathcal{U}} \frac{\exp[jw(s'-s)]}{s'-s}\underline{\mathbf{G}}(s',\mathbf{r}')ds' = 0. \tag{8.13}$$

Since the integrand in (8.13) tends to zero as $\Re(s') \to \infty$, the two vertical paths of the contour do not provide any contribution to the integral, and we let

$$\begin{aligned}
\underline{\mathbf{G}}_w^+(s,\mathbf{r}') &= \frac{1}{2\pi j}\int_{\Re} \frac{\exp[jw(s'-s)]}{s'-s}\underline{\mathbf{G}}(s',\mathbf{r}')ds' \\
&= \frac{1}{2\pi j}\int_{C+} \frac{\exp[jw(s'-s)]}{s'-s}\underline{\mathbf{G}}(s',\mathbf{r}')ds' \\
&= -\frac{\omega\mu}{2\pi}\int_{C+} \frac{\exp[jw(s'-s)+j\psi(s',\mathbf{r}')]}{4\pi(s'-s)R(s',\mathbf{r}')}\underline{\mathbf{N}}(s',\mathbf{r}')ds',
\end{aligned} \tag{8.14}$$

where $\psi(s,\mathbf{r}') = \beta[r(s) - R(s,\mathbf{r}')]$, and the last equality follows from (8.3). Similarly, integration of the second factor along the lower closed contour $\mathcal{L}$ leads to

$$-\frac{1}{2\pi j}\oint_{\mathcal{L}} \frac{\exp[-jw(s'-s)]}{s'-s}\underline{\mathbf{G}}(s',\mathbf{r}')ds' = \underline{\mathbf{G}}(s,\mathbf{r}'), \tag{8.15}$$

where the residue $\underline{\mathbf{G}}(s,\mathbf{r}')$ is due to the presence of the singularity inside $\mathcal{L}$. Since the two vertical paths of the contour again do not provide any contribution to the integral, we let

$$\begin{aligned}
\underline{\mathbf{G}}_w^-(s,\mathbf{r}') &= -\frac{1}{2\pi j}\int_{\Re} \frac{\exp[-jw(s'-s)]}{s'-s}\underline{\mathbf{G}}(s',\mathbf{r}')ds' \\
&= -\frac{1}{2\pi j}\int_{C^-} \frac{\exp[-jw(s'-s)]}{s'-s}\underline{\mathbf{G}}(s',\mathbf{r}')ds' + \underline{\mathbf{G}}(s,\mathbf{r}') \\
&= \frac{\omega\mu}{2\pi}\int_{C^-} \frac{\exp[-jw(s'-s) + j\psi(s',\mathbf{r}')]}{4\pi(s'-s)R(s',\mathbf{r}')}\underline{\mathbf{N}}(s',\mathbf{r}')ds' + \underline{\mathbf{G}}(s,\mathbf{r}').
\end{aligned} \tag{8.16}$$

We now have

$$\Delta\underline{\mathbf{G}}(s,\mathbf{r}') = \underline{\mathbf{G}}_w - \underline{\mathbf{G}} = \underline{\mathbf{G}}_w^+ + \underline{\mathbf{G}}_w^- - \underline{\mathbf{G}}, \tag{8.17}$$

and the problem has been reduced to the computation of the two complex integrals in (8.14) and (8.16). Their evaluation reveals a sharp transition of the bandlimitation error around a critical value of w. The error sharply drops to zero for values of w slightly larger than the critical value, and it quickly rises to its maximum for values slightly smaller than the critical value. The width of the transition can be characterized precisely, and leads to the asymptotic notion of the spatial bandwidth of the radiated field.

8.4.2 Phase Transition of the Bandlimitation Error

The Green's function bandlimitation error (8.17) is evaluated using a limiting argument. This is based on the idea that when computing (8.14) and (8.16) we have a vanishing contribution to the integrals from intervals of s' where the integrands are rapidly oscillating, while most of the contribution comes from the vicinity of points where the phase of the integrands does not change. The evaluation can then be restricted to the stationary points by letting the parameter regulating the velocity of the oscillations away from the stationary points tend to infinity. See also Appendix C for an illustration of this method.

To identify the parameter regulating the velocity of the oscillations, we write

$$\dot{\psi}(s',\mathbf{r}') = \frac{\partial\psi(s',\mathbf{r}')}{\partial s'} = \beta\frac{\partial[r(s') - R(s',\mathbf{r}')]}{\partial s'}, \tag{8.18}$$

and notice that provided that $(r-R)$ tends to a well-defined limit as $s' \to \pm\infty$, we have

$$\lim_{s'\to\pm\infty} \dot{\psi}(s',\mathbf{r}') = 0. \tag{8.19}$$

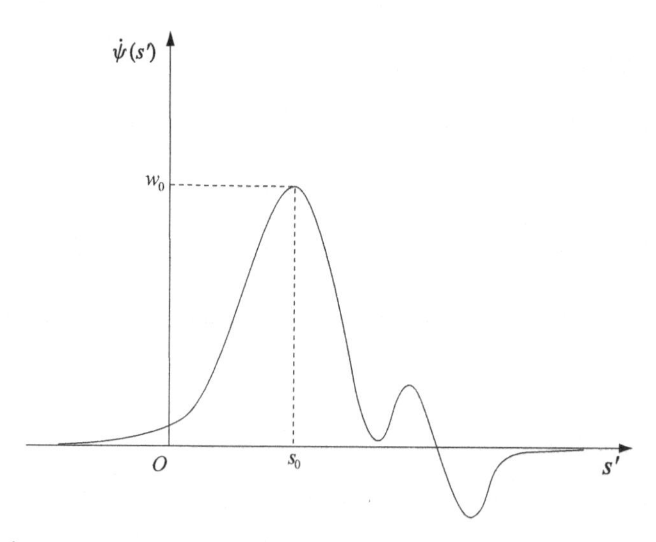

Fig. 8.4 The function $\dot\psi$ has a maximum at s_0.

The proof of this last statement is a cute exercise in analysis – see Problem 8.5. It follows that $\dot\psi$ has at least one extremal point – see Figure 8.4. We then let

$$w_0(\mathbf{r}') = \max_{s'} |\dot\psi(s', \mathbf{r}')| = |\dot\psi(s_0, \mathbf{r}')|, \tag{8.20}$$

and

$$\chi = \frac{w}{w_0}. \tag{8.21}$$

The integrals in (8.14) and (8.16) can now be written in the standard form suitable for asymptotic evaluation:

$$\int_\rho f(s') \exp[w_0 \Phi(s')] ds', \tag{8.22}$$

where

$$\Phi(s') = \pm j\chi(s' - s) + j\frac{\psi(s')}{w_0}, \tag{8.23}$$

the path ρ is either C^+ or C^-, and $f(s')$ is a slowly varying function of s', except in the neighborhood of the singular point $s' = s$. The saddle point method for the asymptotic evaluation of integrals of this type is described in Appendix C.3, and is applied in the next section by letting the parameter regulating the velocity of the oscillations $w_0 \to \infty$.

The result is that the integral (8.22) exhibits a phase transition around the critical point $\chi = 1$, tending to zero as $w_0 \to \infty$ for $\chi > 1$, and to its maximum value for $\chi < 1$. It follows that $w = w_0$ represents the critical value of the bandwidth around which the error's behavior drastically changes. To better visualize the transition, notice that $\exp[w_0 \Phi(s')]$ rapidly oscillates as $w_0 \to \infty$, except at the stationary points where

$$\frac{\partial \Phi(s')}{\partial s'} = 0, \tag{8.24}$$

corresponding to

$$\begin{cases} w = \dot{\psi}(s', \mathbf{r}'), & (8.25a) \\ -w = \dot{\psi}(s', \mathbf{r}'). & (8.25b) \end{cases}$$

If $\chi > 1$ as $w_0 \to \infty$, then $w > |\dot{\psi}(s')|$ for all s' and there are no stationary points. It follows that $\Delta\underline{\mathbf{G}}$ rapidly decays to zero due to the rapid oscillations of the integrands. On the other hand, if $\chi < 1$ then the error $\Delta\underline{\mathbf{G}}$ increases due to the presence of stationary points. The error reaches its maximum value as soon as the contribution of the highest stationary point is taken into account.

8.4.3 Asymptotic Evaluation

Assuming that w_0 corresponds to a maximum of $\dot{\psi}$, we can focus on $\underline{\mathbf{G}}^-$ and asymptotically evaluate the integral in (8.16) using the saddle point method described in Appendix C.3. Expanding $\Phi(s')$ in the neighborhood of s_0, we have

$$\begin{aligned} \Phi(s') &\simeq \Phi(s_0) + \dot{\Phi}(s_0)(s' - s_0) + \frac{1}{2}\ddot{\Phi}(s_0)(s' - s_0)^2 + \frac{1}{3!}\dddot{\Phi}(s_0)(s' - s_0)^3 \\ &= \Phi(s_0) + \dot{\Phi}(s_0)(s' - s_0) + \frac{1}{3!}\dddot{\Phi}(s_0)(s' - s_0)^3 \\ &= -j\chi(s_0 - s) + j\frac{\psi(s_0)}{w_0} - j(\chi - 1)(s' - s_0) - \frac{j}{6}\frac{|\dddot{\psi}(s_0)|}{w_0}(s' - s_0)^3. \end{aligned} \tag{8.26}$$

This third-order expansion is only valid in the neighborhood of s_0; however, as $w_0 \to \infty$ we can substitute it inside (8.16) and integrate along the entire path rather than just around s_0, since away from the stationary point the rapid oscillations let the integral tend to zero. It follows that

$$\begin{aligned} \Delta\underline{\mathbf{G}}(s, \mathbf{r}') \sim\; & \frac{\omega\mu}{2\pi} \exp[j\psi(s_0) + jw(s - s_0)] \frac{\underline{\mathbf{N}}(s_0, \mathbf{r}')}{4\pi R(s_0, \mathbf{r}')} \\ & \int_{C^-} \frac{1}{(s' - s)} \exp\left\{ w_0 \left[-j(\chi - 1)(s' - s_0) \right.\right. \\ & \left.\left. - \frac{j}{6}\frac{|\dddot{\psi}(s_0)|}{w_0}(s' - s_0)^3 \right] \right\} ds'. \end{aligned} \tag{8.27}$$

We now deform the integration path, as we may by Cauchy's theorem, so that it passes through the stationary point s_0 in such a way that the angle approaching the stationary point makes the real part of $\Phi(s')$ exhibit the steepest maximum while keeping the imaginary part stationary. This allows us to capture the largest possible contribution to the integral at the stationary point, and is obtained by letting

$$\tau = -j(s' - s_0)\left(\frac{|\dddot{\psi}(s_0)|}{2}\right)^{1/3}, \tag{8.28}$$

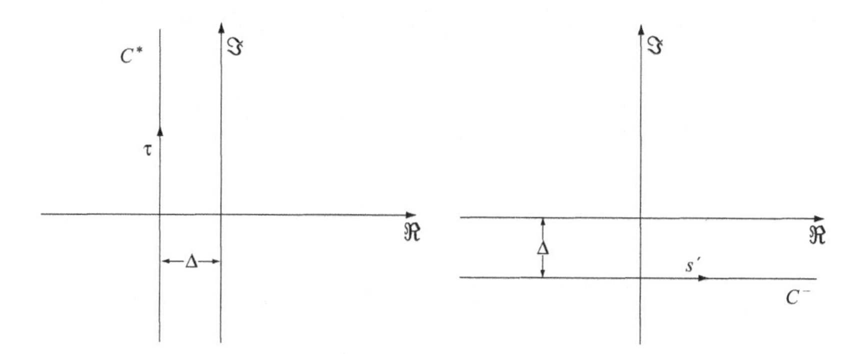

Fig. 8.5 Changing the integration path from C^- to C^*.

so that the integral (8.27) becomes

$$\Delta\underline{\mathbf{G}}(s,\mathbf{r}') \sim \underline{\mathbf{G}}(s_0,\mathbf{r}')\exp[jw(s-s_0)]$$
$$\frac{1}{2\pi j}\int_{C^*}\frac{\exp(\sigma\tau-\tau^3/3)}{\tau+j\left|\frac{\dddot{\psi}(s_0)}{2}\right|^{1/3}(s-s_0)}d\tau, \tag{8.29}$$

where C^* is the vertical integration path in the left complex half-plane shown in Figure 8.5, and

$$\sigma(\mathbf{r}') = \left|\frac{\dddot{\psi}(s_0)}{2}\right|^{-1/3} w_0(\chi-1). \tag{8.30}$$

With a further manipulation, (8.29) can be expressed as a tail integral of the Airy function times a phase term (see Problem 8.2), and we have

$$\Delta\underline{\mathbf{G}}(s,\mathbf{r}') \sim -\underline{\mathbf{G}}(s_0,\mathbf{r}')\exp[jw_0(s-s_0)]$$
$$\int_{\sigma(\mathbf{r}')}^{\infty}\mathrm{Ai}(x)\exp\left[j\left|\frac{\dddot{\psi}(s_0)}{2}\right|^{1/3}(s-s_0)x\right]dx. \tag{8.31}$$

Finally, an application of Parseval's theorem (see Problem 8.3) leads to

$$\int_{-\infty}^{\infty}|\Delta\underline{\mathbf{G}}(s,\mathbf{r}')|^2 ds \sim |\underline{\mathbf{G}}(s_0,\mathbf{r}')|^2 2\pi\left|\frac{\dddot{\psi}(s_0)}{2}\right|^{-1/3}\int_{\sigma(\mathbf{r}')}^{\infty}\mathrm{Ai}^2(x)dx. \tag{8.32}$$

The tail integral on the right-hand side of (8.32) is depicted in Figure 8.6. It is controlled by the value of $\sigma(\mathbf{r}')$, sharply decaying to zero as $\sigma\to\infty$, while growing to its maximum value for $\sigma\to-\infty$.

The behavior of this integral is responsible for the phase transition of the bandlimitation error. Letting $\chi>1$ in (8.30), we show that for all $\mathbf{r}'\in V$,

$$\lim_{w_0\to\infty}\sigma(\mathbf{r}') = \infty. \tag{8.33}$$

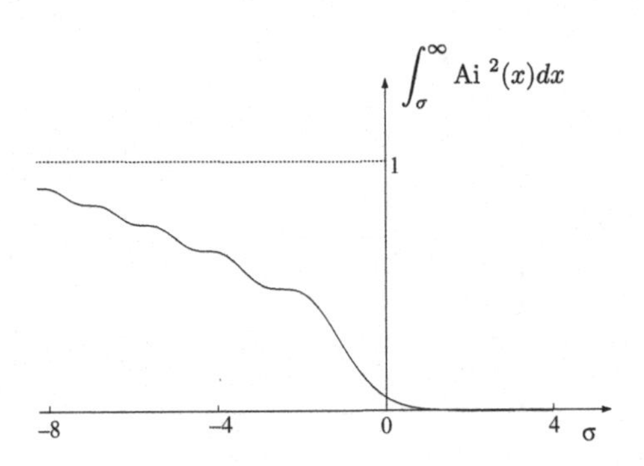

Fig. 8.6 Tail integral behavior governing the Green's function bandlimitation error.

In this regime, the rate of decay of the tail integral on the right-hand side of (8.32) can be computed, and when substituted into (8.32) it shows that the bandlimitation error drops to zero. On the other hand, for $\chi < 1$ the minimum value of $\sigma(\mathbf{r}')$ tends to $-\infty$ and the bandlimitation error sharply increases. By (8.21), the regime $\chi > 1$ corresponds to having values of the bandwidth w slightly larger than the critical value w_0, and in this case the error is negligible for all $\mathbf{r}' \in V$. The regime $\chi < 1$ corresponds to having values of the bandwidth slightly smaller than w_0 and in this case the error may be substantial for some $\mathbf{r}' \in V$.

8.4.4 Critical Bandwidth

To determine the asymptotic behavior of the tail integral on the right-hand side of (8.32), we consider the smallest value,

$$\sigma_{\mathrm{m}} = \inf_{\mathbf{r}' \in V} \sigma(\mathbf{r}'), \tag{8.34}$$

and the largest value of the critical bandwidth,

$$W = \sup_{\mathbf{r}' \in V} w_0(\mathbf{r}'), \tag{8.35}$$

and let

$$\chi_{\mathrm{m}} = \frac{w}{W}, \tag{8.36}$$

$$|\dddot{\psi}|_{\mathrm{M}} = \sup_{\mathbf{r}' \in V} |\dddot{\psi}(s_0, \mathbf{r}')|. \tag{8.37}$$

With these definitions, we have from, (8.30),

$$\sigma_{\mathrm{m}} = \inf_{\mathbf{r}' \in V} \left(\frac{2}{|\dddot{\psi}(\mathbf{r}')|} \right)^{1/3} (w - w_0)$$

$$= \left(\frac{2}{|\dddot{\psi}|_{\mathrm{M}}} \right)^{1/3} (w - W)$$

$$= \left(\frac{2}{|\dddot{\psi}|_{\mathrm{M}}} \right)^{1/3} (\chi_{\mathrm{m}} - 1) W. \tag{8.38}$$

Some geometric considerations yield bounds on $|\dddot{\psi}|_{\mathrm{M}}$ and W. For $s' \in S$ and having normalized the arc length s' to r_{m}, we have

$$\dot{\psi}(s') = \beta \frac{\partial (r - R)}{\partial s'} = \beta r_{\mathrm{m}} (\bar{\mathbf{r}} - \bar{\mathbf{R}}) \cdot \bar{\mathbf{t}}, \tag{8.39}$$

where $\bar{\mathbf{t}}$ is the unit vector tangent to the observation curve – see Figure 8.7. Considering the extremal configuration depicted in Figure 8.8 and corresponding to

$$r' = a, \; r = r_{\mathrm{m}}, \tag{8.40}$$

we have, from (8.39), that this achieves

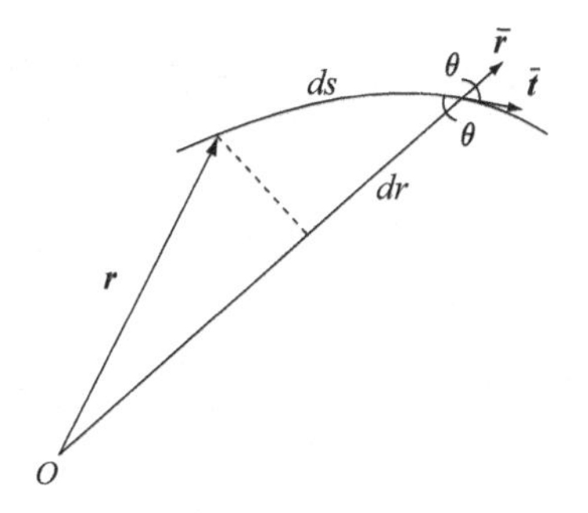

Fig. 8.7 Derivative of a position vector along a curve. $dr = ds \cos\theta = ds\, \bar{\mathbf{r}} \cdot \bar{\mathbf{t}}$.

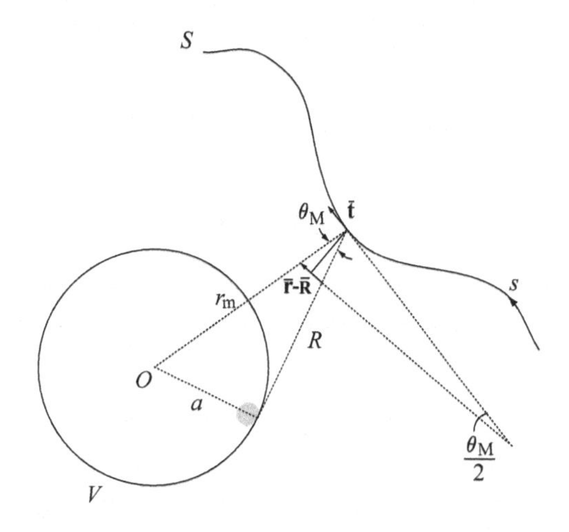

Fig. 8.8 Extremal geometric configuration.

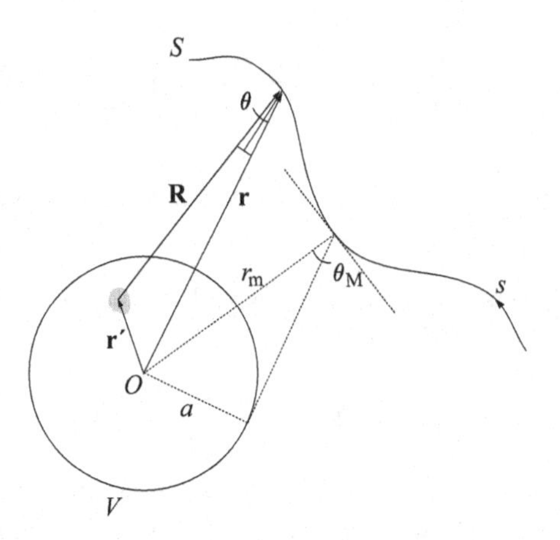

Fig. 8.9 Geometric interpretation.

$$\dot{\psi}(s') = \beta r_m |\bar{\mathbf{r}} - \bar{\mathbf{R}}| \cos\frac{\theta_M}{2}$$
$$= \beta r_m 2 \sin\frac{\theta_M}{2}\cos\frac{\theta_M}{2}$$
$$= \beta a. \tag{8.41}$$

It then follows that

$$W = \sup_{s' \in S, \mathbf{r}' \in V} |\dot{\psi}(s')| \geq \beta a. \tag{8.42}$$

With similar geometric considerations, one can also show that

$$|\dddot{\psi}|_M \leq \kappa \beta a, \tag{8.43}$$

where the positive constant κ depends only on the shape of the observation curve. Substituting (8.42) and (8.43) into (8.38), we obtain

$$\begin{cases} \sigma_m \geq \left(\dfrac{2}{\kappa}\right)^{1/3} (\beta a)^{2/3}(\chi_m - 1) & \text{for } \chi_m > 1, \quad (8.44a) \\ \sigma_m \leq \left(\dfrac{2}{\kappa}\right)^{1/3} (\beta a)^{2/3}(\chi_m - 1) & \text{for } \chi_m < 1. \quad (8.44b) \end{cases}$$

It follows that as $\beta a \to \infty$ we have that $\sigma_m \to \pm\infty$ depending on the value of χ_m. For $\chi_m > 1$, we have $w > W$ and $\sigma_m \to \infty$; for $\chi_m < 1$, we have $w < W$ and $\sigma_m \to -\infty$. From (8.39), we also have

$$|\dot{\psi}(s')| \leq \beta r_m |\bar{\mathbf{r}} - \bar{\mathbf{R}}| = \beta r_m 2 \sin\frac{\theta}{2}$$
$$= \beta r_m \frac{\sin\theta}{\cos\dfrac{\theta}{2}}, \tag{8.45}$$

where θ is the angle between $\bar{\mathbf{r}}$ and $\bar{\mathbf{R}}$ – see Figure 8.9. The right-hand side is maximized when $\theta = \theta_{\mathrm{M}}$, so that by taking the supremum on both sides we get

$$\begin{aligned} W &= \sup_{s' \in S, \mathbf{r}' \in V} |\dot{\psi}| \\ &\leq \beta r_{\mathrm{m}} \frac{\sin \theta_{\mathrm{M}}}{\cos \dfrac{\theta_{\mathrm{M}}}{2}} \\ &\leq \sqrt{2} \beta a, \end{aligned} \tag{8.46}$$

where the last inequality follows from $\sin\theta_{\mathrm{M}} = a/r_{\mathrm{m}}$ and $\theta_{\mathrm{M}} \leq \pi/2$. Putting together (8.42) and (8.46), we have that the critical bandwidth W that determines the diverging behavior of σ_{m} is geometrically related to the size of the scattering system a. Since the propagation constant $\beta = \omega/c = 2\pi/\lambda$, the critical transition point occurs when w is of the order of the circle $2\pi a$, normalized to the transmission wavelength λ.

The critical bandwidth is an intrinsic property of the scattering system, that depends only on its overall dimension, normalized to the transmission wavelength.

In the far field, namely when $r_{\mathrm{m}} \gg a$, the factor $\sqrt{2}$ in (8.46) can be replaced by one, and the critical bandwidth that marks the transition from large to small values of the error becomes essentially βa. In practice, for values of r_{m} slightly larger than a the upper and lower bounds on W are already very close. For example, for $r_{\mathrm{m}} = 1.5a$, the upper bound becomes $W \leq 1.07\beta a$.

8.4.5 Size of the Transition Window

To give a precise characterization of the excess bandwidth required to achieve a small error, we let $w > W$, and evaluate the rate of decay of the bandlimitation error as $\beta a \to \infty$. In this case, by (8.44a) we have $\sigma(\mathbf{r}') \to \infty$ for all $\mathbf{r}' \in V$, and using standard asymptotic expansions of the Airy function, we obtain

$$\int_{\sigma(\mathbf{r}')}^{\infty} \mathrm{Ai}^2(x)dx = O\left[\frac{\exp\left(-\frac{2}{3}\sigma^{3/2}(\mathbf{r}')\right)^2}{\sigma(\mathbf{r}')}\right]. \tag{8.47}$$

By (8.12) and using (8.32), (8.47), and (8.30), we obtain the bound on the bandlimitation error:

$$\begin{aligned} D_{\mathscr{B}_w}(\mathscr{E}) &= \sup_{\mathbf{r}' \in V} \left(\int_{-\infty}^{\infty} |\Delta \underline{\mathbf{G}}(s, \mathbf{r}')|^2 ds \right)^{1/2} \cdot \int_V |\mathbf{I}(\mathbf{r}')| d\mathbf{r}' \\ &= O\left[\frac{\exp\left(-\frac{2}{3}\sigma_{\mathrm{m}}^{3/2}\right)}{\sqrt{w - W}}\right] \\ &= O\left[\frac{\exp\left(-\frac{2}{3}\sigma_{\mathrm{m}}^{3/2}\right)}{\sqrt{(\chi_{\mathrm{m}} - 1)W}}\right]. \end{aligned} \tag{8.48}$$

By (8.44a) and (8.42) it follows that, as $\beta a \to \infty$,

$$D_{\mathscr{B}_w}(\mathscr{E}) = O\left\{\frac{\exp\left[-\frac{2}{3}\left(\frac{2}{\kappa}\right)^{1/2}(\chi_m - 1)^{3/2}\beta a\right]}{\sqrt{(\chi_m - 1)\beta a}}\right\}, \tag{8.49}$$

and the deviation can be made arbitrarily small by choosing

$$\chi_m = 1 + \frac{1}{o(\beta a)}, \tag{8.50}$$

so that $(\chi_m - 1)\beta a \to \infty$. By (8.36), the value of the spatial bandwidth that leads to arbitrarily small deviation is now precisely determined as

$$w = \left(1 + \frac{1}{o(\beta a)}\right)W. \tag{8.51}$$

Using the bounds (8.42) and (8.46), this leads to the following asymptotic notion of spatial bandwidth:

The spatial bandwidth of the field radiated by an arbitrary scattering system of radius a is

$$\beta a + o(\beta a) \le W_0 \le \sqrt{2}\beta a + o(\beta a), \quad \text{as } \beta a \to \infty, \tag{8.52}$$

and in the far field the spatial bandwidth is

$$W_0 = \beta a + o(\beta a), \quad \text{as } \beta a \to \infty. \tag{8.53}$$

Figure 8.2 depicts the transition of the deviation of the space of fields from the space of bandlimited functions. The deviation tends to zero for values of the bandwidth only slightly larger than βa. The transition window viewed at the scale of βa vanishes, and the deviation tends to become a step function.

This result shows that as $\beta a \to \infty$, the space of scattered fields can be approximated with vanishing error by the space of functions essentially bandlimited to βa. In order to bound the degree to which the space of fields can be approximated by finite-dimensional subspaces, we can then apply the results for bandlimited functions developed in Chapters 2 and 3. By projecting $\mathscr{E}$ onto the space of bandlimited fields $\mathscr{B}_w$ and adding this approximation error to the one obtained by a finite interpolation of a strictly bandlimited signal, we obtain a bound on the overall error of the approximation. This technique was illustrated in Section 3.5.1, and we apply it next in the context of electromagnetic signals.

8.5 Degrees of Freedom

In order to bound the number of degrees of freedom of scattered fields, we consider an interval $(-S/2, S/2)$ of interest on the observation curve S. We want to determine the number of degrees of freedom of fields approximately bandlimited to βa, in the sense of (8.53), and observed over this finite interval, as $\beta a \to \infty$. This amounts to determining the dimension of the minimal subspace representing the elements of $\mathscr{E}$ within ϵ accuracy in the interval $(-S/2, S/2)$, namely the minimum number of approximating functions that makes the Kolmogorov N-width $d_N^2(\mathscr{E}) < \epsilon$.

We proceed with the technique developed in Section 3.5.1. By (3.92) we have that the Kolmogorov N-width of the scattered field is bounded by

$$d_N(\mathscr{E}) \leq D_{\mathscr{B}_w}(\mathscr{E}) + d_N(\mathscr{B}_w). \tag{8.54}$$

We can make $d_N^2(\mathscr{E}) < \epsilon$, by ensuring that both the deviation $D_{\mathscr{B}_W}(\mathscr{E})$ and the N-width $d_N(\mathscr{B}_w)$ are arbitrarily small. This can be done by first choosing $w = W_0$, so that $D_{\mathscr{B}_w}(\mathscr{E})$ tends to zero as $\beta a \to \infty$ in the $L^2(-\infty, \infty)$ norm, and hence also in the $L^2(-S/2, S/2)$ norm, and then choosing

$$N = \frac{W_0 S}{\pi} + \frac{1}{\pi} \log\left(\frac{1-\epsilon}{\epsilon}\right) \log\left(\frac{W_0 S}{2}\right) + o(\log W_0 S/\pi), \tag{8.55}$$

so that by (2.132) and (3.13), $d_N^2(\mathscr{B}_w) < \epsilon$ as $\beta a \to \infty$.

We now let

$$N_0 = \frac{\beta a S}{\pi}, \tag{8.56}$$

which corresponds to the space-bandwidth product corresponding to the Nyquist number in our spatial setting. By substituting the upper bound in (8.52) into (8.55) and using (8.56), it follows that we can choose

$$N = \sqrt{2} N_0 + O(\log N_0) \tag{8.57}$$

and make $d_N^2(\mathscr{E}) < \epsilon$. We can now state the final result:

The number of spatial degrees of freedom at level ϵ of the field radiated by an arbitrary scattering system of radius a is

$$N_\epsilon(\mathscr{E}) \leq \sqrt{2} N_0 + O(\log N_0), \quad \text{as } N_0 \to \infty, \tag{8.58}$$

where the dependence on ϵ appears hidden as a pre-constant of the second-order term $O(\log N_0)$ in the phase transition of the number of degrees of freedom. In far field conditions, we also have the tighter bound

$$N_\epsilon(\mathscr{E}) \leq N_0 + O(\log N_0), \quad \text{as } N_0 \to \infty. \tag{8.59}$$

This shows that as $\beta a \to \infty$, the number of eigenfunctions that are needed to interpolate the field within ϵ accuracy is only slightly larger than $N_0 = \beta a S/\pi$. It follows that the effective dimension of the signal's space essentially corresponds to the Nyquist number N_0.

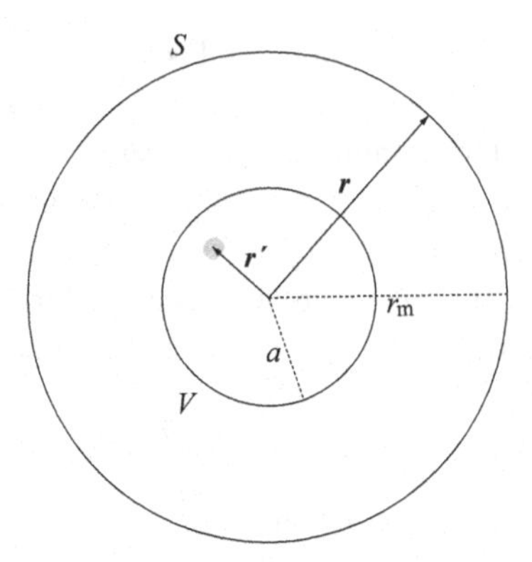

Fig. 8.10 Cylindrical configuration.

The close relationship between effective bandwidth and degrees of freedom must be stressed: it is the existence of an effective bandwidth for the field that makes the number of degrees of freedom practically insensitive to the error level, namely a well-defined quantity.

8.5.1 Hilbert–Schmidt Decomposition

We now attempt a more direct route to determine the number of degrees of freedom of the scattered field, bypassing the intermediate step of computing the bandlimitation error. This entails computing the Hilbert–Schmidt decomposition of the Green's operator (8.7) and then, according to (3.56), studying the behavior of the Nth singular value of this operator to determine the quality of the approximation and thus the degrees of freedom of the field. This direct method relies on the basic result discussed in Chapter 3, that computing the singular values of a Hilbert–Schmidt kernel leads to the optimal decomposition of the kernel into basis functions.

As attractive as it seems, this route is not entirely clear of obstacles. One problem is that we have to compute the spectral decomposition of the Green's operator explicitly, and this computation depends on the geometry under consideration. Nevertheless, in some cases direct computation is possible, and we give one such example next.

We consider the special case of cylindrical propagation, in which V is a disc of radius a, and the observation domain is a circle S concentric and external to V – see Figure 8.10. In cylindrical coordinates, we have $\mathbf{r} = (\rho, \phi_r, z)$, $\mathbf{r}' = (\rho', \phi_{r'}, z)$ with $\rho = r$, $\rho' = r'$, and $z = 0$. The current density is directed along $\bar{\mathbf{z}}$, so that the observed electric field has only the $\bar{\mathbf{z}}$ component, and we can refer to the scalar field only.

By (5.27), the scalar representation of the field is given by

$$E(\mathbf{r}) = (\mathcal{G}I)(\mathbf{r}) = -\frac{\beta^2}{4\omega\epsilon_0} \int_V H_0^{(2)}(\beta|\mathbf{r} - \mathbf{r}'|) I(\mathbf{r}') d\mathbf{r}', \tag{8.60}$$

where $H_0^{(2)}(\cdot)$ is the Hankel function of the second kind and order zero. Using the addition theorem for Hankel functions (see Appendix E.5.3), (8.60) becomes

$$E(\mathbf{r}) = -\frac{\beta^2}{4\omega\epsilon_0}\int_V I(\mathbf{r}')\sum_{n=-\infty}^{\infty} J_n(\beta r')H_n^{(2)}(\beta r)\exp[jn(\phi_r - \phi_{r'})]d\mathbf{r}'. \tag{8.61}$$

We consider the decomposition of (8.60),

$$(\mathcal{G}I)(\mathbf{r}) = \sum_{n=-\infty}^{\infty}\sqrt{\lambda_n}\langle I,\varphi_n\rangle\xi_n(\mathbf{r}), \tag{8.62}$$

where $\langle I,\varphi_n\rangle$ denotes the inner product

$$\langle I,\varphi_n\rangle = \int_V I(\mathbf{r}')\varphi_n^*(\mathbf{r}')d\mathbf{r}'. \tag{8.63}$$

To compute the coefficients of the decomposition, we choose the following basis functions:

$$\xi_n(\mathbf{r}) = -\frac{H_n^{(2)}(\beta r)\exp(jn\phi_r)}{\sqrt{\beta r}|H_n^{(2)}(\beta r)|} \tag{8.64}$$

and

$$\varphi_n(\mathbf{r}) = \frac{J_n(\beta r')\exp(jn\phi_{r'})}{\sqrt{2\pi}\left(\int_0^a |J_k(\beta r')|^2 r'dr'\right)^{1/2}}, \tag{8.65}$$

where $J_n(\cdot)$ is the Bessel function of the first kind of order n, and $H_n^{(2)}(\cdot)$ is the Hankel function of the second kind and of order n. Comparing (8.62) and (8.61), and using (8.64) and (8.65), we obtain

$$\sqrt{\lambda_n} = \frac{\pi\beta^2(\beta r)^{1/2}}{2\omega\epsilon_0\sqrt{2\pi}}\left|H_n^{(2)}(\beta r)\right|\left(\int_0^a |J_n(\beta r')|^2 r'dr'\right)^{1/2}. \tag{8.66}$$

The integral in (8.66) can be computed directly and leads to the final result

$$\sqrt{\lambda_n} = \sqrt{\frac{\mu_0}{\epsilon_0}}\frac{\sqrt{\pi}}{4}\beta a(\beta r)^{1/2}\left|H_n^{(2)}(\beta r)\right|[(J_n(\beta a))^2 - J_{n-1}(\beta a)J_{n+1}(\beta a)]^{1/2}. \tag{8.67}$$

The expansion (8.62) is the bilateral Hilbert–Schmidt decomposition of the Green's operator, and the obtained coefficients $\{\sqrt{\lambda_n}\}$ are the singular values of the operator, while $\{\xi_n\}$ and $\{\varphi_n\}$ are the left and right singular functions. The decomposition can be put in the usual monolateral form used in Section 5.4.1 by using the connection formulas in Appendix E.4 and performing the substitutions

$$\begin{cases} \sqrt{\lambda_n} \rightarrow 2\sqrt{\lambda_n} & \text{for all } n > 0, \\ \xi_n(\mathbf{r}) \rightarrow -\dfrac{H_n^{(2)}(\beta r)\cos(n\phi_r)}{\sqrt{\beta r}|H_n^{(2)}(\beta r)|} & \text{for all } n \geq 0, \\ \varphi_n(\mathbf{r}) \rightarrow \dfrac{J_n(\beta r')\cos(n\phi_{r'})}{\sqrt{2\pi}\left(\int_0^a |J_k(\beta r')|^2 r'dr'\right)^{1/2}} & \text{for all } n \geq 0. \end{cases} \tag{8.68}$$

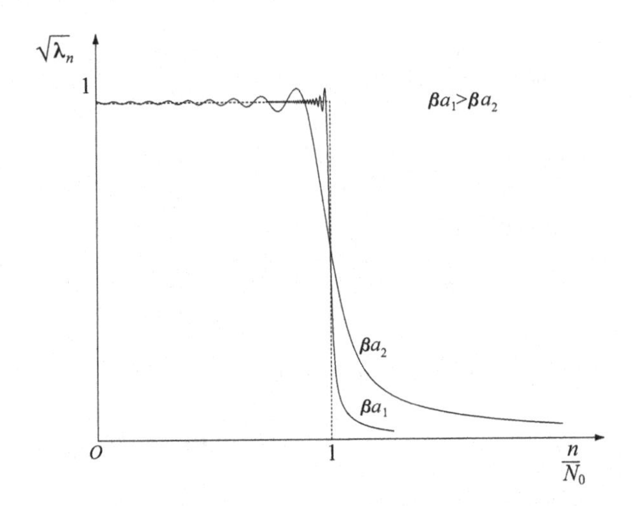

Fig. 8.11 Phase transition of the singular values.

It follows that the electric field on the observation circle lies on a Hilbert space whose dimension depends on the behavior of the singular values (8.67). Figure 8.11 depicts the behavior of the singular values for two different values of βa. The singular values undergo a phase transition around the critical value

$$N_0 = \frac{\beta a S}{\pi} = \frac{\beta a}{\pi} \frac{2\pi r}{r_{\text{m}}}. \tag{8.69}$$

When viewed at the scale of N_0, the transition tends to become a step function as $a \to \infty$. Results analogous to (8.58) and (8.59) can now be derived by studying the asymptotic behavior of $\sqrt{\lambda_n}$ around the critical point $n = N_0$, as $N_0 \to \infty$. This is obtained by some analytic manipulations and using uniform asymptotic expansions for which both the argument and the index of the special functions in (8.61) tend to infinity.

8.5.2 Sampling

For any given field configuration, the coefficients of the optimal field representation leading to the notion of degrees of freedom are functionals of the field, and thus their computation involves knowledge of the field over the whole observation domain. It is of practical interest to consider simpler interpolations using only sampled field values, provided that the loss of optimality in the representation can be tolerated. From the results in Section 2.4.1 it follows that if we use the cardinal series to interpolate strictly bandlimited signals, then any constant increase in the number of sampled values does not lead to an approximation error comparable to the one of the optimal Hilbert–Schmidt representation. Nevertheless, alternative sampling representations can achieve the same error as the optimal representation with a number of extra samples that grows only sublinearly with N_0, and thus provide only a marginal increase in the number of coefficients compared to the optimal Hilbert–Schmidt representation. The advantage of such sampling interpolations is that, while being slightly suboptimal, they require

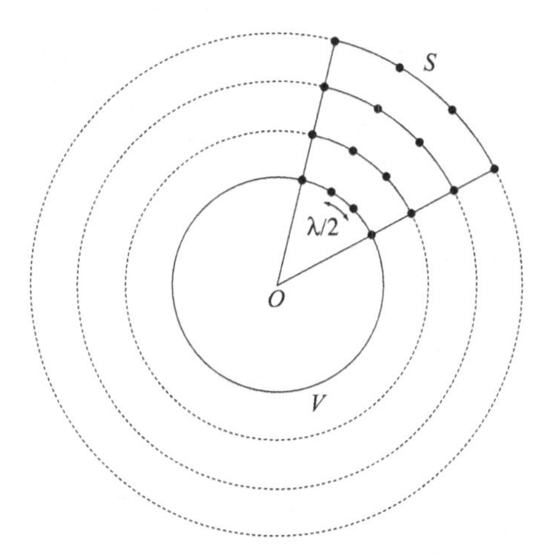

Fig. 8.12 Sampling interpolation over arcs of circumference.

only a discrete set of field values for reconstruction. These interpolation schemes can also be applied to the almost bandlimited signals considered in this chapter, with the same marginal order increase in the number of sampled field values.

In the case of a circular observation domain, the number of sampled field values used for reconstruction depends only on the angular aperture of the observation domain and not on its spatial extension. This follows directly from (8.69), where the arc length S is normalized to r_{m}, so that only the angular aperture of the observation domain plays a role in determining the number of degrees of freedom. It follows that uniform sampling can be performed over the angular domain rather than over the length of the observation domain – see Figure 8.12.

For a full circle, letting λ be the transmission wavelength and recalling that $\beta = 2\pi/\lambda$, we have

$$N_0 = \frac{2\pi}{\lambda}\frac{a}{\pi}\frac{2\pi r}{r_{\text{m}}} = \frac{2\pi a}{\lambda/2}. \tag{8.70}$$

It follows that an angular spacing corresponding to half a wavelength on the perimeter of the radiating disc is the appropriate minimum separation distance for sampling representations of the field. Sampling points can be several wavelengths apart when the receiving domain is far from the radiating system, and more tightly packed together when the receiver is close to the radiating system. The theoretical minimum distance $\lambda/2$ is reached in the proximity of the radiating system and corresponds to the minimum non-redundant spacing given by the Nyquist sampling criterion over the sources' perimeter.

This sampling result can be used to justify the practical design principles of cellular communication systems. Consider a base station equipped with multiple antennas located along S, serving users distributed inside the cell V. The antennas along S can be widely separated when the base station is distant from the users, while they are more closely packed together when the base station is closer to the users, as in the case of microcells. An analogous argument holds for remote sensing systems, where the objective is to construct an image of the sources or scatterers inside V. An optimal placement of detectors occurs at constant angular spacing, so that their optimal separation increases with their distance from the radiating system.

8.6 Cut-Set Integrals

We now provide a refined computation of the number of degrees of freedom, based on the notion of a cut-set integral, that resolves some problems in the physical interpretation of the mathematical results obtained in the previous sections.

From the scaling property of the Fourier transform, as a signal is stretched in the wavenumber domain, spreading its spatial bandwidth, it becomes more concentrated over the angular observation domain. It is then natural to expect, as $a \to \infty$, that sampling points on the observation curve that are far away from the source of radiation provide a vanishing contribution to the information useful for reconstruction. Using sparser sampling in regions far from the scattering system, as illustrated in Figure 8.13, should be sufficient for reconstruction. On the other hand, since the number of degrees of freedom

$$N_0 = \frac{\beta a S}{\pi} \tag{8.71}$$

increases proportionally to the length of the observation curve normalized to r_m, if the minimum distance from the scattering system is kept constant, then the number of spatial samples required for reconstruction increases proportionally to the length of the observation curve. In this case, the information carried by waves propagating through

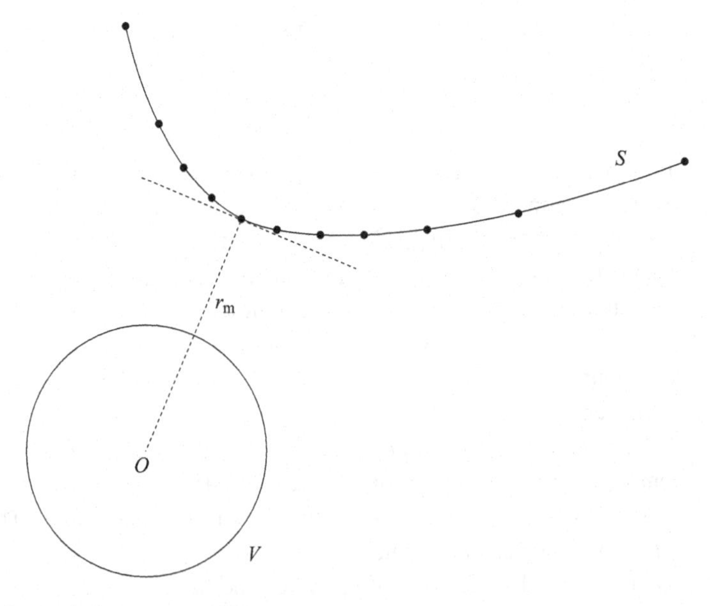

Fig. 8.13 Sampling interpolation over an arbitrary curve.

an indefinite open curve may be larger than the information flow through any circular enclosure of the whole scattering system, given by (8.70). Since one expects a circular enclosure to capture the whole information flow radiating from the scattering system, the degrees of freedom would be a non-conservative information measure – something that should make us slightly uncomfortable, to say the least.

Fortunately, a closer look at the derivation in Section 8.4 makes the model again consistent with our physical intuition, showing that the number of degrees of freedom is indeed a conservative quantity. The crux of the matter is that in our derivation we have considered the largest bandwidth along the observation curve given by (8.20), and then computed the number of degrees of freedom using such a maximum value. As a consequence, unless the spatial bandwidth is constant along the curve, as in the case of a circular observation domain, the samples' spacing can be unnecessarily small in the regions where the spatial bandwidth is significantly smaller than its maximum value, and this can lead to redundancy in the corresponding field representation.

Repeating the derivation in Section 8.4 while accounting for the variability of the spatial bandwidth along an arbitrary observation curve, we obtain a non-redundant representation of the field over the curve, expressed in integral form. The number of degrees of freedom can then be computed as a *cut-set integral* representing the amount of information flowing through any arbitrary observation domain external to the radiating system. As expected, the maximum value of this integral corresponds to a closed circular enclosure of the radiating system, and in this case it reduces to the value $2\beta a = 2\pi a/(\lambda/2)$ obtained previously.

8.6.1 Linear Cut-Set Integral

We consider the field radiated by a scattering system enclosed in a convex domain V bounded by a surface with rotational symmetry obtained by rotating an analytic curve lying in the plane $\phi = 0$ about the z axis. We let $\zeta : S \to \mathbb{R}$ be an arbitrary coordinate along the observation curve, $s : S \to \mathbb{R}$ be the arc length coordinate, and $\bar{\mathbf{t}}$ the unit vector tangent to s – see Figure 8.14. All distance lengths are normalized to the wavelength of transmission.

The reduced field is obtained by extracting an analytical phase factor $\psi(\zeta)$ from the radiated field that depends on the chosen coordinate along the curve. Namely,

$$\mathbf{E}(\zeta) \to \exp(j\psi(\zeta))\mathbf{E}(\zeta). \tag{8.72}$$

Following the same steps as in Section 8.4, we let the local bandwidth at point ζ be

$$w(\zeta) = \max_{\mathbf{r}' \in V} \left| \frac{\partial}{\partial \zeta} (\psi(\zeta) - 2\pi R(\mathbf{r}', \zeta)) \right|. \tag{8.73}$$

The phase factor ψ is chosen so that $w(\zeta)$ is minimized for all ζ. This is accomplished by letting the derivative of ψ be equal to the average between the maximum and minimum values of $2\pi \partial R/\partial \zeta$,

$$\frac{\partial \psi}{\partial \zeta} = \pi \left(\max_{\mathbf{r}' \in V} \frac{\partial R}{\partial \zeta} + \min_{\mathbf{r}' \in V} \frac{\partial R}{\partial \zeta} \right). \tag{8.74}$$

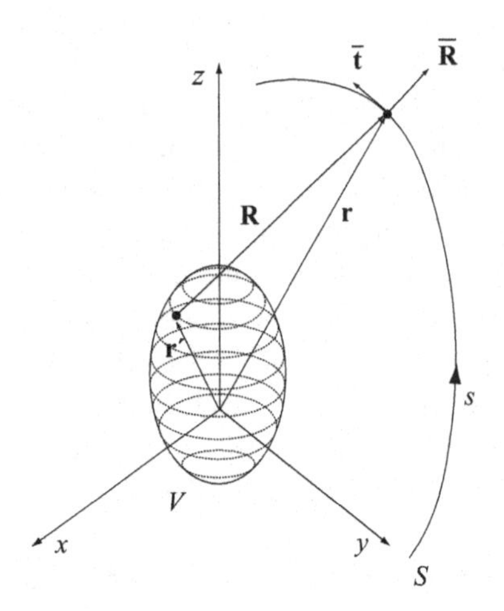

Fig. 8.14 Linear cut-set integral geometry.

Substituting (8.74) into (8.73), we obtain

$$w(\zeta) = \pi \left(\max_{\mathbf{r}' \in V} \frac{\partial R}{\partial \zeta} - \min_{\mathbf{r}' \in V} \frac{\partial R}{\partial \zeta} \right). \tag{8.75}$$

Next, the curvilinear coordinate ζ is chosen so that w is constant along the curve, namely $w(\zeta) = W$ for all $\zeta \in M$. This amounts to imposing the local bandwidth to be constant along the curve by choosing a certain curvilinear coordinate that depends on W.

The number of degrees of freedom is obtained by first substituting into (8.75)

$$\frac{\partial R}{\partial \zeta} = \frac{\partial R}{\partial s} \frac{\partial s}{\partial \zeta}, \tag{8.76}$$

obtaining

$$W d\zeta = \pi \left(\max_{\mathbf{r}' \in V} \frac{\partial R}{\partial s} - \min_{\mathbf{r}'} \frac{\partial R}{\partial s} \right) ds. \tag{8.77}$$

Then, we integrate along the curve S, obtaining

$$\begin{aligned} W \int_S d\zeta &= \pi \int_0^S \left(\max_{\mathbf{r}' \in V} \frac{\partial R}{\partial s} - \min_{\mathbf{r}'} \frac{\partial R}{\partial s} \right) ds \\ &= \pi \int_0^S \left(\max_{\mathbf{r}' \in V} \bar{\mathbf{R}} \cdot \bar{\mathbf{t}} - \min_{\mathbf{r}'} \bar{\mathbf{R}} \cdot \bar{\mathbf{t}} \right) ds. \end{aligned}$$

Finally, we compute the Nyquist number associated with the bandwidth W as

$$N_0 = \frac{W \int_S d\zeta}{\pi} = \int_0^S \left(\max_{\mathbf{r}' \in V} \bar{\mathbf{R}} \cdot \bar{\mathbf{t}} - \min_{\mathbf{r}' \in V} \bar{\mathbf{R}} \cdot \bar{\mathbf{t}} \right) ds. \tag{8.78}$$

This dimensionless quantity corresponds to the space-bandwidth product of the radiated field. It represents the effective dimensionality of the space of signals of bandwidth W, observed over the domain S. The approximation error can be related to the geometric parameters of the scattering system and of the observation curve, and it becomes arbitrarily small when taking appropriate limits of these parameters. An accurate field reconstruction is possible by interpolating only slightly more than N_0 basis functions. As we shall see in the next section, in the case of an arbitrary meridian curve lying on the plane $\phi = 0$, the relevant parameter is the wavelength-normalized meridian length of the scattering system. In the case of an arbitrary circle lying on a plane at constant z, the relevant parameter depends on both the z coordinate and the radius of the circle, but it is always bounded by the wavelength-normalized length of the maximum latitudinal circle of the scattering system.

We call (8.78) the *cut-set integral* associated with the cut S and the radiating system V. It measures the total incremental (maximum minus minimum) variation of the tangential component of $\mathbf{R}$ along the cut S, over all possible scatterer configurations. A physical interpretation is the following:

> The cut-set integral measures the richness of the information content of a multiple scattered field in terms of a variational measure of the scattering system with respect to the cut through which the information must flow.

This interpretation implies that if the combined effect of the scattering environment and the receiver's geometries are such that $\mathbf{R}$ is "highly variable" in the sense of the integral above, then the scattered field has a large information content.

8.6.2 Surface Cut-Set Integral

The results in the previous section can be generalized to surface cuts by considering two-dimensional observation manifolds. The idea is to refer to a set of appropriate coordinate curves on the observation manifold and decompose the field separately along each coordinate. The resulting surface integral, measuring the number of spatial degrees of freedom, is interpreted as a surface boundary through which the information must flow and provides a limit on the amount of information that can radiate from the scattering system.

We consider a cylindrical coordinate system (ρ, ϕ, z) and a scattering system enclosed in a convex domain V bounded by a surface with rotational symmetry obtained by rotating an analytic curve lying in the plane $\phi = 0$ about the z axis. The field is observed over a surface of revolution S external to V and generated by the rotation of an analytic meridian curve S_z lying on the plane $\phi = 0$ about the z axis – see Figure 8.15. We let $\rho'(z')$ be the radial coordinate of V and

$$\max_{z'} \rho'(z') = a. \tag{8.79}$$

Consider the representation of the field on the circumference S_ϕ obtained by intersecting S with a plane at $z = z_s$. The number of degrees of freedom of the field on this circumference can be computed by evaluating the corresponding cut-set integral (8.78). Due to the cylindrical symmetry, the extreme values of $\partial R/\partial s$ are in this case opposite and constant along S_ϕ. It follows that the bandwidth is constant along the curve

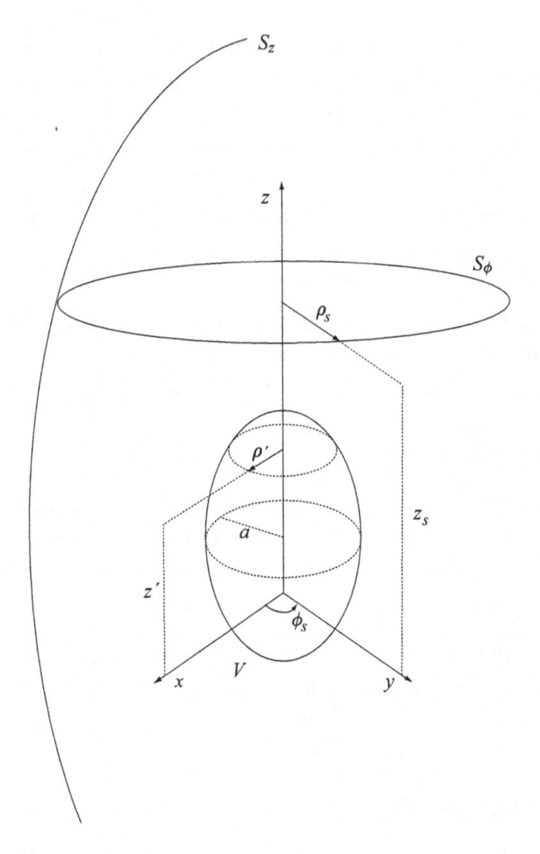

Fig. 8.15 Surface cut-set integral geometry.

and, as in the cylindrical case examined in Section 8.5.1, we can choose the azimuthal angular parametrization of the curve $\zeta = \phi_s$. The cut-set integral (8.78) can then be computed and leads to a Nyquist number that depends on ρ_s and z_s,

$$N_0(\rho_s, z_s) = 2\pi \max_{z'} \left(\sqrt{(z_s - z')^2 + (\rho_s + \rho'(z'))^2} - \sqrt{(z_s - z')^2 + (\rho_s - \rho'(z'))^2} \right) \\ \leq \frac{2\pi a}{\lambda/2}. \tag{8.80}$$

As the circle is moved to the proximity of V and $\rho_s \to \rho'(z')$, we also have the limiting result as $\rho'(z') \to \infty$:

$$N_0(\rho_s, z_s) \sim \frac{2\pi \rho'(z')}{\lambda/2}. \tag{8.81}$$

It follows that the number of degrees of freedom of an azimuthal circle placed near the boundary of V can be approximated as $\rho'(z') \to \infty$, or equivalently $a \to \infty$, by the number of samples equispaced by half a wavelength of the corresponding latitudinal curve on the surface of the radiating system. The upper bound obtained when the circle is in correspondence with the largest radial coordinate of V coincides with (8.70), obtained in the cylindrical case – see also Figure 8.12.

As in the cylindrical case, the optimal representation of the field on S_ϕ is given in terms of cylindrical harmonics,

$$\mathbf{E}(\rho_s,\phi_s,z_s) = \sum_{n=-\infty}^{\infty} \mathbf{E}_n(\rho_s,z_s)e^{jn\phi_s}, \tag{8.82}$$

where the vector coefficients $\{\mathbf{E}_n(\rho_s,z_s)\}$ do not depend on ϕ and represent the field along the meridian curves S_z obtained by intersecting S with any plane at constant ϕ. Comparing (8.82) with the analogous expansion for the cylindrical case given in Section 8.5.1, we see that while the basis functions are the same complex exponentials in the both cases, the coefficients in this case are vectors. This is because the field is not necessarily oriented along z as in the cylindrical case, and we cannot refer to a unique scalar field representation. In addition, the coefficients depend on both ρ and z, while they were independent of z in the cylindrical propagation case.

The number of degrees of freedom of each vector coefficient $\mathbf{E}_n(\rho_S,z_s)$ can also be computed by evaluating the one-dimensional cut-set integral in (8.78) on the meridian curve S_z and choosing an appropriate parametrization $\zeta = \zeta(\rho_s,z_s)$ of the curve. We choose the parametrization for which the spatial bandwidth

$$W = \ell/\lambda \tag{8.83}$$

is constant along the curve and equals the meridian length of the scattering system ℓ, namely the curve obtained by intersecting V with a plane at constant ϕ, normalized by the wavelength. For this parametrization, any meridian closed curve encircling V covers a 2π range, and by (8.78) the corresponding Nyquist number is given by

$$N_O(\ell) = \frac{2\pi\ell}{\pi\lambda} = \frac{\ell}{\lambda/2}. \tag{8.84}$$

It follows that the number of degrees of freedom on any meridian curve encircling the radiating system can be approximated as $\ell \to \infty$, or equivalently $a \to \infty$, by the number of equispaced samples on a meridian of the system's surface.

Now let S be a surface enclosing V; the total number of degrees of freedom on S has a phase transition in the neighborhood of

$$\begin{aligned} N_O &= \frac{1}{2}\int_{S_z} N_O(\ell)N_O(\rho_s,z_s)dz \\ &\sim \frac{\ell}{\lambda/2}\frac{\pi}{\lambda/2}\int_{S_z}\rho'(z')dz' \\ &= \frac{A(S)}{(\lambda/2)^2}, \end{aligned} \tag{8.85}$$

where $A(S)$ is the area of S and the asymptotic equality holds by virtue of (8.81) and (8.84) as S is placed in the proximity of V and is made conformal to its boundary, and as the system's size $a \to \infty$.

This result has an appealing physical interpretation: the number of degrees of freedom of arbitrary scattered fields is essentially coincident with those of an array of radiating elements spaced by $\lambda/2$ and conforming to the surface enclosing the radiating system. It follows that a non-redundant field representation should use a number of parameters at most of the order of (8.85). This is relevant for remote sensing and inverse scattering problems. Using a number of sensors larger than the obtained physical bound is essentially useless. Similarly, in communications, using a number of antennas larger than the obtained physical bound does not lead to additional multiplexing gain. In other words, the obtained bound applies to the spatial decompositions described in Sections 5.3.2 and 5.4.1, providing a physical limit on the rank of the propagation operator. The propagation environment works as a low-pass spatial filter, whose spatial band is ultimately limited by the size of the scattering system.

Although the obtained bound holds for arbitrary scattering systems, in many practical cases the spatial bandwidth can be smaller than what has been computed here, or, in the case of the superposition of a number of occupied sub-bands as described in Section 3.5.2, the field can be represented using a significantly smaller number of basis functions.

8.6.3 Applications to Canonical Geometries

The cut-set integral can be used to bound the effective dimensionality of the field anywhere *outside* a closed surface bounding the radiating system. By symmetry, it also applies to the field anywhere *inside* a closed surface not containing the radiating system. This follows from the equivalence result in Section 4.7 stating that knowledge of the field on any closed surface containing the sources uniquely determines the field outside it. Similarly, for outside sources, knowledge of the field on the surface uniquely determines the field inside. In the following, we illustrate how these results can be applied to different canonical geometries.

Consider an arbitrary radiating environment bounded by a sphere of radius a placed in three-dimensional space. Consider another sphere concentric to the radiating one and of the same order of radius, say $2a$, as depicted in Figure 8.16(a). The number of degrees of freedom of the field measured anywhere inside the annular volume delimited by radii a and $2a$ can be computed in terms of the cut-set integral of the inner annular surface. From the results in Section 8.6.2, as $a \to \infty$ this is given by

$$N_0 = O(a^2/\lambda^2). \tag{8.86}$$

In this spherical geometry the cut-set area, normalized by the wavelength squared, provides a resolution limit for any field observation.

Consider now the half-spherical geometry depicted in Figure 8.16(b). This can represent communication using multiple antennas in an urban environment, where high-rise buildings may be present near the city center, while only lower constructions can be present at the periphery of the city. It is easy to check that (8.86) applies in this case as well. The situation changes when the scattering system does not scale in all three

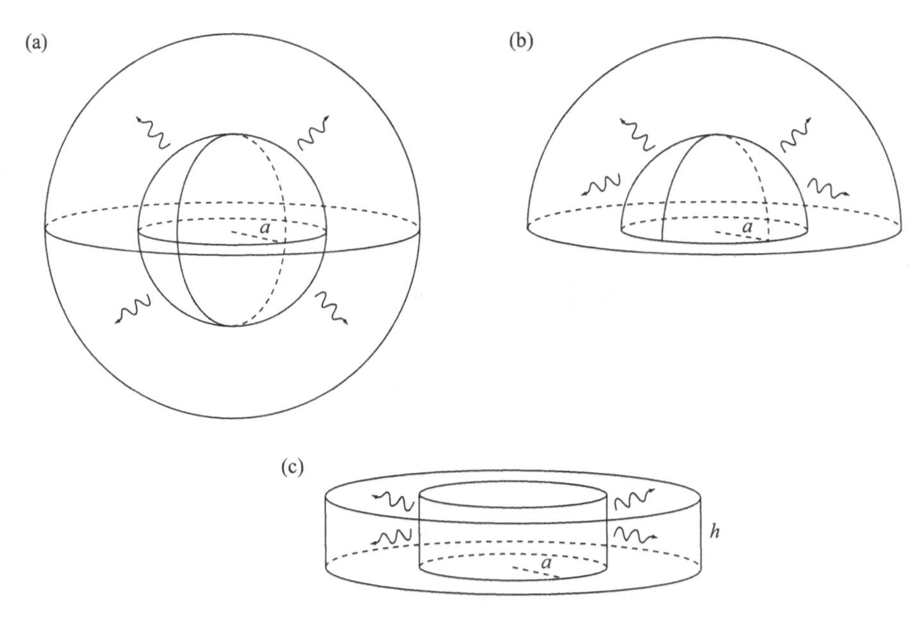

Fig. 8.16 Three canonical geometries.

dimensions. Consider, for example, a model of communication in suburban, residential, and rural areas where the heights of the scattering objects are uniformly bounded. The scattering system can in this case be modeled as a cylinder of radius a and of fixed height h – see Figure 8.16(c). The number of degrees of freedom of the field inside the outer cylindrical annulus of height h and delimited by radii a and $2a$ can be computed in terms of the cut-set integral of its bounding surface to be

$$N_0 = O\left(\frac{ha}{\lambda^2}\right), \tag{8.87}$$

as $a \to \infty$. This shows that since only two-dimensional scaling occurs as $a \to \infty$, the number of degrees of freedom scales in this case as a rather than a^2.

This result follows by first computing the cut-set integral on the perimeter of the rectangle Q depicted on the right-hand side of Figure 8.17, and then extending the integral to a surface cut following the procedure described in Section 8.6.2, namely considering a rotation of the cut-set perimeter along a 2π azimuthal angle. In the following, we assume that the edges of the rectangle are smoothed, so that the observation curve is analytical, and that the rectangle is separated from the radiating domain so that the observation surface is external to the radiating system. With these assumptions, the cut-set integral (8.78) along the perimeter of the rectangle Q can be decomposed into four terms, each corresponding to one side of the rectangle. Since the rectangle has constant height h, the cut-set integral along each vertical side is at most $2h/\lambda$ and, by symmetry, the integrals along the two horizontal sides are equal. It follows that it suffices to compute the integral over the upper horizontal side corresponding to $\rho = s$, $y = 0$, $z = h$, and $s \in [0, a]$. From elementary geometry it is clear that the extreme

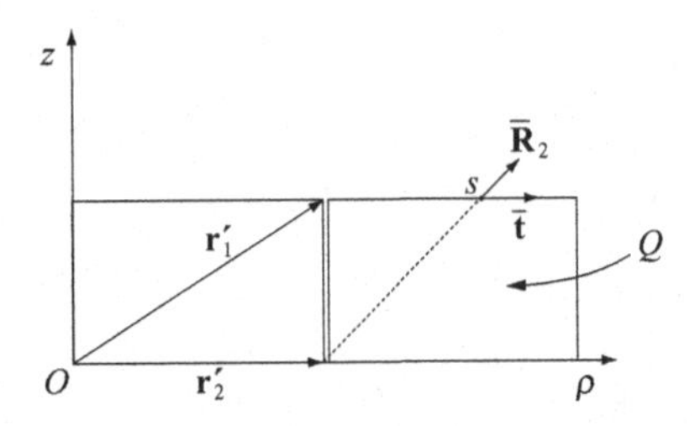

Fig. 8.17 Linear cut-set integral for the cylindrical geometry.

values of $\bar{\mathbf{R}}\cdot\bar{\mathbf{t}}$ correspond to the positions $\mathbf{r}_1'$ and $\mathbf{r}_2'$ illustrated in Figure 8.17, for which we have

$$\bar{\mathbf{R}}_1\cdot\bar{\mathbf{t}} = 1, \tag{8.88}$$

$$\bar{\mathbf{R}}_2\cdot\bar{\mathbf{t}} = \frac{1}{\sqrt{1+(h/s)^2}}. \tag{8.89}$$

Letting the normalized lengths $h^* = h/\lambda$ and $a^* = a/\lambda$, we have

$$\begin{aligned}
&\int_0^{a^*}\left(\max_{\mathbf{r}'}\bar{\mathbf{R}}\cdot\bar{\mathbf{t}} - \min_{\mathbf{r}'}\bar{\mathbf{R}}\cdot\bar{\mathbf{t}}\right)ds\\
&= \int_0^{a^*} 1 - \frac{1}{\sqrt{1+(h/s)^2}}ds\\
&\leq \int_0^{a^*}\frac{(h^*/s)^2}{1+(h^*/s)^2}ds\\
&= h^*\arctan a^*\\
&< 2h^*\\
&= \frac{h}{\lambda/2},
\end{aligned} \tag{8.90}$$

where the first inequality follows from $\sqrt{1+x^2} \leq 1+x^2$ for all x. It follows that the cut-set integral along the perimeter of the rectangle Q is

$$N_0 < \frac{4h}{\lambda/2}. \tag{8.91}$$

Extending the computation along the azimuthal angle as in (8.85) amounts to multiplying by an additional factor of order of $2\pi a/\lambda$, and we obtain (8.87).

8.7 Backscattering

So far, we have considered all sources and scatterers to be enclosed in convex domains bounded by a cut-set surface with rotational symmetry. We now consider the effect of scatterers placed outside the radiating system. These can change the field configuration

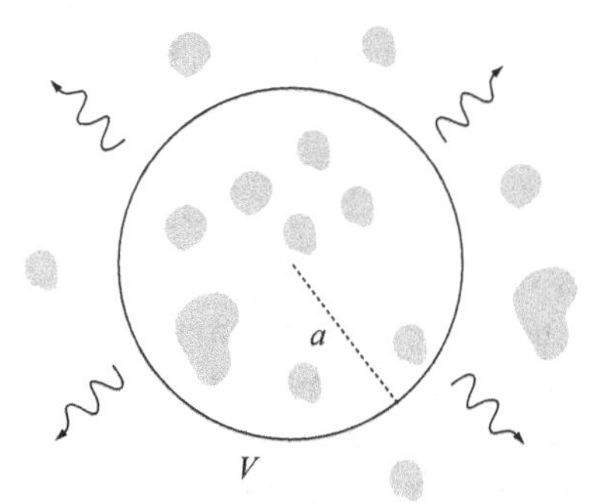

Fig. 8.18 Propagation in the presence of scatterers outside V.

at any observation point outside V. However, the number of distinct field configurations that can be observed by only varying the scattering environment and the currents *inside* V remains the same. Since the number of degrees of freedom refers to the field generated by all possible configurations of sources and scatterers inside the scattering system, it follows that:

> The number of degrees of freedom of the field radiated by V does not change in the presence of external scatterers.

Restated intuitively in terms of information flow, the result is that a closed surface surrounding the radiating system captures the whole information flow coming out of the system and the backscattered field does not provide any new information.

We first provide an heuristic explanation of this result. The radiating and scattering elements that we can vary to produce different field configurations are inside V. By the equivalence result in Section 4.7, the electric field at any point outside V is uniquely determined by the field on the surface of V. This is composed of the field radiated from inside V, which we know has a number of degrees of freedom proportional to a/λ, and the field backscattered from outside V. This latter term changes the field configuration without adding additional degrees of freedom since it is also due to the sources inside V. It follows that the external scatterers can only modulate the spatial modes of the signal, but in the limit $a \to \infty$ the phase transition behavior around a/λ of the number of degrees of freedom is preserved. Next, we provide a more rigorous derivation.

Let the electric field at any point outside V be given by the superposition of the direct contribution due to the currents inside V and the scattered contribution due to the induced currents on the outside scatterers:

$$\mathbf{E}(\mathbf{r}) = \mathbf{E}_{\mathrm{D}}(\mathbf{r}) + \mathbf{E}_{\mathrm{S}}(\mathbf{r}). \tag{8.92}$$

We show that both field vectors in (8.92) have a number of degrees of freedom corresponding to the Nyquist number $N_0 = O(a^2/\lambda^2)$, as $a \to \infty$.

Let us first focus on $\mathbf{E}_{\mathrm{D}}$, the field vector due to the source currents and to the induced currents inside V. The induced currents are due to the scattered field inside V, and also to the field backscattered from outside V. Letting $\mathbf{I}(\mathbf{r}')$, $\mathbf{r}' \in V$, be the total current density, we have

$$\mathbf{E}_\mathrm{D}(\mathbf{r}) = \int_V \underline{\mathbf{G}}(\mathbf{r}-\mathbf{r}')\mathbf{I}(\mathbf{r}')d\mathbf{r}'. \tag{8.93}$$

In the analysis of previous sections the number of degrees of freedom of the field has been determined by assuming an arbitrary current density in V, so that the same analysis applies here and the number of degrees of freedom of the direct field $\mathbf{E}_\mathrm{D}$ still corresponds to the Nyquist number $N_0 = O(a^2/\lambda^2)$.

We now focus on $\mathbf{E}_S(\mathbf{r})$, the field due to the currents induced on the scatterers outside V. Letting V' be the volume occupied by the scattering elements outside V and $\mathbf{I}(\mathbf{r}')$, $\mathbf{r}' \in V'$ be the induced current density in V', we have

$$\mathbf{E}_\mathrm{S}(\mathbf{r}) = \int_{V'} \underline{\mathbf{G}}(\mathbf{r}-\mathbf{r}')\mathbf{I}(\mathbf{r}')d\mathbf{r}'. \tag{8.94}$$

From (4.31) the induced current $\mathbf{I}(\mathbf{r}')$ is not arbitrary but is linearly related to the electric field in V',

$$I(\mathbf{r}') = [\sigma(\mathbf{r}') + j\omega(\epsilon(\mathbf{r}') - \epsilon_0)]\mathbf{E}(\mathbf{r}'), \tag{8.95}$$

where ϵ is the permittivity of the dielectric material of the scatterer and σ its conductivity. In the case of perfect conductors (8.95) cannot be used since $\sigma \to \infty$. In this case, the analysis is completely equivalent by using (4.23) in lieu of (8.95) and referring to the magnetic field. Substituting (8.95) into (8.94), we obtain the scattered field

$$\mathbf{E}_\mathrm{S}(\mathbf{r}) = j\omega \int_{V'} \underline{\mathbf{G}}(\mathbf{r}-\mathbf{r}')[\sigma(\mathbf{r}')(\epsilon(\mathbf{r}') - \epsilon_0]\mathbf{E}(\mathbf{r}')d\mathbf{r}', \tag{8.96}$$

which shows that $\mathbf{E}_\mathrm{S}$ is linearly related to the field on the scatterers.

Substituting (8.96) into (8.92), we obtain the integral equation

$$\mathbf{E}(\mathbf{r}) = \mathbf{E}_\mathrm{D}(\mathbf{r}) + \int_{V'} \underline{\mathbf{G}}(\mathbf{r}-\mathbf{r}')[\sigma(\mathbf{r}')(\epsilon(\mathbf{r}') - \epsilon_0]\mathbf{E}(\mathbf{r}')d\mathbf{r}'. \tag{8.97}$$

This is an inhomogeneous Fredholm integral equation of the second kind, whose resolvent formalism leads to the Liouville–Neumann series – see Appendix A.5.2. We have already encountered the homogeneous version of this equation in the context of Slepian's concentration problem in Chapter 2. The equation forms the basis of the multiple scattering theory examined in Chapter 9. What is important here is that it shows a linear relationship between the direct and the total field. Since by (8.96) the scattered and the total fields are also linearly related, it follows that $\mathbf{E}_\mathrm{S}$ and $\mathbf{E}_\mathrm{D}$ are linearly related and thus have the same number of degrees of freedom. The linear operation cannot increase the spatial bandwidth of the signal and the phase transition behavior is preserved in the usual asymptotic sense. We conclude that the electric field outside V can be expressed as the superposition of two field vectors, $\mathbf{E}_\mathrm{D}(\mathbf{r}')$ and $\mathbf{E}_\mathrm{S}(\mathbf{r}')$, each lying in a Hilbert space whose dimension is limited by the range space of the radiation operator $\mathcal{G}$, yielding as $a \to \infty$ a number of degrees of freedom corresponding to the Nyquist number $N_0 = O(a^2/\lambda^2)$.

8.8 Summary and Further Reading

The question of determining the spatial information content of electromagnetic fields in terms of degrees of freedom dates back to at least Toraldo di Francia (1955, 1969) and Gabor (1961). Our treatment followed the rigorous developments by Bucci and Franceschetti (1987, 1989) on the spatial bandwidth and the degrees of freedom of scattered fields based on the approximation-theoretic arguments developed in Chapter 3. We first showed that the electromagnetic field radiated by an arbitrary scattering system is essentially spatially bandlimited. From this result it follows that the number of degrees of freedom undergoes a phase transition at a critical point that is an intrinsic property of the system, depending only on its size. The same result can be obtained by performing the Hilbert–Schmidt decomposition of the Green's operator, whenever the geometric configuration of the radiation problem permits computation in closed form. These results were placed in an information-theoretic setting by Franceschetti, Migliore, and Minero (2009). An interpretation of the degrees of freedom in terms of information flow leads to the definition of a cut-set integral that was developed by Franceschetti *et al.* (2011), following the work of Bucci, Gennarelli, and Savarese (1998). Near-field effects on the number of degrees of freedom were considered by Janaswamy (2011) and Franceschetti *et al.* (2015).

Related results appeared at different points in the literature, including in optics, by Miller (2000) and Piestun and Miller (2000); in information theory, by Poon, Brodersen, and Tse (2005); and in signal processing, by Kennedy *et al.* (2007).

Sampling interpolations can provide a valid alternative to optimal representations. The truncation errors of these representations have been extensively studied in signal processing. Extensive reviews by Jerri (1977), Unser (2000), and Garcia (2000) contain a compendium of references. Approximate prolate spheroidal sampling functions were introduced by Knab (1979, 1983) and provide order-optimal reconstruction. Their truncation error in the context of approximately bandlimited electromagnetic signals was examined by Bucci and Di Massa (1988).

8.9 Test Your Understanding

Problems

8.1 Consider a sinusoidal source current and use (8.3) to compute the field radiated by an elementary dipole placed at the origin of the space. Show that in the far field the solution coincides with the Fourier transform of (4.92a).

Solution

Considering a sinusoidal current density at coordinate $\mathbf{r}' = 0$, so that $R = r$, we have

$$\mathbf{i}(\mathbf{r}', t) = \Re[I_0 \ell \exp(j\omega t)\delta(\mathbf{r}')]\,\bar{\mathbf{z}}\ [\mathrm{A\,m^{-2}}]. \tag{8.98}$$

We proceed using the complex notation and extract the real part at the end of the computation. For the considered source current, (4.92a) becomes

$$\mathbf{E}(s) = \frac{\zeta}{4\pi}\frac{I_0\ell}{cr} j\omega \exp(j\omega t^*) \sin\theta\, \bar{\boldsymbol{\theta}} = I_0\ell \exp(j\omega t)\frac{j\omega\mu}{4\pi r}\exp(-j\beta r)\sin\theta\, \bar{\boldsymbol{\theta}}. \tag{8.99}$$

We want to show that the same result holds by using (8.3) to compute the electric field when $r \to \infty$. Comparing (8.99) with (8.3) we see that to show the equivalence of the two derivations we need to show that the dyad transformation $\underline{\mathbf{N}}$ amounts to performing a rotation from the $\bar{\mathbf{z}}$ axis to the $\bar{\boldsymbol{\theta}}$ axis and a multiplication by $-\sin\theta$. By applying the dyad $\underline{\mathbf{N}}$ to the column vector

$$\frac{\exp(-j\beta r)}{r}\bar{\mathbf{z}} = \begin{pmatrix} 0 \\ 0 \\ \dfrac{\exp(-j\beta r)}{r} \end{pmatrix},$$

and considering for convenience the electric field radiated in the plane $y = 0$, we get for the dominant field components, as $r \to \infty$,

$$\begin{pmatrix} r\exp(j\beta r)\dfrac{1}{\beta^2}\dfrac{\partial^2}{\partial x \partial z}\dfrac{\exp(-j\beta r)}{r} \\ 0 \\ 1 + r\exp(j\beta r)\dfrac{1}{\beta^2}\dfrac{\partial^2}{\partial z^2}\dfrac{\exp(-j\beta r)}{r} \end{pmatrix} = \begin{pmatrix} -\sin\theta\cos\theta \\ 0 \\ 1-\cos^2\theta \end{pmatrix} = \sin\theta \begin{pmatrix} -\cos\theta \\ 0 \\ \sin\theta \end{pmatrix} = -\sin\theta\, \bar{\boldsymbol{\theta}}.$$

Due to the symmetry along the ϕ coordinate, the above result is valid in any plane containing the z axis.

8.2 Perform the computation to obtain (8.31) from (8.29).

Solution

The Airy integral is defined as

$$\mathrm{Ai}(x) = \frac{1}{2\pi}\int_{-\infty}^{\infty} \exp(jy^3/3 + jxy)dy. \tag{8.100}$$

By the change of variable $jy = \tau$ and moving the integration path to the left complex half-plane, it can also be written as

$$\begin{aligned} \mathrm{Ai}\ (x) &= \frac{1}{2\pi j}\int_{-j\infty}^{j\infty} \exp(-\tau^3/3 + x\tau)d\tau \\ &= \frac{1}{2\pi j}\int_{C^\star} \exp(-\tau^3/3 + x\tau)d\tau. \end{aligned} \tag{8.101}$$

Letting

$$z = \left|\frac{\dddot{\psi}(s_0)}{2}\right|^{1/3}(s - s_0), \tag{8.102}$$

we have

$$\int_{\sigma}^{\infty} \text{Ai}\,(x)\exp(jzx)dx = \frac{1}{2\pi j}\int_{\sigma}^{\infty}\exp(jzx)\int_{C^{\star}}\exp(-\tau^3/3+x\tau)d\tau dx$$
$$= \frac{1}{2\pi j}\int_{C^{\star}}\exp(-\tau^3/3)\int_{\sigma}^{\infty}\exp(jzx+x\tau)dxd\tau$$
$$= -\frac{1}{2\pi j}\int_{C^{\star}}\exp(-\tau^3/3)\frac{\exp(jz\sigma+\tau\sigma)}{jz+\tau}d\tau$$
$$= -\exp[j(s-s_0)(w-w_0)]$$
$$\frac{1}{2\pi j}\int_{C^{\star}}\frac{\exp(-\tau^3/3+\tau\sigma)}{jz+\tau}d\tau, \tag{8.103}$$

where the last equality follows from (8.30) and (8.102). Substituting (8.103) into (8.31), we obtain (8.29).

8.3 Perform the computation to obtain (8.32) from (8.31).

Solution

Letting

$$z = -\left|\frac{\dddot{\psi}(s_0)}{2}\right|^{1/3}(s-s_0), \tag{8.104}$$

we have, from (8.31),

$$\int_{-\infty}^{\infty}|\Delta\underline{\mathbf{G}}(s,\mathbf{r}')|^2 ds \sim |\underline{\mathbf{G}}(s,\mathbf{r}')|^2\left|\frac{\dddot{\psi}(s_0)}{2}\right|^{1/3}$$
$$\int_{-\infty}^{\infty}\left|\int_{-\infty}^{\infty}\text{Ai}\,(x)U(x-\sigma)\exp(-jzx)dx\right|^2 dz, \tag{8.105}$$

where $U(\cdot)$ is Heaveside's step function. By Parseval's theorem, the result now follows.

8.4 Derive (8.47) using the following properties of the Airy function, valid for $x > 0$:

$$\int_{x}^{\infty}\text{Ai}^{\,2}(x)dx = -x\text{Ai}^{\,2}(x) + \left(\frac{\partial \text{Ai}(x)}{\partial x}\right)^2, \tag{8.106}$$

$$\text{Ai}\,(x) \le \frac{\exp[(-2/3)x^{3/2}]}{2\sqrt{\pi}x^{1/4}}, \tag{8.107}$$

$$\left|\frac{\partial \text{Ai}(x)}{\partial x}\right| \le \frac{x^{1/4}\exp[(-2/3)x^{3/2}]}{2\sqrt{\pi}}\left(1+\frac{7}{48x^{3/2}}\right). \tag{8.108}$$

8.5 "A Hardy Old Problem":[2] Suppose that $f(x)$ and $\partial f(x)/\partial x$ are continuous and that their limits for $x \to \infty$ exist. Show that this implies that

$$\lim_{x\to\infty}\frac{\partial f}{\partial x} = 0. \tag{8.109}$$

[2] This problem was posed and discussed by G. H. Hardy in 1908; see Landau and Jones (1983).

Discuss how this result relates to (8.19) and the associated geometrical conditions for the existence of the limits in this case.

Solution

Using L'Hôpital's rule, we have

$$\begin{aligned}\lim_{x\to\infty} f(x) &= \lim_{x\to\infty} \frac{e^x f(x)}{e^x} \\ &= \lim_{x\to\infty} \frac{e^x[f(x) + \partial f(x)/\partial x]}{e^x} \\ &= \lim_{x\to\infty} f(x) + \partial f(x)/\partial x. \end{aligned} \tag{8.110}$$

9 The Time–Frequency Domain

Henceforth space by itself, and time by itself, are doomed to fade away into mere shadows, and only a kind of union of the two will preserve an independent reality.[1]

9.1 Frequency-Bandlimited Signals

In this chapter we extend the information-theoretic treatment of single-frequency sinusoidal waves presented in Chapter 8 to frequency-bandlimited signals, and compute the total number of degrees of freedom in the space–wavenumber and time–frequency domains. This corresponds to determining the total amount of information associated with the spatial and temporal configurations of the waveform, expressed in terms of the effective dimensionality of the corresponding functional space.

We derive the results in a deterministic setting, and refer to scalar space–time waveforms $f(\mathbf{r},t)$. By relying on the theory developed in Chapter 3 for signals of multiple variables, and applying it to the case of radiation from an arbitrary multiple scattering environment, we show that in appropriate asymptotic regimes the total number of degrees of freedom is given by the product of the time–frequency and space–wavenumber degrees of freedom.

Since the space–wavenumber and time–frequency degrees of freedom are not independent of each other, to perform our computation we need to take particular care in scaling the support sets of the signals in the natural and transformed domains to achieve the desired spectral concentration results.

The general conclusion is that the total number of degrees of freedom is proportional to the spatial extension of the cut-set through which the information must flow, which represents the number of space–wavenumber degrees of freedom, and to the time-bandwidth product, representative of the number of time–frequency degrees of freedom.

[1] Herman Minkowski (1908). Address to the society of German natural scientists and physicians at Köln, September 21.

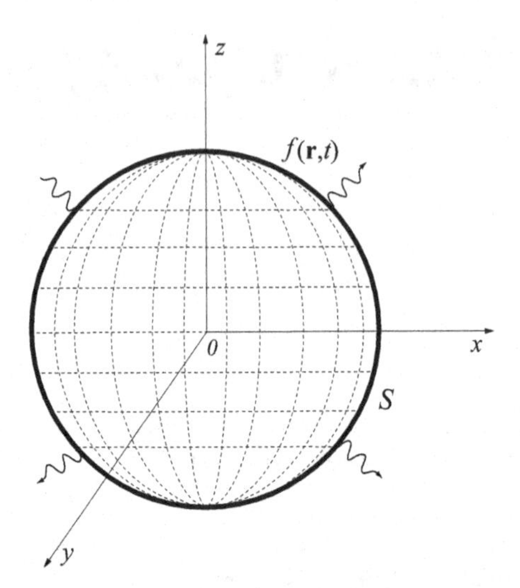

Fig. 9.1 Geometry of the signal model.

9.2 Radiation with Arbitrary Multiple Scattering

Let S be the surface of the sphere of radius r. We consider real space–time waveforms $f : S \times \mathbb{R} \to \mathbb{R}$ on this surface that satisfy the energy constraint

$$\int_{-\infty}^{\infty} \int_{S} f^2(\mathbf{r},t) d\mathbf{r} dt \leq E. \tag{9.1}$$

These waveforms are assumed to have angular frequency support $[-\Omega, \Omega]$ and to be the image of uniformly bounded sources $i(\mathbf{r}', t)$ that are internal to the sphere of radius r, through the convolution integral

$$f(\mathbf{r},t) = \int_{-\infty}^{\infty} \int_{V} g(\mathbf{r} - \mathbf{r}', t - t') i(\mathbf{r}', t') d\mathbf{r}' dt', \tag{9.2}$$

where V is the interior of S containing all the sources, $r' < r$, and g is the free-space Green's function. This geometric setting is depicted in Figure 9.1.

We let $\mathscr{S}$ be the set of square-integrable signals satisfying (9.1) and $\mathscr{B}_{\Omega,S} \subset \mathscr{S}$ the subset of signals frequency-bandlimited to $[-\Omega, \Omega]$ and spatially limited to S satisfying (9.2). We consider the norm

$$\|f\| = \left(\int_{-T/2}^{T/2} \int_{S} f^2(\mathbf{r},t) d\mathbf{r} dt \right)^{1/2}, \tag{9.3}$$

and determine the number of degrees of freedom at level ϵ of $\mathscr{B}_{\Omega,S}$ in $\mathscr{S}$, namely

$$N_\epsilon(\mathscr{B}_{\Omega,S}) = \min\{N : d_N^2(\mathscr{B}_{\Omega,S}, \mathscr{S})/E \leq \epsilon\}, \tag{9.4}$$

where $d_N(\mathscr{B}_{\Omega,S}, \mathscr{S})$ is the Kolmogorov N-width of the space $\mathscr{B}_{\Omega,S}$ in $\mathscr{S}$, introduced in Chapter 3.

This setup extends the one discussed in Chapter 3 for time–frequency signals and the one discussed in Chapter 8 for space–wavenumber signals. In the former case, we considered bandlimited signals $f(t)$ of a single variable, subject to the constraint

$$\int_{-\infty}^{\infty} f^2(t)dt \leq E, \tag{9.5}$$

and equipped with the norm

$$\|f\| = \left(\int_{-T/2}^{T/2} f^2(t)dt\right)^{1/2}, \tag{9.6}$$

and we have shown that the number of degrees of freedom is given by

$$N_\epsilon = \Omega T/\pi + o(T) \quad \text{as } T \to \infty, \tag{9.7}$$

where the ϵ-dependency appears only in the pre-constant of the second-order term $o(T)$. The frequency bandwidth is assumed to be fixed and the observation interval is scaled as $T \to \infty$ to achieve the desired spectral concentration result.

In the case of Chapter 8, we considered time-harmonic signals $f(\mathbf{r})$ on the surface S of a sphere of radius r, subject to the constraint

$$\int_S f^2(\mathbf{r})d\mathbf{r} \leq E, \tag{9.8}$$

and equipped them with the norm

$$\|f\| = \left(\int_S f^2(\mathbf{r},t)d\mathbf{r}\right)^{1/2}. \tag{9.9}$$

In this case, we have shown that the number of degrees of freedom is given by

$$N_\epsilon = 4\pi\left(\frac{\omega r}{c\pi}\right)^2 + o(r^2) \quad \text{as } r \to \infty, \tag{9.10}$$

where once again the ϵ-dependency appears only in the pre-constant of the second-order term $o(r^2)$. The solid angle 4π over which the signal is observed is fixed, and the wavenumber bandwidth is scaled as $r \to \infty$ to achieved the desired spectral concentration result.

We have also shown the additional result that for two-dimensional circular systems radiating a sinusoidal waveform observed over a circumference of radius r,

$$N_\epsilon = 2\pi\frac{\omega r}{c\pi} + o(r) \quad \text{as } r \to \infty. \tag{9.11}$$

We now consider the more general setting where square-integrable signals $f(\mathbf{r},t)$ observed over a finite angular domain are not composed of a single frequency, but have spectral support $[-\Omega,\Omega]$.

9.2.1 Two-Dimensional Circular Domains

We start with a two-dimensional domain of cylindrical symmetry in which an electromagnetic waveform is radiated by a configuration of currents located inside a

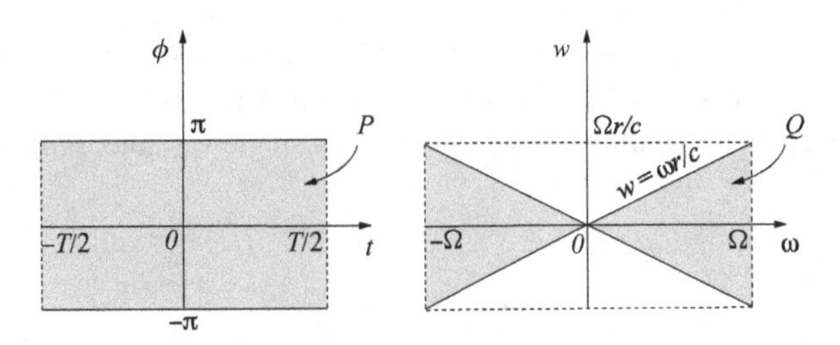

Fig. 9.2 Support sets in the natural and transformed domains.

circular domain of radius r, oriented perpendicular to the domain, and constant along the direction of flow. This can model, for example, an arbitrary scattering environment where a spatial distribution of transmitters is placed inside the circular domain, and communicate with receivers placed outside the domain. The same model applies to a remote sensing system where objects inside the domain are illuminated by an external waveform, and the scattered field is recovered by sensors placed outside the domain. The radiated field away from the scattering system and measured at the receivers is completely determined by the field on the cut-set boundary through which it propagates. On this boundary, we can refer to a scalar field $f(\phi,t)$ that is a function of only two scalar variables: one angular, and one temporal. The geometry is the same as in Figure 1.7, and the corresponding four field representations, linked by Fourier transforms, are depicted in Figure 1.8, where the angular frequency ω indicates the transformed coordinate of the time variable t, the wavenumber w indicates the transformed coordinate of the angular variable ϕ, and the Fourier transform $\mathcal{F}f(\phi,t) = \widehat{F}(w,\omega)$.

We wish to determine the number of degrees of freedom of the space–time field $f(\phi,t)$ on the cut-set boundary. In Section 1.2.3 we gave an informal argument, integrating the number of spatial degrees of freedom along the frequency bandwidth, and neglecting the possible accumulation of the approximation error. To derive rigorous results, we now apply the theory developed in Chapters 2 and 3, and take appropriate scaling limits of the support sets of the field.

To determine the scaling limits to apply, let us examine the geometric constraints imposed by the model. On the one hand, the observation domain is limited to $\phi \in [-\pi,\pi]$. In this case, as discussed in Chapter 8, the wavenumber bandwidth w is related to the frequency of transmission in such a way that for any possible configuration of sources and scatterers inside the circular radiating domain, we have

$$w = \omega r/c + o(\omega r) \quad \text{as } \omega r \to \infty. \tag{9.12}$$

It follows that the wavenumber bandwidth increases linearly with the frequency of radiation up to a maximum value $\Omega r/c$. To apply the scaling results developed in Sections 3.5.3 and 3.5.4, we then need to scale the support sets of the field depicted in Figure 9.2.

We can scale the spectral support Q while keeping P fixed by letting $\Omega \to \infty$. In this case, by (3.119) the number of degrees of freedom over a fixed transmission time and cut-set interval of width 2π, in the wide-band frequency regime, is

$$\begin{aligned} N_\epsilon(\mathscr{T}_P) &= \frac{2\Omega^2 rT2\pi}{c(2\pi)^2} + o(\Omega^2) \\ &= \frac{\Omega T}{\pi}\frac{2\pi r\Omega}{2\pi c} + o(\Omega^2) \quad \text{as } \Omega \to \infty. \end{aligned} \tag{9.13}$$

On the other hand, our geometric configuration does not allow scaling of the support P while keeping Q fixed because the cut-set domain is limited to an angle 2π. In this case we can apply the general result (3.129) to obtain the number of degrees of freedom over a fixed frequency band, by scaling the time coordinate of P and the wavenumber coordinate of Q, and we have

$$N_\epsilon(\mathscr{B}_Q) = \frac{\Omega T}{\pi}\frac{2\pi r\Omega}{2\pi c} + o(rT) \quad \text{as } T, r \to \infty. \tag{9.14}$$

Equations (9.13) and (9.14) confirm the intuition that the number of degrees of freedom is given by the product of two factors, each viewed in an appropriate asymptotic regime: one accounting for the number of degrees of freedom in the time–frequency domain, $\Omega T/\pi$, and the other accounting for the number of degrees of freedom in the space–wavenumber domain, $2\pi r\Omega/(2\pi c)$. The latter factor physically corresponds to the perimeter of the disc of radius r normalized by an interval of wavelengths $2\pi c/\Omega$, and can be interpreted as the spatial cut-set through which the information must flow. The idea is that for any finite-size system, the wavenumber bandwidth is a limited resource. Each parallel channel occupies a certain amount of spatial resource on the cut, proportional to the wavelength of transmission, and these channels must be sufficiently spaced along the cut for the corresponding waveforms to provide independent streams of information. The total number of channels is then given by the total spatial resource, given by the cut length $2\pi r$, divided by the total occupation cost, given by the wavelength interval $2\pi c/\Omega$.

The results are analogous to the ones obtained in Section 1.2.3. Namely, (9.13) and (9.14) correspond to the rigorous derivation of (1.16).

9.2.2 Three-Dimensional Spherical Domains

We now extend the results to three spatial dimensions by considering a spherical radiating system of radius r. In this case, the surface of the sphere is interpreted as a cut-set through which the information must flow and provides a limit on the amount of information that can radiate from the interior of the domain to the outside space. The field $f(\theta, \phi, t)$ on the cut-set boundary is a function of two curvilinear coordinates, identifying a point on the surface, and one temporal one. The measure of the support set P of the field in the natural domain is the same as that of a prism of base 4π, indicating the measure of the solid angle subtended at the center of the sphere, and of height T, indicating the measure of the time observation interval. The measure of the support set Q of the field in the transformed domain is the same as that of the bow-tie shape shown

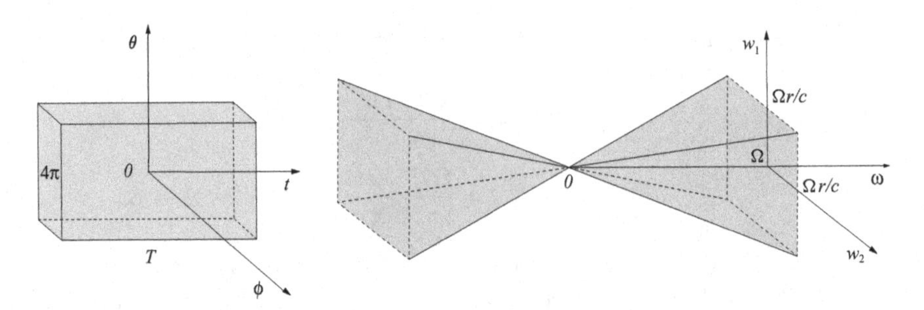

Fig. 9.3 Equivalent volumes of the support sets in the natural and transformed domains.

in Figure 9.3, of base $4(\Omega r/c)^2$ and of height Ω. We then have

$$m(P) = 4\pi T, \tag{9.15}$$

$$m(Q) = \frac{8}{3c^2}\Omega^3 r^2. \tag{9.16}$$

By using (3.128) and (3.129), with

$$\mathrm{A} = \begin{pmatrix} 1 & 0 & 0 \\ 0 & 1 & 0 \\ 0 & 0 & 1 \end{pmatrix}, \ \mathrm{B} = \begin{pmatrix} \rho & 0 & 0 \\ 0 & \rho & 0 \\ 0 & 0 & \rho \end{pmatrix}, \tag{9.17}$$

and using (9.15) and (9.16), it follows that the number of degrees of freedom in the wide-band frequency regime is

$$N_\epsilon(\mathscr{T}_P) = 4\pi r^2 \frac{\Omega^3 T}{3c^2\pi^3} + o(\Omega^3) \ \text{ as } \Omega \to \infty, \tag{9.18}$$

where $\Omega = \rho\Omega'$ with Ω' fixed and $\rho \to \infty$. With the alternative scaling

$$\mathrm{A} = \begin{pmatrix} \tau & 0 & 0 \\ 0 & 1 & 0 \\ 0 & 0 & 1 \end{pmatrix}, \ \mathrm{B} = \begin{pmatrix} 1 & 0 & 0 \\ 0 & \rho & 0 \\ 0 & 0 & \rho \end{pmatrix}, \tag{9.19}$$

we also have that the number of degrees of freedom over a fixed frequency band for large radiating systems and transmission time is

$$N_\epsilon(\mathscr{B}_Q) = 4\pi r^2 \frac{\Omega^3 T}{3c^2\pi^3} + o(r^2 T) \ \text{ as } T, r \to \infty, \tag{9.20}$$

where $T = \tau T'$, $r = \rho r'$, with T', r' fixed and $\tau, \rho \to \infty$.

Once again, the results are analogous to the ones obtained in Section 1.2.3. Namely, (9.18) and (9.20) correspond to the rigorous derivation of (1.19).

9.2.3 General Rotationally Symmetric Domains

The results can be further generalized by considering a radiating system enclosed in a convex domain bounded by a surface with rotational symmetry. Consider a cylindrical coordinate system (r, ϕ, z), a closed analytic curve $\zeta = \zeta(r, z)$ lying in the plane $\phi = 0$

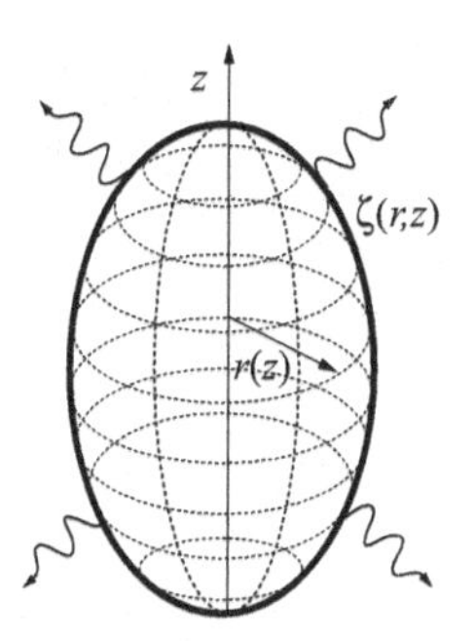

Fig. 9.4 Cut-set boundary of three-dimensional rotationally symmetric domain.

and symmetric with respect to the z axis, and the surface of revolution obtained by rotating the curve about the z axis – see Figure 9.4. In this case, we can choose a pair of coordinates (θ,ϕ) on the surface such that any meridian curve covers a 2π range, and the spatial bandwidth along the meridian is constant and given by

$$w_1 = \frac{\omega\ell}{2\pi c} + o(\omega\ell) \text{ as } \omega\ell \to \infty, \tag{9.21}$$

where ℓ is the Euclidean length of the curve, normalized to the speed of light so that $\omega\ell$ is dimensionless. This should be compared with the geometric constraint for circular curves given in (9.12), where the value of ℓ corresponds to $2\pi r$. In the same fashion, any latitude line is a circle of radius $r(z)$ that also covers a 2π range, and the spatial bandwidth along this line is given by

$$w_2 = \frac{\omega r(z)}{c} + o(\omega r) \text{ as } \omega r \to \infty. \tag{9.22}$$

It follows that in this case, while the support set P covers a solid angle 4π, the set Q varies along z according to the meridian curve parametrization $\zeta = \zeta(r,z)$, and we have

$$m(P) = 4\pi T, \tag{9.23}$$

$$\begin{aligned} m(Q) &= \frac{8\Omega^3}{3c^2}\frac{\ell}{2\pi}\frac{\pi}{2}\int_\zeta r(z)dz \\ &= \frac{8\Omega^3}{3c^2}\frac{\ell}{4\pi}\int_\zeta \pi r(z)dz \\ &= \frac{8\Omega^3}{3c^2}\frac{A}{4\pi}, \end{aligned} \tag{9.24}$$

where $A = A(S)$ is the surface area of the observation domain.

By (3.128) and (3.129), and using (9.23) and (9.24), we have that the number of degrees of freedom in the wide-band frequency regime is

$$N_\epsilon(\mathscr{T}_P) = \frac{A\Omega^3 T}{3c^2\pi^3} + o(\Omega^3) \text{ as } \Omega \to \infty, \tag{9.25}$$

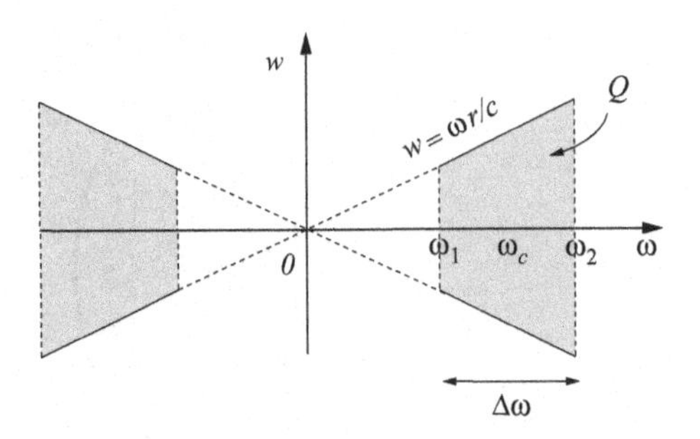

Fig. 9.5 Support set of a modulated signal in the transformed domain.

where $\Omega = \rho\Omega'$ with Ω' fixed and $\rho \to \infty$. The analogous result is obtained by letting the transmission time $T = \tau T'$, with T' fixed and $\tau \to \infty$, and scaling all the coordinates of the radiating volume, so that both $\ell = \rho\ell'$ and $r = \rho r'(z)$ tend to infinity with ℓ' and r' fixed and $\rho \to \infty$. In this case, by (3.128) and (3.129) with the scaling in (9.19), and using (9.23) and (9.24), we have

$$N_\epsilon(\mathscr{B}_Q) = \frac{A\Omega^3 T}{3c^2\pi^3} + o(AT) \quad \text{as } T, A \to \infty, \tag{9.26}$$

where $A = \rho^2 A'$, with A' fixed and $\rho \to \infty$.

9.3 Modulated Signals

We now extend the results to handle the case of modulated signals. Consider the case of a real signal modulating a sinusoid of carrier frequency ω_c, occupying a bandwidth $\Delta\omega = [\omega_1, \omega_2]$ centered around ω_c – see Figure 9.5. In the two-dimensional case, following the same procedure as the previous sections, letting $T = \tau T'$, $r = \rho r'$, and $\rho, \tau \to \infty$, we obtain

$$\begin{aligned} N_\epsilon(\mathscr{B}_Q) &= \frac{T}{\pi} \frac{2\pi r}{c\pi} \frac{(\omega_2^2 - \omega_1^2)}{2} + o(rT) \\ &= \frac{T\Delta\omega}{\pi} \frac{2\pi r \omega_c}{c\pi} + o(rT) \end{aligned} \tag{9.27}$$

as $r, T \to \infty$.

On the other hand, letting $\Delta\omega = \rho\Delta\omega'$, $\omega_c = \rho\omega_{c'}$, and $\rho \to \infty$, we have

$$N_\epsilon(\mathscr{B}_Q) = \frac{T\Delta\omega}{\pi} \frac{2\pi r \omega_c}{c\pi} + o(\omega_c \Delta\omega) \tag{9.28}$$

as $\omega_c, \Delta\omega \to \infty$. The number of degrees of freedom is proportional to the time-bandwidth product and to the cut-set boundary normalized by the radiating carrier.

For spherical domains, letting $T = \tau T'$, $r = \rho r'$, and $\tau, \rho \to \infty$, we have

$$N_\epsilon(\mathscr{B}_Q) = \frac{T}{\pi} \frac{4\pi r^2}{c^2\pi^2} \frac{(\omega_2^3 - \omega_1^3)}{3} + o(Tr^2)$$

$$= \frac{T\Delta\omega}{\pi} \frac{4\pi r^2}{c^2\pi^2} \frac{(\omega_2^2 + \omega_1^2 + \omega_1\omega_2)}{3} + o(Tr^2)$$

$$= \frac{T\Delta\omega}{\pi} \frac{4\pi r^2 \omega_c^2}{c^2\pi^2}(1+\delta) + o(Tr^2) \tag{9.29}$$

as $T, r \to \infty$, where δ is a constant that depends only on the ratio $\Delta\omega/\omega_c$. Analogously, letting $\Delta\omega = \rho\Delta\omega'$, $\omega_c = \rho\omega_{c'}$, $\rho \to \infty$, we have

$$N_\epsilon(\mathscr{B}_Q) = \frac{T\Delta\omega}{\pi} \frac{4\pi r^2 \omega_c^2}{c^2\pi^2}(1+\delta) + o(\omega_c^2 \Delta\omega) \tag{9.30}$$

as $\omega_c, \Delta\omega \to \infty$.

Results (9.29) and (9.30) can also be combined using the scaling matrices

$$\mathrm{A} = \begin{pmatrix} \tau & 0 & 0 \\ 0 & 1 & 0 \\ 0 & 0 & 1 \end{pmatrix}, \quad \mathrm{B} = \begin{pmatrix} \beta & 0 & 0 \\ 0 & \rho\beta & 0 \\ 0 & 0 & \rho\beta \end{pmatrix} \tag{9.31}$$

and letting $\tau\beta, \rho\beta \to \infty$, obtaining

$$N_\epsilon(\mathscr{B}_Q) = \frac{T\Delta\omega}{\pi} \frac{4\pi r^2 \omega_c^2}{c^2\pi^2}(1+\delta) + o(r^2\omega_c^2 \Delta\omega T) \tag{9.32}$$

as $r\omega_c, \Delta\omega T \to \infty$.

Finally, for general rotationally symmetric domains, using (9.31) we have

$$N_\epsilon(\mathscr{B}_Q) = \frac{T\Delta\omega}{\pi} \frac{A\omega_c^2}{c^2\pi^2}(1+\delta) + o(A\omega_c^2 \Delta\omega T) \tag{9.33}$$

as $A\omega_c^2, \Delta\omega T \to \infty$, where $A = \rho^2 A'$ with A' fixed.

For narrow-band signals, $\Delta\omega/\omega_c \ll 1$, and the constant δ can be made arbitrarily small, so that the number of degrees of freedom in three dimensions is essentially given by the first term of (9.33), which is the natural extension of the single-frequency result in Chapter 8, accounting for a non-zero frequency band around frequency ω_c. It follows that the number of degrees of freedom per unit time and per unit angular frequency band is essentially given by

$$N_0 = \frac{A\omega_c^2}{c^2\pi^3}. \tag{9.34}$$

9.4 Alternative Derivations

Analogous results to the ones described in this chapter have been given in a less rigorous setting in Section 1.2.3. In that case, we have simply integrated the number of spatial degrees of freedom over the whole bandwidth. Although non-rigorous, that method provides the right intuition at the basis of the rigorous computation. Each waveform is composed of a spectrum of frequencies, and each frequency carries a number of degrees of freedom that scales with the wavelength-normalized size of the cut-set boundary. Since the wavelength is inversely proportional to the frequency, summing all degrees of freedom in the two-dimensional setting corresponds to integrating a linear function of the frequency, and in a three-dimensional setting corresponds to integrating a quadratic function of the frequency, yielding the expressions in (9.13) and (9.14), (9.18) and (9.20).

9.5 Summary and Further Reading

We have computed the number of degrees of freedom of bandlimited signals propagating in arbitrary time-invariant media using the theory developed in Chapters 3 and 8. For signals observed along a spatial cut-set boundary that separates transmitters and receivers, or between radiating elements and sensing devices in an electromagnetic remote sensing system, the number of degrees of freedom corresponds to the effective number of parallel channels available through the cut-set boundary in the time–frequency and space–wavenumber domains. Thus, they provide a bound on the amount of spatial and frequency multiplexing achievable using arbitrary technologies and in arbitrary scattering environments. The number of degrees of freedom turns out to be proportional to the time-bandwidth product, and to the area of the cut-set boundary expressed in units of wavelengths. Results in this chapter appear in Franceschetti (2015), extending the works of Bucci and Franceschetti (1987, 1989) to frequency-bandlimited signals, and using the mathematical formulation of Landau (1975), described in Chapter 3.

9.6 Test Your Understanding

Problems

9.1 Explain why the equivalent volumes in the natural and transformed domains depicted in Figure 9.3 have measures equal to $m(P)$ and $m(Q)$.

9.2 How are the sets P and Q related to the equivalent volumes in Figure 9.3?

9.3 Check that the value of δ in (9.33) can be made arbitrarily small as $\omega_c \to \infty$.

9.4 Explain the difference between the argument given in Section 1.2.3 to compute the total number of degrees of freedom and the one given in this chapter, and identify what makes the former not completely rigorous.

10 Multiple Scattering Theory

My entire being rebels against order.
But without it I would die, scattered to the winds.[1]

10.1 Radiation with Multiple Scattering

In this chapter we illustrate how the physical constraints imposed by propagation in complex environments limit the amount of information that can be transported by a propagating wave.

We consider the stochastic diversity of signals propagating in a time-invariant random medium and relate it to the parameters of the stochastic model used to describe the medium. We rely on the stochastic frequency representations developed in Chapter 6, and provide additional insights into the design trade-offs discussed in Chapter 7.

From a physical perspective, the general effect of multiple scattering is a damping of the transmitted coherent wave and the creation of an incoherent energy coda – see Figure 10.1. The term "coherent" is used to denote the part of the waveform whose frequency components are not significantly distorted by propagation filtering, so that the waveform retains the original transmitted shape. On the other hand, the term "incoherent" is used to denote the part of the waveform whose frequency components are significantly distorted, so that the original transmitted shape is broadened in time. The transfer of coherent energy into incoherent energy through multiple scattering, as well as the absorption associated with the multiple-scattering process, are responsible for the exponential attenuation of the coherent part of the response. The incoherent response appears delayed and spread, due to the delayed arrival of the different multiple-scattered contributions that combine at the receiver.

From an information-theoretic perspective, as the signal loses coherence due to multiple scattering, it becomes more unpredictable in frequency, increasing the stochastic diversity of the process used to model its frequency variation. From a practical perspective, as the coherence bandwidth decreases, communication using a sequence of short pulses is somewhat inhibited by the multiple scattering process, due to the overlap of the broadened pulses at the receiver. This generally limits the

[1] Albert Camus (1950). *Notebooks, 1942–1951*, Volume 2. English translation by J. O'Brien (2010), Ivan R. Dee.

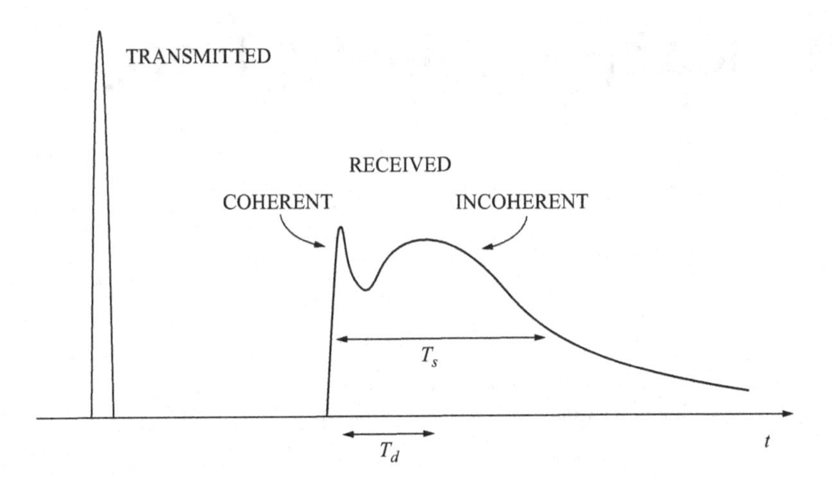

Fig. 10.1 Impulse response consists of a coherent part and an incoherent part. The incoherent part is delayed by T_{d}, the overall response is spread by T_{s}.

performance of time multiplexing technologies. On the other hand, scattering allows for smaller frequency correlation intervals, and favors technologies that use tightly packed frequency pulses that are more spread out in time.

Historically, there are two approaches to studying these phenomena. One is *multiple scattering theory* proper, and the other is *transport theory*. Both theories originated in the 1950s and 60s. At the basis of multiple scattering theory is the Fredholm integral equation of the second kind describing the multiple scattering process for the field. This approach is rigorous, but complete solutions cannot be obtained for complex environments, and approximate solutions that are useful for a specific range of parameters are considered. On the other hand, transport theory deals directly with the transport of energy through a scattering medium using the equation of transfer, which is equivalent to the Maxwell–Boltzmann collision equation used in the kinetic theory of gases, without considering wave propagation effects. In this chapter, we describe these two classical theories and relate them to a more recent probabilistic model of propagation based on random walk theory.

10.1.1 The Basic Equation

We start by considering a single-frequency scalar signal propagating in an environment filled with an arbitrary distribution of scattering objects. At the basis of any multiple scattering formulation is the inhomogeneous Fredholm integral equation of the second kind,

$$E(\mathbf{r}) = E_{\mathrm{D}}(\mathbf{r}) + \int_{V'} K(\mathbf{r},\mathbf{r}')E(\mathbf{r}')d\mathbf{r}', \tag{10.1}$$

where E_{D} is the direct field due to the sources, $K(\mathbf{r},\mathbf{r}')$ is a propagation kernel relating the total field on the scatterers to the reradiated field at any point in the space surrounding them, and the integral is over the space V' occupied by the scattering elements. For an electric field, this is the scalar version of (8.97), showing a linear relationship between the direct and the total field. The equation can be solved iteratively, as described in Appendix A.5.2, yielding the successive approximations

$$\begin{cases} E_0(\mathbf{r}) = E_\mathrm{D}(\mathbf{r}), & (10.2a) \\ E_1(\mathbf{r}) = E_\mathrm{D}(\mathbf{r}) + \int_{V'} K(\mathbf{r},\mathbf{r}')E_\mathrm{D}(\mathbf{r}')d\mathbf{r}', & (10.2b) \\ E_2(\mathbf{r}) = E_\mathrm{D}(\mathbf{r}) + \int_{V'} K(\mathbf{r},\mathbf{r}')E_\mathrm{D}(\mathbf{r}')d\mathbf{r}' & \\ \qquad + \int_{V'}\int_{V'} K(\mathbf{r},\mathbf{r}')K(\mathbf{r}',\mathbf{r}'')E_\mathrm{D}(\mathbf{r}'')d\mathbf{r}''d\mathbf{r}', & (10.2c) \\ E_3(\mathbf{r}) = \cdots & (10.2d) \\ \quad \vdots & \end{cases}$$

In (10.2a)–(10.2c), E_0 accounts for the direct field component only, E_1 approximates the field up to single scattering, E_2 up to double scattering, and so on.

It is in principle possible to obtain a solution using this iterative procedure up to any desired level of accuracy, but this requires complete knowledge of the scatterers' geometry and of the kernel $K(\mathbf{r},\mathbf{r}')$ relating the field at the scatterers to the reradiated field. In principle, this kernel can be obtained using knowledge of the constitutive properties of the scatterers in terms of the conductivity $\sigma(\mathbf{r}')$ and permittivity $\epsilon(\mathbf{r}')$ at any point $\mathbf{r}' \in V'$. By (4.31), one could compute the induced current and then use the free-space Green's function to compute the scattered field at any point in space. This corresponds to solving (8.97). In practice, however, complete knowledge of the scatterers' properties cannot be obtained for all environments, and approximate solutions must be considered. An alternative approach is to model the environment as a random medium, specified by a small set of parameters, and derive results for the average field over multiple spatial realizations. We describe this approach next, starting with the analysis of sinusoidal sources and then extending the treatment to pulse propagation in random media.

10.1.2 Multi-path Propagation

For a single-frequency signal, we have encountered a first model of multiple scattering in Section 6.3, namely the random multi-path model of propagation. The received signal was modeled as the linear superposition of different signals traveling along different multiple-scattered paths from transmitter to receiver. Each path carries a random phase shift and a random attenuation, and summing all of them we argued that the amplitude of the received waveform must be scaled by a Rayleigh random variable of parameter σ, and the phase must be shifted by a uniform random variable in the interval $[-\pi,\pi]$. This leads to a model that multiplies the transmitted signal by a complex, circularly symmetric, Gaussian random variable of standard deviation σ to obtain the received signal. Fixing σ corresponds to fixing the average attenuation, or path loss, between

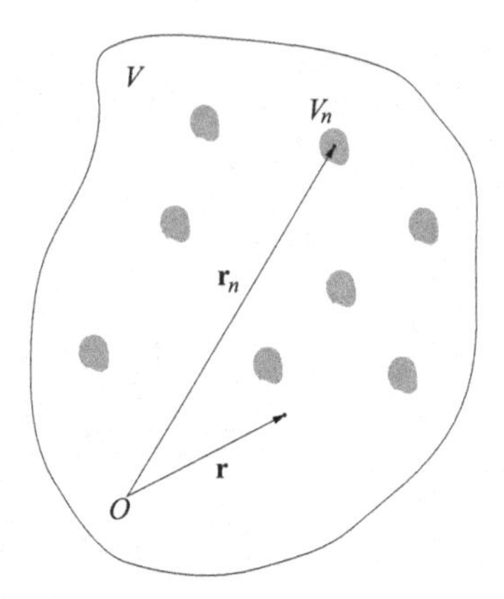

Fig. 10.2 Multiple scattering configuration.

transmitter and receiver. In the presence of absorption at all scattering points, we also argued that the path loss must rapidly decrease with the distance between transmitter and receiver.

We then extended the multi-path model from the sinusoidal regime to propagation of bandlimited signals, introducing the concepts of coherence bandwidth and power delay profile. In this case, we argued that multiple scattering increases the amount of frequency diversity by decreasing the coherence bandwidth, and it increases the time spread of the received waveform. By using multiple scattering theory, it is possible to study these phenomena in more detail, and relate them to the parameters of the stochastic model used to describe the propagation environment.

10.2 Multiple Scattering in Random Media

The theory of multiple scattering in random media starts with a discrete formulation and obtains an integral equation for the average field. Consider a discrete ensemble of N scattering elements $\{V_1, V_2, \ldots, V_N\}$ placed at coordinates $\{\mathbf{r}_1, \mathbf{r}_2, \ldots, \mathbf{r}_N\}$ with $\mathbf{r}_N \in V$, as depicted in Figure 10.2. The total field at coordinate $\mathbf{r}$ in the space surrounding the scatterers is

$$E(\mathbf{r}) = E_{\mathrm{D}}(\mathbf{r}) + \sum_{n=0}^{N} \mathcal{K}_n E_{\mathrm{D}}(\mathbf{r}) + \sum_{n=1}^{N} \sum_{\substack{m=1 \\ m \neq n}}^{N} \mathcal{K}_n \mathcal{K}_m E_{\mathrm{D}}(\mathbf{r})$$

$$+ \sum_{n=1}^{N} \sum_{\substack{m=1 \\ m \neq n}}^{N} \sum_{\substack{l=1 \\ l \neq m}}^{N} \mathcal{K}_n \mathcal{K}_m \mathcal{K}_l E_{\mathrm{D}}(\mathbf{r}) + \cdots, \tag{10.3}$$

where

$$\mathcal{K}_n E_{\mathrm{D}}(\mathbf{r}) = \int_{V_n} K_n(\mathbf{r}, \mathbf{r}') E_{\mathrm{D}}(\mathbf{r}') d\mathbf{r}' \tag{10.4}$$

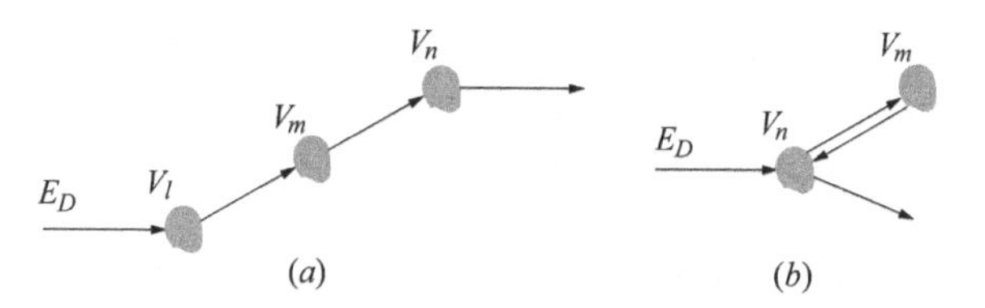

Fig. 10.3 Triple scattering involving (a) three distinct scatterers, and (b) only two scatterers.

indicates the field at point $\mathbf{r}$ due to the scatterer V_n centered at $\mathbf{r}_n$, when a field $E_D(\mathbf{r}')$ due to the sources only is present at the scatterer, and it depends on the position as well as on the characteristics of the scatterer. In (10.3), the first term is the direct field component due to the sources, the second term represents all the single scattering contributions, the third term represents all the double scattering contributions, and so on. In the third summation the terms $m = n$ and $l = m$ are excluded, but the term $l = n$ is not. This means that the summation includes triple scattering, where the propagation path goes through the same particle more than once. We can then rewrite the triple scattering summation to indicate a component that involves three distinct scatterers V_n, V_m, V_l, as depicted in Figure 10.3(a), and a component that involves only two scatterers V_n and V_m, as depicted in Figure 10.3(b), and we have

$$\sum_{n=1}^{N}\sum_{\substack{m=1\\m\neq n}}^{N}\sum_{\substack{l=1\\l\neq m}}^{N}\mathcal{K}_n\mathcal{K}_m\mathcal{K}_l E_D(\mathbf{r}) = \sum_{n=1}^{N}\sum_{\substack{m=1\\m\neq n}}^{N}\sum_{\substack{l=1\\l\neq m\\l\neq n}}^{N}\mathcal{K}_n\mathcal{K}_m\mathcal{K}_l E_D(\mathbf{r}) + \sum_{n=1}^{N}\sum_{\substack{m=1\\m\neq n}}^{N}\mathcal{K}_n\mathcal{K}_m\mathcal{K}_n E_D(\mathbf{r}). \tag{10.5}$$

In general, the complete field can be divided into a contribution of all the multiple-scattered waves involving paths of distinct scatters, and a contribution involving scatterers more than once.

The theory of multiple scattering developed by American mathematical physicist Victor Twersky in the 1950s includes only the first contribution, and neglects the second contribution. This corresponds to considering the field

$$E(\mathbf{r}) = E_D(\mathbf{r}) + \sum_{n=0}^{N}\mathcal{K}_n E_D(\mathbf{r}) + \sum_{n=1}^{N}\sum_{\substack{m=1\\m\neq n}}^{N}\mathcal{K}_n\mathcal{K}_m E_D(\mathbf{r}) + \sum_{n=1}^{N}\sum_{\substack{m=1\\m\neq n}}^{N}\sum_{\substack{l=1\\l\neq m\\l\neq n}}^{N}\mathcal{K}_n\mathcal{K}_m\mathcal{K}_l E_D(\mathbf{r}) + \cdots. \tag{10.6}$$

For large values of N, Twersky's approximation includes almost all multiple-scattering paths. Since an M-fold summation for $M > 2$ in the exact expansion contains $N(N-1)^{M-1}$ terms, while in the approximate expansion contains $N!/(N-M)!$ terms, the ratio between the number of terms in the approximation and in the exact expansion tends to one, and this suggests that the approximation should give accurate results when the number of scatterers is large.

The approximation is also useful to compute the average field over a random location of the scatterers. In this case, we have to average the first sum in (10.6) over the random location of one scatterer, the second sum over the random location of two scatterers, and so on. Letting $p(\mathbf{r})$ be the probability density function of the coordinate where a scatterer is located, and assuming the locations of the scatterers are independent and identically distributed, we have

$$\begin{aligned}\mathbb{E}[E(\mathbf{r})] = {} & E_{\mathrm{D}}(\mathbf{r}) + N \int_V \mathcal{K}_1 E_{\mathrm{D}}(\mathbf{r}) p(\mathbf{r}_1) d\mathbf{r}_1 \\ & + N(N-1) \iint_V \mathcal{K}_1 \mathcal{K}_2 E_{\mathrm{D}}(\mathbf{r}) p(\mathbf{r}_1) p(\mathbf{r}_2) d\mathbf{r}_1 d\mathbf{r}_2 \\ & + N(N-1)(N-2) \iiint_V \mathcal{K}_1 \mathcal{K}_2 \mathcal{K}_3 E_{\mathrm{D}}(\mathbf{r}) p(\mathbf{r}_1) p(\mathbf{r}_2) p(\mathbf{r}_3) d\mathbf{r}_1 d\mathbf{r}_2 d\mathbf{r}_3 \\ & + \cdots, \end{aligned} \tag{10.7}$$

where the integrals are over the whole space where the scatterers are distributed. Letting $p(\mathbf{r}) = \rho(\mathbf{r})/N$, where ρ is a density function indicating the number of scatterers per unit volume and whose integral over the whole space equals N, we have

$$\begin{aligned}\mathbb{E}[E(\mathbf{r})] = {} & E_{\mathrm{D}}(\mathbf{r}) + \int_V \mathcal{K}_1 E_{\mathrm{D}}(\mathbf{r}) \rho(\mathbf{r}_1) d\mathbf{r}_1 \\ & + \frac{(N-1)}{N} \iint_V \mathcal{K}_1 \mathcal{K}_2 E_{\mathrm{D}}(\mathbf{r}) \rho(\mathbf{r}_1) \rho(\mathbf{r}_2) d\mathbf{r}_1 d\mathbf{r}_2 \\ & + \frac{(N-1)(N-2)}{N^2} \iiint_V \mathcal{K}_1 \mathcal{K}_2 \mathcal{K}_3 E_{\mathrm{D}}(\mathbf{r}) \rho(\mathbf{r}_1) \rho(\mathbf{r}_2) \rho(\mathbf{r}_3) d\mathbf{r}_1 d\mathbf{r}_2 d\mathbf{r}_3 \\ & + \cdots, \end{aligned} \tag{10.8}$$

which for large values of N is approximately

$$\begin{aligned}\mathbb{E}[E(\mathbf{r})] = {} & E_{\mathrm{D}}(\mathbf{r}) + \int_V \mathcal{K}_1 E_{\mathrm{D}}(\mathbf{r}) \rho(\mathbf{r}_1) d\mathbf{r}_1 \\ & + \iint_V \mathcal{K}_1 \mathcal{K}_2 E_{\mathrm{D}}(\mathbf{r}) \rho(\mathbf{r}_1) \rho(\mathbf{r}_2) d\mathbf{r}_1 d\mathbf{r}_2 \\ & + \iiint_V \mathcal{K}_1 \mathcal{K}_2 \mathcal{K}_3 E_{\mathrm{D}}(\mathbf{r}) \rho(\mathbf{r}_1) \rho(\mathbf{r}_2) \rho(\mathbf{r}_3) d\mathbf{r}_1 d\mathbf{r}_2 d\mathbf{r}_3 \\ & + \cdots. \end{aligned} \tag{10.9}$$

The obtained relation (10.9) is the iterated version of the integral equation for the average field,

$$\mathbb{E}[E(\mathbf{r})] = E_{\mathrm{D}}(\mathbf{r}) + \int_V \mathcal{K}_{\mathbf{r}'} \, \mathbb{E}[E(\mathbf{r})] \rho(\mathbf{r}') d\mathbf{r}', \tag{10.10}$$

where the operator $\mathcal{K}_{\mathbf{r}'}$ indicates the response due to a scatterer at location $\mathbf{r}'$, and the integral in (10.10) is over all possible locations of the scatterer. This should be compared with the basic equation (10.1) obtained in the deterministic setting. While in (10.1) the integral is over the space V' occupied by the scattering elements, in (10.10) the integral is over the whole space V and it indicates the contribution of the average scatterer. It follows that Twersky's approach corresponds to first determining the response for a single scatterer, and then computing the statistical spatial average for an ensemble of scatterers. Assuming a simple form of the scattering operator $\mathcal{K}_{\mathbf{r}'}$, this approach leads to closed-form solutions. On the other hand, for complex scattering scenarios the method of successive approximations must be employed, neglecting higher-order scattering contributions.

10.2.1 Born Approximation

The Born approximation, named after German physicist Max Born, corresponds to considering only single scattering events. This is equivalent to taking the direct field due to the sources in place of the total field on the right-hand side of (10.1) or (10.10), and is the first-order approximation for the resolvent identity of the integral equation. In the case of (10.1), we have

$$E(\mathbf{r}) = E_{\mathrm{D}}(\mathbf{r}) + \int_{V'} K(\mathbf{r},\mathbf{r}')E_{\mathrm{D}}(\mathbf{r}')d\mathbf{r}'. \tag{10.11}$$

On the other hand, for (10.10) we have

$$\mathbb{E}[E(\mathbf{r})] = E_{\mathrm{D}}(\mathbf{r}) + \int_{V} \mathcal{K}_{\mathbf{r}'}\, \mathbb{E}[E_{\mathrm{D}}(\mathbf{r})]\rho(\mathbf{r}')d\mathbf{r}'. \tag{10.12}$$

Similarly, the nth-order Born approximation corresponds to the nth approximation, and the zeroth order corresponds to $E_{\mathrm{D}}(\mathbf{r})$.

10.2.2 Complete Solutions

The Twersky equation (10.10) can be solved directly by assuming a simple form for the operator $\mathcal{K}_{\mathbf{r}'}$ relating the field at the scatterers to the reradiated field.

Consider a plane wave $E_{\mathrm{D}}(z) = \exp(j\beta z)$, where $\beta = \omega/c$ is the wavenumber, propagating along the z axis and impinging on a random medium composed of a constant density ρ of scattering elements filling the whole space $z > 0$, as depicted in Figure 10.4.

We assume that the observation point at coordinate $\mathbf{r}$ inside the medium is in the far field of all the scatterers. Since the geometry of the medium and of the impinging wave are independent of the x and y coordinates, the average field inside the medium depends on z only. The average field propagates along the $\bar{\mathbf{z}}$ direction, since the average scattering contributions in the positive and negative directions of the x and y axes cancel each other. We then have, from (10.10),

$$\mathbb{E}[E(z)] = \exp(j\beta z) + \int_{z>0} \mathcal{K}_{\mathbf{r}'}\, \mathbb{E}[E(z)]\rho d\mathbf{r}'. \tag{10.13}$$

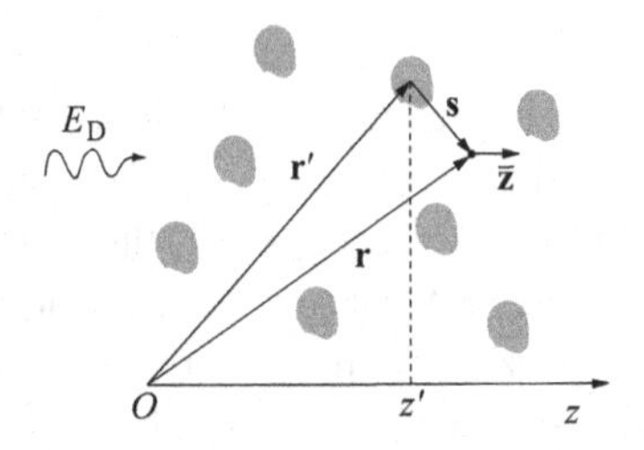

Fig. 10.4 A plane wave impinges on a random medium of a constant density of scatterers.

Since the observation point is assumed to be in the far field of all the scatterers, letting $\bar{\mathbf{s}}$ be the unit vector in the direction of $(\mathbf{r}-\mathbf{r}')$, we can approximate the scattering operator by

$$\mathcal{K}_{\mathbf{r}'}\,\mathbb{E}[E(z)] = \frac{\gamma \exp(j\beta|\mathbf{r}-\mathbf{r}'|)}{|\mathbf{r}-\mathbf{r}'|}\,\mathbb{E}[E(z')], \tag{10.14}$$

where $\gamma(\bar{\mathbf{z}},\bar{\mathbf{s}})$ is a complex scattering coefficient that depends on the characteristics of the scatterer. With this choice, the total average scattered field at depth z inside the medium due to a scatterer at $\mathbf{r}' = (x',y',z')$ is proportional to the average field at depth z'; it undergoes a phase shift proportional to the path length $|\mathbf{r}-\mathbf{r}'|$, a geometric attenuation inversely proportional to $|\mathbf{r}-\mathbf{r}'|$, and is multiplied by an additional scattering coefficient γ accounting for the absorption by the scatterer and for an additional phase shift. Substituting (10.14) into (10.13), we get

$$\mathbb{E}[E(z)] = \exp(j\beta z) + \int_{-\infty}^{\infty} dx' \int_{-\infty}^{\infty} dy' \int_{0}^{\infty} \frac{\gamma \exp(j\beta|\mathbf{r}-\mathbf{r}'|)}{|\mathbf{r}-\mathbf{r}'|}\,\mathbb{E}[E(z')]\rho dz'. \tag{10.15}$$

The integrals with respect to x' and y' can be computed approximately using the method of stationary phase – see Appendix C.[2] The stationary point is given by $x' = x$, $y' = y$, and for these coordinate values the scattering coefficient becomes

$$\begin{cases} \gamma(\bar{\mathbf{z}},\bar{\mathbf{s}}) = \gamma(\bar{\mathbf{z}},\bar{\mathbf{z}}) & \text{if } z > z', \\ \gamma(\bar{\mathbf{z}},\bar{\mathbf{s}}) = \gamma(-\bar{\mathbf{z}},\bar{\mathbf{z}}) & \text{if } z < z'. \end{cases} \tag{10.16}$$

Application of the integration method yields

$$\begin{aligned} \mathbb{E}[E(z)] = \exp(j\beta z) &+ \frac{2\pi j}{\beta}\rho\gamma(\bar{\mathbf{z}},\bar{\mathbf{z}}) \int_0^z \exp(j\beta|z-z'|)\,\mathbb{E}[E(z')]dz' \\ &+ \frac{2\pi j}{\beta}\rho\gamma(-\bar{\mathbf{z}},\bar{\mathbf{z}}) \int_z^\infty \exp(j\beta|z'-z|)\,\mathbb{E}[E(z')]dz'. \end{aligned} \tag{10.17}$$

The second integral accounts for the backward scattering occurring for $z' > z$ and is typically negligible compared to the first integral accounting for the scattering in the

[2] The method described in Appendix C can be extended to the stationary phase evaluation of a multiple integral.

forward direction. Neglecting this term, we obtain

$$\mathbb{E}[E(z)] = \exp(j\beta z)\left[1 + \frac{2\pi j}{\beta}\rho\gamma(\bar{\mathbf{z}},\bar{\mathbf{z}})\int_0^z \exp(j\beta z')\,\mathbb{E}[E(z')]dz'\right], \tag{10.18}$$

which is finally in a form that can be solved directly. Substituting

$$\mathbb{E}[E(z')] = \exp(jKz') \tag{10.19}$$

into (10.18), we obtain

$$K = \beta + \frac{2\pi\rho\gamma(\bar{\mathbf{z}},\bar{\mathbf{z}})}{\beta}. \tag{10.20}$$

Since K is a complex number that depends on $\gamma(\bar{\mathbf{z}},\bar{\mathbf{z}})$, it turns out that the plane wave solution (10.19) attenuates as it propagates through the medium and the rate of attenuation can be expressed in terms of the absorption and scattering cross sections of the scattering elements.

10.2.3 Cross Sections

A wave hitting a scatterer has part of the incident power scattered in different directions, part absorbed by the particle, and part passed through undisturbed. The scattered and absorbed fractions are called the scattering cross section σ_s and absorption cross section σ_a, respectively. The total cross section is $\sigma_t = \sigma_s + \sigma_a$. By expressing the scattering coefficient $\gamma(\bar{\mathbf{z}},\bar{\mathbf{z}})$ in terms of cross sections, the squared amplitude of the average field can be computed from (10.19) and (10.20), and we have

$$|\,\mathbb{E}[E(z)]|^2 = \exp(-\rho\sigma_t z). \tag{10.21}$$

A similar, but more involved, computation than the one in Section 10.2.2 also leads to an approximate expression for the average of the squared amplitude of the field:

$$\mathbb{E}[|E(z)|^2] = \exp(-\rho\sigma_a z). \tag{10.22}$$

By expressing the random field as the sum of an average and a random fluctuating component,

$$E(z) = \mathbb{E}[E(z)] + E'(z), \tag{10.23}$$

assuming $E'(z)$ has zero mean, we have that the total average power is

$$\begin{aligned}\mathbb{E}[|E(z)|^2] &= \mathbb{E}[|\,\mathbb{E}[E(z)] + E'(z)|^2] \\ &= |\,\mathbb{E}[E(z)]|^2 + \mathbb{E}[|E'(z)|^2],\end{aligned} \tag{10.24}$$

and by combining (10.21), (10.22), and (10.24), the average fluctuating power is

$$\begin{aligned}\mathbb{E}[|E'(z)|^2] &= \mathbb{E}[|E(z)|^2] - |\,\mathbb{E}[E(z)]|^2 \\ &= \exp(-\rho\sigma_a z) - \exp(-\rho\sigma_t z).\end{aligned} \tag{10.25}$$

Equations (10.21) and (10.25) represent the coherent and incoherent responses for a given radiated sinusoidal waveform, respectively. Equation (10.22) represents the

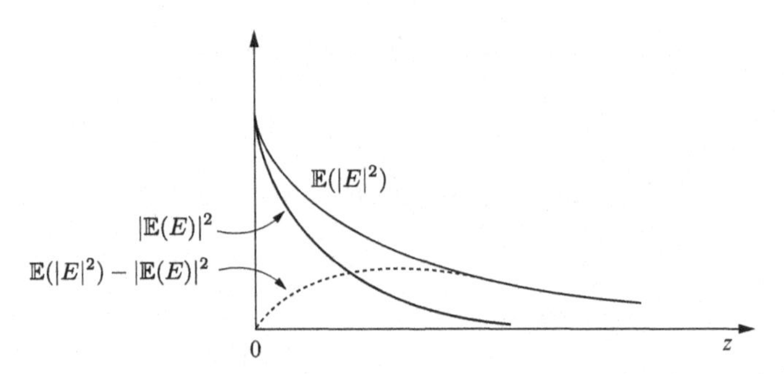

Fig. 10.5 Total (top continuous line), coherent (bottom continuous line), and incoherent (dashed line) response for plane wave propagation in a random medium as a function of the propagation depth.

total response. The coherent response exponentially attenuates due to absorption and scattering. The total response exponentially attenuates due to absorption only, and if $\sigma_a = 0$ there is no power attenuation. These results are consistent with plane wave propagation in a homogeneous medium with absorption as described in Chapter 4. The incoherent response initially increases for short propagation depth due to the transfer of coherent energy into incoherent energy through scattering, but eventually decreases as the waveform penetrates more into the medium and absorption becomes dominant – see Figure 10.5. These effects are also apparent using alternative models of multiple scattering, based on random walk theory and transport theory, and can also be extended to the case of pulse propagation in random media.

10.3 Random Walk Theory

The results obtained using multiple scattering theory essentially predict an exponential attenuation of the average power as a plane wave propagates into the medium. The rate of attenuation depends on the density of the scatterers and on their absorption cross section. The conversion of coherent energy into incoherent energy is governed by the scattering cross section and is responsible for a slower rate of decay of the total response compared to the coherent response. When scatterers are non-absorbing, the rate of decay of the average power tends to zero and the plane wave propagates without attenuation into the medium.

These results can be revisited using a simple model of wave propagation based on continuous random walks that is closely related to transport theory. In this case, we consider a sinusoidal wave of a given frequency radiated at the origin of a space filled by a uniform random distribution of point scatterers. The radiating wave is modeled as a constant stream of photons propagating in the environment, each carrying a quantum of energy. Each radiated photon proceeds along a straight line for a random length, until it hits an obstacle. The photon is then either absorbed by the obstacle, with probability $(1 - \alpha)$, or it is scattered uniformly in a random direction, with probability α. In this

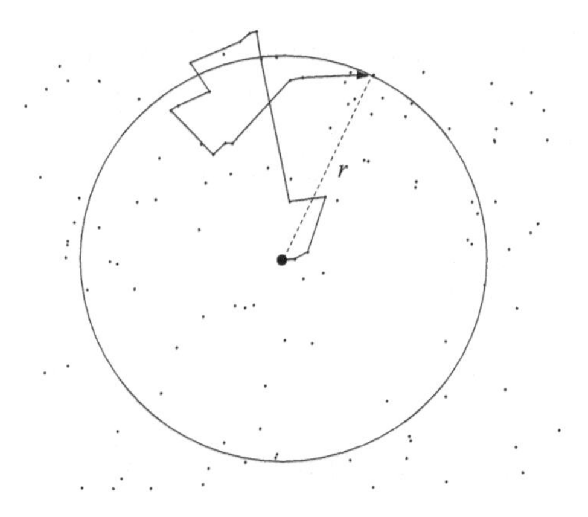

Fig. 10.6 Random walk model. One photon path from a single radiating element is depicted.

way, each photon propagates in the environment according to a random piecewise linear trajectory due to scattering – see Figure 10.6.

For a uniform distribution of obstacles, the path length between successive collisions is an exponential random variable of parameter η, indicating the number of scatterers per unit length, so that $1/\eta$ is the mean free path, and the probability density function (pdf) for the length of each step is

$$q(r) = \eta \exp(-\eta r). \tag{10.26}$$

Since each step is taken in a direction uniform in $[0, 2\pi]$, the pdf of having the first collision at coordinate $\mathbf{r}$ depends only on the distance r from the origin, and we have

$$\begin{cases} p(\mathbf{r}) = p(r) = \dfrac{\eta}{2\pi r} \exp(-\eta r), \\ \\ p(\mathbf{r}) = p(r) = \dfrac{\eta}{4\pi r^2} \exp(-\eta r), \end{cases} \tag{10.27}$$

in two and three dimensions, respectively. These pdfs integrate to one over the plane and over the space:

$$\int_0^{2\pi} d\phi \int_0^{\infty} r \frac{\eta}{2\pi r} \exp(-\eta r) dr = 1, \tag{10.28}$$

$$\int_0^{2\pi} d\phi \int_0^{\pi} \sin\theta d\theta \int_0^{\infty} r^2 \frac{\eta}{2\pi r} \exp(-\eta r) dr d\theta d\phi = 1. \tag{10.29}$$

We now consider the probability density g of the photon's absorption at coordinate $\mathbf{r}$. Once again, this depends only on the distance r from the origin, so that $g(\mathbf{r}) = g(r)$, and it satisfies the integral equation

$$g(r) = (1-\alpha)p(r) + \alpha \int_0^r g(r - r')p(r')dr', \tag{10.30}$$

indicating that the photon is either absorbed in the first step, with probability $(1-\alpha)$, or is scattered in a random direction with probability α. In the latter case, the resulting position is the sum of two random vectors, given by the convolution of the corresponding probability densities. The successive approximations constituting the resolvent formalism for the integral equation are

$$\begin{cases} g_0(r) = (1-\alpha)p(r), & (10.31a)\\ g_1(r) = (1-\alpha)p(r) + \alpha \int_0^r g(r-r')p_0(r')dr', & (10.31b)\\ g_2(r) = (1-\alpha)p(r) + \alpha \int_0^r g(r-r')p_0(r')dr' & \\ \qquad + \alpha^2 \int_0^r \int_0^r p(r-r')g(r'-r'')p_0(r'')dr''dr', & (10.31c)\\ g_3(r) = \cdots, & \\ \quad \vdots & (10.31d) \end{cases}$$

where $g_0(r)$ describes the event that the photon is absorbed at the first step of the random walk, hitting a single obstacle; $g_1(r)$ describes the event that the photon is absorbed either at the first step or at the second; and so on.

For the given kernel (10.27), the integral equation (10.30) can be solved directly by taking the Fourier transform of (10.30). Letting $\widehat{g}(w)$ be the Fourier transform of $g(r)$ and $\widehat{q}(w)$ be the Fourier transform of $q(r)$, we have

$$\widehat{g}(w) = (1-\alpha)\widehat{p}(w) + \alpha\widehat{p}(w)\widehat{g}(w) \tag{10.32}$$

and

$$\widehat{g}(w) = \frac{(1-\alpha)\widehat{p}(w)}{1-\alpha\widehat{p}(w)}. \tag{10.33}$$

The solution $g(r)$ in the spatial domain can then be obtained by taking the inverse transform of (10.33).

Using (10.27), exact expressions for $g(r)$ can be obtained in the two-dimensional case in terms of a series of Bessel polynomials. Useful closed-form approximations can also be obtained in two and three dimensions in terms of exponential functions and a modified Bessel function of the second kind. These are

$$\begin{cases} g(r) = \dfrac{(1-\alpha)\eta}{2\pi r}\left[\exp\left(-(1-\alpha^2)\eta r\right) + \alpha\eta r K_0\left(\sqrt{(1-\alpha^2)}\eta r\right)\right], \\ g(r) = \dfrac{(1-\alpha)\eta}{4\pi r^2}\left[\alpha\eta r \exp(-\sqrt{1-\alpha^2}\eta r) + \exp\left(-(1-\alpha^2)\eta r\right)\right], \end{cases} \tag{10.34}$$

where $K_0(\cdot)$ indicates the modified Bessel function of the second kind and of order zero – see Appendix E.3.

In both cases, as $\alpha \to 0$ and the obstacles become perfectly absorbing, $g(r)$ tends to $q(r)$ given by (10.26) and the random walk reduces to a single exponentially distributed

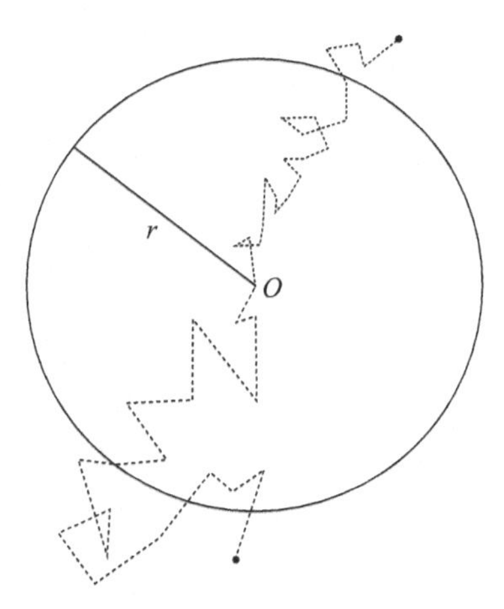

Fig. 10.7 Net power flux in the radial direction is the average number of photons per unit time absorbed outside the sphere of radius r.

step. All expressions are valid probability densities as they integrate to one over the plane, and over the space, respectively.

From these probabilistic results we can compute the average *radiated power density* $S(r)$ and the average *full power density* $P(r)$ at a given distance r from the transmitter.

Referring to the three-dimensional case, integration of $g(r)$ over the whole space outside the sphere of radius r centered at the transmitter gives the probability of a photon's absorption outside the sphere. Multiplying this probability by the total number of radiated photons we obtain the average net energy flux in the radial direction and, when this is taken per unit time, we obtain the average net power flux – see Figure 10.7. The radiated power density in units of W m^{-2} is then obtained by dividing the power flux by the surface of the sphere, and it corresponds to the expected value of the amplitude of the Poynting vector of the radiated field.

Modeling the sinusoidal source as radiating photons at a constant rate A, in two dimensions we have

$$S(r) = \frac{A}{2\pi r} \int_0^{2\pi} d\phi \int_r^{\infty} r' g(r') dr', \tag{10.35}$$

where the factor A represents the transmitted power. Similarly, in three dimensions we have

$$S(r) = \frac{A}{4\pi r^2} \int_0^{2\pi} d\phi \int_0^{\pi} \sin\theta d\theta \int_r^{\infty} r'^2 g(r') dr'. \tag{10.36}$$

On the other hand, the full power density is given by all the photons that enter per unit time an elementary volume placed between radii r and $r + dr$ from the source, and coming from all directions. This takes into account not only the radiated power but also the scattered power reflected towards the observation point from all directions – see Figure 10.8. To compute the full power density we first consider the probability density $g'(r)$ of having a scattering event at distance r from the origin. Since $(1-\alpha)$ is the

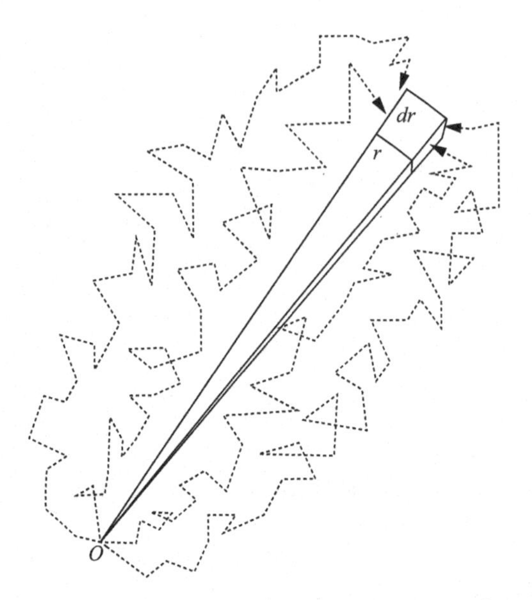

Fig. 10.8 Full power density is the average number of photons per unit time entering an elementary volume between r and $r+dr$, coming from all directions.

conditional density of absorption given that a scattering event occurs, we have

$$g'(r) = \frac{g(r)}{(1-\alpha)}. \tag{10.37}$$

The conditional density of having a scattering event inside the elementary volume, given that the photon reaches it, is given by

$$\lim_{r \to 0} q(r) = \eta. \tag{10.38}$$

It follows that the probability density of a photon reaching distance r from the origin is given by

$$\frac{g'(r)}{\eta} = \frac{g(r)}{(1-\alpha)\eta}. \tag{10.39}$$

Multiplying (10.39) by the total number of radiated photons gives the average energy density at distance r from the origin, and for a constant rate A of radiated photons we have that the average full power density is

$$P(r) = \frac{Ag(r)}{(1-\alpha)\eta}. \tag{10.40}$$

In the absence of scattering, we expect the full power density and the radiated power density to coincide, because all the power propagates in the radial direction. On the other hand, since scattering can provide radiation from different directions, the two quantities can generally be different.

10.3.1 Radiated Power Density

Letting $A = 1$, combining (10.34) and (10.35), in two dimensions we can compute

$$S(r) = \frac{1}{2\pi r}\int_0^{2\pi} d\phi \int_r^{\infty} r' g(r') dr'$$
$$= \frac{1}{2\pi r}\left[\frac{1}{1+\alpha}\exp\left[-(1-\alpha^2)\eta r\right] + \sqrt{\frac{1-\alpha}{1+\alpha}}\alpha\eta r K_1\left(\sqrt{1-\alpha^2}\eta r\right)\right], \quad (10.41)$$

where $K_1(\cdot)$ is the modified Bessel function of the second kind and of order one – see Appendix E.3.

In the limit of no scattering ($\eta \to 0$) or scattering with no absorption ($\alpha \to 1$), we have

$$S(r) = \frac{1}{2\pi r}, \quad (10.42)$$

which is what we expect for free-space propagation of a cylindrical wave, and what we expect by conservation of power for propagation in a scattering environment with no absorption.

In three dimensions, combining (10.34) and (10.36) we have

$$S(r) = \frac{1}{4\pi r^2}\int_0^{2\pi} d\phi \int_0^{\pi} \sin\theta d\theta \int_r^{\infty} r'^2 g(r') dr'$$
$$= \frac{(1-\alpha)\eta}{4\pi r^2}\left\{\eta\alpha\int_r^{\infty} r' \exp\left[-\sqrt{1-\alpha^2}\eta r'\right]dr' + \int_r^{\infty}\exp\left[-(1-\alpha^2)\eta r'\right]dr'\right\}$$
$$= \frac{1}{4\pi r^2}\frac{1}{(1+\alpha)}\left\{\alpha\left(\sqrt{1-\alpha^2}\eta r + 1\right)\exp\left[-\sqrt{1-\alpha^2}\eta r\right]\right.$$
$$\left. + \exp\left[-(1-\alpha^2)\eta r\right]\right\}. \quad (10.43)$$

In the limit of no scattering ($\eta \to 0$) or scattering with no absorption ($\alpha \to 1$), we have $S(r) = 1/(4\pi r^2)$. Again, this is what we expect on physical grounds for propagation of a spherical wave in free space or for propagation in a scattering environment with no absorption.

10.3.2 Full Power Density

Letting $A = 1$, combining (10.34) and (10.40), in two dimensions we have

$$P(r) = \frac{1}{2\pi r}\left[\exp\left(-(1-\alpha^2)\eta r\right) + \alpha\eta r K_0\left(\sqrt{(1-\alpha^2)}\eta r\right)\right], \quad (10.44)$$

and in three dimensions

$$P(r) = \frac{1}{4\pi r^2}\left[\alpha\eta r \exp\left(-\sqrt{1-\alpha^2}\eta r\right) + \exp\left(-(1-\alpha^2)\eta r\right)\right]. \quad (10.45)$$

As expected, in the limit of no scattering ($\eta \to 0$) the full power density coincides with the radiated power density, as it tends to $1/(2\pi r)$ and $1/(4\pi r^2)$ in two and three dimensions, respectively.

Table 10.1 Limiting cases for the full power, and radiated power densities.

	Medium	2D	3D
$\alpha \to 0$	Absorbing	$P(r) = e^{-\eta r}/(2\pi r)$	$P(r) = e^{-\eta r}/(4\pi r^2)$
$\alpha \to 1$	Lossless	$P(r) = \infty$	$P(r) = 1/(4\pi r^2) + \eta/(4\pi r)$
$\eta \to 0$	Free space	$1/(2\pi r)$	$1/(4\pi r^2)$
$\alpha \to 0$	Absorbing	$S(r) = e^{-\eta r}/(2\pi r)$	$S(r) = e^{-\eta r}/(4\pi r^2)$
$\alpha \to 1$	Lossless	$S(r) = 1/(2\pi r)$	$S(r) = 1/(4\pi r^2)$
$\eta \to 0$	Free space	$S(r) = 1/(2\pi r)$	$S(r) = 1/(4\pi r^2)$

10.3.3 Diffusive Regime

The limiting cases of the random walk model of multiple scattering are summarized in Table 10.1. As $\eta \to 0$ the model reduces to free-space propagation and the radiated power density and full power density coincide. On the other hand, for $\eta > 0$ and as $\alpha \to 1$ the radiated power density in three dimensions tends to $1/(4\pi r^2)$, satisfying the conservation of power, while the full power density tends to

$$P(r) = 1/(4\pi r^2) + \eta/(4\pi r). \tag{10.46}$$

This phenomenon of slow decay of the full power density is due to part of the energy being reflected back towards the source; this is responsible for the second term in (10.46), and it indicates that propagation tends to become diffusive.

Under the same conditions of a lossless scattering medium, in two dimensions the full power density diverges. This follows from (10.44) because of the singular behavior of $K_0(\cdot)$ at the origin.

These phenomena can be explained in terms of the recurrence properties of random walks. In two dimensions the random walk is recurrent. This means that in a lossless scattering medium any photon's trajectory revisits infinitely often the same location in space, and this leads to the divergence of the full power density. On the other hand, in three dimensions the probability of revisiting the same location multiple times is less than one, and in this case a lossless medium leads only to a decrease in the rate of decay of the full power density from $1/r^2$ to $1/r$, due to a finite number of recurrent visits and the eventual diffusion of photons in space.

The diffusion effect becomes negligible in a medium with absorption, and power attenuation in three dimensions tends to become of the type $e^{-\eta r}/(4\pi r^2)$, as $\alpha \to 0$.

10.3.4 Transport Theory

The results for the random walk model can be compared with those from transport theory. This theory deals with the transport of energy through a medium with scattering elements. When the scatterers are assumed to be small compared to the wavelength, the energy of the wave impinging on a scatterer is diffused uniformly in all directions. In this case, the results from transport theory are similar to those for the random walk model.

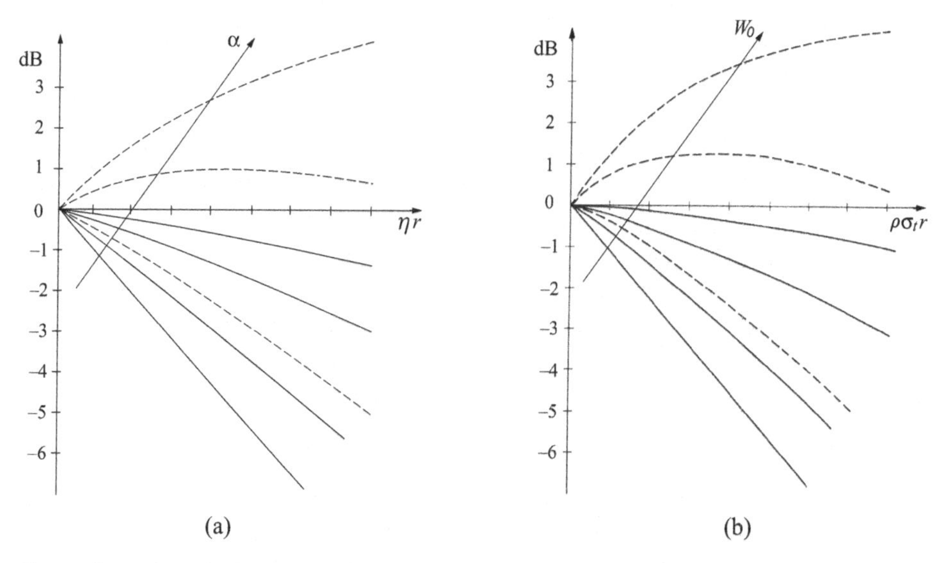

Fig. 10.9 Comparison of results from the random walk model of propagation (a), parametrized by increasing values of the scattering probability α, with results from transport theory (b), parametrized by increasing values of the albedo W_0.

In Figure 10.9(a) we plot $4\pi r^2 S(r)$ (continuous line curves) and $4\pi r^2 P(r)$ (dashed line curves) for the three-dimensional case, and compare them with the corresponding quantities derived in transport theory and plotted in Figure 10.9(b). The parameters of the two theories are different: the random walk model uses the scattering probability α and the average step length $(1/\eta)$, whereas transport theory uses the number of particles per unit volume (ρ) and the absorption (σ_a), scattering (σ_s), and total $(\sigma_t = \sigma_a + \sigma_s)$ particle cross sections. In transport theory, the curves are parametrized by increasing values of the albedo $W_0 = \sigma_s/\sigma_t$, while the curves of the random walk model are parametrized by increasing values of the scattering probability α. The plots are in logarithmic scale (dB).

In both cases, the normalized full power density (dashed line) initially increases for short propagation depths, but eventually decreases as the waveform penetrates more into the medium and absorption become dominant. For strongly absorbing scatterers the power density attenuates exponentially. On the other hand, the normalized radiated power density (continuous line) for lossless scatterers tends to the 0 dB line. Since we have normalized the flux by the geometric attenuation coefficient $1/4\pi r^2$, the situation in this case is similar to that of a plane wave that propagates without attenuation.

10.4 Path Loss Measurements

We now summarize. Propagation losses due to multiple scattering of single-frequency signals can be described by an exponential power attenuation, due to absorption, times a free-space geometric attenuation coefficient. This suggests using an approximate path loss formula,

$$L(r) = \frac{A\exp(-\gamma r)}{4\pi r^2}, \tag{10.47}$$

indicating the attenuation of the power density measured at distance r from the transmitter. The parameter $\gamma = \gamma(\alpha, \eta)$ accounts for both scattering and absorption in the medium. The parameter A accounts for the transmitted power. The model indicates a smooth transition from free-space propagation conditions in the near field ($\gamma r \ll 1$), to exponential attenuation in the far field ($\gamma r \gg 1$). For weakly absorbing scatterers, the model can also be heuristically extended to include a diffusive component,

$$L(r) = \frac{\gamma A\exp(-\gamma r)}{4\pi r^2} + \frac{(1-\gamma)A}{4\pi r}. \tag{10.48}$$

These simplified formulas can be used to predict attenuation when transmitting and receiving antennas are immersed in environments containing a multitude of small scattering objects. The diffusive effect is more evident in indoor environments, due to reverberation effects, and is often negligible outdoors.

Figure 10.10 shows the fit of the theoretical model (10.47), expressed in dB, to outdoor experimental data collected in the center of Rome, Italy, for sinusoidal radiation at 900 MHz over distances up to 350 m and an antenna height of 1.5 m. Data points refer to local space–time averages of the power received at different locations at a given distance from the transmitter. The mean squared error over all sampled points is 3–4 dB. Similar results are obtained by fitting the full power density model (10.45) using values of $\eta = 0.13$ and $\alpha = 0.9$, corresponding to an average inter-obstacle distance of 13 m and an absorption coefficient of 10%. These are typical values for measurements taken with low antennas immersed in an urban scattering environment.

10.5 Pulse Propagation in Random Media

We now extend the results for the random walk model to the case of propagation of a general waveform rather than a sinusoidal one. In this case, rather than a constant stream of photons, we consider a density of photons $f(t)$ radiated at the origin of the space, each carrying a quantum of energy. We wish to determine the expected density of photons $h(r,t)$ at distance r from the origin, according to the random walk model. This corresponds to the expected power response over multiple realizations of the scattering environment, and is obtained by computing a path integral, namely by averaging the number of photons reaching the observation point over all possible multiple-scattered paths of total length $\ell \geq r$.

10.5.1 Expected Space–Time Power Response

We perform the computation in a two-dimensional setting and start considering the joint pdf of having the first scattering event at coordinate $\mathbf{r}$, having traveled a distance ℓ. This

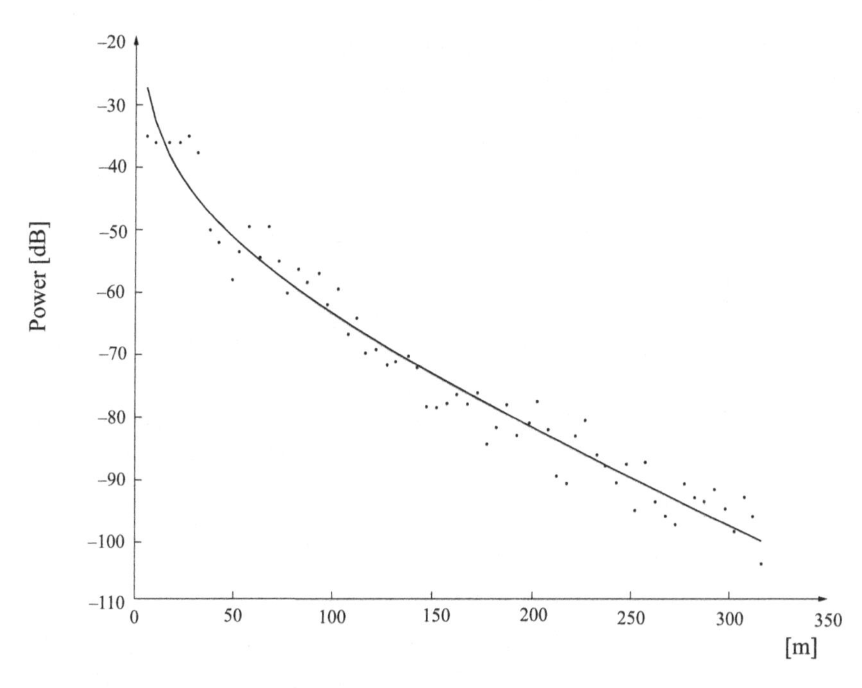

Fig. 10.10 Path loss model fitted to experimental data.

depends only on the distance r from the origin, so that $p(\mathbf{r},\ell) = p(r,\ell)$, and, using (10.27) and Dirac's $\delta(\cdot)$, it is written as

$$p(r,\ell) = \frac{\eta}{2\pi r}\exp(-\eta r)\delta(\ell - r). \tag{10.49}$$

The position after $(n+1)$ steps, and the total path length traveled, are given by the sum of $(n+1)$ i.i.d. random vectors. It follows that the joint pdf of having the $(n+1)$th collision at coordinate $\mathbf{r}$, having traveled a total path length ℓ, can be computed by performing n convolution operations of the single steps, and we have

$$p_n(r,\ell) = \overbrace{p(r,\ell) * p(r,\ell) * \cdots * p(r,\ell)}^{n}. \tag{10.50}$$

The n-fold convolution can be computed by taking the product of $n+1$ Fourier transforms of $p(r,\ell)$,

$$p_n(r,\ell) = \mathcal{F}^{-1}[\mathcal{F}p]^{n+1}, \tag{10.51}$$

where the index $n = 0,1,\ldots$ indicates the number of scattering events, $n+1$ the number of steps, and the Fourier transform in spherical coordinates is

$$\mathcal{F}p(w,\chi) = \int_0^{2\pi} d\theta \int_0^{\infty} rdr \int_0^{\infty} p(r,\ell)e^{-jrw\cos\theta}e^{-j\chi\ell}d\ell, \tag{10.52}$$

where (w,χ) are variables in the transformed domain corresponding to (r,ℓ).

Letting $U(\cdot)$ be the Heaviside step function and $\Gamma(\cdot)$ the Gamma function, it is possible to compute (10.51) in closed form and obtain

$$\begin{cases} p_0(r,\ell) = \dfrac{\eta}{2\pi r}\exp(-\eta r)\delta(\ell - r), \\ \\ p_n(r,\ell) = \dfrac{\eta}{\pi}\exp(-\eta\ell)\dfrac{2\sqrt{\pi}\eta^n}{2^{n-1}\Gamma\left(\dfrac{n+1}{2}\right)\Gamma\left(\dfrac{n}{2}\right)}(\ell^2 - r^2)^{(n-2)/2}U(\ell - r), \end{cases} \tag{10.53}$$

where the first equation coincides with (10.49) and is for one step, and the second equation is for $n > 0$ steps.

We now consider the joint pdf of reaching coordinate $\mathbf{r}$ in any number of steps, having traveled a total path length ℓ, and having a collusion there. This is given by

$$g(r,\ell) = \sum_{n=0}^{\infty}\alpha^n p_n(r,\ell). \tag{10.54}$$

Dividing (10.54) by the conditional density of having a scattering event at coordinate $\mathbf{r}$ given that the photon reaches it, which is given by (10.38), we get the joint pdf of reaching coordinate $\mathbf{r}$ in any number of steps, having traveled a total path length ℓ. Finally, letting c be the velocity of the photons, we compute the expected response:

$$\begin{aligned} h(r,t) &= \int_r^{ct}\frac{g(r,\ell)}{\eta}f(t-\ell/c)d\ell \\ &\int_r^{ct}\sum_{n=0}^{\infty}\frac{\alpha^n}{\eta}p_n(r,\ell)f(t-\ell/c)d\ell \\ &\int_r^{ct}\left[\frac{\alpha^n}{\eta}p_0(r,\ell) + \sum_{n=1}^{\infty}\frac{\alpha^n}{\eta}p_n(r,\ell)\right]f(t-\ell/c)d\ell, \end{aligned} \tag{10.55}$$

where the integration limits are imposed by the geometric constraints $r \le \ell \le ct$. The series in (10.55) can be computed, and we get

$$h(r,t) = \frac{\exp(-\eta r)}{2\pi r}f(t-\ell/c) + \frac{\alpha\eta}{2\pi}\int_r^{ct}f(t-\ell/c)\frac{\exp\left[-\eta\left(\ell - \alpha\sqrt{\ell^2 - r^2}\right)\right]}{\sqrt{\ell^2 - r^2}}d\ell. \tag{10.56}$$

In the case that the transmitted power density is a Dirac impulse, photons reach the observation point at time $t = r/c$ along the line-of-sight path, and at time $t > r/c$ along multiple-scattered paths. It follows that the expected impulse response density is

$$h_\delta(r,t) = \frac{\exp(-\eta r)}{2\pi r}\delta(t - r/c) + \frac{c\alpha\eta}{2\pi}\frac{\exp\left[-\eta\left(ct - \alpha\sqrt{(ct)^2 - r^2}\right)\right]}{\sqrt{(ct)^2 - r^2}}U(t - r/c), \tag{10.57}$$

where the units are photons $\mathrm{m}^{-1}\,\mathrm{s}^{-1}$.

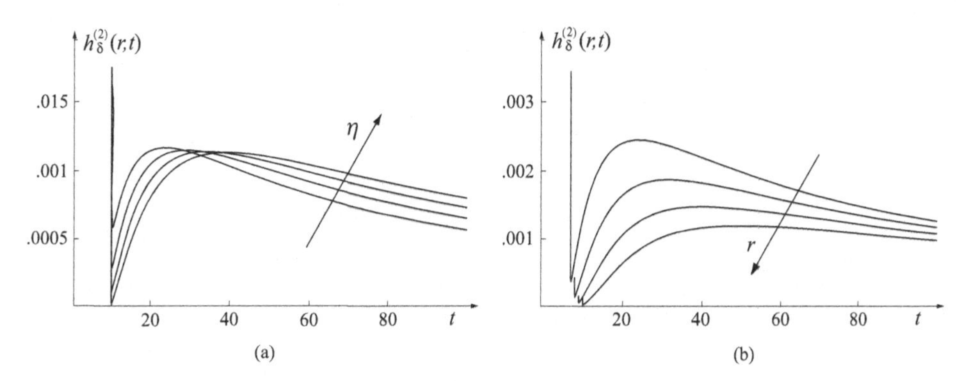

Fig. 10.11 (a) Multiple-scattered part of the impulse response with $c = 1$, $\alpha = 1$, $r = 10$, and $\eta \in [0.5, 0.8]$. (b) Multiple-scattered part of the impulse response with $c = 1$, $\alpha = 1$, $\eta = 1$, and $r \in [6, 9]$.

The first term of (10.57),

$$h_\delta^{(1)}(r,t) = \frac{\exp(-\eta r)}{2\pi r}\delta(t - r/c), \tag{10.58}$$

represents the response due to the single direct line-of-sight path, and the second term,

$$h_\delta^{(2)}(r,t) = \frac{c\alpha\eta}{2\pi}\frac{\exp\left[-\eta\left(ct - \alpha\sqrt{(ct)^2 - r^2}\right)\right]}{\sqrt{(ct)^2 - r^2}}U(t - r/c), \tag{10.59}$$

represents the response due to multiple scattering, including the multiple-scattering straight-line path and paths with different angles and directions. It follows that the direct part of the response is an exponentially attenuated pulse, while the multiple-scattered part, depicted in Figure 10.11, consists of an initial fast-decaying peak, followed by a broadened tail, due to the delayed arrivals of the different multiple-scattering contributions. This tail exhibits a larger spread and is delayed as η or r are increased, and the signal undergoes a larger amount of scattering. The initial, fast-decaying peak in the figure is due to the singularity of (10.59) at $t = r/c$, which is also the time at which the direct line-of-sight contribution (10.58) arrives. Physically, it represents the combined effect of all the multiple-scattering contributions occurring along the straight-line path between the transmitter and receiver.

A closer inspection reveals that the energy associated with the paths occurring along a single straight line tends to disappear as the amount of scattering increases, and the coherent energy associated with the straight-line paths is converted into an incoherent energy coda associated with the non-line-of-sight paths. This is evident from Figure 10.12, showing a steeper decrease of the initial peak of the response as η is increased, indicating that the area under the curve becomes negligible as the amount of scattering increases and the non-line-of-sight paths become dominant.

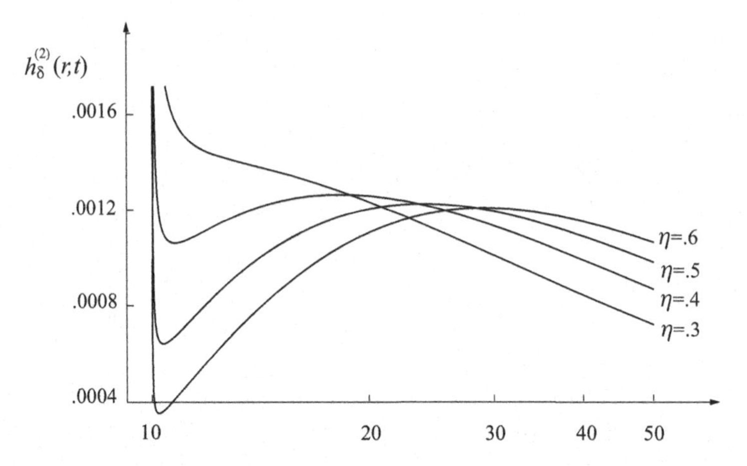

Fig. 10.12 Multiple-scattered part of the impulse response with $c = 1$, $\alpha = 1$, $r = 10$. The line-of-sight contribution disappears as the impulse response becomes more spread out.

10.5.2 Random Walk Interpretation

We now provide an interpretation of the main features of the impulse response in terms of the properties of the random walk. The average distance traveled from the origin in our random walk model grows with the square root of the number of reflections. It follows that after n reflections the distance from the origin is roughly $(1/\eta)\sqrt{n}$, where $1/\eta$ is the typical step length of the random walk. Neglecting absorption, a photon undergoes roughly T/τ reflections in a time interval of duration T, where $\tau = 1/(\eta c)$ is the typical propagation time between successive reflections. Accordingly, the distance from the origin of a multiple-scattered path after time T is of the order of

$$r = \frac{1}{\eta}\sqrt{n} = \sqrt{\frac{cT}{\eta}}. \tag{10.60}$$

The same distance r is reached by the direct line-of-sight path in time r/c. It follows that the time lag between the non-line-of-sight contribution and the line-of-sight one is

$$\begin{aligned}\Delta t &= T - \frac{r}{c}\\ &= \frac{\eta r^2}{c} - \frac{r}{c}\\ &+ \frac{r}{c}(\eta r - 1).\end{aligned} \tag{10.61}$$

This heuristic computation shows that the time delay increases with the distance r and with the amount of scattering η. On the other hand, the time spread is related to the variance of the distance traveled by the random walk, which grows with the number of reflections n. For this reason, the incoherent contribution not only appears delayed, as the distance increases, but also more spread out in time.

10.5.3 Expected Space–Frequency Power Response

Taking the Fourier transform of (10.57), we obtain the expected space–frequency response density

$$H(r,\omega) = \int_{-\infty}^{\infty} h_\delta(r,t)\exp(-j\omega t)dt, \tag{10.62}$$

where the units are photons m^{-1}. This is composed of two factors,

$$H(r,\omega) = H_1(r,\omega) + H_2(r,\omega), \tag{10.63}$$

where

$$H_1(r,\omega) = \frac{\exp[-\eta r(1+j\omega/(\eta c))]}{2\pi r} \tag{10.64}$$

represents the direct line-of-sight part of the response, which is exponentially attenuated and phase shifted by $\omega r/c$, and

$$\begin{aligned} H_2 &= \frac{c\alpha\eta}{2\pi}\int_{-\infty}^{\infty} \frac{\exp\left[-\eta\left(ct-\alpha\sqrt{(ct)^2-r^2}\right)\right]}{\sqrt{(ct)^2-r^2}} U(t-r/c)\exp(-j\omega t)dt \\ &= \frac{\alpha\eta}{2\pi}\int_r^{\infty}\int_{r/c}^{\infty} \delta(t-\ell/c)\frac{\exp\left[-\eta\left(\ell-\alpha\sqrt{\ell^2-r^2}\right)\right]}{\sqrt{\ell^2-r^2}}\exp(-j\omega t)dtd\ell \end{aligned} \tag{10.65}$$

represents the multiple-scattered part of the response. Expanding the exponential function in a Taylor series around $\ell^2 - r^2 = 0$, and integrating term by term, changing the integration variable to $u = \ell/r$, the incoherent response (10.65) can be written in terms of the series

$$\begin{aligned} H_2 &= \frac{\alpha\eta}{2\pi}\int_r^{\infty} \frac{\exp[-\ell(\eta+j\omega/c)]\exp\left[\alpha\eta\sqrt{\ell^2-r^2}\right]}{\sqrt{\ell^2-r^2}}d\ell \\ &= \frac{\alpha\eta}{2\pi}\int_r^{\infty} \exp[-\ell(\eta+j\omega/c)]\sum_{n=0}^{\infty}\frac{(\alpha\eta)^n(\ell^2-r^2)^{\frac{n-1}{2}}}{n!}d\ell \\ &= \frac{\alpha\eta}{2\pi}\sum_{n=0}^{\infty}\frac{(\alpha\eta r)^n}{n!}\int_1^{\infty}\exp[-ur(\eta+j\omega/c)](u^2-1)^{\frac{n-1}{2}}du \\ &= \frac{\alpha\eta}{2\pi}\sum_{n=0}^{\infty}\frac{\alpha^n}{\Gamma(\frac{n}{2}+1)2^{n/2}}\frac{(\eta r)^{n/2}}{(1+j\omega/(\eta c))^{n/2}}K_{n/2}[\eta r(1+j\omega/(\eta c))], \end{aligned} \tag{10.66}$$

where $K_n(\cdot)$ is the modified Bessel function of the second kind of order n, and $\Gamma(\cdot)$ is the Gamma function. This series can be numerically evaluated or analytically approximated using asymptotic expansions of the Gamma function and the modified Bessel function. Its magnitude is a decreasing function of $\omega > 0$, whose rate of decay depends on the scattering density, absorption, and distance from the transmitter.

The frequency response in (10.66) also includes straight-line contributions occurring along multiple-scattered paths. To evaluate the frequency response of the

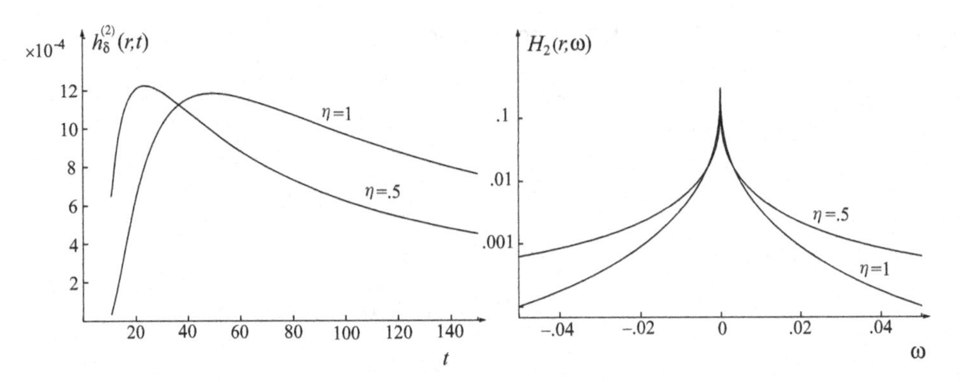

Fig. 10.13 Incoherent part of the impulse response without line-of-sight contribution and with $c = 1$, $\alpha = 1$, $r = 10$.

non-line-of-sight components only, we can numerically compute the Fourier transform of the impulse response with the initial peak removed, as shown in Figure 10.13. As the amount of scattering increases and the signal loses coherence and is broadened in time, the figure shows that the energy concentrates in a smaller frequency band. As we shall see below, the expected space–frequency power response of the random walk model corresponds to the frequency autocorrelation function of a multi-path channel model, and this behavior indicates a decrease of the coherence bandwidth of the signal due to scattering.

10.5.4 Correlation Functions

The time autocorrelation of the received signal in a multi-path channel describing a time-invariant environment having stochastic Green's function $\mathsf{G}(r,\tau)$ is

$$\mathbb{E}[\mathsf{Y}^2(r,t)] = \int_{-\infty}^{\infty}\int_{-\infty}^{\infty} x(r,t-\tau)x(r,t-\tau')\,\mathbb{E}[\mathsf{G}(r,\tau)\mathsf{G}(r,\tau')]d\tau d\tau'. \tag{10.67}$$

In Chapter 6, we defined the uncorrelated scattering multi-path channel to be the one for which the correlation between wave contributions along paths carrying different delays is zero. This also corresponds to a frequency wide-sense stationary model, as described in Section 6.3.1. It follows that for this channel model, the correlation function between the impulse responses $\mathsf{G}(r,\tau)$ and $\mathsf{G}(r,\tau')$ is

$$\mathbb{E}[\mathsf{G}(r,\tau)\mathsf{G}(r,\tau')] = A(r,\tau)\delta(\tau'-\tau), \tag{10.68}$$

indicating zero correlation unless the delays τ and τ' corresponding to different path lengths are equal. In this case, $A(r,\tau)$ represents the average power attenuation at coordinate $\mathbf{r}$ and at time t experienced by the signal along any path carrying a delay $\tau = \ell/c \geq r/c$.

According to the random walk model, the average power attenuation corresponds to the joint density of photons reaching coordinate $\mathbf{r}$ in any number of steps and having traveled a total path length $\ell = c\tau$, and is obtained by dividing the joint density (10.54) by the conditional density (10.38), yielding

$$A(r,\tau) = \sum_{n=0}^{\infty} \frac{\alpha^n}{\eta} p_n(r,\tau). \tag{10.69}$$

Substituting (10.68) into (10.67), taking into account (10.69), and that $r/c \leq \tau \leq t$, we obtain

$$\begin{aligned} \mathbb{E}[\mathsf{Y}^2(r,t)] &= \int_{r/c}^{t} x^2(t-\tau)A(r,\tau)d\tau \\ &= \int_{r/c}^{t} x^2(t-\tau)\sum_{n=0}^{\infty} \frac{\alpha^n}{\eta} p_n(r,\tau)d\tau. \end{aligned} \tag{10.70}$$

It follows that we can interpret the average power attenuation $A(r,\tau)$ in (10.69) as the response to a δ-input power $x^2(t) = \delta(t)$, and comparing (10.70) with (10.55), with the substitution $\tau = \ell/c$, it follows that the density of radiated photons $f(t)$ corresponds to the instantaneous power $x^2(t)$ of the transmitted signal, and the expected response of the random walk model corresponds to the autocorrelation function of a frequency wide-sense stationary, uncorrelated scattering multi-path channel model.

The quantity $h_\delta(r,t)$ in (10.57) represents the power delay profile, and is the inverse transform of the frequency autocorrelation function; we have

$$\mathbb{E}[\mathsf{G}(r,\omega')\mathsf{G}^*(r,\omega'')] = H(r,\omega) = \int_{-\infty}^{\infty} h_\delta(r,t)\exp(-j\omega t)dt, \tag{10.71}$$

which for the given channel depends only on the difference $\omega = \omega'' - \omega'$ and is given by (10.57).

It follows that the spreading properties of the random walk model and the creation of an incoherent energy coda now reflect in the shape of the frequency autocorrelation function and power delay profile of the channel.

10.6 Power Delay Profile Measurements

We now summarize. The main features of the power delay profile predicted by the random walk model are a fast-rising coherent response, followed by the incoherent response that exhibits a decaying tail. The two parts of the response tend to merge as the distance between transmitter and receiver increases or the density of scattering increases, and the incoherent response becomes dominant. Figure 10.14 illustrates this point, showing results for the theoretical model in the case of a transmitted rectangular pulse of width T, obtained by substituting $f(t) = \mathrm{rect}(t/T)$ into (10.56).

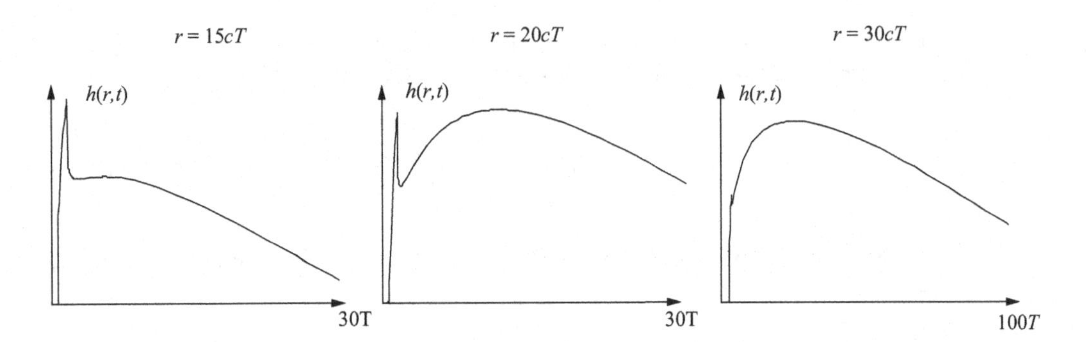

Fig. 10.14 Power response for a rectangular pulse of support T with $\alpha = 0.9$ and $\eta = 0.3cT$.

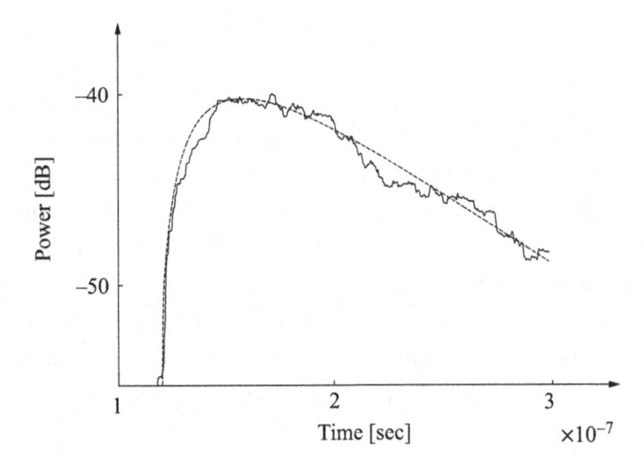

Fig. 10.15 Comparison between the power response of the theoretical model (dashed line) and experimental data (continuous line).

These main features are also apparent in real data. Figure 10.15 shows a comparison between the theoretical power response, expressed in dB, and experimental data collected for waves propagating in a classroom filled with furniture. The average response for a transmitted pulse of width 1 ns was measured over different antenna locations placed at a distance of 6 m from each other and with an occluded line-of-sight path. The average measurement is compared with results from the incoherent response of the random walk model obtained for a rectangular pulse of the same width, an inter-obstacle distance $1/\eta = 1$ m, and a scattering probability of $\alpha = 0.95$. As in the case of the path loss described in Section 10.4, the theoretical model matches well the data for physically reasonable values of the parameters.

10.7 Summary and Further Reading

The main reference for the theory of multiple scattering in random media is the book by Ishimaru (1978). The formulation we provided here was originally developed by Twersky (1957, 1964). A standard reference for transport theory is the book by Chandrasekar (1960). The random walk model of propagation was introduced by Franceschetti, Bruck, and Schulman (2004), and extended by Franceschetti (2004). These works also describe the experimental measurements that we reported here. A general reference for the topic of continuous random walks is the book by Hughes (1995). The version considered here is a variation of the Pearson–Rayleigh random walk where the steps of the walk are independent, identically distributed random vectors of exponential length and uniform orientation, so that the walk is well suited to model an underlying uniform spatial distribution of scatterers. Many variations of the Pearson–Rayleigh continuous random walk have been studied in mathematical physics, with applications ranging from colloid chemistry to stellar dynamics. Additional mathematical results for the distribution of the position of the random walker after a fixed number of steps appear in Stadje (1987), Masoliver, Porrá, and Weiss (1993), and in Franceschetti (2007b). Variations of the problem include walks with different step-length distributions by Le Caër (2010, 2011) and Pogorui and Rodriguez-Dagnino (2011), and walks with drift by Grosjean (1953), Barber (1993), and De Gregorio (2012).

10.8 Test Your Understanding

Problems

10.1 Check that the integrals (10.28) and (10.29) integrate to one.

10.2 Provide a justification for (10.41) using the identity

$$\int xK_0(x)dx = xK_1(x). \tag{10.72}$$

10.3 Compute the limits for $\eta \to 0$ and $\alpha \to 1$ of (10.41), and check that they reduce to (10.42).

Solution

$$\begin{aligned}
\lim_{\eta\to 0} S(r) &= \frac{1}{2\pi r}\left[\frac{1}{1+\alpha} + \lim_{\eta\to 0}\sqrt{\frac{1-\alpha}{1+\alpha}}\alpha\eta r K_1\left(\sqrt{1-\alpha^2}\eta r\right)\right] \\
&= \frac{1}{2\pi r}\left[\frac{1}{1+\alpha} + \sqrt{\frac{1-\alpha}{1+\alpha}}\frac{\alpha}{\sqrt{1-\alpha^2}}\right] \\
&= \frac{1}{2\pi r}.
\end{aligned} \tag{10.73}$$

$$\begin{aligned}
\lim_{\alpha\to 1} S(r) &= \frac{1}{2\pi r}\left[\frac{1}{2} + \frac{1}{2}\right] \\
&= \frac{1}{2\pi r}.
\end{aligned} \tag{10.74}$$

10.4 Check that the approximate solutions in (10.34) are valid probability densities, as they integrate to one over the plane, and over the space, respectively.

Solution

For the first integral, we have

$$\int_0^{2\pi} d\phi \int_0^{\infty} rg(r)dr = (1-\alpha)\eta \left(\frac{1}{\eta(1-\alpha^2)} + \frac{\alpha}{\eta(1-\alpha^2)} \right) = 1. \tag{10.75}$$

For the second integral, we have

$$\begin{aligned} \int_0^{2\pi} d\phi \int_0^{\pi} \sin\theta d\theta \int_0^{\infty} r^2 g(r)dr &= 4\pi \frac{(1-\alpha)\eta}{4\pi} \left[\frac{\alpha}{(1-\alpha^2)\eta} + \frac{1}{(1-\alpha^2)\eta} \right] \\ &= 1. \end{aligned} \tag{10.76}$$

10.5 Go through all the steps to derive (10.53); see Franceschetti (2004).

10.6 Go through all the steps required to compute the series in (10.55) and obtain (10.56); see Franceschetti (2004).

11 Noise Processes

We must break at all cost from this restrictive circle of pure sounds and conquer the infinite variety of noise-sounds.[1]

11.1 Measurement Uncertainty

Every physical apparatus measuring a signal is affected by a measurement error: the measurement appears to fluctuate randomly by a small amount. The signal describing these random fluctuations is called *noise*, and is added to the desired signal. The presence of the noise ensures that the amount of information that can be communicated using signals is always finite. For electromagnetic signals, noise arises from a variety of causes. All of them can be traced back to the basic reason that at the microscopic level the world behaves in a quantized fashion. Despite the complexities associated with microscopic modeling, the net effect is that the noise can, most often, be very precisely modeled as a random signal having values governed by a probability distribution.

11.1.1 Thermal Noise

Thermal noise measured at a receiving antenna is due to free electron charges in the electric conductor material constituting the antenna. At any non-zero temperature these charges move by thermal effects, following random trajectories. Although on average the charges are uniformly distributed and do not affect the received signal, at any given instant of time their non-uniform placement creates a certain random voltage across the terminals of the antenna that varies from instant to instant, creating small fluctuations in the measured signal. These fluctuations result from the combined effect of many independent random trajectories of the electron charges, and affect all frequencies of the signal in the same way. They are modeled as a zero-mean, white Gaussian process $\mathsf{Z}(t)$. Physically, this process delivers an average power $k_B T_K d\omega/(2\pi)$ [W] to any measurement device acting over an elementary angular frequency interval $d\omega$. Here, k_B is the Boltzmann constant, approximately $1.3806488 \times 10^{-23}$ [J K^{-1}], and T_K is the absolute temperature of the electric conductor. Mathematically, the process has a power

[1] Luigi Russolo (1913). *The Art of Noises*. Reprinted 1986, Something Else Press.

spectral density

$$S_Z = k_B T_K \quad \text{for all } \omega \in \mathbb{R}. \tag{11.1}$$

By the Wiener–Khinchin theorem, taking the inverse Fourier transform of (11.1) we obtain the (generalized) autocorrelation

$$s_Z(\tau) = \mathbb{E}(Z(t)Z(t+\tau)) = k_B T_K \delta(\tau), \tag{11.2}$$

which corresponds physically to having a coupling between the values of the noise that is essentially zero at any two instants of time separated by more than a tiny interval.

What is described above is clearly an idealized situation, and particular care should be used when talking about random processes with impulsive correlations. The mathematical meaning of having a random process with an infinite variance given by $s_Z(0)$ should raise an eyebrow, to say the least. Nevertheless, the situation can be backed up by a rigorous formulation. The white Gaussian process can also be obtained in an appropriate limiting sense as the generalized derivative of the Wiener process. This process is defined as having independent Gaussian increments over disjoint intervals. Using an appropriate smoothing function and letting the derivative at any point be a functional rather than a real number, a rigorous mathematical treatment of white Gaussian noise in a distributional sense can be developed. Despite these mathematical sophistications, in many applications it is possible to think about the white Gaussian process as simply producing essentially independent Gaussian random variables at distinct time intervals. In practice, white Gaussian noise is always observed as a convolution integral through a linear system of finite bandwidth that acts as a smoothing function, and this makes any real measurement consistent with the rigorous mathematical formulation.

From the physical perspective, however, a simple thermal noise model with constant power spectral density remains problematic. Even if our linear system measurement may not show it, the total expected power of the noise theoretically diverges if its power spectral density were to be integrated over all frequencies. This divergence is unnatural to describe the thermal agitations occurring in nature at finite temperatures. As we shall see below, this inconsistency can be solved by taking a closer look at the quantum mechanical nature of the thermal agitation that leads to the noise process. This provides a more precise expression for (11.1) that avoids divergence of the expected power and can also explain the Gaussian form of the distribution in a manner consistent with the second law of thermodynamics.

11.1.2 Shot Noise

As with thermal noise, the shot noise at a receiving antenna arises because of the quantization of electric charges. Any current is composed of a stream of charges that cross a given surface. At the microscopic level, these charges do not cross the surface at regular intervals, but have random spatial separations between them. Hence, the number of charges that pass through the surface fluctuates randomly over time around a given value. A mathematical model to describe the situation is given by a Poisson process

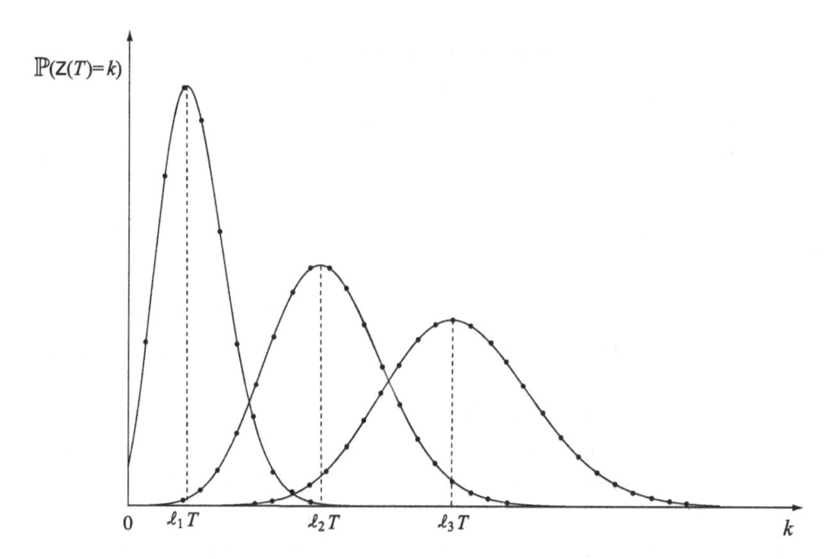

Fig. 11.1 Convergence of the Poisson distribution to the Gaussian distribution.

$Z(t)$ of intensity ℓ. This is a continuous process that counts random events occurring uniformly over time. It is defined by two properties: for mutually disjoint intervals of time the numbers of crossing charges are independent, and the probability that the number of charges crossing in an interval of length T is equal to k is given by the Poisson formula

$$\mathbb{P}(Z(T) = k) = \frac{(\ell T)^k}{k!} \exp(-\ell T). \tag{11.3}$$

It follows from this definition that in any finite time interval of length T there are only finitely many crossing charges, on average ℓT, and these are distributed uniformly inside the interval. For a large average number of crossings, the deviation from the mean $Z(T) - \ell T$ approaches a zero-mean Gaussian random variable of variance ℓT. This is somewhat evident from Figure 11.1, and Problem 11.1 takes a closer look at the mathematical derivation. It follows that if a large number of charges is counted in the measurement process, then the shot noise is essentially Gaussian. In this case, the shot noise is mathematically no different than the thermal noise considered earlier. Physically, however, it is due to impressed sources and is proportional to the current, while the thermal noise occurs in equilibrium in the absence of sources and is proportional to the temperature.

When the value of the current is increased, in addition to becoming Gaussian, the effect of the shot noise tends to vanish when seen at the scale of the measured signal level. Since the standard deviation of the shot noise is $\sqrt{\ell T}$, while its average is ℓT, it follows that as $\ell T \to \infty$ the fluctuations in the measured signal around the average value become negligible, and the effect of the noise disappears compared to the overwhelming signal.

On the other hand, with very small currents and considering observations at shorter time scales, the effect of the shot noise and its deviation from the Gaussian distribution can be significant. In conductors, however, even at the very small scale the shot noise is mostly suppressed by the repulsive action of the Coulomb force that acts against charge build-ups and smoothes out accumulation points, reducing the fluctuations around the mean that one would expect assuming independent interactions.

11.1.3 Quantum Noise

Poisson shot noise also arises in electromagnetic radiation in the framework of individual photon detection when appropriate care is taken to reduce the thermal noise that is otherwise dominant, for example by cooling the receiving apparatus and transmitting at high frequency with limited energy.

Electromagnetic waves are composed of a stream of photons propagating in the environment, and at high frequency only a limited number of highly energetic photons can be transmitted with a given energy budget. Fluctuations can then be observed in the numbers of received photons over a given time interval. The statistics of these fluctuations are similar to those occurring in the case of current charges, and are due to individual photons not being evenly spaced when crossing a given surface, but having random spatial separations among them. In this case, the shot noise models quantum effects due to the quantized nature of propagation, and is called *quantum noise*. For a fixed radiation energy this noise increases linearly with frequency, since at high frequency a smaller number of photons are radiated and quantum fluctuations become more evident.

Quantum noise is closely related to the uncertainty principle examined in Chapter 2. The uncertainty in the number of photons received in a given time interval is due to the quantum fluctuations of the time of arrival of each photon composing the signal, and any reduction of this uncertainty must be balanced by a corresponding quantum uncertainty in the frequency of the radiated photons. Since the energy of a photon is proportional to its frequency of radiation, this results in an additional energy uncertainty. Furthermore, since the signal received at a given distance from the transmitter is phase shifted by an amount proportional to the frequency of radiation, this results in a corresponding phase uncertainty. It follows that time–frequency, time–energy, as well as number of photons–phase, are all pairs subject to the principle of *quantum complementarity*, as they cannot be simultaneously measured accurately. Quantum noise represents the irreducible amount of noise imposed by these quantum constraints. At the macroscopic scale, for a large number of radiated photons, it reduces to thermal noise, having a Gaussian distribution. At the microscopic scale, for a small number of radiated photons, it is better described by a Poisson distribution.

11.1.4 Radiation Noise

Another kind of noise that is also of thermal origin but not intrinsic to the transmitting or receiving apparatus is due to the propagation environment. Even in free-space propagation conditions, a certain amount of background radiation from distant objects contributes to the noise at the receiver. Mathematically, this kind of noise follows the same statistics, and is indistinguishable from the intrinsic thermal noise of conductors.

In idealized free space there is no radiation noise; however, in practice there are a large number of sources of radiation that affect any antenna measurement. Even pointing an antenna towards the cold sky it is possible to detect cosmic background radiation from the boundary of our universe consistent with the possible remains of the Big Bang.

Radiation noise occurs by natural emission of electromagnetic waves due to thermal effects inside radiating bodies. Any body at a given temperature radiates energy to, and absorbs energy from, the environment. The *black body* is an idealized version of a real body that is in perfect thermal equilibrium, so that the radiating and absorbing energies coincide at all frequencies. Loosely speaking, all the incoming radiation is "thermalized" inside the body and re-emitted in the form of electromagnetic waves. Due to the random mechanism of radiation associated with the random thermal motions of the elementary constituents of the body, the radiated electromagnetic field has a Gaussian intensity, and the expected power is proportional to the temperature and distributed across a wide range of frequencies.

11.2 The Black Body

The radiation law of a black body is at the basis of all thermal noise models and its maximum entropy properties provide insight into the largest amount of information associated with electromagnetic radiation.

11.2.1 Radiation Law, Classical Derivation

The radiation law can be derived by first determining the energy absorbed per unit volume inside the body, and then imposing the condition that at equilibrium the absorbed energy equals the radiated energy. Since at the boundary of the body with the outside space the incoming power flux is equal to the emitted one, the net flux is zero and we have

$$\mathbf{e} \times \mathbf{h} \cdot \bar{\mathbf{n}} = 0, \tag{11.4}$$

where $\mathbf{e}$ is the electric field, $\mathbf{h}$ is the magnetic field, and $\bar{\mathbf{n}}$ is the normal to the boundary. It follows that we can assume that at the boundary the tangential electric field is zero, or the tangential magnetic field is zero. Assuming the former, the electric field inside the body can be written as the superposition of standing waves, that we call modes, with zero tangential component at the boundaries. Counting all possible configurations of these modes leads to an expression for the internal energy absorbed by the body that must be equal to the radiated energy.

From Maxwell's equations, the electric field must satisfy a wave equation that is the three-dimensional generalization of (4.44):

$$\frac{\partial e}{\partial x^2} + \frac{\partial e}{\partial y^2} + \frac{\partial e}{\partial z^2} = \frac{1}{c^2} \frac{\partial e}{\partial t^2}. \tag{11.5}$$

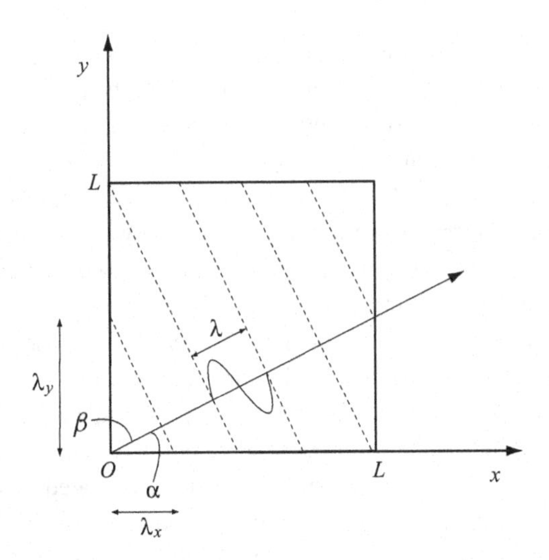

Fig. 11.2 Standing wave forming angles α and β with respect to the coordinate axes inside the black body.

The solution can be obtained as a superposition of sinusoidal components, each corresponding to a standing wave of wavelength λ inside the body. We consider a cubical volume V of side length L. The cubical geometry is taken for convenience, but the result easily generalizes to different geometries. Waves traveling orthogonally to the faces of the cube exhibit an electric field that is polarized along the coordinate axes tangential to the faces of the cube. These sinusoidal waves must satisfy the zero boundary condition, and we have

$$n\lambda = 2L, \tag{11.6}$$

for any positive integer n.

For waves that travel at angles α, β, γ with respect to the coordinate axes, we have three sets of standing waves along the three coordinate axes that form a complete modal expansion of the field inside the cube with the prescribed boundary condition, provided that

$$\begin{cases} \lambda_x = \dfrac{\lambda}{\cos\alpha}, & (11.7a) \\ \lambda_y = \dfrac{\lambda}{\cos\beta}, & (11.7b) \\ \lambda_z = \dfrac{\lambda}{\cos\gamma}, & (11.7c) \end{cases}$$

where $\lambda_x, \lambda_y, \lambda_z$ are the wavelength spacings measured along the three axes – see Figure 11.2. It then follows that

$$\begin{cases} n_x\lambda = 2L\cos\alpha, & (11.8a) \\ n_y\lambda = 2L\cos\beta, & (11.8b) \\ n_z\lambda = 2L\cos\gamma, & (11.8c) \end{cases}$$

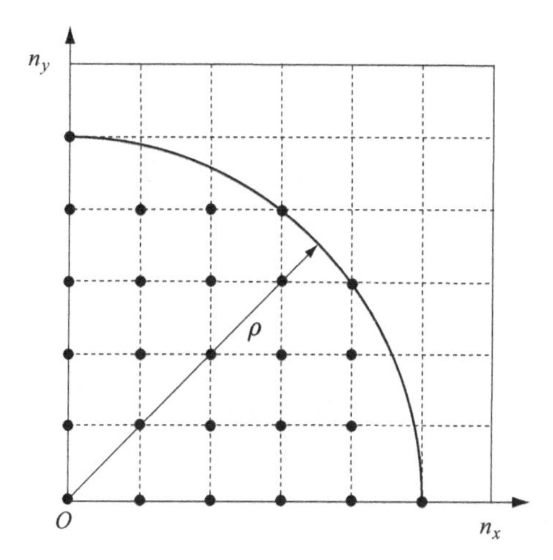

Fig. 11.3 Approximating the number of lattice points in the positive quadrant within radius ρ from the origin with the area of the positive quadrant subtended by the circle of radius ρ.

for positive integers n_x, n_y, n_z. Squaring and summing the three equations in (11.8) and using $\cos^2\alpha + \cos^2\beta + \cos^2\gamma = 1$, we have that the permitted wavelengths of the standing waves are

$$\lambda = \frac{2L}{\sqrt{n_x^2 + n_y^2 + n_z^2}}. \tag{11.9}$$

We now count the number $M(\lambda)$ of modes that can meet this condition. This amounts to counting all possible combinations of the positive integer values n_x, n_y, n_z leading to the different values of λ. Consider the n-space defined by the integer cubical lattice with coordinates (n_x, n_y, n_z). In this space, the quantity

$$\rho = \sqrt{n_x^2 + n_y^2 + n_z^2} \tag{11.10}$$

represents the magnitude of a vector in the positive octant of the space, and each value of ρ gives a corresponding value for λ. In the scaling limit of $L \to \infty$, the discrete space of possible vectors can be seen as being essentially continuous, and the counting process corresponds to computing the volume of the three-dimensional positive octant of the sphere of radius ρ. The error in treating the radius ρ as a continuous variable is of the order of the surface of the sphere, and is thus negligible compared to its volume; see Figure 11.3 for a two-dimensional picture of the situation. Accounting for an additional factor of two for the two possible mode polarizations, we have

$$M(\lambda) = 2\frac{1}{8}\frac{4\pi}{3}(n_x^2 + n_y^2 + n_z^2)^{3/2} = \frac{\pi}{3}(n_x^2 + n_y^2 + n_z^2)^{3/2}, \tag{11.11}$$

and substituting (11.9) into (11.11), we get

$$M(\lambda) = \frac{8\pi L^3}{3\lambda^3}. \tag{11.12}$$

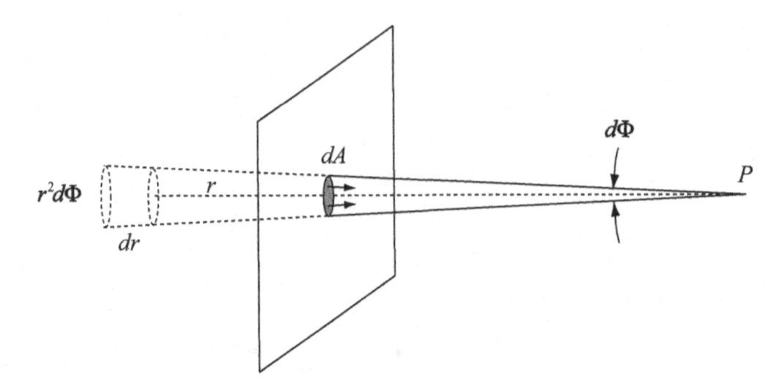

Fig. 11.4 Geometry of the radiated energy through the surface of the black body as seen by a detector at P.

The number of modes per unit wavelength is then given by

$$\frac{\partial M(\lambda)}{\partial \lambda} = -\frac{8\pi L^3}{\lambda^4}, \tag{11.13}$$

where the negative number shows that the number of modes decreases with increasing wavelength.

Next, we wish to assign energy to the standing waves inside the body. First, we assume equipartition of energy among the modes in thermal equilibrium, namely all standing waves have the same expected energy value, proportional to the temperature. Second, according to classical statistical mechanics, we assume that all energy values due to random thermal oscillations are possible, but higher energy values have an exponentially decaying probability of occurrence. As we shall see below, for the given expected energy constraint, this distribution maximizes the entropy of the system. Hence, we model each standing wave as having random energy E, distributed continuously according to an exponential probability density function of average $k_B T_K$:

$$f_{\mathsf{E}}(e) = \frac{1}{k_B T_K} \exp\left(-\frac{e}{k_B T_K}\right). \tag{11.14}$$

Letting u be the total expected energy inside the body, it follows from (11.13) and (11.14) that the expected energy per unit wavelength is given by

$$\frac{\partial u(\lambda, V)}{\partial \lambda} = -k_B T_K \frac{\partial M(\lambda)}{\partial \lambda} = \frac{8\pi L^3 k_B T_K}{\lambda^4} = \frac{8\pi V k_B T_K}{\lambda^4}. \tag{11.15}$$

The expected energy per unit wavelength and per unit volume is then given by

$$\frac{\partial^2 u(\lambda, V)}{\partial \lambda \partial V} = \frac{8\pi k_B T_K}{\lambda^4}. \tag{11.16}$$

Consider now an elementary volume $dV = r^2 dr d\Phi$ inside the cube, subtended by solid angle $d\Phi$ centered at point P outside the cube – see Figure 11.4. Since in thermal equilibrium absorption and emission are balanced, and we assume isotropic radiation of energy in all directions with velocity c, it follows that a $1/(4\pi)$ fraction of energy is radiated from this volume at velocity $c = dr/dt$ along the direction of point P and a $c/(4\pi)$ fraction of energy per unit time passes through the elementary area dA subtended by the solid angle $d\Phi$ on the surface of the cube. The expected radiated power per unit wavelength, per unit area dA, and into the angle $d\Phi$ along the direction orthogonal to the surface of the cube is then given by

$$P(\lambda) = \frac{c}{4\pi}\frac{\partial^2 u(\lambda, V)}{\partial\lambda\,\partial V} = \frac{8\pi c k_B T_K}{4\pi\lambda^4} = \frac{2c}{\lambda^4}k_B T_K. \tag{11.17}$$

The analogous expression per unit angular frequency is immediately obtained using the relation $\omega = 2\pi c/\lambda$, so that

$$\frac{d\omega}{d\lambda} = -\frac{2\pi c}{\lambda^2}, \tag{11.18}$$

where the minus sign reminds us that an increase in frequency corresponds to a decrease in wavelength and vice versa. From the differential relation

$$P(\lambda)\,|d\lambda| = P(\omega)\,|d\omega|, \tag{11.19}$$

we then have

$$P(\omega) = \frac{S_Z(\omega)}{2\pi} = \frac{\omega^2}{4\pi^3 c^2}k_B T_K, \tag{11.20}$$

where $S_Z(\omega)$ is the power spectral density per unit area and per unit solid angle of the noise process. It follows that the expected power per unit frequency is proportional to the temperature, and is also proportional to the square of the angular frequency of radiation: classical statistical mechanics predicts that higher power is delivered at higher frequencies by the radiating body.

11.2.2 Thermal Noise, Classical Derivation

Thermal noise is a manifestation of black body radiation in a single dimension. By performing the same steps leading to the radiation law in one dimension, we obtain (11.1). Consider two identical resistors connected by a lossless transmission line (an ideal pair of parallel wires) of length L. One of the resistors models the receiving conductive antenna, and we wish to determine the thermal voltage measured across its terminals. Assume the transmission line has characteristic impedance equal to the resistance of the resistors, so that power is efficiently coupled between them. In equilibrium, the transmission line can only support standing waves having zero voltages at its ends. All other modes are lossy and dissipate over the resistors, violating equilibrium – see Figure 11.5.

It follows that the standing waves at equilibrium must satisfy

$$2L = n\lambda, \; n = 1, 2, 3, \ldots, \tag{11.21}$$

where $\lambda = 2\pi v/\omega$ is the wavelength of the electric signal propagating along the line, and v is the propagation velocity of the signals along the line. We then have

$$n = \frac{L\omega}{\pi v}, \tag{11.22}$$

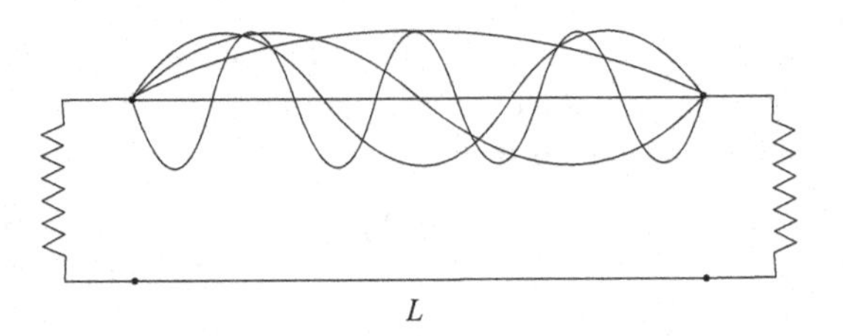

Fig. 11.5 Thermal noise between coupled resistors.

and the number of modes per unit frequency is

$$\frac{\partial n(\omega)}{\partial \omega} = \frac{L}{\pi v}. \tag{11.23}$$

In equilibrium, each mode has average energy equal to $k_B T_K$, so the average energy per unit frequency and per unit length in the transmission line is

$$\frac{\partial^2 u(\omega)}{\partial \omega \partial L} = \frac{k_B T_K}{\pi v}. \tag{11.24}$$

Finally, we need to convert the energy density inside the transmission line into power flow from one end to the other of the transmission line. This is simply done multipying by v, and obtaining the total average power per unit frequency generated by two resistors. It follows that the average power per unit frequency of one resistor is

$$P(\omega) = \frac{S_Z(\omega)}{2\pi} = \frac{k_B T_K}{2\pi}. \tag{11.25}$$

The average power per unit frequency of the noise is in this case proportional to the temperature and constant across the frequency spectrum. The similarity between this result and (11.20) should be clear. The space of the transmission line has dimension one rather than three, so that the frequency dependence is ω^0 in the case of (11.25), rather than ω^2 as in the case of (11.20).

11.2.3 Quantum Mechanical Correction

There is a problem with the results obtained by the classical derivation of the thermal and radiation noises: the total expected power obtained by integrating the power spectral density over all frequencies diverges. This is clearly a paradox, as the expected power is in both cases due to thermal agitations in equilibrium at finite temperature, so it cannot be infinite. The phenomenon is clear from the divergence of (11.20) at high frequencies, and is called the "ultraviolet catastrophe" prediction of classical physics. Similarly, the power spectral density of white noise in (11.25), although not diverging, remains constant across all frequencies, and its integral across the whole spectrum also gives an infinite expected power.

Fortunately, quantum mechanics comes to the rescue and brings the model closer to reality. The quantum mechanical correction of the formulas (11.1) and (11.20) is based on the fundamental observation that the electromagnetic modes cannot have arbitrary energy values, but these are quantized and radiant energy can only exist in discrete quanta $n\hbar\omega$, $n = 1,2,3,\ldots$, where $\hbar$ is the reduced Planck's constant. This effectively means that the energy is proportional to a discrete number of emitted photons, each carrying a quantum of energy $\hbar\omega$. In this case, the exponential distribution (11.14) for the energy becomes discrete, taking values with probability

$$\mathbb{P}(\mathsf{E} = n\hbar\omega) = \frac{\exp(-n\hbar\omega/(k_B T_K))}{\sum_{n=0}^{\infty} \exp(-n\hbar\omega/(k_B T_K))}, \tag{11.26}$$

and the average energy per mode is then given by

$$\mathbb{E}(\mathsf{E}) = \frac{\sum_{n=0}^{\infty} n\hbar\omega \exp(-n\hbar\omega/(k_B T_K))}{\sum_{n=0}^{\infty} \exp(-n\hbar\omega/(k_B T_K))}. \tag{11.27}$$

In order to perform the computation in (11.27) we write $x = \exp(-\hbar\omega/(k_B T_K))$, so that we have

$$\begin{aligned}
\mathbb{E}(\mathsf{E}) &= \hbar\omega \frac{\sum_{n=0}^{\infty} n x^n}{\sum_{n=0}^{\infty} x^n} \\
&= \hbar\omega x \frac{\sum_{n=0}^{\infty} n x^{n-1}}{\sum_{n=0}^{\infty} x^n} \\
&= \hbar\omega x \frac{(1-x)^{-2}}{(1-x)^{-1}} \\
&= \frac{\hbar\omega x}{1-x} \\
&= \frac{\hbar\omega}{1/x - 1} \\
&= \frac{\hbar\omega}{\exp(\hbar\omega/(k_B T_K)) - 1}.
\end{aligned} \tag{11.28}$$

By using (11.28) in place of $k_B T_K$ in the derivations of the power spectral densities given before, we obtain in one dimension the power spectral density per unit angular frequency, and in three dimensions the power spectral density per unit angular frequency, per unit area, and per unit solid angle:

$$\begin{cases}
S_{\mathsf{Z}}^{(1)}(\omega) = \dfrac{\hbar\omega}{\exp(\hbar\omega/(k_B T_K)) - 1}, & \text{(11.29a)} \\[2ex]
S_{\mathsf{Z}}^{(3)}(\omega) = \dfrac{\hbar\omega^3}{2\pi^2 c^2} \dfrac{1}{\exp(\hbar\omega/(k_B T_K)) - 1}. & \text{(11.29b)}
\end{cases}$$

The analogous expressions per unit wavelength are immediately obtained using the chain rule with $\omega = 2\pi c/\lambda$:

$$\begin{cases}
S_{\mathsf{Z}}^{(1)}(\lambda) = \dfrac{(2\pi)^2 c^2 \hbar}{\lambda^3} \dfrac{1}{\exp(2\pi\hbar c/(\lambda k_B T_K)) - 1}, & \text{(11.30a)} \\[2ex]
S_{\mathsf{Z}}^{(3)}(\lambda) = \dfrac{(2\pi)^2 2 c^2 \hbar}{\lambda^5} \dfrac{1}{\exp(2\pi\hbar c/(\lambda k_B T_K)) - 1}. & \text{(11.30b)}
\end{cases}$$

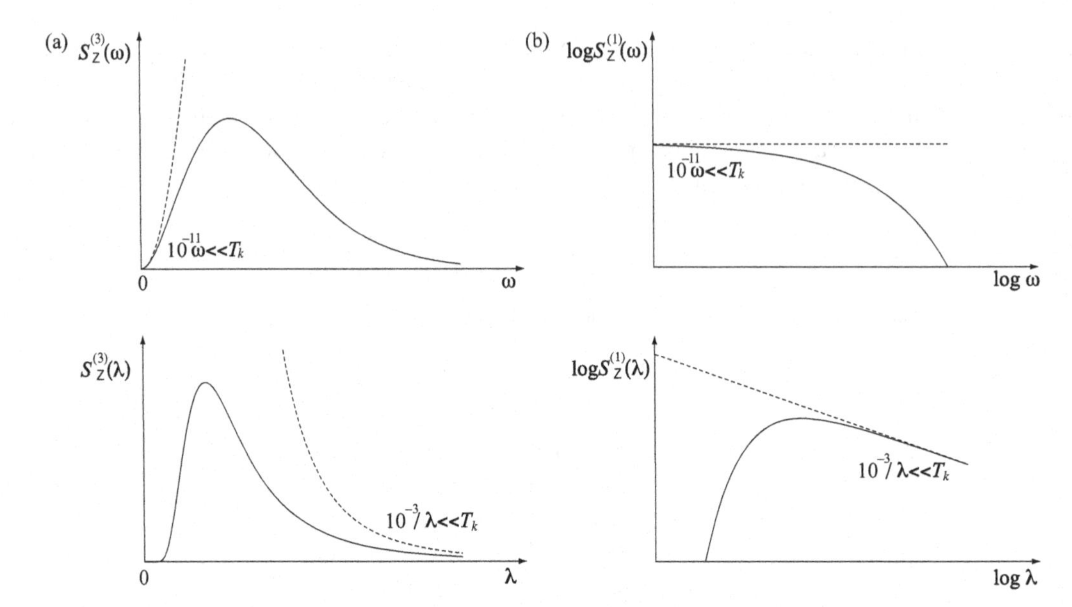

Fig. 11.6 Power spectral densities of radiation noise (a) and ideal thermal noise (b), with corresponding classical approximations (dashed lines).

The results from classical mechanics obtained earlier are recovered by letting the energy of a photon $\hbar\omega \to 0$ and using the Maclaurin expansion

$$\exp(\hbar\omega/(k_B T_K)) = 1 + \frac{\hbar\omega}{k_B T_K} + O((\hbar\omega)^2). \tag{11.31}$$

Exact formulas and their low-frequency approximations are depicted in Figure 11.6. The left-hand side of the figure depicts the power spectral density in the three-dimensional case, and the right-hand side depicts the one-dimensional case. When communication occurs in the range of frequencies $\hbar\omega/k_B \approx 10^{-11}\omega \ll T_K$, the white noise approximation applies. For example, for communication up to the GHz range at an ambient temperature of 300 K, the approximation is satisfied with four orders of magnitude. On the other hand, at higher frequencies and lower temperatures, quantum effects become relevant. Figure 11.7 depicts the power spectral density in the three-dimensional case for different values of the temperature, showing that the peak of the curve shifts to the right and the curve becomes spread out over a larger wavelength interval as the temperature decreases.

11.3 Equilibrium Configurations

In the derivations above, we have assumed that thermally excited electromagnetic modes have exponentially distributed energy. We justified this assumption by stating that at equilibrium it allows all energy values due to random thermal oscillations, but weights higher energy values by an exponentially decaying probability of occurrence. A

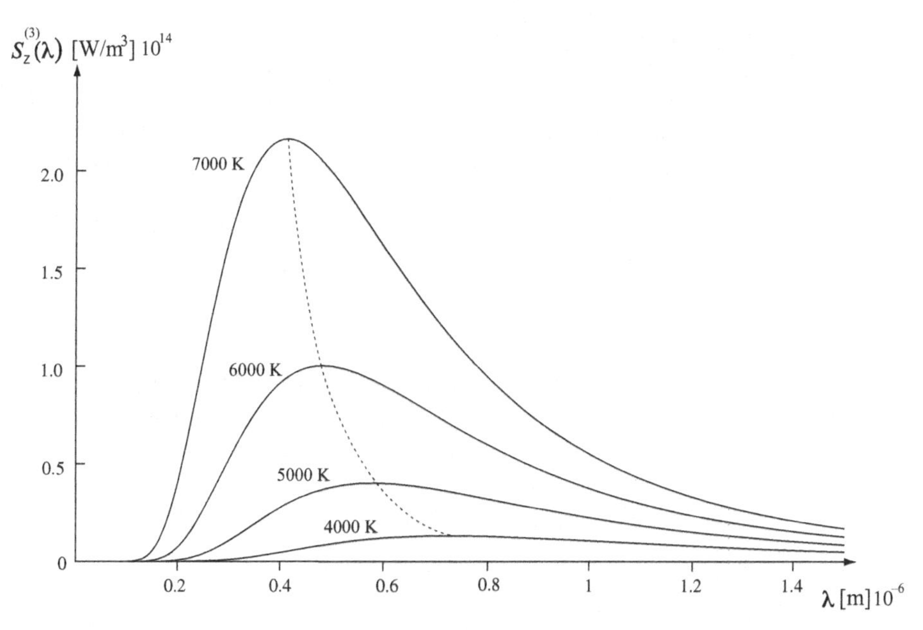

Fig. 11.7 Power spectral density of three-dimensional radiation noise as a function of the wavelength for different temperature values.

consequence of the exponential assumption is that the electromagnetic radiated signal has a Gaussian distributed amplitude at all frequencies. This can also be justified by noting that thermal excitations are the result of a large number of independent random trajectories of the elementary constituents of the body. For example, in a resistor each conduction band electron produces a random elementary current wave as it thermally agitates. The total current is then the result of the addition of these elementary independent components, and one can invoke the central limit theorem to justify the resulting Gaussian distribution. The main idea is that by adding many independent random variables and hence convolving their distributions, a Gaussian shape is obtained in the limit. This fixed point distribution has the property that additional convolutions with Gaussians have only scaling effects but do not change its overall shape. Although the argument can be made mathematically rigorous, its physical foundation based on the random elementary effects still appears somewhat heuristic.

An alternative justification for the choice of the probability distribution of the noise can be given in terms of the second law of thermodynamics, according to which at equilibrium the energy of the thermally excited modes inside the body assumes a configuration that maximizes the entropy of the radiating system. This also provides an interpretation of the noise as the most random process, for a given average energy constraint, namely the one for which we have the least amount of information regarding its actual physical state. Instead of the central limit theorem, the mathematical tool for this physical interpretation is the law of large numbers.

11.3.1 Statistical Entropy

To relate our probabilistic modeling to the second law, we recall some statistical mechanical definitions introduced in Chapter 1. Consider a system that can assume any one of N discrete states, where the nth state occurs with probability p_n. The Shannon entropy of this system is

$$H = -\sum_{n=1}^{N} p_n \log p_n, \tag{11.32}$$

where the logarithm is base two, we define by continuity $0 \log 0 = 0$, and the entropy is measured in bits. A simple change of base leads to a change of units, as $\log_b p_n = \log_b a \log_a p_n$. The entropy formula was proposed by Shannon in 1948 as a measure of information: the entropy of a system can be thought of as the amount of information we lack about it.

In statistical mechanics, the Boltzmann (1896–8) entropy of a system with N states is defined as

$$H_{\mathrm{B}} = k_{\mathrm{B}} \log N. \tag{11.33}$$

Apart from the scaling factor k_{B}, this corresponds to letting $p_n = 1/N$ in (11.32), so that $H = \log N$. This choice of distribution yields a maximization of (11.32) subject to the constraint on the support of the interval where the state distribution is defined to be of size N. The proof follows easily from Jensen's inequality:

$$\begin{aligned} H &= \sum_{n=1}^{N} p_n \log 1/p_n \\ &\leq \log \sum_{n=1}^{N} p_n/p_n \\ &= \log N, \end{aligned} \tag{11.34}$$

where it can be immediately verified that the equality occurs if and only if $p_n = 1/N$. The normalization factor k_{B} in Boltzmann's formula gives to the entropy the dimensions of energy over temperature and, as we shall see below, allows us to interpret the statistical definition as the thermodynamic entropy of an isolated system in thermal equilibrium, where all states are equally probable and occur independently.

Another statistical mechanical definition of entropy, that is analogous to Shannon's formula, was introduced by Gibbs in 1902 as a generalization of Boltzmann's entropy:

$$H_{\mathrm{G}} = -k_{\mathrm{B}} \sum_{n=1}^{N} p_n \log p_n. \tag{11.35}$$

This is aimed at describing systems in thermal equilibrium whose states are not necessarily uniformly distributed. For example, a black body is a closed system in thermal equilibrium that is not isolated, as it continuously exchanges energy with the outside world. The energy is not uniformly distributed among all the electromagnetic modes inside the body, but only the average energy per mode is kept constant, while higher energy modes are exponentially less likely than the lower ones.

Finally, in a system having a continuous density of states, an integral expression analogous to (11.32) is given by the differential entropy,

$$h(f) = -\int_S f(x) \log f(x) dx, \tag{11.36}$$

where x is the realization of a continuous random variable representing the state of the system having support S and probability density function f. In this case, the uniform distribution is the maximum entropy distribution over a given support interval, the exponential distribution is the maximum entropy distribution among all distributions with given non-zero support subject to a first moment constraint, and the Gaussian distribution has maximum entropy among all distributions subject to a second moment constraint (see Problems 11.2, 11.3, and 1.15).

Given the definitions above, it is now clear that the probability distributions employed in the description of thermal noise are consistent with the principle of statistical entropy maximization: each noise mode is assumed to have uniform phase, exponentially distributed energy subject to an average constraint, and Gaussian intensity subject to a variance constraint, corresponding to the average power of the noise. The probability distributions for the shot noise are also consistent with the same principle. Fixing the intensity of the Poisson process corresponds to fixing the average time between successive arrivals, and under this constraint the inter-arrival time is exponentially distributed. Fixing the length of any time interval, the distribution of the arrivals in the given interval is uniform, and this maximizes the entropy given the support of the distribution.

We conclude that using the maximum entropy principle to model the noise has two appealing features. On the one hand, by maximizing the Shannon entropy subject to physical constraints, the noise maximizes the lack of information regarding its actual physical state. On the other hand, maximization of entropy is consistent with the second law of thermodynamics, according to which a system in thermal equilibrium always reaches a configuration where entropy is maximized. To better understand this latter connection, we now take a closer look at the relationship between the statistical and thermodynamic definitions of entropy.

11.3.2 Thermodynamic Entropy

The phenomenological definition of thermodynamic entropy is stated in variational terms as:

> A variation of thermodynamic entropy is given by the ratio of energy absorbed by the system in the form of heat and the macroscopic temperature of the system.

This definition was proposed by Clausies (1850–65). Expressed mathematically, we have

$$dH_C = \frac{dE}{T_K}, \tag{11.37}$$

where the entropy has the dimensions of energy divided by temperature.

The fundamental insight of statistical mechanics is that phenomenological observations reflect the average of the microscopic configurations of the system. Once equilibrium is reached, microscopic energy fluctuations are smoothed out and only average values over the possible microscopic energy configurations are observed. Consider a black body in thermal equilibrium containing M electromagnetic modes, each of average energy $k_B T_K$. The body has total average energy

$$E = M k_B T_K. \tag{11.38}$$

Differentiating (11.38) and dividing both sides by the temperature T_K, we have

$$\frac{dE}{T_K} = k_B dM. \tag{11.39}$$

Putting together (11.37), (11.38), and (11.39), it follows that the quantity $k_B M$ corresponds to the thermodynamic entropy. Namely, the entropy is proportional to the number of elementary modes of the system. This equivalence holds only on average, with statistical variations that tend to disappear as the number of elementary modes increases, or equivalently according to (11.12) in the scaling limit of large bodies. A larger body allows a larger number of possible electromagnetic modes inside it and, as the number of modes grows, this number also converges to the thermodynamic entropy. The equivalence is limited to thermodynamic equilibrium, as a macroscopic notion of temperature can only be defined at equilibrium.

11.3.3 The Second Law of Thermodynamics

The second law of thermodynamics can be stated in many different ways. At the end of his presentation to the Philosophical Society of Zurich in 1865, Rudolf Clausius stated that the entropy of the universe tends to a maximum. More precisely, we have that:

> The entropy of a system that moves from one state of thermal equilibrium to another can only increase.

It is important to emphasize that the increase in entropy is measured only between states of thermal equilibrium. During the time it takes for the system to reach equilibrium, thermodynamic entropy is not defined and one can only talk of an increase (if any) in statistical entropy.

Statistical mechanics provides an explanation of the second law in terms of microscopic configurations that the system can attain at equilibrium. Consider M modes, each having two equally probable binary energy levels. They generate $N = 2^M$ equally probable states of the system, and this distribution maximizes statistical entropy subject to a constraint on the number of states being equal to N. In this case, we immediately have

$$\begin{aligned} H_{\mathrm{B}} &= k_{\mathrm{B}} \log N \\ &= k_{\mathrm{B}} \log 2^M \\ &= k_{\mathrm{B}} M \\ &= H_{\mathrm{C}}. \end{aligned} \tag{11.40}$$

The binary cardinality of the energy levels of the elementary modes does not really matter, as long as it remains finite and yields a finite cardinality N of the number of states on which to maximize the entropy. Letting the set of permitted energy levels be $\mathscr{E}$ and its cardinality be $|\mathscr{E}|$ rather than 2, we also have

$$H_{\mathrm{B}} = k_{\mathrm{B}} \log |\mathscr{E}|^M = k_{\mathrm{B}} M \log |\mathscr{E}|, \tag{11.41}$$

where the factor $\log |\mathscr{E}|$ can be eliminated by changing the units with which we measure entropy by an appropriate choice of the base of the logarithm. It follows that for systems in equilibrium composed of elementary modes whose energy takes discrete values in a set of finite cardinality, and where all states are equally probable, the statistical and thermodynamic entropies coincide. Gibbs' definition (11.35) attains Boltzmann's form (11.33), corresponding to its maximum, and Shannon's definition (11.34) corresponds to the analogous dimensionless quantity.

11.3.4 Probabilistic Interpretation

So far, we have considered discrete systems in thermal equilibrium composed of elementary modes that can have a finite number of energy levels. In this case, the statistical and thermodynamic definitions of entropy coincide. In the scaling limit of large systems, however, the cardinality of the permitted energy levels of the elementary modes increases, so that the factor $\log|\mathscr{E}|$ in (11.41) diverges, and the relationship between statistical and thermodynamic entropy becomes unclear. In addition, systems in equilibrium may also have states that are not equally probable, and in this case Gibbs' and Shannon's definitions of entropy differ from Boltzmann's. It follows that to obtain a satisfactory probabilistic interpretation of the second law, we need a more elaborate description that takes into account these discrepancies.

At the basis of this generalization is the *asymptotic equipartition property* for a statistical ensemble of microstates. This property ensures that as the number of modes $M \to \infty$ and the system converges to equilibrium, only a subset of the infinitely many possible states of the system are typical, and these are essentially equally probable. The cardinality of the typical states is an exponential function of the statistical entropy of the elementary modes and of the number of modes, and (11.40) becomes

$$\begin{aligned} H_{\mathrm{B}} &= k_{\mathrm{B}} \log 2^{MH(p)} \\ &= k_{\mathrm{B}} MH(p), \end{aligned} \tag{11.42}$$

where $p = p_{\mathsf{E}}(e)$ is the common distribution of the energy levels of the modes, and $H(p)$ is its Shannon entropy. The cardinality of the energy levels being finite or not does not matter anymore; as long as their entropy is finite, the number of typical states $2^{MH(p)}$ is well defined. Since the typical states are essentially equiprobable, Gibbs', Shannon's, and Boltzmann's definitions once again assume the same form.

Comparing (11.40) and (11.42), it follows that the quantity $MH(p)$ in (11.42) can be interpreted as the number of equiprobable binary modes of an equivalent thermodynamic system in equilibrium, and maximizing the statistical entropy $H(p)$ subject to energy constraints corresponds to maximizing the thermodynamic entropy of this system.

11.3.5 Asymptotic Equipartition Property

In mathematical terms, the asymptotic equipartition property states that the probability of a sequence of M i.i.d. random variables $\{\mathsf{E}_k\}$ with common distribution $p_{\mathsf{E}}(e)$ over the set $\mathscr{E}$ is close to $2^{-MH(p)}$, and the number of such "typical sequences" is approximately $2^{MH(p)}$.

Asymptotic equipartition follows from the law of large numbers. The (weak) law of large numbers states that the sample average $(1/N)\sum_{k=1}^{N} \mathsf{E}_k$ converges in probability to its expected value. Consider now another sequence of random variables $\{\mathsf{W}_k\}$, such that each W_k takes the value $w(e) = -\log p_{\mathsf{E}}(e)$ for $\mathsf{E}_k = e$. The independence of the E_k's implies that the W_k's are also independent, with mean

$$\mathbb{E}(\mathsf{W}_k) = -\sum_{e \in \mathscr{E}} p_{\mathsf{E}}(e) \log p_{\mathsf{E}}(e) = H(p). \tag{11.43}$$

By applying the law of large numbers to the sample average

$$\frac{\mathsf{W}_1 + \cdots + \mathsf{W}_M}{M} = \frac{1}{M} \sum_{k=1}^{M} -\log p_{\mathsf{E}}(e_k), \tag{11.44}$$

we have

$$\lim_{M \to \infty} \mathbb{P}\left(\left| \frac{1}{M} \sum_{k=1}^{M} -\log p_{\mathsf{E}}(e_k) - H(p) \right| > \epsilon \right) = 0. \tag{11.45}$$

This result essentially states that given a long sequence of i.i.d. random variables, there is a typical set of possible realizations defined by

$$2^{-M(H(p)+\epsilon)} \leq \mathbb{P}(\mathsf{E}_1 = e_1, \mathsf{E}_2 = e_2, \ldots, \mathsf{E}_M = e_M) \leq 2^{-M(H(p)-\epsilon)}, \tag{11.46}$$

whose aggregate probability is almost one. All the elements of the typical set are nearly equiprobable, each having probability roughly $2^{-MH(p)}$, and the number of elements in the typical set is roughly $2^{MH(p)}$ – see Figure 11.8.

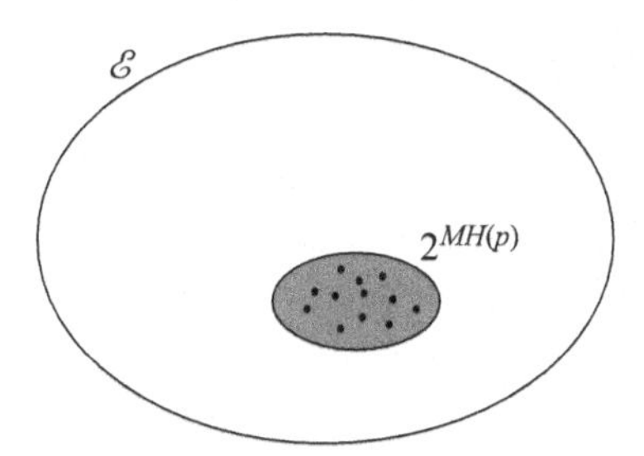

Fig. 11.8 Asymptotic equipartition of the set $\mathcal{E}$. The typical set is composed of approximately $2^{MH(p)}$ elements.

Consider now a system in thermal equilibrium constituted by a large number M of elementary modes, each having energy E_k. The entropy of each mode is

$$H(p) = -\sum_{e \in \mathcal{E}} p_{\mathsf{E}}(e) \log p_{\mathsf{E}}(e). \tag{11.47}$$

By asymptotic equipartition, all states are typical only when the energy is uniformly distributed among them, while in general only a small number $2^{MH(p)}$ of states have non-negligible probability and these are nearly equally probable. It follows that choosing a distribution for the energy of each mode that maximizes the entropy subject to physical constraints corresponds to choosing the distribution of system states that leads to the largest number of nearly equiprobable states. Physically, this is the most natural distribution that is generated by the largest number of possible configurations of the elementary constituents of the system. By (11.42), the Boltzmann entropy of the equipartitioned system composed of the typical states is maximized by choosing an appropriate energy distribution for the modes maximizing their Shannon entropy. Since the Boltzmann entropy corresponds to the thermodynamic entropy of an equivalent isolated system in thermal equilibrium where all states are equally probable, it follows that this maximization is consistent with the second law of thermodynamics.

11.3.6 Entropy and Noise

We now summarize the basic results on thermodynamic and statistical entropy, asymptotic equipartition, and their relationship with the noise model. For systems that take a discrete number of states, statistical entropy coincides with thermodynamic entropy at thermal equilibrium. The equivalence is immediate for isolated systems, where states are equally probable and the statistical entropy attains Boltzmann's form $H = \log N$. For systems that are not isolated, the asymptotic equipartition property shows that as the state space grows, it concentrates around the typical set, and the statistical entropy of this set again attains Boltzmann's form $H = \log 2^{MH(p)}$. By choosing the maximum entropy distribution for the mode energies $H(p)$, we can maximize the thermodynamic entropy of a system in equilibrium having $MH(p)$ equiprobable binary modes. Defining the noise process in this way is consistent with the classical statement of the second law.

11.4 Relative Entropy

Of course, one can wonder what happens to the statistical entropy as the system *reaches* equilibrium. Classical thermodynamics has little to say in this case, since thermodynamic entropy can only be defined at equilibrium. However, for systems whose states evolve according to a Markov chain, and whose equilibrium distribution is uniform over a finite state space, it can be shown that statistical entropy increases over time.

This result follows from some properties of the *relative entropy* between probability distributions $p(x)$ and $q(x)$,

$$D(p||q) = \sum_{x \in \mathscr{X}} p(x) \log \frac{p(x)}{q(x)}. \tag{11.48}$$

The first property is the information inequality,

$$D(p||q) \geq 0, \tag{11.49}$$

where equality holds if and only if $p = q$. The second property is the chain rule,

$$D[p(x,y)||q(x,y)] = D[p(x)||q(x)] + D[p(y|x)||q(y|x)], \tag{11.50}$$

where

$$D(p(y|x)||q(y|x)) = \sum_x p(x) \sum_y p(y|x) \log \frac{p(y|x)}{q(y|x)}. \tag{11.51}$$

From (11.50), it follows that given any two joint probability distributions $p(x_n, x_{n+1})$ and $q(x_n, x_{n+1})$ on the state space of the Markov chain, where x_n indicates the state at time n and x_{n+1} the state at time $n+1$, we have

$$\begin{aligned} &D[p(x_n, x_{n+1})||q(x_n, x_{n+1})] \\ &= D[p(x_n)||q(x_n)] + D[p(x_{n+1}|x_n)||q(x_{n+1}|x_n)] \\ &= D[p(x_{n+1})||q(x_{n+1})] + D[p(x_n|x_{n+1})||q(x_n|x_{n+1})]. \end{aligned} \tag{11.52}$$

From (11.49), and since $p(x_{n+1}|x_n) = q(x_{n+1}|x_n)$, it follows that

$$D[p(x_n)||q(x_n)] \geq D[p(x_{n+1})||q(x_{n+1})]. \tag{11.53}$$

In the case that there is a unique stationary distribution, and letting $q(x_n) = q(x_{n+1})$ be such a distribution, the result implies that any state distribution gets closer and closer to the stationary one as time goes by.

> The relative entropy of a Markov chain evolving towards a unique stationary distribution is non-increasing in time.

Finally, if the stationary distribution is the uniform one, we have, by (11.48),

$$D[p(x_n)||q(x_n)] = \sum_{x_n \in \mathscr{X}} p(x_n) \log p(x_n) + \sum_{x_n \in \mathscr{X}} p(x_n) \log |\mathscr{X}|$$

$$= -H(p_{X_n}) + \log|\mathscr{X}|, \tag{11.54}$$

so that the monotonic decrease in relative entropy implies a monotonic increase in entropy.

We now return to the physics. For systems in thermal equilibrium, each state is identified by a sequence of M i.i.d. random variables $\{\mathsf{E}_k\}$ representing the values of the energy of its elementary modes. When the number of modes is large then asymptotic equipartition ensures that only a typical set of states have non-negligible probability of occurrence and these are essentially equiprobable. Assuming that the system evolves towards thermal equilibrium in a Markovian fashion, it follows that the stationary distribution is uniform and the statistical entropy increases over time.

In practice, however, the equiprobable approximation can lead to entropy decrease over short time scales. From (11.46), it follows that the quality of the approximation improves as $\epsilon \to 0$. However, small values of ϵ require larger values of M for the probability of the typical set (11.45) to be close to one. When M is large but fixed, the state distribution admits non-typical states that have small, but non-zero, aggregate probability. As the system evolves towards equilibrium and the state distribution approaches the almost-uniform stationary one, small fluctuations exhibiting a decrease in entropy are possible – see Problem 11.7. In the study of non-equilibrium thermodynamics, fluctuation theorems bound the probability of such events.

11.5 The Microwave Window

In the design of communication systems, it is desirable to keep the noise to a minimum level. The laws of radiation show that the noise from different sources affecting the transmitted signal depends on the temperature and on the frequency of radiation. It turns out that in the microwave range of frequencies between 1 and 60 GHz the environmental radiation noise is at a minimum, which makes it an ideal candidate band for communication using electromagnetic signals.

Clearly, radiation noise is significant in the visible part of the spectrum. For example, the sun's surface temperature is around 5800 K and its emission peaks at a wavelength of around 0.5×10^{-6} m. The emission appears as white light because around its peak the power spectral density is approximately constant. This peak makes it difficult to communicate at visible optical frequencies during daytime. In comparison, the earth's surface temperature at around 300 K peaks in the infrared part of the spectrum at a wavelength of around 10^{-5} m. The intensity of radiation is much lower, about a 10^{-6} scale factor compared to the sun. The cosmic background radiation is isotropic across the whole sky and corresponds to a universal temperature of 2.73 K, which constitutes a noise floor for the microwave window.

Figure 11.9 shows the combined effect of the different forms of natural radiation noise in space. It plots the equivalent noise temperature spectral density, namely

$$T_{\mathrm{N}} = \frac{S_{\mathsf{Z}}}{k_{\mathrm{B}}}. \tag{11.55}$$

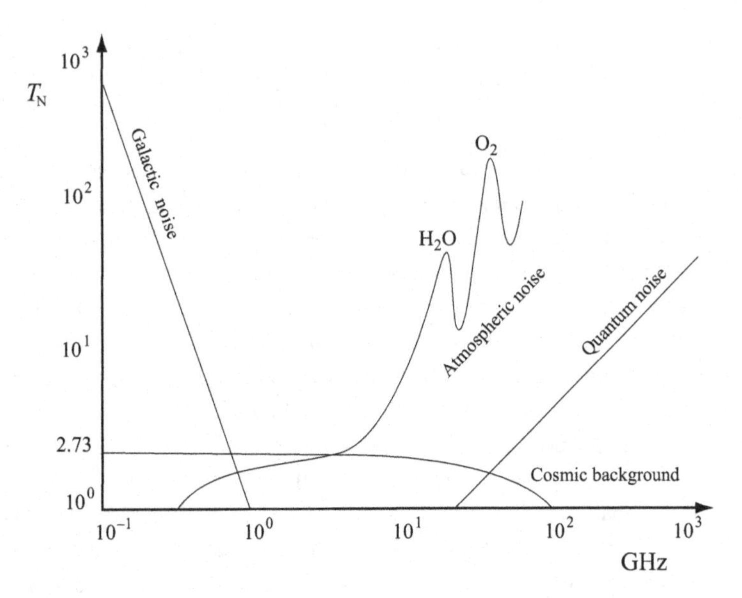

Fig. 11.9 Radiation noise contributions. The peaks in the atmospheric noise are due to the resonant absorption frequencies for water and oxygen.

We now identify the different contributions. First there is "galactic noise," the combined radiation noise from the Milky Way. This noise rises steeply below 1 GHz. Second, there is the 2.73 K cosmic background, the relic radiation emitted at the beginning of the universe and reaching us today from the boundaries of space. Third, there is "quantum noise," representing a fundamental quantum limitation on the receiver's sensitivity. This reflects vacuum fluctuations and represents the uncertainty of detecting a single photon of energy $\hbar\omega$. Letting

$$k_B T_N = \hbar\omega, \tag{11.56}$$

we get

$$T_N = \frac{\hbar\omega}{k_B}. \tag{11.57}$$

It follows that the quantum noise temperature increases linearly with frequency and becomes the dominant form of noise at high frequencies.

Attenuation in the atmosphere reduces the received galactic noise level, but it replaces it with the noise from the hotter atmosphere that in first approximation acts as a black body itself, emitting the absorbed radiation. For example, a galactic black-body noise source of temperature T_b viewed through a lossy medium, such as the atmosphere, of loss γ and temperature T_a, yields an effective temperature of

$$T_e = \gamma T_b + (1-\gamma)T_a. \tag{11.58}$$

Atmospheric noise peaks at the resonant absorption frequencies for the different elements in the atmosphere, at which absorption is maximum. The atmospheric noise contribution also depends on the angle between the direction of the antenna and the zenith. Looking to the zenith, the antenna is facing the smallest possible thickness of the atmosphere. Looking towards the horizon, the antenna signal path needs to pass through a much longer section of atmosphere. Above 1 GHz, the galactic noise falls below the isotropic cosmic background, and above about 60 GHz the quantum noise exceeds the cosmic background. Thus, the frequency range of 1–60 GHz is ideal for communication as natural noise is at a minimum. This is the microwave window in free space.

With the development of personal communication systems and the myriad man-made devices operating in the microwave range, the collective interference from other devices can also become a source of noise. The noise at the terminals of the antenna becomes a weighted average of the natural noise sources, plus the man-made noise. Man-made noise, emanating from electrical motors and power devices, typically plays little part in the microwave range and appears mostly in the low frequencies. However, some household devices also operate in the microwave range and can cause unwanted noise effects on communication.

11.6 Quantum Complementarity

Thermal and radiation noise fall off as $\omega \to \infty$. If these were the only kinds of noise present, then one could mitigate their effect by simply transmitting signals at increasing frequencies. Quantum limitations do not allow such a scenario. They arise from the quantization of charges or photons that propagate in the environment and they are another manifestation of the uncertainty principle illustrated in Chapter 2.

Recall that the uncertainty principle states that it is impossible to achieve simultaneous concentration in time and frequency. The more a signal is localized in time, the less it is localized in frequency, and vice versa. For a unit energy signal $f(t)$ of spectrum $F(\omega)$, this means that

$$\sigma_t^2 \sigma_\omega^2 \geq \frac{1}{4}, \tag{11.59}$$

where

$$\sigma_t^2 = \int_{-\infty}^{\infty} t^2 f^2(t) dt \tag{11.60}$$

$$\sigma_\omega^2 = \frac{1}{2\pi} \int_{-\infty}^{\infty} \omega^2 |F(\omega)|^2 d\omega. \tag{11.61}$$

For a single photon, Planck's equation,

$$E_{\mathrm{p}} = \hbar\omega, \tag{11.62}$$

shows that localization in frequency is equivalent to the determination of its energy, and we have the time–energy indetermination inequality

$$\sigma_t^2 \sigma_{E_{\mathrm{p}}}^2 \geq \frac{\hbar^2}{4}. \tag{11.63}$$

The more accurately we know the energy of a photon, the more monochromatic its wave function becomes, and the less accurately we can measure the time of arrival. On the other hand, in a small time interval the detected energy is uncertain and it is impossible to determine the presence or absence of a photon in time: the vacuum is subject to quantum fluctuations over small time scales.

The illustrated relations are examples of the general principle of quantum complementarity, according to which not all aspects of a system can be viewed simultaneously, as complementary quantities cannot be simultaneously measured accurately. This principle leads to the fundamental limitation labeled as quantum noise in Figure 11.9. For any given energy level, quantum noise increases linearly with frequency, as quantum fluctuations become more evident when a smaller number of highly energetic photons are radiated in space.

11.7 Entropy of a Black Body

Noise signals maximize entropy, subject to energy constraints. It follows that thermal radiation can be used to compute the largest amount of information, in terms of entropy, associated with a radiating body of given average energy. To illustrate this point, we note that by (11.12) the number of internal states of a black body is proportional to its volume expressed in units of wavelength. These states, however, are not all equally probable as higher frequency modes are weighted by a lower probability of occurrence that also depends on the temperature, according to (11.26). By the results described in Section 11.3, only a smaller number of states are typical, and these are essentially equally probable. The amount of information inside the body, integrated over all frequencies of radiation, can then be computed by taking the logarithm of the number of typical states that are thermally excited, which, for large systems in thermal equilibrium, coincides with the thermodynamic entropy. Finally, the thermodynamic entropy can be derived using (11.37) from the total average energy of the system, considering the variation of energy occurring at a given temperature due to an amount of heat absorbed by the system of constant volume. In the following, we describe these computations.

11.7.1 Total Energy

To compute the total average energy inside the body, we first compute the energy per unit volume lying between ω and $\omega + d\omega$. Using (11.29b) and recalling (11.17), we have the expected energy per unit frequency and per unit volume:

$$\frac{4\pi}{c}\frac{S_Z^{(3)}(\omega)}{2\pi} = \frac{\hbar\omega^3}{\pi^2 c^3}\frac{1}{\exp(\hbar\omega/(k_B T_K)) - 1}. \tag{11.64}$$

Letting $x = \hbar\omega/(k_B T_K)$, the total average energy per unit volume is then given by

$$\int_0^\infty \frac{\hbar\omega^3}{\pi^2 c^3}\frac{d\omega}{\exp(\hbar\omega/(k_B T_K)) - 1} = \frac{(k_B T_K)^4}{\pi^2\hbar^3 c^3}\int_0^\infty \frac{x^3}{\exp(x) - 1}dx. \tag{11.65}$$

The integral on the right-hand side can be computed to be $\pi^4/15$, so that the total average energy of a black body of volume V and temperature T_K is given by

$$E = \frac{\pi^2}{15\hbar^3 c^3}(k_B T_K)^4 V = \alpha V T_K^4, \tag{11.66}$$

where we have defined

$$\alpha = \frac{\pi^2}{15\hbar^3 c^3} k_B^4. \tag{11.67}$$

11.7.2 Thermodynamic Entropy

The total average energy of a black body depends on both its volume and its temperature. To derive an expression for the entropy, we consider a process where the volume is held constant and the energy variation is due to a temperature variation only. By (11.66), we have

$$dH_C = \frac{dE}{T_K} = 4\alpha V T_K^2 \, dT_K. \tag{11.68}$$

Integrating both sides and imposing that the entropy vanishes at zero temperature, we get

$$H_C = \frac{4}{3}\alpha V T_K^3. \tag{11.69}$$

Since at equilibrium the thermodynamic entropy corresponds to the logarithm of the number of typical states of the system, it follows that the number of degrees of freedom of the system is an exponential function of the volume and of the cube of the temperature. This functional law is somehow expected: by (11.12), the total number of modes of a given wavelength that can be thermally excited inside the body is proportional to the volume and inversely proportional to the cube of the wavelength. Smaller wavelengths yield a larger number of modes, but these require larger energy values to be excited. Since the average energy is proportional to the temperature, larger values of the temperature increase the number of degrees of freedom by decreasing the wavelength of the typical modes of radiation.

By combining (11.66) and (11.69), we obtain

$$H_C = \frac{4}{3} E^{3/4} V^{1/4} \alpha^{1/4}. \tag{11.70}$$

It follows that the entropy of a black body is proportional to the total average energy of the body to the power 3/4. By increasing the energy, the typical modes of radiation occur at higher frequency and carry a larger amount of information.

A natural question then arises of whether information can grow arbitrarily large by increasing the energy inside the body. It turns out that the high-frequency thermal states require high energy to be excited and can eventually lead to gravitational instability of the source of the radiation. Even if we have assigned an exponentially low probability of occurrence to the high-frequency modes, the typical states responsible for the entropy in (11.69) are still too massive for large volumes and large temperatures, and can lead to gravitational collapse of the body into a black hole. On the other hand, if we keep the average energy inside the body below a value that leads to gravitational collapse, we obtain an entropy bound that depends only on the size of the system expressed in a certain unit of length, called the Planck length. In the following, we first introduce this length based on quantum uncertainty relations, and then turn to the computation of the entropy bound.

11.7.3 The Planck Length

The main idea for defining the Planck length is to introduce a unit of length so small that quantum fluctuations of energy within its radius may be so large that they could create a small black hole. By the time–energy uncertainty principle for a single photon (11.63), we have

$$\sigma_t \sigma_{E_\mathrm{p}} \geq \frac{\hbar}{2}, \tag{11.71}$$

from which we have the localization–energy uncertainty

$$\sigma_\ell \sigma_{E_\mathrm{p}} \geq \frac{\hbar c}{2}, \tag{11.72}$$

expressing the general principle that photons of precise energy, or equivalently precise frequency, cannot be localized accurately due to the uncertainty on their time of arrival at any given point in space. A larger energy uncertainty allows for a smaller distance uncertainty, and vice versa. A bound on the reduction in distance uncertainty is imposed by general relativity. For distances smaller than the Schwarzschild radius

$$r_\mathrm{S} = \frac{2Gm}{c^2} = \frac{2Gmc^2}{c^4} = \frac{2G\sigma_{E_\mathrm{p}}}{c^4}, \tag{11.73}$$

where m is the mass of the radiating body and G is Newton's gravitational constant, the laws of general relativity predict that such a precise localization requires energy fluctuations so large that they can lead to gravitational collapse and the formation of a small black hole of radius r_S. It follows that the distance uncertainty can be reduced only up to

$$\sigma_\ell \geq \frac{2G\sigma_{E_\mathrm{p}}}{c^4} \geq \frac{G\hbar c}{2c^4 2\sigma_\ell}, \tag{11.74}$$

from which it follows that

$$\sigma_\ell^2 \geq \frac{G\hbar}{c^3}. \tag{11.75}$$

We define the square root of the right-hand side of (11.75) as the Planck length:

$$\ell_\mathrm{P} = \sqrt{\frac{G\hbar}{c^3}}. \tag{11.76}$$

The Planck length represents the smallest standard deviation of the distance that can be appreciated before the energy required to achieve such resolution becomes so large as to create a black hole at the measurement point.

11.7.4 Gravitational Limits

We now determine a bound on the entropy of a black body in terms of its size expressed in Planck length units. We consider a spherical black body of radius r, and we impose the condition that the total average energy is below the Schwarzschild energy value leading to gravitational collapse. It follows that we need

$$r \geq r_S = \frac{2Gm}{c^2}. \tag{11.77}$$

Multiplying both sides by c^2 and using $E = mc^2$, we have that the energy of the black body must satisfy

$$E \leq \frac{c^4 r}{2G}. \tag{11.78}$$

Using (11.78) as a bound for the average energy in (11.70), it follows that an upper bound on the entropy is

$$H_C \leq \frac{4}{3} \frac{c^3 r^{3/4}}{(2G)^{3/4}} \left(\frac{4}{3} \pi r^3 \right)^{1/4} \alpha^{1/4}. \tag{11.79}$$

Finally, substituting (11.67) into (11.79), we obtain

$$H_C \leq k_B \alpha' \left(\frac{r}{\ell_p} \right)^{3/2}, \tag{11.80}$$

where α' is a dimensionless constant close to one. The gravitational bound (11.80) states that the entropy of a black body can be at most proportional to the square root of its volume, expressed in Planck length units. This contrasts with the original estimate (11.69), for which the entropy is proportional to the volume of the body and to the cube of the temperature. For bodies of moderate size and temperature, however, (11.69) can be much smaller than (11.80). The entropy can be proportional to the volume and to the cube of the temperature until the body is so large and so energetic that the additional constraint of not collapsing into a black hole expressed by (11.80) becomes dominant – see Problem 11.11.

11.8 Entropy of Arbitrary Systems

So far, we have provided maximum entropy bounds for black bodies filled with thermal electromagnetic radiation. These correspond to systems radiating at all frequencies and subject to average energy and quantum radiation constraints, as described in

Section 11.2.3. Similar bounds can be obtained for general matter systems and for electromagnetic systems radiating over a finite frequency bandwidth. The holographic information bound introduced in Chapter 1 is one such example. Another example is the universal entropy bound that we introduce below and discuss further in Chapter 13.

11.8.1 The Holographic Bound

The argument for the holographic bound given in Chapter 1 is that the entropy of any arbitrary system cannot be larger than the entropy of a black hole of the same size, and this is proportional to the area of the event horizon, measured in Planck length units. Expressing the results in Section 1.5.4 in terms of Boltzmann's entropy, we have

$$H_{\mathrm{B}} \leq \frac{k_{\mathrm{B}} \pi r^2}{\ell_{\mathrm{P}}^2} \log e. \tag{11.81}$$

Compared to the gravitational entropy bound for a black body (11.80), the bound (11.81) is somewhat more generous, allowing entropy to scale with the area, rather than the area to the power 3/4, expressed in Planck length units. The bound (11.81) holds for arbitrary systems, and a black body, like many other systems, falls short of saturating it. Other trivial examples that fall short of reaching the holographic bound are a sphere surrounding empty space or a crystal at zero temperature, which both have zero entropy.

For systems far from gravitational collapse, gravitational bounds such as (11.81) and (11.80) are generally loose. Most systems of "reasonable" energy have a much smaller entropy that can be bounded directly in terms of their energy.

11.8.2 The Universal Entropy Bound

A general bound for matter systems of reasonable size and energy is the universal entropy bound, introduced by Jacob Bekenstein in 1981. This states that the entropy of any matter system enclosed in a sphere of radius r with energy at most E is

$$H_{\mathrm{B}} \leq \frac{k_{\mathrm{B}} 2\pi r E}{\hbar c} \log e. \tag{11.82}$$

Imposing gravitational stability, we have

$$E \leq \frac{c^4 r}{2G}, \tag{11.83}$$

which substituted into (11.82) recovers the holographic bound (11.81). The Bekenstein bound, however, is intended for weak self-gravitating systems. These are systems far from gravitational collapse for which $E \ll c^4 r/(2G)$, and for those it provides a tighter entropy bound than the holographic one.

As we shall see in Chapter 13, while the black body entropy formula (11.70) applies to a maximum-entropy electromagnetic system radiating at all frequencies according to Planck's law (11.29b), the universal entropy bound (11.82) can be achieved by a maximum-entropy electromagnetic system radiating photons over a fixed bandwidth. In both cases the entropy is proportional to the volume of the body, but the energy scalings

Table 11.1 Entropy limits.

Black body classical	Black body gravitational	General holographic	General universal
$H_C = \left(\frac{4}{3}\right)^{5/4}(\alpha\pi)^{1/4}(rE)^{3/4}$	$H_C \le \alpha'\left(\frac{r}{\ell_p}\right)^{3/2}$	$H \le \pi\left(\frac{r}{\ell_p}\right)^2 \log e$	$H \le \frac{2\pi rE}{\hbar c}\log e$

are different. While the energy of a black body grows with the cube of the radius of the radiating body, in the case of bandlimited radiation achieving the universal entropy bound, the energy grows only with the square of the radius.

The different entropy bounds are summarized in Table 11.1. Compared to what is achieved with current technologies, these bounds are all extremely lenient. For example, a current flash memory card of size of the order of a centimeter, and weight of a few grams, can store about one terabyte of data, that is, of the order of 10^{12} bits. In comparison, expressing the universal bound (11.82) in bits, rather than energy over temperature, and using $E = mc^2$, we obtain the entropy bound

$$H \le \frac{2\pi rE}{\hbar c}\log e = \frac{2\pi rmc}{\hbar}\log e \approx 10^{38} \text{ bits.} \tag{11.84}$$

What allows this bound to be so large is that it requires conversion of all matter into radiation carrying information, and by doing so it considers all available degrees of freedom at the most fundamental level.

If we were to increase the mass of our memory card and bring it to the verge of becoming a black hole, the entropy would be limited by the holographic bound,

$$H \le \frac{\pi r^2}{\ell_p^2}\log e \approx 10^{65} \text{ bits,} \tag{11.85}$$

which is 53 orders of magnitude more than the current technology and 27 orders of magnitude more than the universal bound for objects of reasonable energy. Clearly, these bounds, although important for matters of principle, are of no great practical use for current technologies, and are not likely to be for many years to come. To view the whole information content of our hypothetical memory card on the verge of gravitational collapse, we would need to convert all matter into radiation and observe a waveform on a cut-set boundary surrounding our object carrying 10^{65} bits of information. This would require radiation at a wavelength of

$$\lambda \approx 1/\sqrt{10^{65}} \approx 10^{-33} \text{ m.} \tag{11.86}$$

11.9 Entropy of Black Holes

The Bekenstein–Hawking formula for the entropy of a black hole is at the basis of the derivation of the holographic entropy bound. This formula states that the entropy of a black hole is proportional to the surface area of its event horizon, measured in Planck length units. The holographic bound then states that the entropy of any system cannot be larger than the entropy of a black hole of the same size.

The Bekenstein–Hawking result can be obtained from a clever physical argument based on quantum uncertainty relations. This was Bekenstein's original argument and we describe it below. From the time–frequency uncertainty principle for a single photon (11.59), we have

$$\sigma_t \sigma_\omega \geq \frac{1}{2}, \tag{11.87}$$

from which we have the localization–frequency uncertainty

$$\sigma_\ell \sigma_\omega \geq \frac{c}{2} \tag{11.88}$$

and the localization–wavelength uncertainty

$$2\pi \frac{\sigma_\ell}{\sigma_\lambda} \geq \frac{1}{2}. \tag{11.89}$$

It follows that a lower bound on the uncertainty of the localization is

$$\sigma_\ell \geq \frac{\sigma_\lambda}{4\pi}. \tag{11.90}$$

We now choose a wavelength of radiation $\lambda \approx 4\pi r_S$ for a photon directed at a black hole of radius r_S, so that when the photon is localized within the radius of the black hole, it disappears inside the hole. In this way, when the photon disappears, knowledge of its existence within a radius r_S of the center of the hole is lost. We assume that after the photon falls in, we have no information about whether the photon exists or not inside the hole, and we have lost one bit of information.

The energy increment of the black hole due the photon is

$$\Delta E = \hbar\omega = \frac{\hbar 2\pi c}{4\pi r_S}. \tag{11.91}$$

The new radius of the black hole resulting from adding this amount of energy is

$$\begin{aligned} r'_S &= r_S + \Delta r \\ &= \frac{2Gm}{c^2} + \frac{2G}{c^2}\frac{\Delta E}{c^2} \\ &= \frac{2Gm}{c^2} + \frac{2G\hbar 2\pi}{c^3 4\pi r_S}. \end{aligned} \tag{11.92}$$

The new area of the horizon due to the increase in radius is

$$\begin{aligned} A' &= A + \Delta A \\ &= 4\pi (r_S + \Delta r)^2 \\ &= 4\pi r_S^2 (1 + \Delta r / r_S)^2 \\ &= 4\pi r_S^2 + 4\pi r_S^2 \left[2\Delta r / r_S + (\Delta r / r_S)^2 \right]. \end{aligned} \tag{11.93}$$

It follows, by combining (11.93), (11.92), and (11.76), that

$$\Delta A = 4\pi r_S^2 \left[\frac{4G\hbar 2\pi}{c^3 4\pi r_S^2} + \left(\frac{2G\hbar 2\pi}{c^3 4\pi r_S^2} \right)^2 \right] = 8\pi \ell_p^2 + 4\pi \ell_p^4 / r_S^2. \tag{11.94}$$

The key observation now is that the first term in (11.94) is independent of the size of the black hole. Ignoring the second term, which is much smaller than the first one, it follows that adding one bit of information increases the area of the horizon of a black hole of any size by an amount proportional to ℓ_p^2. This suggests that, adding information bit by bit, the total amount of entropy in our black hole is

$$H \approx \frac{4\pi r_S^2}{8\pi \ell_p^2} = \frac{A}{8\pi \ell_p^2} \text{ bits}, \tag{11.95}$$

which is proportional to the area of the event horizon, measured in Planck length units. The proportionality constant $1/(8\pi)$ is the one originally obtained by Bekenstein's (1973) calculation, and is only approximate. It is based on the assumption that the localization of the photon occurs precisely within a radius r_S, while in reality this cannot be so sharply determined because it is only expressed in terms of standard deviation. Bekenstein (1973) noticed that the approximation is the price to pay for having given a semi-classical argument and not a full quantum one. A more detailed calculation by Hawking (1975) obtained

$$H = \frac{4\pi r_S^2}{4\ell_p^2} \log e \text{ bits}. \tag{11.96}$$

This result leaves open the question of what the actual physical states present in a black hole are. In statistical mechanics, physical states are identified as configurations of elementary particles, or radiating modes in a black body. Providing a similar interpretation for black holes is a topic of current research in physics, with some recent proposals made in the areas of string theory.

11.10 Maximum Entropy Distributions

We conclude this chapter by deriving the statistics of the noise process as maximum entropy distributions subject to energy constraints. Since the energy of a signal is related to its average power, to its magnitude, and to its waveform shape, we show that the maximum entropy distributions associated with the noise process are all related to each other through elementary transformations of random variables. We consider a frequency regime $\hbar\omega \ll k_B T_K$, so that quantum effects can be ignored and we can assume continuous probability densities. We then show the required modifications to obtain the analogous results valid for all frequencies, taking into account quantum effects. We do not provide a fully rigorous treatment in terms of generalized functions, and instead resort to a less rigorous but more intuitive argument that emphasizes physical aspects.

Let E be the energy of a thermally excited standing wave of angular frequency ω composing the noise signal. The average radiated power at angular frequency ω, which has the dimensions of energy per unit time, is proportional to the energy of the standing wave via a coupling coefficient indicating the elementary bandwidth over which it is radiated, and we have

$$P = \frac{E d\omega}{2\pi}. \tag{11.97}$$

In this way, the energy of the standing wave can be interpreted as the average radiated power per unit frequency, namely the average power of a single sinusoid,

$$x(t) = A\cos(\omega t + \phi),\ A \geq 0, \tag{11.98}$$

which is given by

$$P = \frac{A^2}{2}. \tag{11.99}$$

From (11.99) and (11.97), we have the differential relationships

$$dP = AdA, \tag{11.100}$$

$$dEd\omega = 2\pi\, dP. \tag{11.101}$$

We now use these relationships to derive the distribution of the noise starting from the exponential distribution of the energy of each mode given by (11.14). It follows from (11.101) that the average power is distributed as

$$f_{\mathsf{P}}(p) = \frac{2\pi}{k_{\mathrm{B}} T_{\mathrm{K}} d\omega} \exp(-2\pi p/(k_{\mathrm{B}} T_{\mathrm{K}} d\omega)), \tag{11.102}$$

and from (11.100) that the amplitude is distributed according to the Rayleigh probability density function

$$g_{\mathsf{A}}(a) = \frac{2\pi a}{k_{\mathrm{B}} T_{\mathrm{K}} d\omega} \exp(-2\pi a^2/(2k_{\mathrm{B}} T_{\mathrm{K}} d\omega)),\ a \geq 0. \tag{11.103}$$

We now turn to considering the distribution of the random phasor representing the noise signal,

$$\underline{\mathsf{A}} = \mathsf{A}\exp(j\phi). \tag{11.104}$$

Since the thermal excitations of different modes are not synchronized in time, ϕ can be assumed to be distributed uniformly at random according to maximum entropy with the only constraint being that it lies inside the interval $[0, 2\pi]$. The probability that the tip of the phasor lies in an elementary area dS at distance between a and $a + da$ from the origin is

$$q_{\underline{\mathsf{A}}}(m) dS = \frac{dS}{2\pi a da} g_{\mathsf{A}}(a) da = \frac{g_{\mathsf{A}}(a)}{2\pi a} dS; \tag{11.105}$$

see Figure 11.10. It follows that the probability density function of the phasor over the plane is

$$q_{\underline{\mathsf{A}}}(a) = \frac{g_{\mathsf{A}}(a)}{2\pi a} = \frac{1}{k_{\mathrm{B}} T_{\mathrm{K}} d\omega} \exp(-2\pi a^2/(2k_{\mathrm{B}} T_{\mathrm{K}} d\omega)), \tag{11.106}$$

which is a two-dimensional, circularly symmetric Gaussian distribution. This density

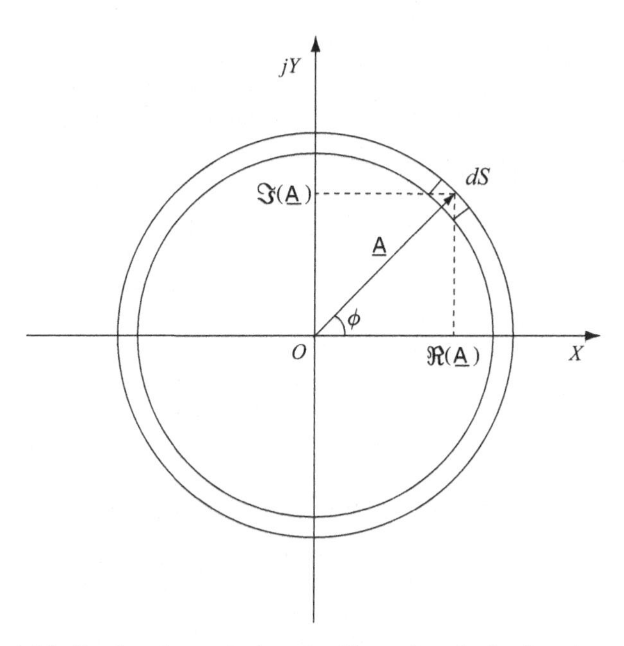

Fig. 11.10 A phasor of Rayleigh-distributed magnitude and uniform phase is distributed as a circularly symmetric complex Gaussian over the complex plane.

depends only on the magnitude A and not on the phase ϕ of the signal. When viewed in the complex plane, it corresponds to a zero-mean vector oriented uniformly in $[0, 2\pi]$ with a Rayleigh-distributed magnitude and variance:

$$\mathbb{E}(\mathsf{A}^2) = 2\,\mathbb{E}(\mathsf{P}) = k_\mathrm{B} T_\mathrm{K} d\omega/\pi. \tag{11.107}$$

Since $\mathsf{A}^2 = \Re(\mathsf{A})^2 + \Im(\mathsf{A})^2$, the two-dimensional distribution factorizes into two mono-dimensional Gaussians,

$$\begin{aligned} q_{\underline{\mathsf{A}}}(a) &= \frac{1}{k_\mathrm{B} T_\mathrm{K} d\omega} \exp(-2\pi(\Re(a)^2 + \Im(a)^2)/(2k_\mathrm{B} T_\mathrm{K} d\omega)) \\ &= \frac{1}{\sqrt{k_\mathrm{B} T_\mathrm{K} d\omega}} \exp(-2\pi(\Re(a)^2/(2k_\mathrm{B} T_\mathrm{K} d\omega)) \\ &\quad \frac{1}{\sqrt{k_\mathrm{B} T_\mathrm{K} d\omega}} \exp(-2\pi(\Im(a)^2/(2k_\mathrm{B} T_\mathrm{K} d\omega)), \end{aligned} \tag{11.108}$$

that are in the form $(1/\sqrt{2\pi\sigma^2})\exp(-x^2/(2\sigma^2))$, with $\sigma = \sqrt{k_\mathrm{B} T_\mathrm{K} d\omega/2\pi}$, $x \in \{\Re(a), \Im(a)\}$, and correspond to the distributions of the real and imaginary parts of the phasor.

We now consider the distribution of the instantaneous real signal produced by the random thermal excitation of a single mode,

$$\mathsf{X}(t) = \Re(\underline{\mathsf{A}} \exp(j\omega t)) = \mathsf{A} \cos(\omega t + \phi). \tag{11.109}$$

Multiplication of the phasor by $\exp(j\omega t)$ corresponds to a rotation in the complex plane and does not change its distribution, which depends only on the magnitude. We then obtain the Gaussian probability density function for the real signal $\mathsf{X}(t)$ representing the instantaneous amplitude of the noise,

$$q_{\mathsf{X}}(x) = \frac{1}{\sqrt{k_B T_K d\omega}} \exp(-2\pi x^2/(2k_B T_K d\omega)). \tag{11.110}$$

Finally, if the noise is composed of the superposition of $2\Omega/d\omega$ thermally excited modes, each distributed as $\mathsf{X}(t)$,

$$\mathsf{Z}(t) = \sum_{n=1}^{2\Omega/d\omega} \mathsf{X}_n(t). \tag{11.111}$$

Assuming independence of modes separated in frequency by an elementary interval $d\omega$, the resulting distribution is still Gaussian, and we have

$$q_{\mathsf{Z}}(z) = \frac{1}{\sqrt{k_B T_K 2\Omega}} \exp(-2\pi x^2/(2k_B T_K 2\Omega)), \tag{11.112}$$

where the variance has been scaled by the number of modes $2\Omega/d\omega$. The expected power of the noise, accounting for the contribution from all modes, is then given by

$$\mathbb{E}(\mathsf{Z}^2) = k_B T_K 2\Omega/2\pi. \tag{11.113}$$

Throughout, we have assumed $\hbar\omega/(k_B T_K) \ll 1$, so that we could deal with continuous distributions. This results in the two drawbacks already mentioned: an average power that increases with the bandwidth, and an impulsive autocorrelation function that allows independence of modes separated by an arbitrarily small frequency interval. Using the discrete distribution described in Section 11.2.3 and going through the same steps substituting the quantum mechanical correction (11.28) for the frequency-dependent expected energy of a mode in place of $k_B T_K$ leads to a colored Gaussian distribution for the noise, with expected power

$$\mathbb{E}(\mathsf{Z}^2) = \frac{1}{2\pi} \int_{-\Omega}^{\Omega} \frac{\hbar\omega}{\exp(\hbar\omega/(k_B T_K)) - 1} d\omega, \tag{11.114}$$

rather than the one in (11.113).

11.11 Summary and Further Reading

Noise provides a fundamental limitation on the resolution at which we can detect electromagnetic signals. Thermal radiation occurs due to random agitations of charges inside media. The precise spectrum of the noise signal can be determined by considering the superposition of elementary radiating electromagnetic modes with appropriate boundary conditions, and accounting for the quantized levels of the electromagnetic radiation. The stochastic distribution of the noise follows from basic thermodynamics considerations. The second law provides an interpretation of the noise signal in terms of maximum uncertainty subject to imposed physical constraints. Shot noise is related to quantum limitations arising from Heisenberg's uncertainty principle, and provides an ultimate irreducible uncertainty limit.

A classic reference for the treatment of black body radiation is the book by Kittel and Kroemer (1980). The first volume of the Feynman lectures on physics (Feynman, Leighton, and Sands, 1964), of course, also covers thermal radiation. A theoretical treatment of thermal and quantum noises from an engineering perspective is given in the classic paper of Oliver (1965). The statistical mechanics interpretation of thermal noise is due to Nyquist (1928). Connections between statistical entropy and thermodynamic entropy have a long and rich history, dating back to Boltzmann (1872); see Uffink (2008) for a historical account. Modern theories allow for a temporary decrease of entropy in systems in dynamical evolution, while the increase in entropy always holds macroscopically, as the system evolves from one state of thermal equilibrium to another. In this context, fluctuation theorems give quantitative estimates for the probability of a decrease in statistical entropy in systems in dynamical evolution. In our treatment we considered equilibrium states only, where thermodynamic entropy and statistical entropies coincide.

A different point of view was taken by Cover (1994), who studied, independent of the physics, the class of finite state Markov stochastic processes for which the second law holds. He showed that statistical entropy increases over time only for Markov processes for which the equilibrium distribution is uniform over the finite state space. He argued that an appropriate generalized statement of the second law in the context of all stochastic processes should be that the relative entropy of the current distribution with respect to the stationary distribution always decreases.

The connection between statistical entropy and thermodynamic entropy at equilibrium via the asymptotic equipartition property was argued by Jaynes (1965). The asymptotic equipartition property was first stated by Shannon (1948), who proved the result for i.i.d. processes and stated it for stationary ergodic processes. Several extensions are provided in the literature, including for ergodic countable processes and ergodic, real-valued ones. Jaynes (1982) argued for the use of the maximum entropy principle in a broader context. More recently, connections between thermodynamic and statistical notions of entropy led to many advances in engineering and computer science, described, for example, in the book of Mézard and Montanari (2009). Further connections between statistical physics and information theory appear in Merhav (2010).

For a detailed introduction to the holographic information bound see Susskind (1995). A review of information limits due to gravitation with a focus on the holographic information bound and its extensions appears in Bousso (2002). The universal entropy bound is due to Bekenstein (1981). The calculation of the entropy of black holes is due to Bekenstein (1973) and Hawking (1975), who settled the proportionality constant at $1/4$. The microscopic origin of black hole entropy has been investigated from a string-theoretic perspective by Strominger and Vafa (1996).

11.12 Test Your Understanding

Problems

11.1 Show that the deviation from the mean $\mathsf{Y} = \mathsf{N} - \alpha$ of a Poisson random variable N of mean α approaches, for large values of α, a zero-mean Gaussian of variance α.

Solution

We have the Poisson probability mass function

$$p_{\mathsf{N}}(n) = \frac{\alpha^n \exp(-\alpha)}{n!}. \tag{11.115}$$

Substituting $n = \alpha + y$ into (11.115), we get the probability density function

$$p_{\mathsf{Y}}(y) = \frac{\alpha^{(\alpha+y)} \exp(-\alpha)}{(\alpha+y)!}. \tag{11.116}$$

By using Stirling's formula, for large values of α this is approximately

$$\begin{aligned}\frac{\alpha^{(\alpha+y)} \exp(-\alpha)}{(\alpha+y)^{(\alpha+y)} \exp[-(\alpha+y)]\sqrt{2\pi(\alpha+y)}} &= \frac{\exp(y)}{\sqrt{2\pi(\alpha+y)}} \frac{\alpha^{(\alpha+y)}}{(\alpha+y)^{(\alpha+y)}} \\ &= \frac{\exp(y)}{\sqrt{2\pi(\alpha+y)}} \left(1 + \frac{1}{\alpha/y}\right)^{-(\alpha+y)} \\ &= \frac{\exp(y)}{\sqrt{2\pi(\alpha+y)}} \left[\left(1 + \frac{1}{\alpha/y}\right)^{-\alpha/y}\right]^{\frac{y^2+\alpha y}{\alpha}}.\end{aligned} \tag{11.117}$$

For large values of α, the right-most term tends to $\exp(-y)\exp(-y^2/\alpha)$, and $(\alpha+y) \approx \alpha$. It follows that the probability density function of the deviation from the mean becomes approximately

$$p_{\mathsf{Y}}(y) = \frac{\exp(-y^2/\alpha)}{\sqrt{2\pi\alpha}}. \tag{11.118}$$

A more rigorous derivation can be performed considering the characteristic function of the random variable $\mathsf{Y}_\alpha = (\mathsf{N} - \alpha)/\sqrt{\alpha}$,

$$\phi_{\mathsf{Y}_\alpha}(t) = \exp[-jt\sqrt{\alpha} + \alpha(\exp(jt/\sqrt{\alpha}) - 1)], \tag{11.119}$$

and showing that

$$\lim_{\alpha\to\infty} \phi_{\mathsf{Y}_\alpha}(t) = \exp(-t^2/2), \tag{11.120}$$

which is the characteristic function of the standard zero-mean, unit-variance, Gaussian.

11.2 Show that among all continuous random variables defined over a real interval and having finite differential entropy, the differential entropy is maximized only for those having uniform distribution over that interval.

11.3 Show that among all continuous random variables defined over the positive reals and having finite differential entropy and mean μ, the differential entropy is maximized only for exponentials.

11.4 Prove the information inequality (11.49) using Jensen's inequality.

11.5 Prove the chain rule (11.50).

11.6 Show that the ratio of typical states to the total number of states in a system approaches zero exponentially in the number of states. Thus, for large systems, the vast majority of states are atypical. Their probabilistic contribution, however, is negligible.

11.7 In the asymptotic equipartition property, the typical set depends on the choice of ϵ. By (11.46), as $\epsilon \to 0$, the elements in the typical set become equiprobable, but for the probability of deviation from the typical set in (11.45) to be negligible, we require a larger number M of random variables. Show at what rate ϵ can be decreased as $M \to \infty$ to tighten the bounds in (11.46).

Solution

Apply the Chebyshev inequality to bound the deviation in (11.45) for all M, and express ϵ as a function of M.

11.8 Consider a system composed of a continuous set of elementary modes that are stochastic signals of N_0 degrees of freedom. Use the continuum version of the asymptotic equipartition property described in Section 1.4.2 to choose a distribution for the signals' coefficients that is consistent with the second law. Compare your results with the probabilistic interpretation of the second law given in Section 11.3.4.

Solution

The system of signals is described by N_0 i.i.d. stochastic coefficients. The typical volume of the state space is then roughly

$$V = 2^{N_0 h}, \tag{11.121}$$

where $h = h_{\mathsf{X}}(f)$ is the differential entropy of each coefficient, having density function f. Assuming signals are observed quantized at level ϵ, we obtain a number of observable typical states

$$N = \frac{2^{N_0 h}}{\epsilon^{N_0}}, \tag{11.122}$$

that are all roughly equiprobable. It follows that the cardinality of the observable typical set is an exponential function of the differential entropy h and of the number N_0 of elementary modes of the system. The entropy of the observable typical set is

$$H = N_0 \log \frac{2^h}{\epsilon}. \tag{11.123}$$

We now choose a Gaussian probability distribution for the signals' coefficients that maximizes the differential entropy subject to the constraint $\mathbb{E}(\mathsf{X}^2) \leq P$, so that

$$h = \log(\sqrt{2\pi e P}), \tag{11.124}$$

and substituting into (11.123) this choice maximizes the entropy of the observable typical set,

$$H = N_0 \log \frac{\sqrt{2\pi eP}}{\epsilon}. \tag{11.125}$$

For any given signal to quantum resolution constraint, the chosen distribution maximizes the entropy of an equivalent thermodynamic system in equilibrium formed by the typical quantized states, and in this case the entropy is proportional to the number of degrees of freedom. This probabilistic interpretation is consistent with the one provided in Section 11.3.4 for a discrete ensemble of elementary modes. Comparing (11.123) and (11.125) with (11.42), it follows that N_0 plays the role of the number M of elementary modes of the system, and $h - \log \epsilon$ plays the role of the entropy $H(p)$ of each quantized mode of radiation. In Chapter 13 we relate the quantum resolution ϵ to the frequency of radiation, and obtain an expression for the maximum entropy of radiation for a system of given size over a fixed frequency band.

11.9 Complete the computation of the integral in (11.65).

11.10 Compute the value of the constant α' in (11.80), using $k_B = 1.381 \times 10^{-23}$ J K^{-1}, $\ell_p = 1.62 \times 10^{-35}$ m, $\hbar = 1.055 \times 10^{-34}$ J s, and verify that it is dimensionless.

11.11 Verify that for a black body of the size of the sun, $r \approx 6.96 \times 10^8$ m, and temperature $T \approx 5800$ K, the entropy in (11.69) proportional to the volume is much smaller than the gravitational bound (11.80). Provide an example of a body of given temperature and radius that reaches the gravitational bound (11.80).

11.12 Wien's displacement law states that the black body radiation curve for different temperatures peaks at a wavelength inversely proportional to the temperature. Provide the precise formula.

11.13 Compute the expected power of a black body per unit wavelength $d\lambda$, per unit surface area dA, and radiated in all directions out from dA. This corresponds to the irradiance of a black body and requires extending (11.17) to all outgoing directions from the surface of the body.

Solution

Consider radiation of the black body per unit wavelength, per unit surface area, and in a direction forming an angle θ with the normal to the surface, as depicted in Figure 11.11. Letting dA be the area subtended by the angle $d\Phi$ on the surface of the body, the projected area orthogonal to the direction of observation is $dA \cos\theta$. By (11.17), the radiation per unit wavelength in the direction of P, and per unit projected area, is

$$S_Z(\lambda) = \frac{c}{4\pi} \frac{\partial^2 u(\lambda)}{\partial\lambda \partial V}. \tag{11.126}$$

It follows that the radiation per unit wavelength in the direction of P, and per unit area dA on the surface of the body, is

$$S_Z(\lambda, \theta) = S_Z(\lambda) \cos\theta. \tag{11.127}$$

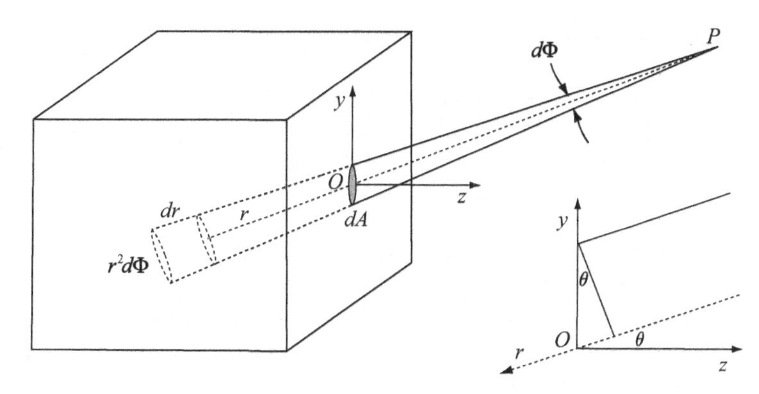

Fig. 11.11 Geometry of the radiated energy at an angle θ with the normal to the surface of the body as seen by a detector at P.

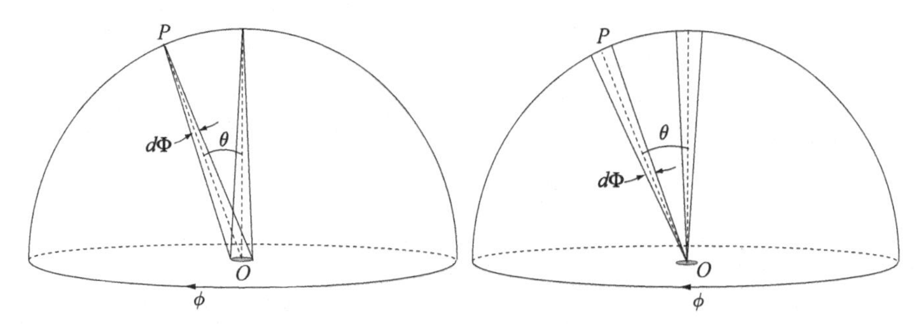

Fig. 11.12 Conservation of radiance. The detector observes the source through an elementary solid angle $d\Phi$ and at an angle θ with the normal to the surface of the body. The source radiates through an elementary solid angle $d\Phi$ towards a detector at an angle θ with the normal to the surface of the body.

We now compute the total radiation per unit wavelength in all directions through dA by letting P vary over the hemisphere U surrounding the area dA – see Figure 11.12. By conservation of radiance, the radiation emitted by the source at an angle θ into an elementary solid angle $d\Phi$ is the same as that received by a detector observing the source at the same angle θ, through an elementary solid angle, and since

$$d\Phi = \sin\theta d\theta d\phi, \tag{11.128}$$

we obtain the irradiance

$$\int_U S_Z(\lambda,\theta)d\Phi = S_Z(\lambda)\int_0^{2\pi}\int_o^{\pi/2}\cos\theta\sin\theta d\phi d\theta = \pi S_Z(\lambda). \tag{11.129}$$

This leads to Planck's formula per unit area on the surface of the body and the associated Rayleigh–Jeans limit,

$$\pi S_Z(\lambda) = \frac{(2\pi c)^2}{\lambda^5}\frac{\hbar}{\exp\left(\dfrac{2\pi\hbar c}{\lambda k_B T_K}\right)-1} \approx \frac{2\pi c k_B T_K}{\lambda^4}, \tag{11.130}$$

which differ from (11.30b) and (11.20) by the factor π.

11.14 The Stefan–Boltzmann law states that the power radiated per unit surface area by a black body in all directions and over all frequencies is proportional to the fourth power of the temperature. Provide the precise formula.

Solution

Assuming isotropic radiation of energy in all directions, a $c/4\pi$ fraction of the energy per unit volume passes per unit time through the elementary area dA subtended by a solid angle $d\Phi$ in the direction of the observer on the surface of the body. Adding a factor of π to account for radiation in all directions, as explained in Problem 11.13, and using (11.66), we obtain

$$\frac{\partial P}{\partial A} = \frac{E}{V}\frac{c}{4\pi}\pi = \frac{c}{4}\alpha T^4 = \frac{\pi^2}{15\hbar^3 4c^2}(k_B T_K)^4. \tag{11.131}$$

12 Information-Theoretic Quantities

Nec scire fas est omnia.[1]

12.1 Communication Using Signals

The act of communication occurs by *encoding* a message we wish to send at the transmitter, and *decoding* it at the receiver. The encoding process translates the message into a form suitable for transmission over a given channel. For us, the channel allows transmission and reception of real electromagnetic signals, so that the translation process amounts to representing the message by a point in the signals' space described in Section 2.2.4. By transmitting the signal, we identify one point in the space corresponding to a given message, and communicate a certain amount of information that is at most equal to the entropy of the space.

A continuum signals' space, however, contains an uncountable number of points and a single signal can represent infinitely many possible messages. It follows that the amount of information transferred in the communication process is unbounded.

To limit the amount of information, we impose a regularity constraint on the space of transmitted signals. This constraint is expressed in terms of bandlimitation of the spectral support of the signal, which is reflected in a limitation on the number of coordinates required to identify it. As discussed in Chapter 2, any signal in $\mathscr{B}_\Omega$ can be identified by essentially $N_0 = \Omega T/\pi$ real numbers as $N_0 \to \infty$. It follows that the encoding process translates the message into a signal's *codeword* composed of roughly N_0 real numbers. The decoding process translates the received signal back into the corresponding message by observing the N_0 real numbers of the codeword. By (3.16) it follows that the communication rate achieved by transmitting a bandlimited signal of finite energy, and observing it with arbitrary accuracy in the interval $[-T/2, T/2]$, is given by

$$\lim_{T\to\infty} \frac{N_0 + O(\log N_0)}{T} = \frac{\Omega}{\pi} \text{ reals per second.} \tag{12.1}$$

Measuring information using real numbers may seem, however, somewhat artificial, because a real number carries infinite precision and can be specified by using an infinite number of bits. If we insist on making binary digits suitable units of information, we

[1] Horace, *Carmina* IV (circa 30 BC).

also need to introduce a limit on the resolution at which signals can be observed. A remarkable intuition of Shannon was that the noise poses a resolution limit at which different signals can be reliably distinguished, and ensures that only a finite number of bits per second can be transmitted reliably.

In the presence of the noise, only a "corrupt" version of the signal can be received, so that the decoding process can only "guess" what the transmitted signal was. A first-order model of the noise process is additive, white Gaussian. Without delving into the mathematical sophistications required to develop a rigorous spectral theory of random functions, in Chapter 11 we have argued that it is possible to think about the white Gaussian process as producing essentially independent Gaussian random variables at distinct time intervals. This physically corresponds to having a coupling between the values of the noise that is essentially zero at any two instants of time separated by more than a tiny interval. As shown in Figure 2.3, each point of the signals' space is perturbed by a certain random distance that is roughly proportional to the standard deviation of the noise, and this produces a certain region of uncertainty, providing a resolution limit at which one can distinguish different signals. In order to avoid decoding errors, signals must be separated enough that the corresponding uncertainty regions do not overlap. However, because of their limited energy, signals cannot be separated arbitrarily far in space. It then follows that for any channel with a given noise variance, and an available energy budget, one can only communicate at most at a finite bit rate: the capacity of the channel.

12.2 Shannon Capacity

The Shannon capacity is defined as the largest rate of communication that can be achieved between a transmitter and a receiver with vanishing probability of error. When presenting his theory in the context of continuum waveforms, Shannon considered transmitting a bandlimited signal $f(t) \in \mathscr{B}_\Omega$, subject to an energy constraint that scales linearly with the number of dimensions,

$$\int_{-T/2}^{T/2} f^2(t)dt \leq PN_0. \tag{12.2}$$

He also assumed the transmitted signal to be essentially timelimited, in the sense that only a negligible fraction of its energy falls outside the interval $[-T/2, T/2]$. Finally, he assumed the received signal $g(t)$ to be corrupted by additive noise $z(t)$, so that

$$g(t) = f(t) + z(t). \tag{12.3}$$

Shannon then asked what the largest amount of information is that can be transferred by sending $f(t)$ and receiving $g(t)$, over an interval of T seconds.

To answer this question, he considered a codebook composed of a subset of waveforms in the space, each corresponding to a given message. He let a transmitter select any one of these signals, and a receiver observe the same signal corrupted by the noise. He then noticed that for many reasonable noise models one could choose signals in the codebook sufficiently separated from each other that the receiver could recover the message with an arbitrarily small probability of error.

Shannon defined the capacity, measured in bits, as the logarithm base two of the largest number of messages $M^\delta(P)$ that can be communicated with probability of error at most $\delta > 0$:

$$C(\delta) = \log M^\delta(P) \text{ bits.} \tag{12.4}$$

Clearly, this depends on the choice of the noise model and on the value of δ. For a Gaussian noise model, he showed that the capacity measured per unit time over a large time interval,

$$C = \lim_{T\to\infty} \frac{\log M^\delta(P)}{T} \text{ bits per second,} \tag{12.5}$$

is proportional to the bandwidth and to the logarithm of the energy of the signal divided by the energy of the noise; most surprisingly, it is independent of δ.

Shannon's result is striking. *A priori*, it is not even clear whether it is possible to communicate at a non-zero rate with arbitrarily small error probability. It may very well be that to decrease the error, the amount of spacing we need to add between the codewords in the signals' space to combat the noise drives the rate to zero. Indeed, in the early days of communication theory it was believed that the only way to decrease the probability of error was to proportionally reduce the rate. Shannon showed this belief to be incorrect: by accurately choosing the encoding and decoding processes, one can communicate at a strictly positive rate, and at the same time with as small a probability of error as desired. Furthermore, there is a highest achievable critical rate C, called the capacity of the channel, for which this can be done. If one attempts to communicate at rates above the channel capacity, then it is impossible to do so with arbitrarily low error probability.

To give a more precise mathematical description of this result, we need to provide more details on the communication model. We consider the noise $z(t)$ to be the realization of a white Gaussian noise process $\mathsf{Z}(t)$ having power spectral density

$$S_{\mathsf{Z}}(\omega) = \begin{cases} N & |\omega| \leq \Omega \\ 0 & |\omega| > \Omega. \end{cases} \tag{12.6}$$

This stochastic process has finite average power

$$\mathbb{E}(\mathsf{Z}^2(t)) = s_{\mathsf{Z}}(0) = \frac{1}{2\pi}\int_{-\infty}^{\infty} S_{\mathsf{Z}}(\omega)d\omega = \frac{N\Omega}{\pi}, \tag{12.7}$$

where s_{Z} is the autocorrelation of the noise. The average energy of the noise within the observation interval $[-T/2, T/2]$ is given by

$$\int_{-T/2}^{T/2} \mathbb{E}(\mathsf{Z}^2(t))dt = \frac{N\Omega T}{\pi} = NN_0. \tag{12.8}$$

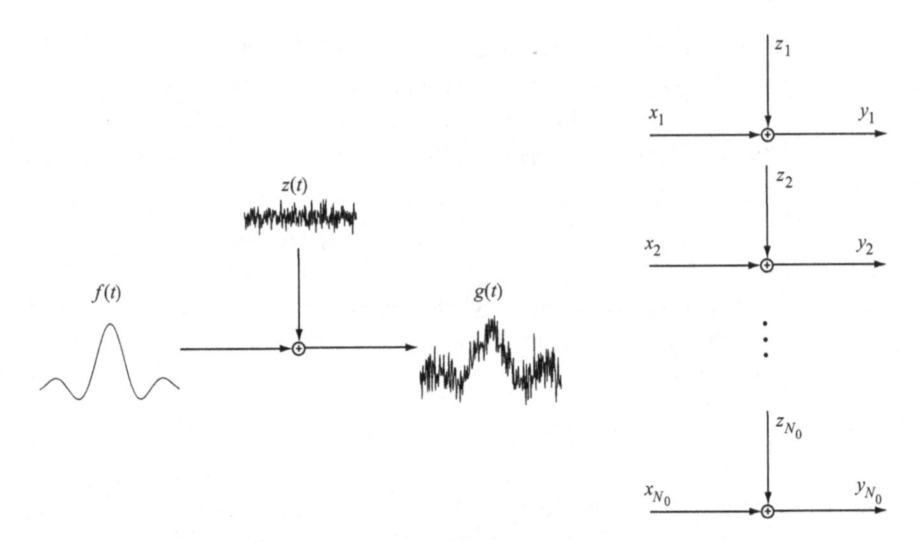

Fig. 12.1 The additive Gaussian channel and its discrete representation.

By combining (12.2) and (12.8), we define the signal-to-noise ratio in Shannon's model to be

$$\mathrm{SNR_S} = P/N. \tag{12.9}$$

Since signals are essentially timelimited, the input can also be expressed as a linear combination of N_0 orthonormal bandlimited basis functions,

$$f(t) = \sum_{n=1}^{N_0} x_n \psi_n(t). \tag{12.10}$$

In this case, the energy constraint (12.2) essentially corresponds to

$$\sum_{n=1}^{N_0} x_n^2 \leq PN_0, \tag{12.11}$$

and the projection of the noise onto the space spanned by the basis functions forms a discrete i.i.d. Gaussian process that adds an independent Gaussian random variable of zero mean and variance N to each signal's coordinate x_n, so that the continuum time communication model (12.3) becomes the discrete superposition of N_0 parallel communication channels,

$$y_n = x_n + z_n, \tag{12.12}$$

where $n \in \{1, 2, \ldots, N_0\}$; see Figure 12.1.

Using this model, Shannon showed that:

It is possible to communicate information at the rate

$$C < \frac{\Omega}{2\pi} \log\left(1 + \mathsf{SNR_S}\right) \text{ bits per second} \tag{12.13}$$

with the probability of error tending to zero as $T \to \infty$, while it is not possible to communicate at a higher rate than C with vanishing probability of error.

The result is obtained by exploiting two main arguments called *sphere packing* and *random coding*. First, one observes that the noise turns the coded signal point into a cloud sphere of an essentially fixed radius. Then, by counting how many disjoint noise spheres fit into the whole signals' space, one argues that no reliable transmission at a rate larger than the capacity is possible. The second argument relies on the probabilistic method, which gained prominence in combinatorial mathematics around the same time that Shannon's information theory came about. This is a non-constructive method for proving the existence of a prescribed kind of mathematical object. It works by showing that if one randomly chooses objects from a specified class, the probability of picking the "right" object of the prescribed kind is larger than zero. Although the proof uses probability, the final conclusion of existence is certain.

In our case, the mathematical object picked at random is a particular encoding system. A given selection of points in the signals' space depicted in Figure 2.3 corresponds to a particular encoding system. In order to show the existence of an encoding system capable of transmitting C bits per second with vanishing probability of error, it is enough to show that by performing the selection of $M = 2^{CT}$ points at random inside the sphere, the probability of error in the transmission of the messages associated with such points vanishes as $T \to \infty$. It then follows that using a random encoding system it is possible to communicate with a vanishing probability of error at rate $(\log M)/T = C$ bits per second, so that there must exist a particular coding scheme that allows transmission at such a rate and with probability of error tending to zero. In Shannon's (1949) own words:

> It turns out, rather surprisingly, that it is possible to choose our M signal functions at random from the points inside the sphere of radius $\sqrt{PN_0}$, and achieve the most that is possible.

In fact, not only does this argument show that there exists a capacity-achieving code, but since a random selection leads to a capacity-achieving code, almost all codes must share this property.

12.2.1 Sphere Packing

The main idea of the sphere packing argument is that the noise places a resolution limit on the possible signals that can be distinguished at the receiver. In the geometrical representation of signals, each signal point is surrounded by a small region of uncertainty due to the noise. The perturbations of the different coordinates in the

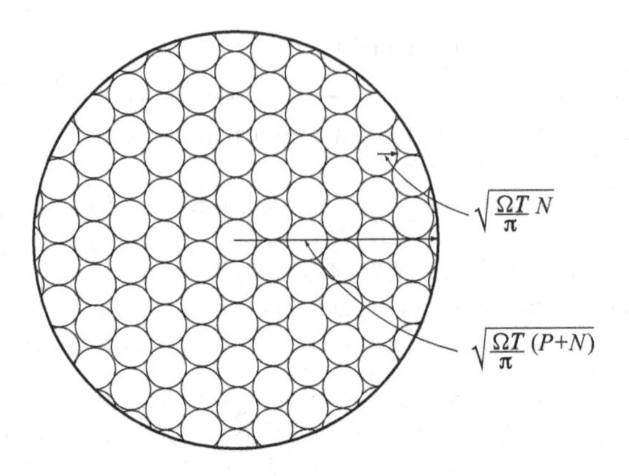

Fig. 12.2 The geometry of Shannon's outer bound.

signals' space are Gaussian and independent. Let us focus on the difference between the transmitted signal and the received one, namely on the differences between the coordinates of the corresponding points. The joint probability density function of the coefficients,

$$(y_1 - x_1, y_2 - x_2, \dots, y_{N_0} - x_{N_0}) = (z_1, z_2, \dots, z_{N_0}), \tag{12.14}$$

is given by the product of N_0 Gaussian densities, which depends only on the sum $\sum_{i=1}^{N_0} z_i^2$. This shows that the region of uncertainty about each received point in the signals' space is spherical. Furthermore, for large T, the noise perturbation about each point probabilistically concentrates around its typical value $\sqrt{N_0 N}$. This means that while fluctuations of the value of the perturbation are possible, these tend to vanish as $T \to \infty$, so that the received signal lies with high probability on the surface of a hard sphere of radius $\sqrt{N_0 N}$ centered at the transmitted point. By a similar argument, since the received signals have an average power at most $P + N$, they all lie with high probability inside the sphere of radius $\sqrt{N_0(P+N)}$. In order to recover different transmitted codewords without error, we need to space them sufficiently far apart so that their noise-perturbed versions are still distinguishable. An upper bound on the number of distinguishable signals is then given by the volume of the sphere of radius $\sqrt{N_0(P+N)}$ divided by the volume of the sphere of radius $\sqrt{N_0 N}$, since overlap of the noise spheres results in message confusion. A geometric picture is given in Figure 12.2. It follows that an upper bound on the number of distinguishable signals is given by

$$M \leq \left(\sqrt{\frac{P+N}{N}} \right)^{N_0}, \tag{12.15}$$

and the rate is bounded by

$$C = \frac{\log M}{T} \leq \frac{\Omega}{2\pi} \log \left(1 + \frac{P}{N} \right) \text{ bits per second.} \tag{12.16}$$

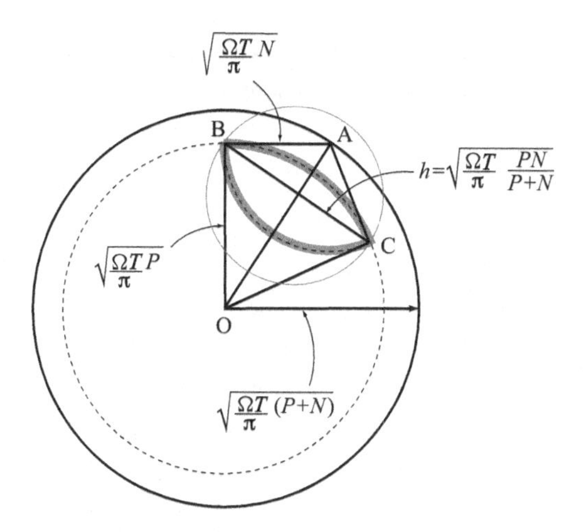

Fig. 12.3 The geometry of Shannon's inner bound.

12.2.2 Random Coding

The main idea of the random coding argument is to let the M encoded signals be picked uniformly at random inside the sphere of radius $\sqrt{N_0 P}$. The ratio of the volume of the shell between r and $r-\epsilon$, for $\epsilon \in (0,r)$, and the whole sphere is given by

$$\frac{r^{N_0}-(r-\epsilon)^{N_0}}{r^{N_0}}=1-\left(1-\frac{\epsilon}{r}\right)^{N_0}, \tag{12.17}$$

and tends to one as $N_0 \to \infty$. It follows that in high dimensions nearly all the volume is very close to the surface, so that the transmitted signals are, with high probability, very close to the surface of the sphere and by symmetry they all have the same probability of error. Figure 12.3 shows a cross section through the high-dimensional sphere defined by a typically transmitted signal B, received signal A, and the origin O. The high-dimensional lens-shaped region whose boundary is highlighted in gray, identified by the intersection of the surface of the signals' space sphere of radius $\sqrt{N_0 P}$ and the surface of the sphere of radius $\sqrt{N_0 N}$ centered in A, contains all possible signals that might have caused A, since the distance between a transmitted and a received signal is concentrated around $\sqrt{N_0 N}$. The volume of this region is smaller than that of the sphere of radius $\rho = \overline{BC}/2$. The radius ρ can be computed by solving the following equation, where the area of the triangle OAB appears on both sides:

$$\frac{1}{2}\rho\sqrt{N_0(P+N)}=\frac{1}{2}\sqrt{N_0 P}\sqrt{N_0 N}, \tag{12.18}$$

yielding

$$\rho=\sqrt{N_0\frac{PN}{P+N}}. \tag{12.19}$$

Since the M encoded signals are chosen uniformly at random among the points inside the signals' space of radius $\sqrt{N_0 P}$, it follows that the probability of a signal point being inside the lens-shaped region is less than the ratio of the volume of the signals' space sphere and the volume of the sphere of radius ρ, and it is given by

$$\left(\frac{\sqrt{N_0 \dfrac{PN}{P+N}}}{\sqrt{N_0 P}}\right)^{N_0} = \left(\frac{N}{P+N}\right)^{N_0/2}. \tag{12.20}$$

The transmitted signal B is decoded correctly when the remaining $M-1$ possible encoded signals are all outside the lens-shaped region. By the union bound, the probability of this latter event can be made greater than $(1-\delta)$ by appropriately choosing M, such that

$$1-(M-1)\left(\frac{N}{P+N}\right)^{N_0/2} > (1-\delta), \tag{12.21}$$

from which it follows that we need

$$(M-1) < \left(\frac{P+N}{N}\right)^{N_0/2} \delta. \tag{12.22}$$

By taking logarithms on both sides and dividing by T, we have

$$\frac{\log(M-1)}{T} < \frac{\Omega}{2\pi}\log\left(1+\frac{P}{N}\right)+\frac{\log\delta}{T}. \tag{12.23}$$

It follows that as $T \to \infty$ any communication rate $R = (\log M)/T$ arbitrarily close to the capacity $C = (\Omega/2\pi)\log(1+P/N)$ can be achieved.

12.2.3 Capacity and Mutual Information

Shannon also showed that the capacity of the channel coincides with the largest *mutual information* between a transmitted signal and the corresponding received one corrupted by the noise. This result is known as Shannon's channel coding theorem, and links an operational quantity, such as the largest achievable communication rate, with an information-theoretic one, such as the largest mutual information between transmitted and received signals.

The mutual information between a single transmitted random coefficient of the signal and the corresponding received one is defined as the residual uncertainty associated with the transmitted coefficient, given the received one. Viewing the uncertainty as the "surprise" that a transmitter can generate at the receiver by selecting one of the possible signals in the constellation, this also corresponds to the amount of information transferred in the communication process. It follows that the capacity can be obtained by maximizing the mutual information over all possible distributions of the coefficients compatible with the given power constraint.

To rephrase this intuition in the language of mathematics, we consider a bandlimited signal $f(t)$ corrupted by additive white Gaussian noise. By (12.10), we can view this signal as a point in high-dimensional space subject to the constraint (12.11). To define the mutual information, we introduce a probability distribution for the input coefficients, and we let $\{x_n\}$ be realizations of i.i.d. random variables $\{\mathsf{X}_n\}$ having probability density function $p_{\mathsf{X}}(x)$. As discussed in Chapter 1, probabilistic concentration ensures that the constraint (12.11) is essentially equivalent, as $N_0 \to \infty$, to

$$\mathbb{E}(\mathsf{X}^2) \leq P, \tag{12.24}$$

so that we can maximize the mutual information over all input distributions subject to the second moment constraint (12.24).

The mutual information between random variables X and Y is defined in terms of the differential entropy,

$$h_{\mathsf{X}} = -\int_{-\infty}^{\infty} p_{\mathsf{X}}(x) \log p_{\mathsf{X}}(x) dx, \tag{12.25}$$

and conditional differential entropy,

$$h_{\mathsf{X}|\mathsf{Y}} = -\int_{-\infty}^{\infty} p_{\mathsf{X},\mathsf{Y}}(x,y) \log(p_{\mathsf{X}|\mathsf{Y}}(x,y)) dx dy, \tag{12.26}$$

as

$$\mathcal{I}(\mathsf{X};\mathsf{Y}) = h_{\mathsf{X}} - h_{\mathsf{X}|\mathsf{Y}}, \tag{12.27}$$

and can be interpreted as the residual uncertainty associated with X once Y is known. Shannon's coding theorem is stated as follows:

> The information capacity of the channel $\mathsf{Y} = \mathsf{X} + \mathsf{Z}$ with average power constraint P is
>
> $$C = \sup_{p_{\mathsf{X}}(x):\mathbb{E}(\mathsf{X}^2) \leq P} \mathcal{I}(\mathsf{X};\mathsf{Y}) \text{ bits.} \tag{12.28}$$

Consider now the parallel channels used to transmit N_0 coefficients of a bandlimited signal. For each elementary channel, we have

$$\begin{aligned} \mathcal{I}(\mathsf{X};\mathsf{Y}) &= h_{\mathsf{X}} - h_{\mathsf{X}|\mathsf{Y}} \\ &= h_{\mathsf{Y}} - h_{\mathsf{X}+\mathsf{Z}|\mathsf{X}} \\ &= h_{\mathsf{Y}} - h_{\mathsf{Z}|\mathsf{X}} \\ &= h_{\mathsf{Y}} - h_{\mathsf{Z}}, \end{aligned} \tag{12.29}$$

where the last equality follows from Z being independent of X. Since X and Y are also independent, we have

$$\mathbb{E}(\mathsf{Y}^2) = \mathbb{E}(\mathsf{X}+\mathsf{Z})^2 = \mathbb{E}(\mathsf{X}^2) + N \leq P + N. \tag{12.30}$$

Recalling from Problem 1.15 that a zero-mean Gaussian random variable maximizes the differential entropy for a given variance constraint, and

$$h_{\mathsf{Z}} = \frac{1}{2}\log(2\pi e N), \tag{12.31}$$

we have, from (12.29),

$$\begin{aligned} \mathcal{I}(\mathsf{X};\mathsf{Y}) &\leq \frac{1}{2}\log[2\pi e(P+N)] - \frac{1}{2}\log(2\pi eN) \\ &= \frac{1}{2}\log\left(1+\frac{P}{N}\right) \text{ bits,} \end{aligned} \tag{12.32}$$

where equality is attained when X is a zero-mean Gaussian random variable with variance P. This expression can be interpreted as the largest amount of information transferred in a realization of one of the parallel channels in Figure 12.1.

Transmitting a bandlimited signal corresponds to having $N_0 = \Omega T/\pi$ parallel realizations, each with energy per degree of freedom P and noise variance per degree of freedom N, yielding a capacity

$$C = \frac{N_0}{2}\log\left(1+\mathsf{SNR}_\mathsf{S}\right) \text{ bits,} \tag{12.33}$$

or, in terms of bits per degree of freedom,

$$C = \frac{1}{2}\log\left(1+\mathsf{SNR}_\mathsf{S}\right) \text{ bits per degree of freedom,} \tag{12.34}$$

and in bits per unit time,

$$C = \frac{\Omega}{2\pi}\log\left(1+\mathsf{SNR}_\mathsf{S}\right) \text{ bits per second,} \tag{12.35}$$

which coincides with the expression of the operational capacity given earlier.

12.2.4 Limiting Regimes

The Shannon capacity shows a "law of diminishing returns" as SNR_S is increased. While for low values of SNR_S the capacity grows linearly with SNR_S, for large values it grows only logarithmically with SNR_S; see Figure 12.4. In the first case we are in a power-limited regime and a small increase in the transmitted power results in a corresponding increase in capacity, while in the second case the capacity is practically insensitive to increases in power.

The low-SNR_S regime was investigated in Section 1.5.1. In that case, rather than (12.11), we assumed the fixed energy constraint

$$\sum_{n=1}^{N_0} x_n^2 \leq E, \tag{12.36}$$

yielding a signal-to-noise ratio

$$\mathsf{SNR}_\mathsf{S} = \frac{E}{N_0 N}, \tag{12.37}$$

which tends to zero as $N_0 \to \infty$. It follows that the capacity in this low-SNR_S regime is proportional to the energy, and vanishes as $N_0 \to \infty$:

$$C = \frac{1}{2}\log\left(1+\frac{E}{N_0 N}\right)$$

$$\simeq \frac{E}{2N_0N} \log e \text{ bits per degree of freedom.} \tag{12.38}$$

To obtain a low-$\mathsf{SNR_S}$ regime with a non-vanishing capacity, we can replace the energy constraint (12.36) with

$$\int_{-T/2}^{T/2} f^2(t)dt \leq \bar{P}T, \tag{12.39}$$

where $\bar{P}$ has the units of energy per unit time. This new constraint essentially corresponds to

$$\sum_{n=1}^{N_0} x_n^2 \leq \bar{P}T, \tag{12.40}$$

and we have

$$\mathsf{SNR_S} = \frac{\bar{P}T}{N_0N} = \frac{\pi\bar{P}T}{\Omega TN} = \frac{\pi\bar{P}}{\Omega N}. \tag{12.41}$$

It follows that in this case $\mathsf{SNR_S}$ tends to zero as $\Omega N \to \infty$, and if we keep N fixed and let $\Omega \to \infty$, the capacity in the wide-band regime tends to a constant proportional to the signal's average power, namely

$$\begin{aligned} C &= \frac{\Omega}{2\pi} \log\left(1 + \frac{\pi\bar{P}}{\Omega N}\right) \\ &\simeq \frac{\Omega}{2\pi} \frac{\pi\bar{P}}{\Omega N} \log e \\ &= \frac{\bar{P}}{2N} \log e \text{ bits per second.} \end{aligned} \tag{12.42}$$

All of the above results differ in the applied energy constraint. With the constraint (12.11), $\mathsf{SNR_S}$ is independent of N_0, and the formula (12.35) holds. With the constraint

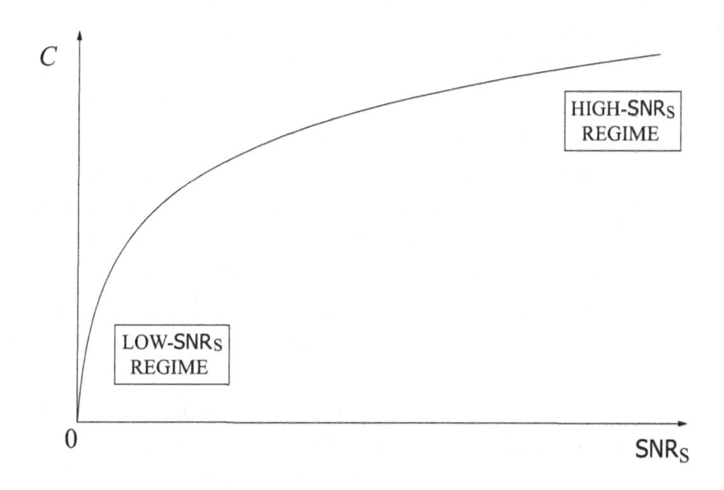

Fig. 12.4 Capacity of the additive white Gaussian channel.

(12.36), SNR_S vanishes as $N_0 \to \infty$, and leads to the capacity formula (12.38). With the constraint (12.40), SNR_S vanishes as $\Omega N \to \infty$ and leads to the capacity formula (12.42).

12.2.5 Quantum Constraints

The capacity results described above hold only in the regime where the quantum nature of radiation can be neglected. To illustrate additional quantum constraints, recall from the discussion in Chapter 11 that for the white noise assumption to hold we need $k_B T_K \gg \hbar\omega$ for all radiated frequencies, where k_B is the Boltzmann constant, $\hbar$ is the reduced Planck constant, and T_K is the absolute temperature of the noise, so that (11.31) can be used as a valid approximation for the noise. It follows that if we transmit over an interval of frequencies $[-\Omega, \Omega]$, we need

$$N = k_B T_K \gg \hbar\Omega, \tag{12.43}$$

and this leads to the noise-to-bandwidth constraint,

$$\frac{N}{\Omega} \gg \hbar. \tag{12.44}$$

Assuming the energy constraint (12.39), and plugging (12.44) into (12.35), we get

$$\begin{aligned} C &= \frac{\Omega}{2\pi} \log\left(1 + \frac{\pi \bar{P}}{\Omega N}\right) \\ &\le \frac{\bar{P}}{2N} \log e \\ &\ll \frac{\bar{P}}{2\hbar\Omega} \log e \text{ bits per second,} \end{aligned} \tag{12.45}$$

where the first inequality follows from $\log(1+x) \le x \log e$ for all $x > 0$.

In the wide-band regime, the first inequality in (12.45) is tight because $\Omega N \gg P$, and we have

$$C \simeq \frac{\bar{P}}{2N} \log e \ll \frac{\bar{P}}{2\hbar\Omega} \log e. \tag{12.46}$$

It follows that in the wide-band regime the capacity can be proportional to the average power only when the power spectral density of the noise is much larger than $\hbar\Omega$. When this physical constraint is violated, (12.42) no longer holds and, as we shall see below, the capacity becomes proportional to the square root of the average power.

For radio communications, and typical values of the frequency and temperature, the white noise assumption holds and quantum mechanical effects can indeed be neglected. For example, for transmission up to the GHz range at a temperature of 300 K, we have that $\hbar\Omega \approx 10^{-25}$, while $k_B T_K$ is of the order of 10^{-21}. This value of $k_B T_K$ can be regarded as a minimum noise attainable by an ideal receiver at a temperature around 300 K. Noise is usually greater by some factor expressed in decibels and known as the noise figure. It follows that (12.43) is typically satisfied in many operating conditions. For low temperature and high frequency of radiation, however, quantum effects kick in, noise distributions different from the Gaussian one become more appropriate, including the Poisson shot-noise models discussed in Chapter 11, and different capacity computations are required.

12.2.6 Capacity of the Noiseless Photon Channel

For small values of the noise compared to the largest frequency of radiation times the reduced Planck constant, (12.43) is violated and the white Gaussian noise model must be replaced with a more appropriate quantum noise model. In the limiting case of the absence of noise, the signal itself is subject to quantized radiation constraints; in this case the capacity reduces to the entropy per unit time of the waveform reaching the receiver, and is proportional to the square root of the average radiated power.

To illustrate this result, let us model the signal as a stochastic process $\mathsf{f}(t)$ of given average power

$$\mathbb{E}(\mathsf{f}^2(t)) = \frac{1}{2\pi}\int_{-\infty}^{\infty}(S_{\mathsf{Z}}(\omega))d\omega = \bar{P}. \tag{12.47}$$

The maximum entropy of the quantized process is achieved by black body radiation in one dimension, having a power spectral density of the form (11.29a), and discrete energy per mode of distribution (11.26). Assuming a single polarization state, the capacity can then be computed to be

$$C = \sqrt{\frac{\bar{P}\pi}{3\hbar}}\log e \text{ bits per second.} \tag{12.48}$$

A simple derivation of the functional form in (12.48) is given as follows. Assume that each radiated photon has a characteristic frequency ω_c and carries an energy $\hbar\omega_c$. In a modulated communication system, ω_c corresponds to the carrier frequency of the signal and $\hbar\omega_c$ to the minimum detectable energy of a single mode of radiation at that frequency. Assume that the detection of each photon carries one bit of information, and occurs at the time scale of one signal oscillation. Then, we have

$$\bar{P}/(\hbar\omega_c) \approx C \approx \omega_c; \tag{12.49}$$

namely, the capacity is roughly of the same order as the average number of radiated photons per second and the number of oscillations per second. By transmitting one quantum of energy per second, the signal carries at most one bit per second, and evolves at a rate of one cycle per second. Eliminating ω_c from (12.49) leads to

$$\bar{P} \approx \hbar C^2, \tag{12.50}$$

from which it follows that

$$C \approx \sqrt{\frac{\bar{P}}{\hbar}}, \tag{12.51}$$

which has the same functional form as (12.48).

This simple derivation indicates the optimal transmission strategy, captured by the slogan: "one quantum–one bit–one mode."

To achieve capacity one can modulate the signal once a period, by transmitting a sequence of sinusoidal modes, each of which lasts roughly one period. The effective bandwidth of the modulated signal corresponds to an interval of frequencies ranging from zero, corresponding to transmitting a fixed oscillation at frequency ω_c, to a maximum value ω_c, corresponding to transmitting an oscillation varying at the same time scale as the carrier. It follows that achieving capacity requires transmission over a bandwidth that is of the same order as the characteristic frequency of radiation.

A more rigorous derivation of the capacity of the noiseless photon channel (12.48) can be obtained using a statistical mechanical argument to maximize the number of possible discrete energy configurations that the one-dimensional signal can have, by allocating a given number of quanta of energy $\hbar\omega$ at each frequency, subject to an average power constraint. This approach was used in Chapter 11 to compute the maximum entropy of three-dimensional radiating bodies. In the one-dimensional case, it leads to a capacity of the form (12.51) that corresponds to the entropy rate of black body radiation in one dimension.

12.2.7 Colored Gaussian Noise

In some frequency regimes a colored Gaussian distribution that decays at high frequencies may be a more appropriate model of the thermal noise, and the capacity results should be modified accordingly.

In this case, by dividing the band into small sub-bands where the spectral density of the noise is approximately constant, the total capacity is given by

$$C = \int_0^{\Omega} \frac{1}{2\pi} \log\left(1 + \frac{P(\omega)}{N'(\omega)}\right) d\omega, \tag{12.52}$$

where N' is the one-sided power spectral density of the noise, and $P(\omega)$ is the power allocation of the signal, chosen to maximize (12.52) subject to the constraint

$$\int_0^{\Omega} P(\omega) d\omega = \bar{P}. \tag{12.53}$$

The problem is standard in the calculus of variations, and amenable to a Lagrangian formulation. Letting $(\cdot)_+ = \max\{0, \cdot\}$, the optimal solution corresponds to a piecewise constant value of the power,

$$P(\omega) = [\kappa - N'(\omega)]_+, \tag{12.54}$$

with κ satisfying

$$\int_0^{\Omega} [\kappa - N'(\omega)]_+ d\omega = \bar{P}. \tag{12.55}$$

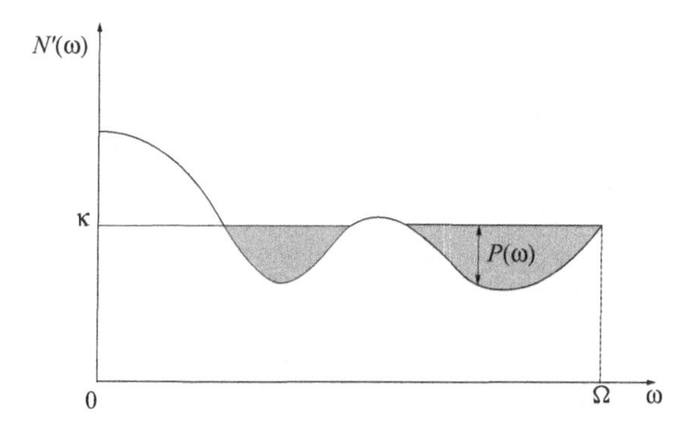

Fig. 12.5 Water filling interpretation of the optimal power allocation.

Substituting the optimal power allocation (12.54) into (12.52), we obtain the capacity of the additive (colored) Gaussian channel,

$$C = \int_0^{\Omega} \frac{1}{2\pi} \log\left(1 + \frac{[\kappa - N'(\omega)]_+}{N'(\omega)}\right) d\omega. \tag{12.56}$$

The solution has the graphical interpretation given in Figure 12.5. We can think of the total energy of the signal as being poured into a reservoir whose bottom is shaped by the power spectral density of the noise. The level to which the water rises is κ, while $P(\omega)$ is the depth of the water at various locations in the reservoir.

If one varies the shape of $N(\omega)$, while keeping its integral across the spectrum constant, and adjusts $P(\omega)$ to obtain the maximum transmission rate, it turns out that the worst shape is indeed the constant one corresponding to white Gaussian noise. Namely, white Gaussian is the "worst noise" among all possible Gaussian noises from a capacity standpoint. This result holds in greater generality, as the white Gaussian noise is the worst kind of additive noise among all noises of a given variance.

12.2.8 Minimum Energy Transmission

We have so far considered the highest reliable transmission rate under a given energy constraint. We now assume a given transmission rate and determine the minimum energy per bit needed to achieve it. This leads to the classic Shannon limit on the energy needed to reliably communicate one bit of information over a white Gaussian channel.

Consider transmission of bits at rate R, and average energy per bit E_b. The average energy per unit time is $\bar{P} = RE_b$. Using (12.35), we have

$$\begin{aligned} R &\leq \frac{\Omega}{2\pi} \log\left(1 + \frac{\pi \bar{P}}{\Omega N}\right) \\ &= \frac{\Omega}{2\pi} \log\left(1 + \frac{2\pi R E_b}{\Omega N'}\right), \end{aligned} \tag{12.57}$$

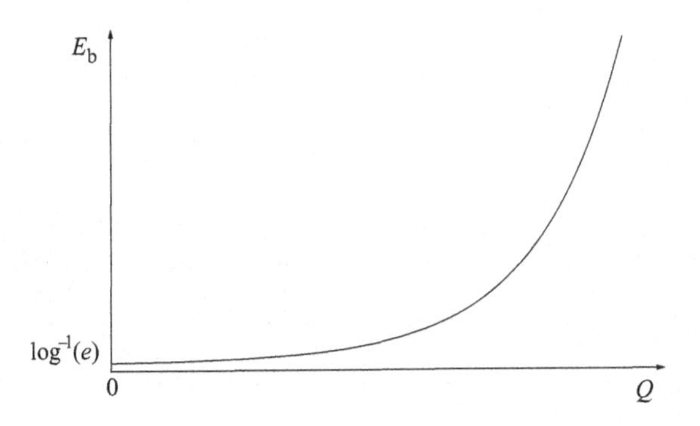

Fig. 12.6 Shannon's minimum energy limit.

where $N' = 2N$ is the one-sided power spectral density. From (12.57), we have

$$\frac{2\pi R}{\Omega} \leq \log\left(1 + \frac{2\pi R E_b}{\Omega N'}\right). \tag{12.58}$$

The quantity

$$Q = 2\pi R/\Omega \tag{12.59}$$

is the spectral efficiency, namely the rate of transmission per unit frequency. Substituting (12.59) into (12.58), it follows that

$$\frac{E_b}{N'} \geq \frac{2^Q - 1}{Q}. \tag{12.60}$$

This gives a bound on the normalized energy per bit of information when transmitting at rate R over the Gaussian channel using signals in $\mathscr{B}_\Omega$. As $Q \to 0$, the function converges to its minimum,

$$\frac{E_b^*}{N'} = \log^{-1}(e) = 0.6931\ldots, \tag{12.61}$$

where E_b^* represents Shannon's energy limit in the classical regime: the minimum energy necessary to transmit one bit of information is a factor of about 0.7 times the one-sided noise spectral density; see Figure 12.6.

12.3 A More Rigorous Formulation

All the results described so far swept under the carpet a number of issues required to provide a completely rigorous mathematical formulation. We have argued that communication occurs over a time interval of length T, and that results hold in the limit of $T \to \infty$, provided that a negligible amount of the signal's energy falls outside the interval $[-T/2, T/2]$. However, any strictly bandlimited signal has an infinite time duration, and in the computation of the capacity we have not accounted for the degrees of freedom present in the vanishing tail of the signal. A rigorous derivation requires a more careful account, and use of the notions of time concentration and frequency concentration developed in Chapter 2.

Fortunately, the phase transition of the number of degrees of freedom ensures that Shannon's results are robust, and do not change significantly when a more precise derivation is performed. On the other hand, a more detailed account reveals that more accurate noise models must be developed to provide a sound physical interpretation of the results.

12.3.1 Timelimited Signals

Consider first a signal strictly timelimited to $[-T/2, T/2]$, and subject to the energy constraint

$$\int_{-\infty}^{\infty} f^2(t)dt = \int_{-T/2}^{T/2} f^2(t)dt \leq \bar{P}T. \tag{12.62}$$

This signal is spread over an infinite bandwidth, but its frequency concentration over the interval $[-\Omega, \Omega]$, as defined in (2.53), is given by $\beta^2(\Omega) \geq 1 - \epsilon_\Omega^2$; see Figure 12.7. By considering white Gaussian noise of constant power spectral density

$$S_Z(\omega) = N \tag{12.63}$$

to be added to the signal over all frequencies, we obtain a more precise version of Shannon's formula:

$$C = \frac{\Omega}{2\pi} \log\left(1 + (1 - \epsilon_\Omega^2)\frac{\pi \bar{P}}{\Omega N}\right) + \epsilon_\Omega^2 \frac{\bar{P}}{2N} \log e \text{ bits per second.} \tag{12.64}$$

By letting $\epsilon_\Omega \to 0$, the signal's energy tends to be concentrated inside the band and the capacity reduces to the familiar form (12.35), where SNR_S is given by (12.41). As $\epsilon_\Omega \to 1$ and more and more of the signal's energy falls outside the band, the capacity approaches that obtained in the wide-band regime (12.42). On the other hand, for any $0 < \epsilon_\Omega < 1$ we have two contributions to the capacity: a contribution due to the concentrated part of the signal that scales logarithmically with the average power $(1 - \epsilon_\Omega^2)\bar{P}$, and a contribution due to the unconcentrated part that scales linearly with the average power $\epsilon_\Omega^2 \bar{P}$. As $\epsilon_\Omega \to 0$, the latter part disappears and the whole contribution is due to the concentrated part of the signal only.

The generalized formula (12.64) is obtained by expanding the signal into a prolate spheroidal basis set and then summing the contribution to capacity due to the first N_0 coefficients associated with the concentrated part of the signal and the contribution of the remaining coefficients associated with the unconcentrated part of the signal. The first contribution occurs over N_0 degrees of freedom and is of the form of (12.35). The second contribution occurs over an infinite number of degrees of freedom and is of the form of (12.42). A completely rigorous derivation requires using the phase transition results presented in Chapters 2 and 3. We provide here a sketch of the main steps.

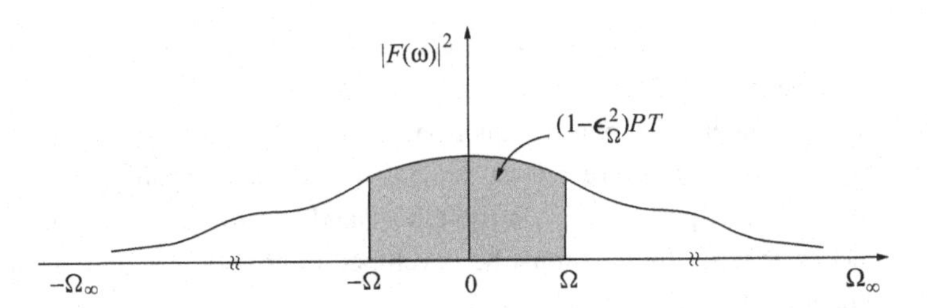

Fig. 12.7 Spectral concentration of a timelimited signal. The area of the shaded region is at most $(1-\epsilon_\Omega^2)\bar{P}T$.

By allocating the energy $(1-\epsilon_\Omega^2)\bar{P}T$ over the first $N_0 = \Omega T/\pi$ degrees of freedom, the first contribution to the capacity is

$$\begin{aligned} C_1 &= \frac{\Omega}{2\pi}\log\left(1+(1-\epsilon_\Omega^2)\frac{\bar{P}T}{N_0 N}\right) \\ &= \frac{\Omega}{2\pi}\log\left(1+(1-\epsilon_\Omega^2)\frac{\pi\bar{P}}{\Omega N}\right) \text{ bits per second.} \end{aligned} \tag{12.65}$$

Letting $\Omega_\infty > \Omega$, by allocating the remaining $\epsilon_T^2\bar{P}T$ energy over $(\Omega_\infty - \Omega)T/\pi$ degrees of freedom, a second contribution to the capacity is

$$C_2^\infty = \frac{\Omega_\infty - \Omega}{2\pi}\log\left(1+\frac{\epsilon_\Omega^2\pi\bar{P}}{N(\Omega_\infty - \Omega)}\right) \text{ bits per second.} \tag{12.66}$$

Finally, letting $\Omega_\infty \to \infty$, we have

$$\begin{aligned} C_2 &= \lim_{\Omega_\infty\to\infty} C_2^\infty \\ &= \frac{\Omega_\infty - \Omega}{2\pi}\frac{\epsilon_\Omega^2\pi\bar{P}}{N(\Omega_\infty - \Omega)}\log e \\ &= \epsilon_\Omega^2\frac{\bar{P}}{2N}\log e \text{ bits per second.} \end{aligned} \tag{12.67}$$

The total capacity is now given by the sum of (12.65) and (12.67), which coincides with (12.64).

12.3.2 Bandlimited Signals

An analogous result is obtained for bandlimited signals of infinite time duration, of energy constraint

$$\int_{-\infty}^{\infty} f^2(t)dt \le \bar{P}T, \tag{12.68}$$

and of time concentration $\alpha^2(T) \ge 1-\epsilon_T^2$, as defined in (2.52) and depicted in Figure 12.8. In this case, we need to distinguish two different time scales. The first time scale is representative of the concentrated part of the signal and corresponds to the

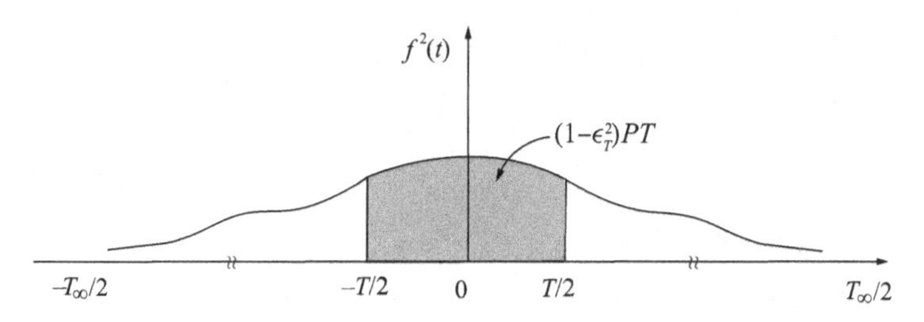

Fig. 12.8 Time concentration of a bandlimited signal. The area of the shaded region is at most $(1-\epsilon_T^2)\bar{P}T$.

time duration T, which is chosen large enough so that with the first part of the signal we can communicate

$$C_1 = \frac{\Omega T}{2\pi} \log\left(1 + (1-\epsilon_T^2)\frac{\pi \bar{P}}{\Omega N}\right) \text{ bits} \tag{12.69}$$

with an arbitrarily small error probability. The second time scale is representative of the infinite tail of the signal and corresponds to the interval $(T_\infty - T)$. Letting $T_\infty \to \infty$ and having fixed T, at this time scale we can communicate

$$\begin{aligned} C_2 &= \lim_{T_\infty \to \infty} \frac{\Omega(T_\infty - T)}{2\pi} \log\left(1 + \epsilon_T^2 \frac{\pi \bar{P} T}{\Omega(T_\infty - T)N}\right) \\ &= \frac{\Omega(T_\infty - T)}{2\pi} \frac{\pi \epsilon_T^2 \bar{P} T}{\Omega(T_\infty - T)N} \log e \\ &= \frac{\epsilon_T^2 \bar{P} T}{2N} \log e \text{ bits.} \end{aligned} \tag{12.70}$$

It follows that the total number of bits communicated by transmitting a single bandlimited signal of infinite duration is given by the sum of (12.69) and (12.70).

A problem now arises in the definition of communication rate. If we simply sum (12.69) and (12.70) and divide by T, we obtain

$$C = \frac{\Omega}{2\pi} \log\left(1 + (1-\epsilon_T^2)\frac{\pi \bar{P}}{\Omega N}\right) + \epsilon_T^2 \frac{\bar{P}}{2N} \log e \text{ bits per second,} \tag{12.71}$$

which is the analog of (12.64) for bandlimited, rather than timelimited, signals. This formula, however, has a limited physical meaning. Since any bandlimited signal has infinite duration, the interpretation of communication rate over any interval of fixed length T is problematic. When viewed at the scale of T_∞ the signal carries, strictly speaking, zero capacity. This follows by summing (12.69) and (12.70), dividing by T_∞, and taking the limit for $T_\infty \to \infty$. Since the energy constraint does not scale with T_∞, similar to the case in (12.38), the number of bits per unit time tends to zero. On the other hand, in the timelimited signal model leading to (12.64), where the linear contribution to the capacity is due to the infinite tail of the signal in the frequency domain rather than in the time domain, a meaningful notion of communication rate can be defined.

12.3.3 Refined Noise Models

We now summarize. A communication model based on bandlimited signals does not allow a rigorous definition of communication rate, since each transmitted signal has an infinite time duration. A communication model based on timelimited signals is perhaps more natural, as it allows an operational interpretation of the capacity in terms of communication rate using signals of finite time, leading to (12.64). However, a problem arises in this case as well: since noise of constant power spectral density is added over all frequencies of a signal of infinite bandwidth, the average power of the noise must be infinite. As we have discussed in Chapter 11, this is not a good physical model and it leads to the ultraviolet catastrophe paradox.

It follows that to carefully account for the spectral concentration properties of signals and fully generalize Shannon's formula, we need to develop more accurate models of the noise process. Ideally, we would like to work with timelimited signals and bandlimited noise, so that we could have a meaningful definition of communication rate as well as avoiding physical impossibilities in the noise model. Any noise model that is zero in some frequency range, however, would immediately lead to unbounded capacity. By allocating some signal power over this frequency range, SNR_S and hence the capacity would be arbitrarily high.

One way to circumvent these difficulties has been anticipated in Sections 12.2.5 and 12.2.6. We could assume a Gaussian noise model that vanishes outside a certain range of frequencies, as well as a quantum noise model that kicks in at high frequencies. The Gaussian model may account for the amount of information carried by the concentrated part of the signal, while the quantum model may describe the amount of information carried by radiation at high frequencies, where quantum constraints are dominant. When the Gaussian noise vanishes and only quantum constraints are present, the capacity must become proportional to the square root of the average radiated power, and reduce to (12.48).

Another possibility that retains a continuous signal modeling approach but still provides valuable physical insight is to assume that the noise is given by the superposition of a bandlimited Gaussian component acting within the frequency interval $[-\Omega, \Omega]$, as well as a small Gaussian residual noise acting out of band. The bandlimited component may approximate thermal noise, while the small noise floor may represent an inherent measurement uncertainty acting on the degrees of freedom of the signal occurring within the unconcentrated part the spectrum.

If the model of the noise is the one depicted in Figure 12.9, given by

$$S_Z(\omega) = \begin{cases} N & |\omega| \leq \Omega, \\ \nu_\Omega^2 N & |\omega| > \Omega, \end{cases} \tag{12.72}$$

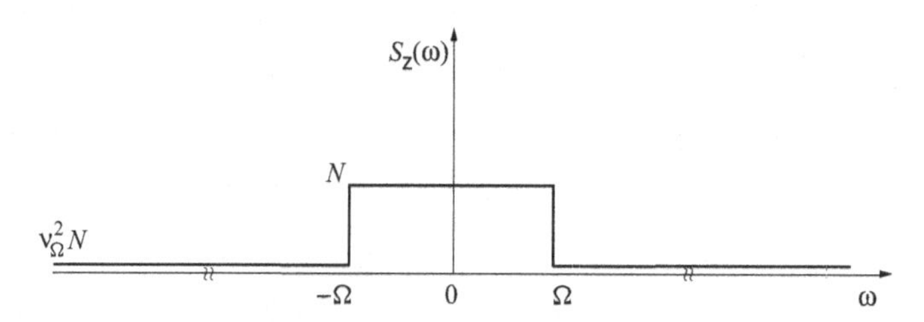

Fig. 12.9 Bandlimited noise with vanishing floor.

then we have

$$C = \frac{\Omega}{2\pi}\log\left(1+(1-\epsilon_\Omega^2)\frac{\pi\bar{P}}{\Omega N}\right)+\left(\frac{\epsilon_\Omega}{\nu_\Omega}\right)^2\frac{\bar{P}}{2N}\log e \text{ bits per second.} \tag{12.73}$$

If we now assume $\nu_\Omega \to 0$, then the noise becomes essentially bandlimited and physically well defined, but the capacity diverges because the fraction ϵ_Ω^2 of unconcentrated energy of the signal lies in a portion of the spectrum where there is vanishing noise and thus can be used to communicate at infinite rate. However, if ϵ_Ω approaches zero as well, then the capacity can remain bounded.

By interpreting ν_Ω and ϵ_Ω as limits on the measurement accuracy of the high frequency components, the former measuring the noise and the latter measuring the signal, it is reasonable to assume that they tend to zero at the same rate, as $T \to \infty$. In this case, a first contribution to capacity is due to the concentrated part of the signal that is affected by essentially constant noise on each degree of freedom, and as $T \to \infty$ it tends to Shannon's original result obtained for a bandlimited white Gaussian noise process. A second contribution is due to the unconcentrated part of the signal and, assuming that both the out-of-band signal and noise contributions tend to zero at the same rate, this contribution tends to a constant value, as $T \to \infty$.

Shannon's classic formula and its extension are compared in Table 12.1. The second term in the extended formula quantifies the ability to capture the higher-order components in the phase transition of the number of degrees of freedom of the signal, and use them to communicate information. If the signal-to-noise ratio in the tail of the transition window of the number of degrees of freedom is large, this constant term can be substantial. In the context of spatial signals, an analogous term arises in the computation of the capacity of imaging systems where sensors attempt to capture the vanishing tail of the number of degrees of freedom of the signal in the space–wavenumber domain to improve the quality of the constructed image, and it leads to the *super-resolution* gain that we discussed in Section 1.2.4.

It is important to point out that all technologies developed in this framework must obey the bounds imposed by the laws of physics and information theory that are based on spectral concentration occurring in the appropriate asymptotic regimes. While the

Table 12.1 Shannon capacity.

Classical	Extended
$C = \frac{\Omega}{2\pi}\log\left(1+\frac{\pi\bar{P}}{\Omega N}\right)$	$C = \frac{\Omega}{2\pi}\log\left(1+(1-\epsilon_\Omega^2)\frac{\pi\bar{P}}{\Omega N}\right) + (\epsilon_\Omega/\nu_\Omega)^2\frac{\bar{P}}{2N}\log e$

amount of super-resolution depends on the properties of the noise model, any reasonable model dictates that unbounded super-resolution is a physical impossibility, since the second term of (12.71) must always remain bounded, even if $\nu_\Omega \to 0$.

12.4 Shannon Entropy

We now turn to consider another information-theoretic notion introduced by Shannon: the *entropy*, which captures the "surprise" associated with a given stochastic realization, as discussed in Section 1.4.

While the Shannon capacity is closely related to the problem of packing the signals' space with balls of a given radius, the entropy is closely related to the geometric problem of covering the space with balls of a given radius. Each source signal, modeled as a stochastic process, corresponds to a random point in the space, and by quantizing all coordinates at a given resolution, the Shannon entropy corresponds to the number of bits needed on average to represent the quantized signal. Thus, the entropy depends on both the probability distribution of the process and the quantization step along the coordinates of the space.

A quantizer, however, does not need to act independently on each coordinate, and can more generally viewed as a discrete set of balls covering the space. The source signal is represented by the closest center of a ball covering it, and the distance to the center of the ball represents the distortion measure associated with this representation. In this context, the Shannon *rate–distortion function* provides the minimum number of bits that must be specified per unit time to represent the source process with a given average distortion.

12.4.1 Rate–Distortion Function

We consider a zero-mean white Gaussian stochastic process $\mathsf{f}(t)$ of constant power spectral density P of support $[-\Omega, \Omega]$:

$$S_{\mathsf{f}}(\omega) = \begin{cases} P & |\omega| \leq \Omega, \\ 0 & |\omega| > \Omega. \end{cases} \tag{12.74}$$

This stochastic process has finite average power

$$\mathbb{E}(\mathsf{f}^2(t)) = s_{\mathsf{f}}(0) = \frac{1}{2\pi}\int_{-\infty}^{\infty} S_{\mathsf{f}}(\omega)d\omega = \frac{P\Omega}{\pi}, \tag{12.75}$$

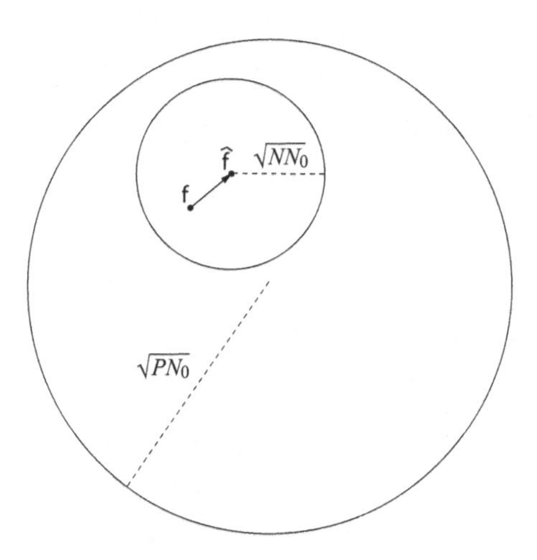

Fig. 12.10 Codebook representation of a stochastic process.

where s_f is the autocorrelation of the process. When observed over the interval $[-T/2, T/2]$, the process can be viewed as a random point having essentially N_0 independent Gaussian coordinates of zero mean and variance P, and distributed inside the ball of radius

$$\left(\int_{-T/2}^{T/2} \mathbb{E}(f^2(t))dt\right)^{1/2} = \left(\frac{P\Omega T}{\pi}\right)^{1/2} = \sqrt{PN_0}. \tag{12.76}$$

A source codebook is composed of a subset of points in the space, and each codebook point is a possible representation of the stochastic process. The distortion associated with the representation of $f(t)$ using codebook point $\hat{f}(t)$ is defined in terms of mean-squared energy,

$$d(f, \hat{f}) = \int_{-T/2}^{T/2} \mathbb{E}[f(t) - \hat{f}(t)]^2 dt. \tag{12.77}$$

We let $L_N(P)$ be the smallest number of codebook points that can be used to represent the source process with distortion at most NN_0. This roughly corresponds to the minimum number of balls of radius $\sqrt{NN_0}$ that cover the whole ball of radius $\sqrt{PN_0}$; see Figure 12.10. The definition of $L_N(P)$ should be compared with that of $M^\delta(P)$ used for capacity, representing the largest number of messages that can be communicated with probability of error at most δ, which roughly corresponds to the maximum number of balls of radius $\sqrt{NN_0}$ that can be packed inside the ball of radius $\sqrt{PN_0}$.

The rate–distortion function is then defined as

$$R_N = \lim_{T \to \infty} \frac{\log L_N(P)}{T} \text{ bits per second,} \tag{12.78}$$

and it represents the minimum number of bits per unit time needed to represent the stochastic process with distortion at most NN_0.

Letting $\mathsf{SNR}_\mathsf{S} = P/N$ be the signal-to-distortion ratio, Shannon's formula for the rate–distortion function of this Gaussian source is

$$R_N = \begin{cases} \dfrac{\Omega}{2\pi} \log(\mathsf{SNR}_\mathsf{S}) \text{ bits per second} & \text{if } P \geq N, \\ 0 & \text{if } P < N. \end{cases} \tag{12.79}$$

The rate–distortion formula (12.79) should be compared with the capacity formula (12.35). Both formulas scale linearly with the bandwidth and logarithmically with the signal-to-noise ratio. It should also be compared with the entropy formula (1.46). While the entropy considers the number of bits needed on average to represent a stochastic process quantized at fixed resolution along each dimension of the space, the rate–distortion function considers a more general quantization model where signals are represented by codebook points that are arbitrarily located inside the signals' space, and represents the number of bits needed to approximate the stochastic process with a given distortion value.

The geometric insight behind (12.79) should be clear. The minimum number of balls needed to cover the whole space is given by at least the volume of the ball of radius $\sqrt{N_0 P}$ divided by the volume of the ball of radius $\sqrt{N_0 N}$, and we have

$$L_N(P) \geq \left(\sqrt{\frac{P}{N}}\right)^{N_0}. \tag{12.80}$$

By taking the logarithm and dividing by T, a lower bound matching (12.79) follows. A rigorous proof complements the lower bound with a random coding argument, similar to the one given for the capacity, showing the achievability of this minimum rate. Namely, random coding shows that there exists a collection of balls of cardinality at most $(P/N)^{N_0/2}$ that covers the whole space with high probability, as $T \to \infty$.

12.4.2 Rate–Distortion and Mutual Information

The operational definitions of capacity and rate–distortion have geometric interpretations in terms of maximum-cardinality sphere packing and minimum-cardinality sphere covering of the space. The corresponding information-theoretic interpretations are given in terms of the largest mutual information between transmitted and received signals, and the smallest mutual information between a signal and its perturbed representation. We discussed the information-theoretic interpretation of capacity in Section 12.2.3. We now give an analogous interpretation for the rate–distortion.

We begin by considering the smallest mutual information between a Gaussian random variable X and its codebook representation $\hat{\mathsf{X}}$,

$$R_N(\mathsf{X};\hat{\mathsf{X}}) = \inf_{p_{\hat{\mathsf{X}}|\mathsf{X}} : \mathbb{E}[(\hat{\mathsf{X}}-\mathsf{X})^2] \leq N} \mathcal{I}(\mathsf{X};\hat{\mathsf{X}}) \text{ bits}, \tag{12.81}$$

where $p_{\hat{X}|X}$ is the conditional density of the codebook representation $\hat{X}$, given X. This naturally extends to a sequence $\{X_n\}$ of N_0 independent random variables distributed as X, and we have

$$R_N(\{X_n\};\{\hat{X}_n\}) = N_0 R_N(X;\hat{X}) \text{ bits.} \tag{12.82}$$

As $N_0 \to \infty$ the mutual information $\mathcal{I}(\{X_n\};\{\hat{X}_n\})$ can be identified with the mutual information between the stochastic processes $f(t)$ and $\hat{f}(t)$, and (12.82) represents the growth rate of the corresponding rate–distortion function. When this is measured per unit time it agrees with (12.79).

To prove this latter statement, we first compute a lower bound on the mutual information:

$$\begin{aligned}\mathcal{I}(X;\hat{X}) &= h_X - h_{X|\hat{X}} \\ &= \frac{1}{2}\log(2\pi eP) - h_{X-\hat{X}|\hat{X}} \\ &\geq \frac{1}{2}\log(2\pi eP) - h_{X-\hat{X}} \\ &\geq \frac{1}{2}\log(2\pi eP) - h_Z,\end{aligned} \tag{12.83}$$

where Z is a Gaussian random variable of zero mean and variance $\mathbb{E}[(X-\hat{X})^2]$, which maximizes the differential entropy $h_{X-\hat{X}}$. Since $\mathbb{E}[(X-\hat{X})^2] \leq N$, we get

$$\begin{aligned}\mathcal{I}(X;\hat{X}) &\geq \frac{1}{2}\log(2\pi eP) - \frac{1}{2}\log(2\pi eN) \\ &\geq \frac{1}{2}\log\frac{P}{N}.\end{aligned} \tag{12.84}$$

Next, to find the conditional density $p_{\hat{X}|X}$ that achieves this lower bound, assuming $N \leq P$, we choose

$$X = \hat{X} + Z, \tag{12.85}$$

where $\hat{X}$ is a Gaussian random variable of mean zero and variance $P-N$, Z is another Gaussian of mean zero and variance N, and $\hat{X}$ and Z are independent of each other. The resulting "test channel" that gives the original signal from the estimate by adding noise is depicted in Figure 12.11. It follows that we have $\mathbb{E}[(X-\hat{X})^2] = N$, and

$$\begin{aligned}\mathcal{I}(X;\hat{X}) &= h_X - h_{X|\hat{X}} \\ &= h_X - h_Z \\ &= \frac{1}{2}\log(2\pi eP) - \frac{1}{2}\log(2\pi eN) \\ &= \frac{1}{2}\log\frac{P}{N},\end{aligned} \tag{12.86}$$

achieving the lower bound (12.84). If $N \leq P$, we can choose $\hat{X} = 0$ with probability one, so that the rate–distortion function is zero.

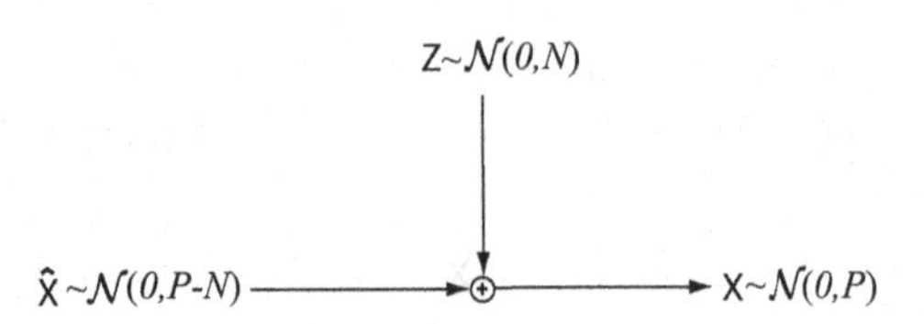

Fig. 12.11 Density achieving the lower bound for the rate–distortion function.

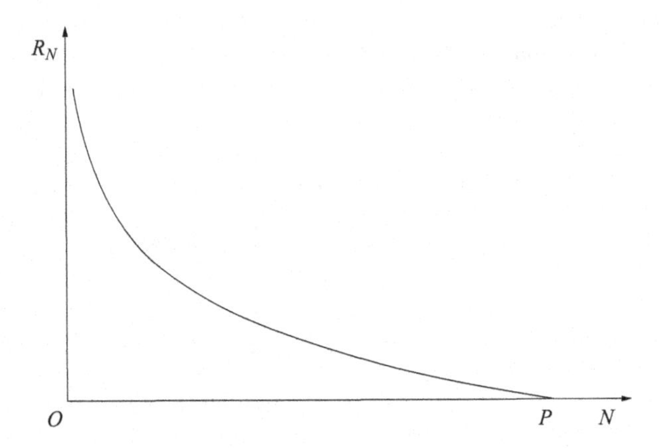

Fig. 12.12 Rate–distortion function for a Gaussian source.

Substituting (12.86) into (12.82), dividing by T, and taking the limit for $T \to \infty$, we obtain (12.79). A plot of the rate–distortion function for a Gaussian source is depicted in Figure 12.12. We can also invert (12.79) and obtain the distortion in terms of rate,

$$N = P2^{-2R_N\pi/\Omega}, \tag{12.87}$$

showing that each additional bit of description of the stochastic process reduces the distortion per degree of freedom by a factor of four. This exponential decrease of the distortion with the rate should be compared to the exponential decrease of the probability of error for rates below the capacity in the computation performed in Section 12.2.2.

12.5 Kolmogorov's Deterministic Quantities

The stochastic notions of entropy and capacity put forth by Shannon can be related to analogous ones developed by the Russian school of Kolmogorov in a deterministic setting.

A first deterministic information-theoretic notion is the Kolmogorov N-width that we introduced in Chapter 3. For bandlimited signals, the existence of a steep transition point in the N-width leads to the definition of the number of degrees of freedom that describes the effective "size" or "massiveness" of the space. This number can be computed in a precise asymptotic setting by solving Slepian's concentration problem. The relationship between the number of degrees of freedom and the largest achievable information rate when using bandlimited signals for communication is evident in Shannon's formulas,

$$C = \lim_{T\to\infty} \frac{N_0}{T} \log(1 + \mathsf{SNR_S}) \text{ bits per second,} \tag{12.88}$$

$$R_N = \lim_{T\to\infty} \frac{N_0}{T} \log(\mathsf{SNR_S}) \text{ bits per second,} \tag{12.89}$$

where the number of degrees of freedom appears in front of the logarithmic term. The number of degrees of freedom corresponds to the effective number of parallel channels available in the space, and the logarithmic term reflects the power constraint.

In addition to the N-width, Kolmogorov introduced notions of capacity and entropy in the purely deterministic setting of functional approximation. This led to the development of a functional theory of information that, as mentioned in Chapter 1, has several points of contact with the probabilistic theory of information developed in the West.

12.5.1 ϵ-Coverings, ϵ-Nets, and ϵ-Entropy

Let A be a subset of a metric space $\mathscr{X}$, and let $\epsilon > 0$. A family $\{U_1, U_2, \ldots, U_Q\}$ of subsets of $\mathscr{X}$ is called an *ϵ-covering* of A if the largest distance between any two points in U_k is at most 2ϵ and if the family covers A, namely $A \subset \cup_{k=1}^{Q} U_k$. The minimum number of sets in an ϵ-covering is an invariant of the set A, which depends only on ϵ, and is denoted by $Q_\epsilon(A)$. The *ϵ-entropy* of A is defined as the base two logarithm

$$H_\epsilon(A) = \log Q_\epsilon(A). \tag{12.90}$$

This is also called the *metric entropy*, in contradistinction to the statistical entropy used in Shannon's theory.

A set of points $\{x_1, x_2, \ldots, x_P\}$ of $\mathscr{X}$ is called an *ϵ-net* for A if for each $x \in A$ there is at least an x_k of the net at a distance from x not exceeding ϵ, namely $\|x - x_k\| \leq \epsilon$. The minimum number of points in an ϵ-net is denoted by $Q_\epsilon^{\mathscr{X}}(A)$ and may also depend, as the notation indicates, on the larger space $\mathscr{X}$ in which A is contained. The *ϵ-entropy* of A relative to $\mathscr{X}$ is defined as the base two logarithm

$$H_\epsilon^{\mathscr{X}}(A) = \log Q_\epsilon^{\mathscr{X}}(A). \tag{12.91}$$

For *centerable spaces*, ϵ-coverings naturally induce ϵ-nets and the two entropies defined above coincide. A set $U \subset \mathscr{X}$ is centerable in $\mathscr{X}$ if there exists a center point $x_0 \in \mathscr{X}$ such that $\|x - x_0\| \leq \epsilon$ for all $x \in U$. If all sets U_k are centerable, then we can think about them as "balls" and their centers form an ϵ-net for A. It follows that, in this case,

$$H_\epsilon^{\mathscr{X}}(A) = H_\epsilon(A). \tag{12.92}$$

Figure 12.13 provides an illustration of an ϵ-covering and the associated ϵ-net.

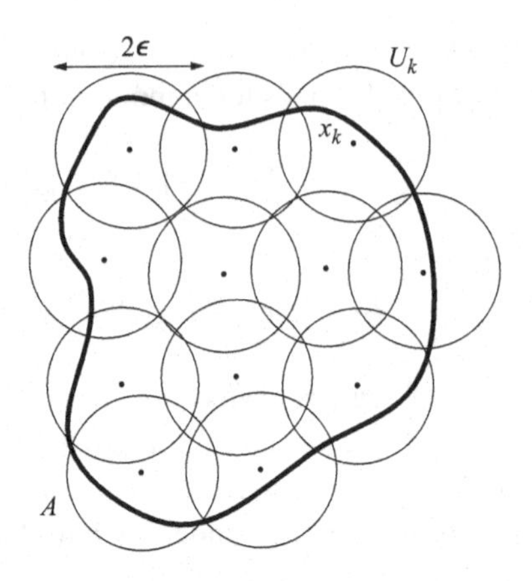

Fig. 12.13 An ϵ-covering of the set A and its associated ϵ-net.

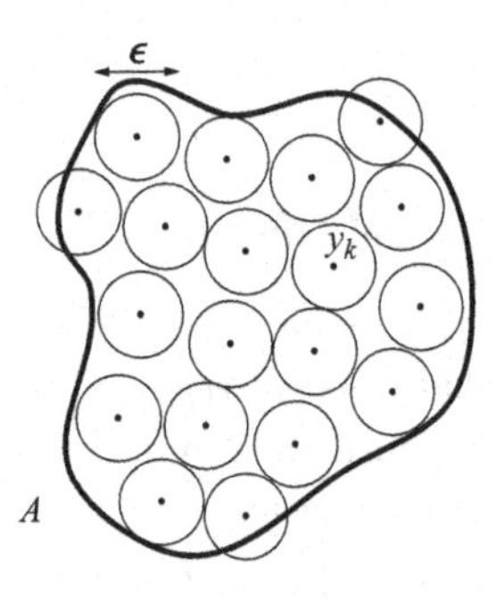

Fig. 12.14 A set of ϵ-distinguishable points for the set A.

12.5.2 ϵ-Distinguishable Sets and ϵ-Capacity

Points $y_1, y_2, \ldots, y_M$ of $\mathscr{X}$ are called *ϵ-distinguishable* if the distance between any two of them exceeds ϵ, namely $\|y_i - y_k\| > \epsilon$ for all $i \neq k$. The maximum number of ϵ-distinguishable points is an invariant of A that depends only on ϵ, and is denoted by $M_\epsilon(A)$. The *ϵ-capacity* of A is defined as the base two logarithm

$$C_\epsilon(A) = \log M_\epsilon(A). \tag{12.93}$$

Figure 12.14 provides an illustration of distinguishable points for a set A.

12.5.3 Relation Between ϵ-Entropy and ϵ-Capacity

A general relation between entropies and capacities is

$$C_{2\epsilon}(A) \leq H_\epsilon(A) \leq H_\epsilon^{\mathscr{X}}(A) \leq C_\epsilon(A). \tag{12.94}$$

It follows that a typical technique to estimate entropy and capacity is to find a lower bound for $C_{2\epsilon}$ and an upper bound for H_ϵ, and if these are close to each other, then they are good estimates for both capacity and entropy.

The basic relation (12.94) is obtained as follows. Since every ϵ-net also induces an ϵ-covering, we have

$$Q_\epsilon(A) \leq Q_\epsilon^{\mathscr{X}}(A). \tag{12.95}$$

Given a set of 2ϵ-distinguishable points of A of cardinality m and an ϵ-covering of A of cardinality n, then we have $m \leq n$, otherwise two distinguishable points would be contained in the same element of the covering set, which is impossible. This situation is illustrated in Figure 1.13.

It follows that

$$M_{2\epsilon}(A) \leq Q_\epsilon(A). \tag{12.96}$$

Finally, a maximal set of ϵ-distinguishable points of A of cardinality $m = M_\epsilon(A)$ also form an ϵ-net for A, so that

$$Q_\epsilon^{\mathscr{X}}(A) \leq M_\epsilon(A). \tag{12.97}$$

Taking logarithms of (12.95), (12.96), and (12.97), we obtain (12.94).

12.6 Basic Deterministic–Stochastic Model Relations

To investigate the relationship between the Shannon and Kolmogorov entropies and capacities, we consider an ϵ-bounded noise channel and revisit Shannon's results in this setting.

12.6.1 Capacity

Consider a probabilistic communication setting in which the input signal belongs to a subset A of a metric space $\mathscr{X}$ and the output signal belongs to the same metric space, but is perturbed by random noise that has bounded support. Namely, with probability one the output signal is at most at distance ϵ from the input signal. This means that given a class $\mathscr{P}$ of probability distributions of the input signals,

$$\mathscr{P} \subseteq \{P_{\mathsf{X}} : \mathbb{P}_{\mathsf{X}}(B) = \mathbb{P}(\mathsf{X} \in B)\}, \tag{12.98}$$

where $B \subset A$, we can use the probabilistic kernel

$$K(x, Q) = \mathbb{P}(\mathsf{Y} \in Q | \mathsf{X} = x) \tag{12.99}$$

to represent the channel in Figure 12.15 with the output distribution

$$\mathbb{P}(\mathsf{Y} \in Q) = \int_A K(x, Q)\, \mathbb{P}_{\mathsf{X}}(dx) \tag{12.100}$$

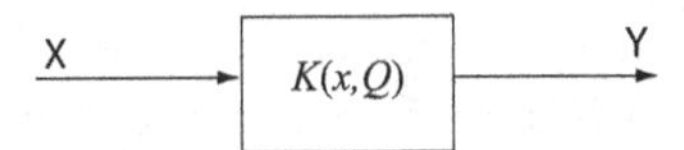

Fig. 12.15 A probabilistic channel.

and the joint distribution

$$\mathbb{P}[(\mathsf{X},\mathsf{Y}) \in U] = \mathbb{P}_{\mathsf{XY}}(dx,dy) = \iint_U K(x,dy)\,\mathbb{P}_{\mathsf{X}}(dx). \tag{12.101}$$

For this probabilistic communication channel, the restriction on the support of the noise is expressed as $K \in K_\epsilon$, where

$$K_\epsilon = \{K : K(x,\{y : \|x-y\| \le \epsilon\}) = 1\}. \tag{12.102}$$

The mutual information between input and output in this general model is

$$\begin{aligned}\mathcal{I}(\mathsf{X};\mathsf{Y}) &= h_{\mathsf{X}} - h_{\mathsf{X}|\mathsf{Y}} \\ &= \int_A \int_Q \mathbb{P}_{\mathsf{XY}}(dx,dy) \log \frac{\mathbb{P}_{\mathsf{XY}}(dx,dy)}{\mathbb{P}_{\mathsf{X}}(dx)\,\mathbb{P}_{\mathsf{Y}}(dy)},\end{aligned} \tag{12.103}$$

and the Shannon capacity is

$$C(K) = \sup_{\mathbb{P}_{\mathsf{X}} \in \mathscr{P}} \mathcal{I}(\mathsf{X};\mathsf{Y}), \tag{12.104}$$

which clearly depends on the noise model through the kernel K.

We now consider the smallest Shannon capacity over all noise distributions of bounded support ϵ, namely

$$C^*_\epsilon = \inf_{K \in K_\epsilon} C(K). \tag{12.105}$$

The largest amount of information in this worst-case condition is then related to the ϵ-entropy and ϵ-capacity as follows:

$$C_{2\epsilon}(A) \le C^*_\epsilon \le H_\epsilon(A) \le C_\epsilon(A). \tag{12.106}$$

The smallest Shannon capacity for an ϵ-noise perturbed channel falls between the ϵ and the 2ϵ Kolmogorov capacities of the corresponding functional space.

To show the validity of the bounds in (12.106), we notice that there exists a noise distribution kernel $\tilde{K}$ such that

$$C(\tilde{K}) = H_\epsilon(A). \tag{12.107}$$

This kernel simply assigns with probability one to each $x \in A$ one of the points $y \in Q$ of an optimal ϵ-net defining the ϵ-entropy of A. Clearly, this choice implies (12.107), since for all input distributions we have

$$\begin{aligned}\mathcal{I}(\mathsf{X};\mathsf{Y}) &= h_{\mathsf{X}} - h_{\mathsf{X}|\mathsf{Y}} \\ &= h_{\mathsf{Y}} - h_{\mathsf{Y}|\mathsf{X}}\end{aligned}$$

$$
\begin{aligned}
&= h_{\mathsf{Y}} \\
&\leq \log(Q_\epsilon) \\
&= H_\epsilon(A),
\end{aligned}
\tag{12.108}
$$

and the equality in the derivation above is achieved by choosing the input corresponding to a uniform distribution of the output over the ϵ-net. By combining (12.105) and (12.107), and using the basic relation (12.94), it now follows that

$$C^*_\epsilon \leq H_\epsilon(A) \leq C_\epsilon(A). \tag{12.109}$$

The lower bound on the left-hand side of (12.106) follows by choosing one input signal uniformly at random among the $M_{2\epsilon}(A)$ input signals described by the points of a maximal 2ϵ distinguishable set defining the 2ϵ capacity of A. In this way, the probability of error at the output is zero, and

$$
\begin{aligned}
\mathcal{I}(\mathsf{X};\mathsf{Y}) &= h_{\mathsf{X}} - h_{\mathsf{X}|\mathsf{Y}} \\
&= h_{\mathsf{X}} \\
&= \log(M_{2\epsilon}) \\
&= C_{2\epsilon}.
\end{aligned}
\tag{12.110}
$$

12.6.2 Rate–Distortion

To obtain a relation between the Shannon capacity and the Kolmogorov capacity, we have considered the largest mutual information compatible with a class of input probability distributions and then minimized it among all bounded probabilistic perturbations. We now consider the smallest mutual information compatible with any bounded probabilistic perturbation and maximize it among the class of possible input distributions. We let the Shannon rate–distortion function

$$R_\epsilon(P_{\mathsf{X}}) = \inf_{K \in K_\epsilon} \mathcal{I}(\mathsf{X};\mathsf{Y}), \tag{12.111}$$

which clearly depends on the model for the input distribution $P_{\mathsf{X}} \in \mathscr{P}$. If we maximize it over all input probability distributions, we obtain

$$R^*_\epsilon(A) = \sup_{P_{\mathsf{X}} \in \mathscr{P}} R_\epsilon(P_{\mathsf{X}}). \tag{12.112}$$

The quantities (12.105) and (12.112) are duals of each other, and we have the following inequalities:

$$C^*_{2\epsilon}(A) \leq R^*_\epsilon(A) \leq C^*_\epsilon. \tag{12.113}$$

The largest ϵ-bounded rate–distortion function falls between the smallest Shannon capacities of an ϵ- and a 2ϵ-noise channel.

The inequality on the right-hand side follows immediately from the general duality formula

$$\sup_x \inf_y f(x,y) \le \inf_y \sup_x f(x,y). \tag{12.114}$$

The inequality on the left-hand side follows from $C_{2\epsilon}(A) \ge C^*_{2\epsilon}(A)$ by (12.109) and the existence of a $P_X \in \mathscr{P}$ for which

$$R_\epsilon(P_X) \ge C_{2\epsilon}(A). \tag{12.115}$$

This distribution is simply the uniform one assigning to each of the $Q_{2\epsilon}$ points of a maximal 2ϵ-separated set the probability $1/Q_{2\epsilon}$.

12.7 Information Dimensionality

Kolmogorov's entropy and capacity measure the amount of information associated with functional spaces. The larger their values, the more complicated the structure of the space is. This suggests the introduction of a notion of information dimensionality based on these information-theoretic quantities. One approach defines the fractal dimension as the order of growth of the entropy and capacity at increasingly higher levels of discretization of the space. For many functional spaces, however, entropy and capacity diverge for any fixed value of ϵ. In this case, a second approach considers a sequence of finite-dimensional approximating subspaces, and determines the order of growth of the dimensionality of successive elements of this sequence.

We start by investigating the first approach, and distinguish the two cases of *informationally meager* functional sets, having a slower rate of growth of the ϵ-entropy and ϵ-capacity, and *informationally abundant* functional sets, having a faster rate of growth as $\epsilon \to 0$.

12.7.1 Metric Dimension

Let A be a bounded subset of $\mathbb{R}^d$ with non-zero interior; then we have, as $\epsilon \to 0$,

$$H_\epsilon(A) = d \log \frac{1}{\epsilon} + O(1), \tag{12.116}$$

$$C_\epsilon(A) = d \log \frac{1}{\epsilon} + O(1). \tag{12.117}$$

The derivation of (12.116) and (12.117) follows from the computation of the entropy and the capacity in the one-dimensional case. For an interval $A = [a,b]$ and Euclidean metric, $Q_\epsilon(A)$ is the smallest $n \ge (b-a)/(2\epsilon)$, and $M_\epsilon(A)$ is the largest $m \le (b-a)/\epsilon$. It follows that

$$Q_\epsilon(A) = \frac{b-a}{2\epsilon} + O(1), \tag{12.118}$$

$$M_\epsilon(A) = \frac{b-a}{\epsilon} + O(1), \tag{12.119}$$

and by taking logarithms,

$$H_\epsilon(A) = \log(1/\epsilon) + O(1), \tag{12.120}$$

$$C_\epsilon(A) = \log(1/\epsilon) + O(1). \tag{12.121}$$

The entropy and capacity for the higher-dimensional case can now be estimated by considering a Cartesian product of intervals defining a parallelepiped and using the results for the individual factors. Let $A = \prod_{k=1}^{d} A_k \subset \mathbb{R}^d$ and each A_k be an interval of the kth coordinate axis; then

$$H_\epsilon(A) \leq \sum_{k=1}^{d} H_{\epsilon/\sqrt{d}}(A_k), \tag{12.122}$$

$$C_{2\epsilon}(A) \geq \sum_{k=1}^{d} C_{2\epsilon/\sqrt{d}}(A_k). \tag{12.123}$$

Using (12.120) and (12.121) to express the ϵ-entropy of the monodimensional sets $\{A_k\}$ and the basic relation (12.94), the result (12.116) and (12.117) now follows in the case that $A \subset \mathbb{R}^d$ is a parallelepiped. The extension to more general compact sets with interior points follows from the monotonicity of entropy and capacity, in conjunction with an approximation argument.

The exact determination of the factor $O(1)$ in (12.116) corresponds to determining the most economical covering of the space $\mathbb{R}^d$, and in (12.117) the tightest packing by balls of unit radius. The exact solution to these geometric problems is not known for $d > 3$. The solution to the packing problem in three dimensions was first conjectured by Kepler in 1611. A computer-aided proof, which has been accepted by the wider mathematical community, was only provided in 1998 by Thomas Hales. Without having to solve this difficult combinatorial problem, the first-order characterization in (12.116) shows that as $\epsilon \to 0$ the behavior of the ϵ-entropy and ϵ-capacity for subsets of the finite-dimensional Euclidean space is principally given by the dimension d of the space. Their rate of divergence as $\epsilon \to 0$ is logarithmic, indicating that subsets of this kind are "informationally meager."

To characterize the massiveness of other kinds of sets, it is then natural to compare them to the massiveness of subsets of the finite-dimensional Euclidean space. This is easily done by defining the *metric dimension* of a subset $A \subset \mathscr{X}$ as

$$\mathcal{D}(A) = \lim_{\epsilon \to 0} \frac{H_\epsilon(A)}{\log(1/\epsilon)}, \tag{12.124}$$

or

$$\mathcal{D}(A) = \lim_{\epsilon \to 0} \frac{C_\epsilon(A)}{\log(1/\epsilon)}, \tag{12.125}$$

provided that these limits exist, or otherwise defining the corresponding upper and lower metric dimensions using lim sup and lim inf, respectively. These definitions are due to Kolmogorov, and (12.124) is also known as the Minkowski–Bouligand, or fractal, dimension used to determine the "degree of fractality" of a set. We have encountered this quantity in Section 3.6.2 in the context of the reconstruction of multi-band signals from linear measurements. This deterministic quantity should also be compared with the analogous information dimension (3.174) introduced by Alfred Rényi in a stochastic setting, based on the Shannon entropy, and discussed in Section 3.7.3.

It is easy to see that, by virtue of (12.116) and (12.117), for any bounded subset of $\mathbb{R}^d$ we have

$$\mathcal{D}(A) = d, \tag{12.126}$$

and the metric dimension coincides with the dimension of the space. On the other hand, for infinite-dimensional convex sets the order of growth of $H_\epsilon(A)$ and $C_\epsilon(A)$ always exceeds $\log(1/\epsilon)$, and the metric dimension diverges. Infinite-dimensional sets are "informationally abundant." For this reason, Kolmogorov concludes:

> Metric dimension is useless for distinguishing the massiveness of sets of this sort. (Kolmogorov and Tikhomirov, 1959)

To quantify the amount of information of informationally abundant sets, Kolmogorov then suggested considering alternative notions of metric dimension.

12.7.2 Functional Dimension and Metric Order

One example of an informationally abundant set is an infinite-dimensional compact set of finite smoothness. The growth of the ϵ-entropy and capacity is typically a power of $(1/\epsilon)$ with an exponent that decreases as the supply of smoothness of A increases. This confirms the intuition that the faster $H_\epsilon(A)$ tends to infinity as $\epsilon \to 0$, the more complicated the structure of A is.

Stated more precisely, for any infinite-dimensional set A of functions of s variables defined on a unit hypercube with uniformly bounded partial derivatives up to order r, Kolmogorov showed that

$$H_\epsilon(A) = O\left(\frac{1}{\epsilon}\right)^{s/r} \quad \text{as } \epsilon \to 0, \tag{12.127}$$

$$C_\epsilon(A) = O\left(\frac{1}{\epsilon}\right)^{s/r} \quad \text{as } \epsilon \to 0. \tag{12.128}$$

On the other hand, he also showed that the order of magnitude of the ϵ-entropy and ϵ-capacity for analytic functions is essentially different, being only

$$H_\epsilon(A) = O[\log(1/\epsilon)]^{s+1} \quad \text{as } \epsilon \to 0, \tag{12.129}$$

$$C_\epsilon(A) = O[\log(1/\epsilon)]^{s+1} \quad \text{as } \epsilon \to 0. \tag{12.130}$$

When viewed together, these results show that:

The information content of analytic functions is much less than that of functions of finite smoothness.

For these functional classes, the most natural measures of information growth are what Kolmogorov calls the *functional dimension*,

$$\mathcal{F}(A) = \lim_{\epsilon \to 0} \frac{\log H_\epsilon(A)}{\log \log(1/\epsilon)}, \tag{12.131}$$

and the *metric order*,

$$\mathcal{M}(A) = \lim_{\epsilon \to 0} \frac{\log H_\epsilon(A)}{\log(1/\epsilon)}. \tag{12.132}$$

By (12.129) it follows that the functional dimension of an analytic function is roughly the number of variables, or, geometrically, the dimension of the manifold where the function is supported. For smooth functions, by (12.127) we have that the metric order corresponds to the exponent of smoothness s/r.

A derivation of a slightly weaker upper bound than (12.127) for smooth functions can be obtained easily. Consider a function of one variable with bounded derivatives up to order r defined on the unit interval $[0, 1]$. Divide the interval into equal subintervals of size $\epsilon^{1/r}$ and compute the value of the function and its derivatives up to order r with accuracy ϵ at the endpoints of each subinterval. One can then compute the value of the function at any point with accuracy of the order of ϵ by finding the subinterval containing the given point and using the rth-order Taylor expansion about its left end point. In this way, the remainder of the Taylor expansion is of the order of ϵ for any point in $[0, 1]$. Now, to specify one derivative at a given point within accuracy ϵ we need of the order of $1/\epsilon$ numbers, and $(1/\epsilon)^{r+1}$ to specify all derivatives. Since the total number of points is $\epsilon^{-1/r}$, we need a total number of $(1/\epsilon)^{(r+1)\epsilon^{1/r}}$ numbers to compute any value of the function in $[0, 1]$ within accuracy ϵ. Taking the logarithm, this leads to an upper bound on the ϵ-entropy of the order of $(1/\epsilon)^{1/r} \log 1/\epsilon$. The correct upper bound, without the additional logarithmic factor, can be obtained using a more accurate calculation.

12.7.3 Infinite-Dimensional Spaces

For spaces where the ϵ-entropy and ϵ-capacity are infinite, we can consider a sequence of finite-dimensional approximating subspaces, and provide asymptotic results on the order of growth of the dimensionality as the accuracy of the approximation increases. Following this approach, we argued in Chapter 3 that for bandlimited signals the effective dimensionality corresponds to the number of degrees of freedom of the space, as the size of the observation interval $T \to \infty$. We then extend this notion to multi-band signals, considering the Minkowski–Bouligand fractal dimension in place of the number of degrees of freedom. We now follow the approximation approach to determine the order of growth of the ϵ-entropy and ϵ-capacity of bandlimited signals as $T \to \infty$.

12.8 Bandlimited Signals

We consider the entropy and capacity of the infinite-dimensional space of bandlimited signals. In this case, all the results obtained in Shannon's probabilistic setting have direct analogs in a deterministic setting. We first summarize the geometric insight that forms the basis of the two models, and then give the details of the mathematical results.

12.8.1 Capacity and Packing

Shannon's capacity is closely related to the problem of geometric packing of "billiard balls" in high-dimensional space. Roughly speaking, each transmitted signal, represented by the coefficients of an orthonormal basis expansion, corresponds to a point in the space, and balls centered at the transmitted points represent the probability density of the uncertainty of the observation performed at the receiver. A certain amount of overlap between the balls is allowed to construct dense packings corresponding to codebooks of high capacity, as long as the overlap does not include typical noise concentration regions, and this allows us to achieve reliable communication with vanishing probability of error.

Similarly, in Kolmogorov's deterministic setting, communication between a transmitter and a receiver occurs without error, balls of fixed radius ϵ representing the uncertainty introduced by the noise about each transmitted signal are not allowed to overlap, and his notion of 2ϵ-capacity corresponds to the Shannon capacity of the ϵ-bounded noise channel with the probability of error equal to zero.

We now wish to introduce a notion of vanishing error in a deterministic setting as well. With this goal in mind, we allow a certain amount of overlap between the ϵ-balls. In a deterministic setting, a codebook is composed of a subset of waveforms in the space, each corresponding to a given message. A transmitter can select any one of these signals, which is observed at the receiver with perturbation at most ϵ. If signals in the codebook are at distance less than 2ϵ of each other, then a decoding error may occur due to the overlap region between the corresponding ϵ-balls. The total volume of the error region, normalized by the total volume of the ϵ-balls in the codebook, represents a measure of the fraction of the space where the received signal may fall and result in a communication error. We then define the (ϵ, δ)-capacity as the logarithm base two of the largest number of signals that can be placed in a codebook having a normalized error region of size at most δ. It turns out that the (ϵ, δ)-capacity grows linearly with the number of degrees of freedom, but only logarithmically with the signal-to-noise ratio, and, when taken per unit time, as in the case of Shannon's capacity, it is independent of the value of δ. This was Shannon's original insight, expressed by (12.35), which remains invariant when the model is subject to a deterministic formulation.

12.8.2 Entropy and Covering

Shannon's rate–distortion function is closely related to the geometric problem of covering a high-dimensional space with balls of a given radius. Roughly speaking, each source signal, modeled as a stochastic process, corresponds to a random point in the space, and a set of covering balls provides a possible representation of any signal in the space. By representing the signal with the closest center of a ball covering it, the distance to the center of the ball represents the distortion measure associated with this representation. Shannon's rate–distortion function provides the minimum number of bits that must be specified per unit time to represent the source process with a given average distortion.

In Kolmogorov's deterministic setting, the ϵ-entropy is the logarithm of the minimum number of balls of radius ϵ needed to cover the whole space and, when taken per unit time, it corresponds to the Shannon rate–distortion function, as it also represents the minimum number of bits that must be specified per unit time to represent any source signal with distortion at most ϵ. It turns out that the ϵ-entropy of bandlimited functions grows linearly with the number of degrees of freedom and logarithmically with the ratio of the norm of the signal to the norm of the distortion. Once again, this was Shannon's key insight, expressed by (12.79), which remains invariant when the model is subject to a deterministic formulation.

12.8.3 ϵ-Capacity of Bandlimited Signals

We consider waveforms $f \in \mathscr{B}_\Omega$, namely bandlimited, one-dimensional, real, scalar waveforms of a single scalar variable and supported over an angular frequency interval $[-\Omega, \Omega]$. We assume that waveforms are square-integrable, and satisfy the energy constraint

$$\int_{-\infty}^{\infty} f^2(t)dt \leq E. \tag{12.133}$$

These bandlimited waveforms have unbounded time support, but are observed over a finite interval $[-T/2, T/2]$ and equipped with the norm

$$\|f\| = \left(\int_{-T/2}^{T/2} f^2(t)dt \right)^{1/2}. \tag{12.134}$$

By the results in Chapter 3 it follows that any signal can be expanded in terms of a suitable set of orthonormal basis functions, and for T large enough it can be seen as a point in a space of essentially

$$N_0 = \Omega T/\pi \tag{12.135}$$

dimensions, corresponding to the number of degrees of freedom of the waveform, and of radius $\sqrt{E}$.

We consider an uncertainty sphere of radius ϵ centered at each signal point, representing the energy of the noise that is added to the observed waveform. In this model, due to Kolmogorov, the signal-to-noise ratio is

$$\mathsf{SNR}_\mathsf{K} = E/\epsilon^2. \tag{12.136}$$

A codebook is composed of a subset of waveforms in the space, each corresponding to a given message. A transmitter can select any one of these signals, which is observed at the receiver with perturbation at most ϵ. By choosing signals in the codebook to be at a distance of at least 2ϵ from each other, the receiver can decode the message without error. The 2ϵ-capacity is the logarithm base two of the maximum number $M_{2\epsilon}(E)$ of distinguishable signals in the space. This corresponds geometrically to the maximum number of disjoint balls of radius ϵ with their centers situated inside the signals' space, and it is given by

$$C_{2\epsilon} = \log M_{2\epsilon}(\mathscr{B}_\Omega) \text{ bits.} \tag{12.137}$$

We also define the capacity per unit time:

$$\bar{C}_{2\epsilon} = \lim_{T\to\infty} \frac{\log M_{2\epsilon}(\mathscr{B}_\Omega)}{T} \text{ bits per second.} \tag{12.138}$$

Comparing this communication model with Shannon's as described in Section 12.2, it is clear that the geometric insight on which each is built is the same. However, while in Kolmogorov's deterministic setting packing is performed with "hard" spheres of radius ϵ and communication in the presence of arbitrarily distributed noise over a bounded support is performed without error, in Shannon's stochastic model packing is performed with "soft" spheres of effective radius $\sqrt{NN_0}$ and communication in the presence of Gaussian noise of unbounded support is performed with arbitrarily low probability of error δ.

Shannon's energy constraint (12.2) scales with the number of dimensions, rather than being a constant. The reason for this should be clear: if the noise is assumed to act independently on each signal's coefficient, the statistical spread of the output, given the input signal, corresponds to an uncertainty ball of radius $\sqrt{NN_0}$. It follows that the norm of the signal should also be proportional to $\sqrt{N_0}$, to avoid a vanishing signal-to-noise ratio as $N_0 \to \infty$. In contrast, in the case of Kolmogorov the capacity is computed assuming an uncertainty ball of fixed radius ϵ and the energy constraint is constant. In both cases, spectral concentration ensures that the size of the signals' space is essentially of N_0 dimensions. Probabilistic concentration ensures that the noise in Shannon's model concentrates around its standard deviation, so that the functional form of the capacity is similar in the two cases, and we have

$$\begin{cases} \bar{C}_{2\epsilon} \le \dfrac{\Omega}{\pi} \log\left(1 + \sqrt{\mathsf{SNR}_\mathrm{K}/2}\right) \text{ bits per second,} & (12.139) \\[2ex] \bar{C}_{2\epsilon} \ge \dfrac{\Omega}{\pi} \left(\log\sqrt{\mathsf{SNR}_\mathrm{K}} - 1\right) \text{ bits per second.} & (12.140) \end{cases}$$

When comparing (12.139) and (12.140) with (12.35), it is important to keep in mind that while Shannon's formula refers to communication with vanishing probability of error over a Gaussian channel, Kolmogorov's formula refers to communication with zero error over a bounded noise channel. To provide a more precise comparison, we extend the deterministic model introducing the possibility of having a decoding error by allowing signals in the codebook to be at a distance of less than 2ϵ from each other.

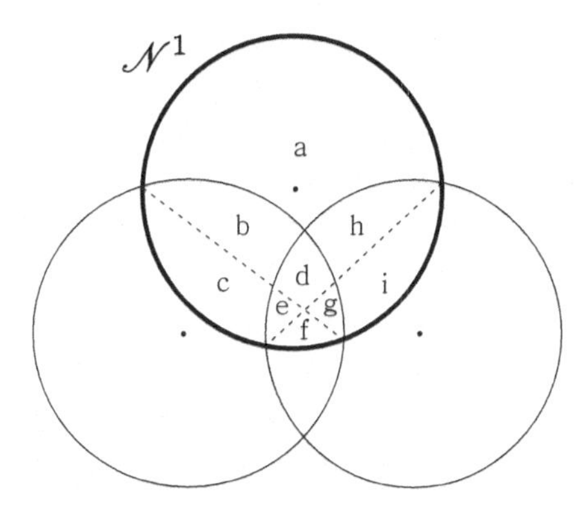

Fig. 12.16 Illustration of the error region for a signal in the space. The letters indicate the volumes of the corresponding regions of the ball $\mathscr{N}^1$, and $\Delta_1 = (c+e+f+g+i)\,/\,(a+b+c+d+e+f+g+h+i)$.

This yields the notion of (ϵ,δ)-capacity that is the analog of Shannon's capacity in a deterministic setting.

12.8.4 (ϵ,δ)-Capacity of Bandlimited Signals

For any $y_i \in \{y_1, y_2, \ldots, y_m\} \subset \mathscr{B}_\Omega$, we let the noise ball be

$$\mathscr{N}^i = \{x : \|x - y_i\| \leq \epsilon\}, \tag{12.141}$$

where ϵ is a positive real number, and we let error region be

$$\mathscr{D}^i = \{x \in \mathscr{N}^i : \exists j \neq i : \|x - y_j\| \leq \|x - y_i\|\}. \tag{12.142}$$

We then define the error measure for the ith signal,

$$\Delta_i = \frac{\mathrm{vol}(\mathscr{D}^i)}{\mathrm{vol}(\mathscr{N}^i)}, \tag{12.143}$$

where $\mathrm{vol}(\cdot)$ indicates volume in $\mathscr{X}$, and the cumulative error measure

$$\Delta = \frac{1}{M}\sum_{i=1}^{M} \Delta_i. \tag{12.144}$$

Figure 12.16 provides an illustration of the error region for a given signal in the space.

Clearly, we have $0 \leq \Delta \leq 1$. For any $\delta > 0$, a set of points in $\mathscr{B}_\Omega$ is an (ϵ,δ)-distinguishable set if $\Delta \leq \delta$. The maximum cardinality of an (ϵ,δ)-distinguishable set is an invariant of the space, which depends only on ϵ and δ, and is denoted by $M_\epsilon^\delta(\mathscr{B}_\Omega)$. The (ϵ,δ)-capacity of $\mathscr{B}_\Omega$ is defined as the base two logarithm

$$C_\epsilon^\delta = \log M_\epsilon^\delta(\mathscr{B}_\Omega) \text{ bits}; \tag{12.145}$$

see Figure 12.17. We also define the (ϵ,δ)-capacity per unit time:

$$\bar{C}_\epsilon^\delta = \lim_{T\to\infty} \frac{\log M_\epsilon^\delta(\mathscr{B}_\Omega)}{T} \text{ bits per second}. \tag{12.146}$$

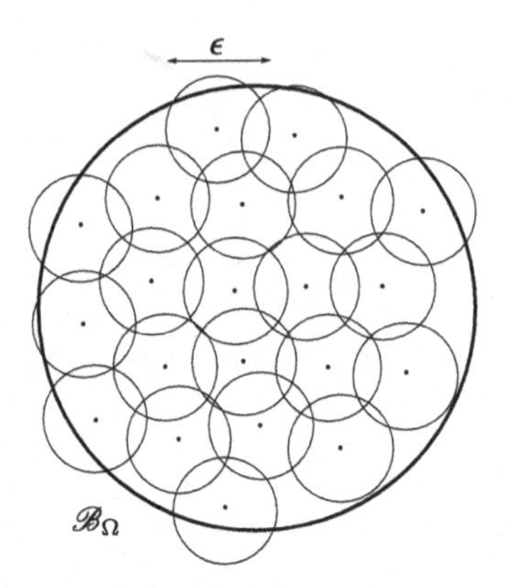

Fig. 12.17 Illustration of the (ϵ,δ)-capacity of bandlimited functions. An overlap among the ϵ-balls is allowed, provided that the cumulative error measure $\Delta \leq \delta$.

In this case, we have, for any $\epsilon, \delta > 0$,

$$\begin{cases} \bar{C}_\epsilon^\delta \leq \dfrac{\Omega}{\pi} \log\left(1+\sqrt{\mathsf{SNR}_\mathrm{K}}\right) \text{ bits per second,} & (12.147) \\ \bar{C}_\epsilon^\delta \geq \dfrac{\Omega}{\pi} \log\sqrt{\mathsf{SNR}_\mathrm{K}} \text{ bits per second.} & (12.148) \end{cases}$$

As in Shannon's case, these results do not depend on the size of the error region δ, and are the deterministic analog of (12.35).

12.8.5 ϵ-Entropy of Bandlimited Signals

In the case of ϵ-entropy, a source codebook is composed of a subset of points in the space, and each codebook point is a possible representation for the signals that are within radius ϵ of itself. If the union of the ϵ balls centered at all codebook points covers the whole space, then any signal in the space can be encoded by its closest representation. The radius ϵ of the covering balls provides a bound on the largest estimation error between any source $f(t)$ and its codebook representation $\hat{f}(t)$. When signals are observed over a finite time interval $[-T/2, T/2]$, this corresponds to

$$d(f,\hat{f}) = \int_{-T/2}^{T/2} [f(t) - \hat{f}(t)]^2 dt \leq \epsilon^2. \qquad (12.149)$$

The signal-to-distortion ratio in this source coding model is $\mathsf{SNR}_\mathrm{K} = \sqrt{E}/\epsilon$.

The Kolmogorov ϵ-entropy is the logarithm base two of the minimum number $Q_\epsilon(E)$ of ϵ-balls covering the whole space, and it is given by

$$H_\epsilon = \log Q_\epsilon(E) \text{ bits.} \qquad (12.150)$$

Table 12.2 Comparison of stochastic and deterministic models.

	Stochastic	Deterministic
Signal constraint	$\int_{-T/2}^{T/2} f^2(t)dt \leq PN_0$	$\int_{-\infty}^{\infty} f^2(t)dt \leq E$
Additive noise	$\mathbb{E}\sum_{i=1}^{N_0} \mathsf{Z}_i^2 = NN_0$	$\sum_{i=1}^{\infty} z_i^2 \leq \epsilon^2$
Signal-to-noise ratio	$\mathsf{SNR_S} = P/N$	$\mathsf{SNR_K} = E/\epsilon^2$
Capacity	$C = \frac{\Omega}{\pi}\log(\sqrt{1+\mathsf{SNR_S}})$	$\bar{C}_\epsilon^\delta \leq \frac{\Omega}{\pi}\log(1+\sqrt{\mathsf{SNR_K}})$
		$\bar{C}_\epsilon^\delta \geq \frac{\Omega}{\pi}\log\sqrt{\mathsf{SNR_K}}$
Source constraint	$\int_{-T/2}^{T/2} \mathbb{E}(\mathsf{f}^2(t))dt = PN_0$	$\int_{-\infty}^{\infty} f^2(t)dt \leq E$
Distortion	$d(\mathsf{f},\hat{\mathsf{f}}) \leq N_0 N$	$d(\mathsf{f},\hat{f}) \leq \epsilon^2$
Rate–distortion function	$R_N = \frac{\Omega}{\pi}\log\sqrt{\mathsf{SNR_S}}$	$\bar{H}_\epsilon = \frac{\Omega}{\pi}\log\sqrt{\mathsf{SNR_K}}$

We also define the ϵ-entropy per unit time:

$$\bar{H}_\epsilon = \lim_{T\to\infty} \frac{\log Q_\epsilon(E)}{T} \text{ bits per second.} \tag{12.151}$$

In this case, we have

$$\bar{H}_\epsilon = \begin{cases} \dfrac{\Omega}{\pi}\log\left(\sqrt{\mathsf{SNR_K}}\right) \text{ bits per second} & \text{if } E \geq N, \\ 0 & \text{if } E < N, \end{cases} \tag{12.152}$$

which is the deterministic analog of (12.79).

12.8.6 Comparison with Stochastic Quantities

Table 12.2 provides a comparison between the results in the deterministic and stochastic settings. In the computation of capacity, a transmitted signal subject to a given energy constraint is corrupted by additive noise. Due to spectral concentration, the signal has an effective number of dimensions N_0. In a deterministic setting, the noise represented by the deterministic coordinates $\{z_i\}$ can take any value inside a ball of radius ϵ. In a stochastic setting, due to probabilistic concentration, the noise represented by the stochastic coordinates $\{\mathsf{Z}_i\}$ can take values essentially uniformly at random inside a ball of effective radius NN_0. In both cases, the maximum cardinality of the codebook used for communication depends on the error measure $\delta > 0$, but the capacity in bits per unit time does not, and it depends only on the signal-to-noise ratio. The special case $\delta = 0$ does not appear in the table and corresponds to the Kolmogorov 2ϵ-capacity, the analog of the Shannon zero-error capacity of an ϵ-bounded noise channel.

In the computation of the rate–distortion function, a source signal is modeled as either an arbitrary or a stochastic process of given energy constraint. The distortion measure corresponds to the estimation error incurred when this signal is represented by an element of the source codebook. The minimum cardinality of the codebook used for representation depends on the distortion constraint, and so does the rate–distortion function.

The functional form of the results indicates that in both Kolmogorov's and Shannon's settings, capacity and entropy grow linearly with the number of degrees of freedom, but only logarithmically with the signal-to-noise ratio. This was Shannon's original insight, which transcends the details of the stochastic or deterministic description of the information-theoretic model.

12.9 Spatially Distributed Systems

We now extend the capacity results to communication with spatially distributed transmitters and receivers. These results can be obtained using the singular value decomposition described in Section 5.3.3 that provides the number of spatial degrees of freedom per unit frequency in terms of the rank of the Green's matrix $\mathrm{G} = \{\mathrm{G}_{i,k}(\omega)\}$ connecting transmitters and receivers. To obtain the total capacity, we sum the rate contributions of all spatial channels expressed in bits per time–frequency degree of freedom and given by Shannon's formula (12.34). The derivation follows the usual steps, using an orthogonal decomposition of the signal in space rather than in frequency, and with each spatial channel contributing to the capacity by a value given by (12.34).

12.9.1 Capacity with Channel State Information

We assume a deterministic, time-invariant setting, where the transmitters have knowledge of the Green's function, so that they can optimally allocate energy over the different spatial channels. By (5.41), each parallel channel corresponds to a non-zero singular value of the matrix $\mathrm{G}(\omega)$, and the total number of channels per unit frequency of transmission is physically limited by the number of spatial degrees of freedom per unit frequency $N_0(\omega)$ of the radiating system, as studied in Chapter 8.

We consider adding an independent, zero-mean, Gaussian noise random variable of standard deviation N on each spatial channel in (5.41), obtaining

$$\widetilde{Y}_n = \sigma_n \widetilde{X}_n + z_n, \quad n = 1, \ldots, N_0(\omega). \tag{12.153}$$

We also assume a communication system with a number of transmitter–receiver pairs equal to the number of spatial degrees of freedom $N_0(\omega)$, and that the transmitters are subject to an energy constraint that scales linearly with the number of degrees of freedom:

$$\sum_{n=1}^{N_0(\omega)} \widetilde{X}_n^2 \leq P N_0(\omega). \tag{12.154}$$

The signal-to-noise ratio per unit frequency of transmission is then given by

$$\mathsf{SNR}_\mathsf{S} = \frac{P N_0(\omega)}{N N_0(\omega)} = \frac{P}{N}. \tag{12.155}$$

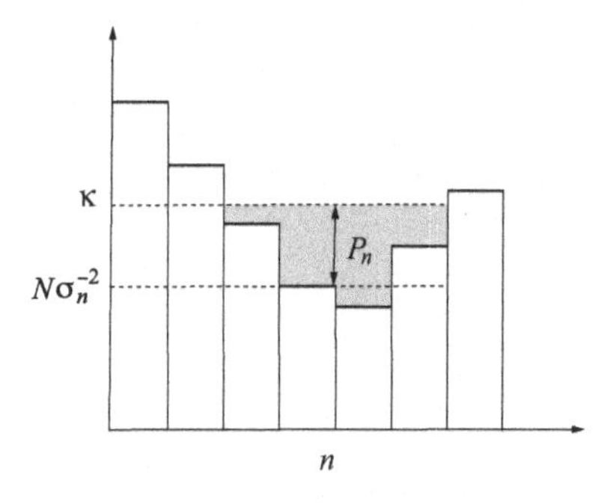

Fig. 12.18 Illustration of water filling in the discrete spatial setting.

In this case, the Shannon capacity, in bits per (time–frequency) degree of freedom, equals the supremum of the sum of the achievable rates from the transmitters to the receivers, exploiting all the spatial degrees of freedom available for communication at any given frequency.

To compute the capacity, we observe that the scaling of the signal $\widetilde{X}_n$ by σ_n in each channel of (12.153) is equivalent to adding an amount of noise that is inversely proportional to the corresponding singular value. The capacity can then be obtained using an optimal power allocation similar to the one described in Section 12.2.7, where water filling is performed in space rather than in the frequency domain. The solution corresponds to a value of the power analogous to (12.54),

$$P_n = [\kappa - N\sigma_n^{-2}]_+, \tag{12.156}$$

with κ satisfying

$$\sum_{n=1}^{N_0(\omega)} [\kappa - N\sigma_n^{-2}]_+ = PN_0(\omega), \tag{12.157}$$

and the capacity in bits per (time–frequency) degree of freedom is obtained by summing over all spatial channels,

$$\begin{aligned} C &= \sum_{n=1}^{N_0(\omega)} \frac{1}{2}\log\left(1 + \frac{\sigma_n^2 P_n}{N}\right) \\ &= \sum_{n=1}^{N_0(\omega)} \frac{1}{2}\log\left(1 + \frac{\sigma_n^2[\kappa - N\sigma_n^{-2}]_+}{N}\right), \end{aligned} \tag{12.158}$$

which is the analog of (12.56), where the noise here is modulated by the singular values of the channel over space rather than over frequency. An illustration of the water filling solution is depicted in Figure 12.18.

An upper bound on the capacity is obtained using Jensen's inequality,

$$\frac{\sum_{n=1}^{N_0(\omega)} \log x_n}{N_0(\omega)} \le \log\left(\frac{\sum_{n=1}^{N_0(\omega)} x_n}{N_0(\omega)}\right), \tag{12.159}$$

yielding

$$\begin{aligned} C &= \sum_{n=1}^{N_0(\omega)} \frac{1}{2} \log\left(1 + \frac{P_n \sigma_n^2}{N}\right) \\ &\leq \frac{N_0(\omega)}{2} \log\left[\frac{1}{N_0(\omega)}\left(N_0(\omega) + \frac{\sum_{n=1}^{N_0(\omega)} P_n \sigma_n^2}{N}\right)\right] \\ &= \frac{N_0(\omega)}{2} \log\left(1 + \frac{\sum_{n=1}^{N_0(\omega)} P_n \sigma_n^2}{N N_0(\omega)}\right). \end{aligned} \tag{12.160}$$

By comparing(12.160) with (12.34), we observe a *rate gain* due to the $N_0(\omega)$ spatial channels, while the second term inside the logarithm corresponds to the signal-to-noise ratio at the receivers, with the power on each channel being scaled by the corresponding singular value. The bound is tight when the singular values are all equal, and in this case the optimal power allocation corresponds to having equal power on all channels. This suggests the general principle that the less spread out the singular values are, the larger the capacity. By defining the condition number of the matrix G as the ratio between the maximum and the minimum singular values, we have that:

> Multiple antenna systems have higher capacities in well-conditioned propagation environments.

On the other hand, by the results in Chapter 8 it follows that the number of degrees of freedom in any propagation environment is bounded by the size of the minimum cut-set boundary containing all the transmitters, and the capacity cannot increase beyond this value imposed by the physics of the propagation process. Any model of the Green's function that violates this constraint is a physical impossibility. Furthermore, since by (5.38) the number of degrees of freedom is at most equal to the minimum between the number of transmitter and receiver antennas, increasing the number of antennas cannot increase capacity beyond the number of degrees of freedom. Finally, if the singular values are all close to one, then the capacity becomes

$$C = \frac{N_0(\omega)}{2} \log\left(1 + \mathsf{SNR}_S\right), \tag{12.161}$$

and the rate gain of $N_0(\omega)$ parallel channels is now directly comparable with the single point-to-point Shannon formulas (12.33) and (12.34). In (12.33), N_0 refers to the number of time–frequency degrees of freedom of a point-to-point channel, and the capacity is measured in bits; (12.34) gives the corresponding formula in bits per time–frequency degree of freedom. On the other hand, in (12.161), $N_0 = N_0(\omega)$ refers to the number of spatial–wavenumber degrees of freedom available at every frequency, and the capacity is measured in bits per time–frequency degree of freedom. It follows that the spatial system provides a rate gain equal to the number of spatial–wavenumber degrees of freedom.

All of the above results depend on the power constraint (12.154), which has been assumed to grow proportionally to the number of degrees of freedom. On the other hand, in the presence of a fixed energy constraint

$$\sum_{n=1}^{N_0(\omega)} \widetilde{X}_n^2 \leq E, \tag{12.162}$$

we need to choose κ satisfying

$$\sum_{n=1}^{N_0(\omega)} [\kappa - N\sigma_n^{-2}]_+ = E, \tag{12.163}$$

and for equal power allocation over unitary channels, we get

$$C = \frac{N_0(\omega)}{2} \log\left(1 + \frac{E}{N_0(\omega)N}\right). \tag{12.164}$$

In this case, as $N_0(\omega) \to \infty$ we enter a vanishing signal-to-noise ratio regime, and the capacity tends to the limiting value

$$C = \frac{E}{2N} \log_2 e. \tag{12.165}$$

Comparing (12.165) with (12.161), it follows that, consistent with the physics, increasing the capacity by increasing the number of degrees of freedom requires energy expenditure. For fixed energy systems, as the number of degrees of freedom increases, what matters eventually is only the total energy expenditure.

12.9.2 Capacity without Channel State Information

In a time-invariant setting, the Green's function can easily be estimated by sending a pilot signal and measuring it at the receivers. The estimate can then be sent back to the transmitters to perform power allocation as discussed above. When the Green's function is modeled as randomly time-varying, however, this estimation process may not be completed in a time where the channel is constant and all we can hope for is to have channel state information at the receivers only. The best strategy in this case is to allocate equal power on all channels, and the capacity, computed averaging over the channel variations over time, is

$$\begin{aligned} C &= \frac{1}{2} \sum_{n=1}^{N_0(\omega)} \mathbb{E}\left[\log\left(1 + \mathsf{SNR}_\mathsf{S} \sigma_n^2\right)\right] \\ &\leq \frac{N_0(\omega)}{2} \mathbb{E}\left[\log\left(1 + \frac{\mathsf{SNR}_\mathsf{S} \sum_{n=1}^{N_0} \sigma_n^2}{N_0}\right)\right]. \end{aligned} \tag{12.166}$$

Clearly, the capacity now depends on the statistical properties of the singular values. If the stochastic Green's function is statistically well conditioned, then the rate gain equals the number of degrees of freedom, and the system achieves its full multiplexing capability. In this case, the model chosen for the Green's function is crucial to determining the value of the capacity, and the use of random matrix models in the analysis of communication systems has become widespread. Probabilistic estimates of the eigenvalues of these matrices are a key tool for capacity predictions in different scenarios. However, one should keep in mind that the chosen stochastic model must always be consistent with physical reality, predicting a multiplexing gain no larger than what can be physically supported by the channel.

12.10 Summary and Further Reading

Shannon's model of communication using bandlimited signals was introduced in Shannon (1949). The rigorous formulation of his capacity result for continuous signals is given by Wyner (1965). Close examination of these results shows that refined noise models are required to account for additional quantum constraints and super-resolution effects occurring at high frequencies. A more accurate physical model assumes Gaussian noise that vanishes at high frequencies, where quantum noise kicks in. In this case, there is a contribution to capacity associated with the continuous wave nature of the signal that follows Shannon's classic formula, and a contribution associated with the discrete particle nature of the signal that depends on the details of the quantum noise model, which is dominant at high frequencies. The book by Papen and Blahut (2018) gives a compendium of additional results on the capacity of electromagnetic channels subject to quantum constraints in different regimes.

In the absence of noise, and in the presence of quantized radiation constraints, the formula (12.48) has been obtained by Lebedev and Levitin (1966), Bowen (1967), and Pendry (1983), following previous work by Gordon (1962) and Stern (1960). Lachmann, Newman, and Moore (2004) point out that the formula corresponds to the entropy rate of black body radiation in one dimension. We studied black body radiation in Chapter 11 using a statistical physics approach. Different derivations of the capacity formula (12.48) are reviewed by Caves and Drummond (1994). A general derivation considers all possible representations of the radiated signal that include non-orthogonal basis sets and arbitrary quantum measurement operators. Using a theorem by Holevo (1973), extended to infinite-dimensional systems by Yuen and Ozawa (1993), it uses the full machinery of quantum mechanics to confirm the capacity bound (12.48) in a general setting.

Parallel to Shannon's theory, there is a deterministic theory of information that originates from the Russian school of Kolmogorov. The notions of ϵ-entropy and ϵ-capacity were introduced by Kolmogorov (1956). His functional theory of information is reviewed in Kolmogorov and Tikhomirov (1959). The basic relations with Shannon's theory of information discussed in Section 12.6 appear in Appendix II of that paper. Bounds on the ϵ-entropy and ϵ-capacity of bandlimited functions were first given by Jagerman (1969, 1970) and Wyner (1973). We presented some improved versions obtained by Lim and Franceschetti (2017b), who also introduced the concept of (ϵ,δ)-capacity. The connection between deterministic and stochastic models is discussed by Donoho (2000).

The capacity of systems composed of multiple antennas is a well-studied topic in communication theory. Comprehensive treatments can be found in the books by Biglieri (2005), Goldsmith (2005), and Tse and Viswanath (2005). Relationships with random matrix theory are described by Tulino and Verdú (2004).

12.11 Test Your Understanding

Problems

12.1 Check that SNR_S in (12.37) and (12.41) is in dimensionless units.

12.2 Check that the capacity in (12.46) is in bits per second.

12.3 Write the expression for the Shannon capacity of the additive white Gaussian noise channel subject to the following power constraints:

$$\int_{-T/2}^{T/2} f^2(t) \leq PN_0, \tag{12.167}$$

$$\int_{-T/2}^{T/2} f^2(t) \leq \bar{P}T, \tag{12.168}$$

$$\int_{-T/2}^{T/2} f^2(t) \leq E, \tag{12.169}$$

where E, P, and $\bar{P}$ are constants.

12.4 Rewrite (12.38) in bits per second and compare it with (12.42).

12.5 Consider a signal subject to the energy per degree of freedom constraint (12.167). Show the following quantum limitation and compare it to (12.45):

$$C \ll \frac{P}{2\pi\hbar}\log e \text{ bits per second.} \tag{12.170}$$

12.6 In Sections 12.2 and 12.2.3 we stated that the energy constraints (12.2) and (12.39) "essentially" correspond to (12.11) and (12.40). Discuss the kind of approximation that has been made in these statements.

12.7 Provide a lower bound on the probability of error in an additive white Gaussian noise channel for any achievable rate $R < C$ in terms of T, C, and R.

12.8 Explain the difference between a probabilistic proof for the existence of a capacity-achieving code by showing that a random codebook selection implies the selection of a capacity-achieving codebook with positive probability, and the proof given in Section 12.2.2.

12.9 Check that the units of (12.45) and (12.48) are bits per second.

12.10 Provide a geometric argument that justifies (12.122) and (12.123).

12.11 Show that for high values of SNR_K the Kolmogorov 2ϵ-capacity is

$$(\Omega/\pi)(\log\sqrt{\mathsf{SNR}_\mathrm{K}} - 1) \le \bar{C}_{2\epsilon} \le (\Omega/\pi)(\log\sqrt{\mathsf{SNR}_\mathrm{K}} - 1/2). \tag{12.171}$$

12.12 Show that if $\sqrt{\mathsf{SNR}_\mathrm{K}} \ge \sqrt{2}/(\sqrt{2}-1)$ then $\bar{C}^\delta_\epsilon > \bar{C}_{2\epsilon}$.

12.13 Provide an intuitive justification for the factor $1+\sqrt{\mathsf{SNR}_\mathrm{K}}$ inside the logarithm in (12.147), compared to $\sqrt{1+\mathsf{SNR}_\mathrm{S}}$ in Shannon's setting.

12.14 Verify that the capacity in (12.166) can also be written as

$$C = \frac{1}{2}\,\mathbb{E}\big[\log\,\det\left(\mathrm{I} + \mathsf{SNR}_\mathrm{S}\mathrm{G}\mathrm{G}^\dagger\right)\big]. \tag{12.172}$$

13 Universal Entropy Bounds

Lo! thy dread Empire, CHAOS! is restor'd;
Light dies before thy uncreating word:
Thy hand, great Anarch! lets the curtain fall;
And Universal Darkness buries All.[1]

13.1 Bandlimited Radiation

In Chapter 11, we modeled black body radiation as a stochastic process radiating at all frequencies and having a Gaussian distribution whose power spectral density has a vanishing tail according Planck's law. The entropy of black body radiation is proportional to $(r\bar{E})^{3/4}$, where $\bar{E}$ is the average energy of the radiating body and r is its radius. This result follows by imposing quantized radiation at all frequencies, while keeping the total average energy of the body finite. The average energy is proportional to the volume, so that radiation from a black body carries an amount of information proportional to its volume. In the one-dimensional case, black body radiation achieves the capacity of the noiseless photon channel expressed by (12.48).

Radiation over all frequencies, however, is an idealization as it allows emission of photons of arbitrarily high energy. Even though highly energetic photons are weighted by a small probability of emission to avoid the "ultraviolet catastrophe," transient radiation of unbounded energy is still possible. A different point of view adopted in this chapter starts with the premise that any radiated signal must be a strictly bandlimited function, and computes entropy bounds by imposing quantized radiation constraints in this setting.

We consider both a deterministic and a stochastic model of radiation. Deterministic signals of energy at most E can be arbitrarily distributed over a finite frequency bandwidth, and stochastic signals of expected energy at most $\bar{E}$ are assumed to have a Gaussian distribution whose power spectral density is constant over a finite bandwidth. By combining the results on the degrees of freedom in Chapter 9 with quantized radiation constraints, and using the entropy formulas derived in Chapter 12, we show that in a three-dimensional setting of radiation from a sphere of radius r, the entropy is proportional to (rE), and it achieves the universal bound (11.82). This bound is tight when the energy is proportional to r^2, so that the maximum entropy of bandlimited

[1] Alexander Pope (1743). *The Dunciad*, Bk. IV, L. 649.

radiation is also proportional to the volume of the radiating body. However, the energy expenditure to achieve this bound scales with the surface of the radiating body, while for black body radiation the expected energy scales with the volume of the body.

13.2 Deterministic Signals

Our derivation follows from the three constraints highlighted at the beginning of the book: an energy constraint, a dimensionality (or bandlimitation) constraint, and a resolution (or quantization) constraint.

We first consider arbitrary square-integrable, bandlimited waveforms propagating with velocity c from the interior of a sphere of radius r to the outside space. The waveform on the surface of the sphere is identified by one temporal coordinate t and two angular ones (θ, ϕ). The propagating signal is assumed to be the image of sources internal to the sphere, impressed or induced through multiple scattering, through the Green's free space propagation operator, and is observed on the external surface of the sphere for a time interval of duration $T = r/c$. We consider the norm

$$\|f\|^2 = \int_{-T/2}^{T/2} \int_S f^2(\theta, \phi, t) d\theta d\phi dt, \tag{13.1}$$

and we assume that the signal is subject to the energy constraint

$$\int_{-\infty}^{\infty} \int_S f^2(\theta, \phi, t) d\theta d\phi dt \leq E. \tag{13.2}$$

As the sphere expands, the entropy of the observation increases and our objective is to find its scaling law as a function of r. This geometric setup is depicted in Figure 9.1.

We consider the Fourier transform

$$f(\theta, \phi, t) \longleftrightarrow \widehat{F}(w_1, w_2, \omega), \tag{13.3}$$

and two sets, P and Q, corresponding to the coordinate points of the space–time region where the norm is computed, and of the support set of the Fourier transform of the signal. By (9.15), the observation domain P has measure $4\pi T$. By (9.16), the spectral support Q has measure of the order of $8\Omega^3 r^2/(3c^2)$ as $r \to \infty$. By letting $T = r/c$ in (9.18), and using the norm (9.3), it follows that the Nyquist number representing the effective dimensionality of the space is

$$N_0 = \frac{4\pi}{3} \left(\frac{\Omega r}{c\pi} \right)^3; \tag{13.4}$$

in other words, the number of degrees of freedom grows with the volume of the radiating system, normalized to the cube of the bandwidth expressed in wavelength units. By the results in Chapter 12 we also have that the ϵ-entropy is

$$H_\epsilon = \begin{cases} N_0 \log(\sqrt{E}/\epsilon) + o(N_0) \text{ bits} & \text{if } \sqrt{E}/\epsilon \geq 1, \\ 0 & \text{if } \sqrt{E}/\epsilon < 1 \end{cases} \tag{13.5}$$

as $r \to \infty$. The ϵ-entropy grows linearly with the number of degrees of freedom and logarithmically with the signal-to-quantization ratio.

On the other hand, for radiation of single-frequency, time-harmonic signals subject to the constraint (9.8), and considering the norm (9.9), using (9.10) we have an effective number of dimensions per unit frequency of

$$N_0(\omega) = 4\pi \left(\frac{\omega r}{c\pi}\right)^2; \tag{13.6}$$

in other words, the number of spatial degrees of freedom grows with the surface of the radiating system, normalized by half a wavelength squared.

13.2.1 Quantization Error

We now introduce the notion of quantization error, and provide a lower bound on the resolution at which signals can be observed due to the physical nature of the propagation process. Due to the quantized nature of radiation, there is a minimum resolution at which each frequency component of the signal can be observed on any spatial dimension, and by virtue of (13.6) this leads to a cumulative quantization error per unit frequency that grows with the square of the radius of the system. It follows that to keep a non-vanishing signal-to-quantization ratio, the energy of the signal must also be subject to the same scaling law. When these constraints are taken into account in (13.5), they lead to a maximum entropy expression matching the universal entropy bound (11.82).

We let $q(t,\mathbf{r})$ be the error committed when a signal f is identified with the closest center $\hat{f}$ of an ϵ-ball of a minimum-cardinality covering of the space $\mathscr{Q}$, namely

$$q(t,\mathbf{r}) = f - \hat{f}, \tag{13.7}$$

where

$$\hat{f} = \arg\min_{f^* \in \mathscr{Q}} \|f - f^*\|. \tag{13.8}$$

Clearly, for all signals in the space we have

$$\int_{-T/2}^{T/2} \int_S q^2(t,\mathbf{r}) d\mathbf{r} dt \le \epsilon^2. \tag{13.9}$$

Consider the largest possible error occurring for signals located on the boundary of an ϵ-ball and approximated by the center of the ball. For all $\omega \in \mathbb{R}$, we decompose the frequency spectrum of the error using a basis set $\{\psi_n(\mathbf{r})\}$, orthonormal over the surface S of the sphere of radius r, obtaining

$$\mathcal{F}q(\omega,\mathbf{r}) = \sum_{n=1}^{\infty} a_n(\omega)\psi_n(\mathbf{r}), \tag{13.10}$$

where $\mathcal{F}$ indicates the Fourier transform operator with respect to the variable t. Using orthonormality, we have that the energy of the error per unit angular frequency is

$$\epsilon^2(\omega) = \int_S |\mathcal{F}q(\omega,\mathbf{r})|^2 d\mathbf{r} \ge \sum_{n=1}^{N_0(\omega)} |a_n(\omega)|^2, \tag{13.11}$$

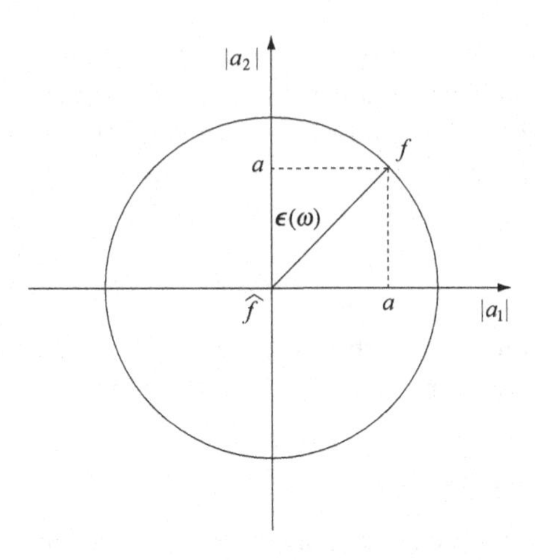

Fig. 13.1 Illustration of the lower bound on the quantization error.

where $N_0(\omega)$ is given by (13.6). Since radiation is quantized in photonic units of energy $\hbar\omega > 0$, we can model the energy of the error per unit angular frequency occurring on each of the $N_0(\omega)$ spatial dimensions as taking discrete values that are multiples of $\hbar/2$, namely

$$|a_n(\omega)|^2 = k\hbar/2, \quad k \in \{0, 1, 2, \dots\} \tag{13.12}$$

In this way, each pair of opposite frequencies of any real signal located on the boundary of the sphere is subject to a quantization error that is a multiple of $\hbar$. Some of the components in (13.12) can be zero, but the cumulative error must be proportional to the number of dimensions, namely

$$\epsilon^2(\omega) = \sum_{n=1}^{N_0(\omega)} |a_n(\omega)|^2 \geq N_0(\omega)\hbar/2. \tag{13.13}$$

This follows by noticing that all signals located on the boundary of the ϵ-ball have the same cumulative error per unit angular frequency $\epsilon^2(\omega) > 0$. Consider one such signal, for which $|a_n(\omega)| = a(\omega) > 0$ for all $0 \leq n \leq N_0(\omega)$; see Figure 13.1. In this case, we have

$$\epsilon^2(\omega) = N_0(\omega)a^2(\omega). \tag{13.14}$$

Since $a^2(\omega)$ is strictly positive, and by (13.12) it is also quantized, we have $a^2(\omega) \geq \hbar/2$. It follows that $\epsilon^2(\omega) \geq N_0(\omega)\hbar/2$, which is the desired lower bound on the energy of the error per unit angular frequency. This result shows that the quantization error per unit angular frequency has energy proportional to the number of dimensions per unit angular frequency of the radiating signal.

13.2.2 Kolmogorov Entropy Bound

The universal entropy bound for the ϵ-entropy of bandlimited radiation is now stated.

The ϵ-entropy of the space of bandlimited signals radiated from the sphere of radius $r \to \infty$ is

$$H_\epsilon \leq \frac{2\pi r E}{\hbar c} \frac{1}{2\pi e} \log e + o(rE) \text{ bits} \tag{13.15}$$

if $\sqrt{E}/\epsilon \geq 1$, and zero otherwise. The bound is tight for signals carrying $(1/2)\log e$ bits per degree of freedom.

The derivation of (13.15) goes as follows. By (13.9) and Parseval's theorem, we have

$$\begin{aligned}\epsilon^2 &\geq \int_{-T/2}^{T/2} \int_S q^2(t,\mathbf{r}) d\mathbf{r} dt \\ &= \frac{1}{2\pi} \int_{-\Omega}^{\Omega} \int_S |\mathcal{F}q(\omega,\mathbf{r})|^2 d\mathbf{r} d\omega \\ &\quad - \epsilon_T^2 \int_{-\infty}^{\infty} \int_S q^2(t,\mathbf{r}) d\mathbf{r} dt \\ &\geq \frac{1}{2\pi} \int_{-\Omega}^{\Omega} \int_S |\mathcal{F}q(\omega,\mathbf{r})|^2 d\mathbf{r} d\omega - \epsilon_T^2 E, \end{aligned} \tag{13.16}$$

where

$$\epsilon_T^2 = 1 - \frac{\int_{-T/2}^{T/2} \int_S q^2(\mathbf{r},t) d\mathbf{r} dt}{\int_{-\infty}^{\infty} \int_S q^2(\mathbf{r},t) d\mathbf{r} dt}. \tag{13.17}$$

We now consider a signal located on the boundary of an ϵ-ball and the corresponding error occurring by approximating it with the center of the ball. Integrating (13.11) over the bandwidth and using (13.13), we get

$$\begin{aligned}\int_{-\Omega}^{\Omega} \int_S |\mathcal{F}q(\omega,\mathbf{r})|^2 d\mathbf{r} d\omega &\geq \int_{-\Omega}^{\Omega} \sum_{n=1}^{N_0(\omega)} |a_n(\omega)|^2 d\omega \\ &\geq \int_0^{\Omega} \hbar N_0(\omega) d\omega. \end{aligned} \tag{13.18}$$

Substituting (13.18) into (13.16) and using (13.6), we obtain the lower bound

$$\begin{aligned}\epsilon^2 &\geq \frac{1}{2\pi} \int_0^{\Omega} 4\pi \left(\frac{\omega r}{c\pi}\right)^2 \hbar d\omega - \epsilon_T^2 E \\ &= \frac{\hbar}{2\pi} \frac{4\pi r^2}{c^2 \pi^2} \frac{\Omega^3}{3} - \epsilon_T^2 E. \end{aligned} \tag{13.19}$$

Letting the signal-to-quantization ratio

$$E/\epsilon^2 = \mathsf{SNR}_\mathrm{K} \geq 1, \tag{13.20}$$

and using (13.19), we have

$$E \geq \frac{\hbar}{2\pi} \frac{4\pi r^2}{c^2 \pi^2} \frac{\Omega^3}{3} \mathsf{SNR}_\mathrm{K} - E\epsilon_T^2 \mathsf{SNR}_\mathrm{K}. \tag{13.21}$$

From this, it follows that

$$\epsilon_T^2 E \geq \frac{\epsilon_T^2}{(1+\epsilon_T^2 \mathsf{SNR}_\mathrm{K})} \frac{\hbar}{2\pi} \frac{4\pi r^2}{c^2\pi^2} \frac{\Omega^3}{3} \mathsf{SNR}_\mathrm{K}. \tag{13.22}$$

Since ϵ_T^2 tends to zero as $T = r/c \to \infty$, it follows that for any fixed value of $\mathsf{SNR}_\mathrm{K} \geq 1$ the right-hand side is $o(r^2)$. Plugging this result into (13.19), we have the asymptotic relation

$$\epsilon^2 \geq \frac{\hbar}{2\pi} \frac{4\pi r^2}{c^2\pi^2} \frac{\Omega^3}{3} - o(r^2). \tag{13.23}$$

Finally, applying (13.20) and (13.23) to (13.5), we obtain

$$\begin{aligned} H_\epsilon &= \frac{4\pi}{3}\left(\frac{\Omega r}{c\pi}\right)^3 \log\sqrt{\mathsf{SNR}_\mathrm{K}} + o(r^3) \\ &\leq \frac{2\pi r E}{\hbar c} \frac{\log \mathsf{SNR}_\mathrm{K}}{2\pi \mathsf{SNR}_\mathrm{K}} + o(rE). \end{aligned} \tag{13.24}$$

Since the function $(\log \mathsf{SNR}_\mathrm{K})/\mathsf{SNR}_\mathrm{K}$ is maximized for $\mathsf{SNR}_\mathrm{K} = e$, (13.15) follows.

13.2.3 Saturating the Bound

The bound (13.15) is tight when the energy of the quantization error is a minimum, namely (13.23) holds with equality, and the signal-to-quantization ratio is $\mathsf{SNR}_\mathrm{K} = e$. In this case, roughly $N_0 \log\sqrt{e}$ bits are needed to identify any signal in the space, and the universal entropy bound is saturated. It follows that in this case the quantized signals' space is equivalent to a system of

$$N_0 \log\sqrt{\mathsf{SNR}_\mathrm{K}} = N_0 \log\sqrt{e} \tag{13.25}$$

binary spins, forming

$$N = 2^{(N_0/2)\log e} \tag{13.26}$$

states, and having Boltzmann entropy

$$\begin{aligned} H_\mathrm{B} &= \log N \\ &= (N_0/2)\log e \text{ bits}, \end{aligned} \tag{13.27}$$

where N_0 is given by

$$N_0 = \frac{4\pi}{3}\left(\frac{\Omega r}{c\pi}\right)^3. \tag{13.28}$$

By (13.21), the bound (13.15) also requires the energy to be proportional to r^2 and to the cube of the bandwidth. It follows that the entropy of bandlimited radiation can be proportional to the volume of the radiating body and increases with the cube of the bandwidth of the radiation. This result should be compared with the analogous one for black body radiation over an infinite spectrum of frequencies. In that case, the expected energy is proportional to r^3 and to the temperature raised to the fourth power, and the entropy is proportional to $(\bar{E}r)^{3/4}$. It follows that in both cases the entropy can scale with the volume of the radiating body, although the energy expenditures are different. Black body radiation reflects an equilibrium configuration where the expected energy per unit volume remains constant as the radiating system expands. On the other hand, bandlimited radiation achieves the universal entropy bound with an energy per unit volume that vanishes as the radiating system expands. For bandlimited radiation, the entropy increases with the bandwidth, while for thermal radiation the temperature plays the role of an "effective bandwidth" in its entropy expression.

13.3 Stochastic Signals

Analogous results are obtained by considering the rate–distortion function of stochastic signals having a Gaussian distribution with constant power spectral density over the bandwidth. In this case, we replace the energy constraint (13.2) with the average one,

$$\int_{-T/2}^{T/2} \int_{S} \mathbb{E}(\mathsf{f}^2(\theta,\phi,t)) d\theta d\phi dt \leq PN_0. \tag{13.29}$$

By the results in Section 12.4, we have that letting the quantization error associated with the representation of $\mathsf{f}(t)$ using codebook point $\hat{\mathsf{f}}(t)$ be

$$\mathsf{q}(t,\mathbf{r}) = \mathsf{f} - \hat{\mathsf{f}}, \tag{13.30}$$

and imposing the average energy per degree of freedom constraint

$$\frac{1}{N_0} \int_{-T/2}^{T/2} \int_{S} \mathbb{E}[\mathsf{q}^2(t,\mathbf{r})] d\mathbf{r} dt \leq \sigma^2, \tag{13.31}$$

then the rate–distortion function, measured in bits, is given by

$$\begin{cases} R_{\sigma^2} = N_0 \log(\sqrt{P}/\sigma) + o(N_0) \text{ bits} & \text{if } \sqrt{P}/\sigma \geq 1, \\ 0 & \text{if } \sqrt{P}/\sigma < 1. \end{cases} \tag{13.32}$$

We now impose a lower bound on the quantization energy σ^2. In the deterministic setting, dividing the lower bound (13.19) by N_0 and using (13.28), we obtain a lower bound on the empirical mean energy per signal dimension of the quantization error:

$$\frac{\epsilon^2}{N_0} \geq \frac{\hbar c}{2r} - o(1/r). \tag{13.33}$$

As shown in Section 13.2.1, the error on any given signal dimension can be zero, but since the cumulative error has energy proportional to the number of dimensions, the positive lower bound on the empirical mean (13.33) follows. We now impose the same constraint on the stochastic mean energy per signal dimension of the quantization error,

$$\sigma^2 \geq \frac{\hbar c}{2r} - o(1/r); \tag{13.34}$$

in other words, the average energy per signal dimension cannot be smaller than one photonic unit of radiation.

13.3.1 Shannon Rate–Distortion Bound

The universal bound on the rate–distortion function is now stated.

> The Shannon rate–distortion function of the space of Gaussian signals of constant power spectral density over a fixed bandwidth radiated from the sphere of radius $r \to \infty$ is
>
> $$R_{\sigma^2} \leq \frac{2\pi r\bar{E}}{\hbar c}\frac{1}{2\pi e}\log e + o(r\bar{E}) \text{ bits}, \tag{13.35}$$
>
> where $\bar{E} = PN_0$. This result holds if $\sqrt{P}/\sigma \geq 1$, otherwise the rate distortion is zero. The bound is tight for signals carrying $(1/2)\log e$ bits per degree of freedom.

The derivation follows by substituting (13.34) into (13.32) and going through the same steps as in the deterministic case. We have, as $r \to \infty$,

$$\begin{aligned} R_{\sigma^2} &\leq N_0 \log\sqrt{\frac{P}{\dfrac{\hbar c}{2r} - o(1/r)}} \\ &\leq \frac{N_0 P}{\dfrac{\hbar c}{2r}[1 - o(1)]}\frac{\log e}{2e} \\ &= \frac{2\pi r\bar{E}}{\hbar c}\frac{1}{2\pi e}\log e[1 + o(1)] \\ &= \frac{2\pi r\bar{E}}{\hbar c}\frac{1}{2\pi e}\log e + o(r\bar{E}), \end{aligned} \tag{13.36}$$

which coincides with (13.35).

The bound is tight if the energy of the quantization error is a minimum, so that the first inequality holds with equality, and if P is of the order of $1/r$, so that the second inequality is also tight. By (13.28), this means that the energy of the signal must be of the order of r^2, which is consistent with the analogous result in the deterministic setting. The results in the two settings are completely analogous and follow from the minimum quantization error constraints (13.33) and (13.34), which impose an energy of the error increasing as r^2 or, equivalently, an energy of the error per degree of freedom decreasing as $1/r$. The essential steps in the derivation are to use the minimum quantization allowed by quantum indetermination constraints in the entropy and rate–distortion formulas, scale the energy to maintain a constant signal-to-quantization ratio, and write the result in terms of the energy and radius of the system.

13.3.2 Shannon Entropy Bound

By comparing (13.15) and (13.35) with (11.82), we observe a discrepancy of a factor $1/(2\pi e)$ between the bounds for radiated signals and the universal entropy bound for the thermodynamic entropy of arbitrary systems. This is due to the way we have performed quantization in the signal model. In the computation of the ϵ-entropy and rate–distortion we considered the minimum cardinality covering of the space, and then performed quantization of each signal dimension with respect to this covering. If instead we compute the Shannon entropy of an equipartitioned signals' space where quantization of the energy is performed separately on each dimension, we obtain an entropy bound matching (11.82).

The Shannon entropy of the space of Gaussian signals of constant power spectral density over a fixed bandwidth radiated from the sphere of radius $r \to \infty$ and whose energy is quantized at level σ on each dimension of the signals' space is

$$H \leq \frac{2\pi r\bar{E}}{\hbar c} \log e + o(r\bar{E}) \text{ bits,} \tag{13.37}$$

where $\bar{E} = PN_0$ and $\sqrt{P}/\sigma \geq 1$.

This result follows by noticing that the number of typical quantized signals is in this case given by the volume of the typical set divided by the volume of the box of side length σ,

$$N = \left(\frac{\sqrt{2\pi eP}}{\sigma}\right)^{N_0}. \tag{13.38}$$

These signals are roughly equally probable, each having probability of occurrence $p_n = 1/N$ for all $n \in [1,N]$. This corresponds to an equilibrium configuration where the Shannon entropy attains Boltzmann's form:

$$\begin{aligned} H &= \sum_{n=1}^{N} -p_n \log p_n \\ &= \log N \\ &= N_0 \log \sqrt{2\pi eP/\sigma^2}. \end{aligned} \tag{13.39}$$

We now apply the lower bound on the energy of the quantization error per degree of freedom and substitute (13.34) into (13.39). Letting $\bar{E} = PN_0$ and following the usual steps, we obtain

$$\begin{aligned} H &\leq \frac{2\pi eP2r}{\hbar c}\frac{N_0}{2e} \log e + o(rN_0) \\ &= \frac{2\pi r\bar{E}}{\hbar c} \log e + o(rE) \text{ bits,} \end{aligned} \tag{13.40}$$

which coincides with (13.37).

The geometric reason for the disappearance of the factor $1/(2\pi e)$ should be clear: in the case of the Kolmogorov entropy we cover a space of signals of radius E with balls of radius ϵ, so that the number of quantized signals is roughly

$$N = \left(\frac{E}{\epsilon}\right)^{N_0}. \tag{13.41}$$

Similarly, in the case of the rate–distortion function, the volume of the typical set is $(\sqrt{2\pi eP})^{N_0}$ and the volume of the typical error region is $(\sqrt{2\pi e\sigma})^{N_0}$, so that the number of quantized signals is roughly

$$N = \left(\frac{\sqrt{2\pi eP}}{\sqrt{2\pi e\sigma}}\right)^{N_0} = \left(\frac{P}{\sigma}\right)^{N_0}. \tag{13.42}$$

On the other hand, in the case of the Shannon entropy the volume of the typical set is $(\sqrt{2\pi eP})^{N_0}$, but quantization occurs independently on each coordinate of the space, which geometrically corresponds to covering the space using a grid, with each cell having volume σ^{N_0}, and we have

$$N = \left(\frac{\sqrt{2\pi eP}}{\sigma}\right)^{N_0}. \tag{13.43}$$

It is important to note, however, that from the discussion in Section 1.4.4, we have that (13.43) holds only for $\sigma^2 \ll P$, namely in the high signal-to-quantization ratio regime. Since tightness of the bound requires P and σ^2 to be of the same order, the accuracy of the derivation, which is precise for high values of the signal-to-quantization ratio, is traded off by the tightness of the obtained bound.

13.4 One-Dimensional Radiation

We now revisit results in the one-dimensional case, where $N_0 = \Omega T/\pi$. Given a minimum quantization energy

$$\epsilon^2 \geq \hbar\Omega, \tag{13.44}$$

we obtain the entropy bound

$$\begin{aligned} H_\epsilon &= N_0 \log(\sqrt{E}/\epsilon) + o(N_0) \\ &\leq N_0 \frac{E}{\epsilon^2}\frac{\log e}{2e} + o(N_0) \\ &\leq \frac{TE}{\hbar}\frac{\log e}{2\pi e} + o(T) \text{ bits.} \end{aligned} \tag{13.45}$$

Similarly, in the stochastic setting using the lower bound on the quantization error per signal dimension

$$\sigma^2 \geq \frac{\hbar\Omega}{N_0}, \tag{13.46}$$

we obtain the rate–distortion bound

$$R_\sigma \leq \frac{T\bar{E}}{\hbar}\frac{\log e}{2\pi e} + o(T) \text{ bits,} \tag{13.47}$$

and also the entropy bound

$$H \leq \frac{T\bar{E}}{\hbar} \log e + o(T) \text{ bits.} \tag{13.48}$$

These bounds are tight when the energy of the quantization error is a minimum and the signal's energy is a constant of the same order as the error. As in the three-dimensional case, the largest entropy signal carries $(1/2)\log e$ bits per degree of freedom, and the size of the observation interval T plays the role of the volume of the space.

13.5 Applications

We now discuss in what settings the universal entropy bounds computed in the previous sections are applicable. First, we observe that they ignore gravitational constraints, so they must be intended for weak self-gravitating systems, that are far from gravitational collapse. Second, by identifying information with energy, they consider all available degrees of freedom at the most fundamental level. Exploiting all of these degrees of freedom would require incredible technological advancements and the emergence of novel quantum communication architectures.

13.5.1 High-Energy Limits

From a physical perspective, the main conclusion that can be drawn from the computations above is that information can be identified with energy, but also that space can be considered as an information-bearing object. For a single "burst" of energy of vanishing density propagating inside a spherical volume and occupying a finite bandwidth, the entropy is proportional to the volume of the sphere and to the energy, which scales with the square of the radius. Larger volumes allow for a larger number of waveform configurations, and these in turn lead to larger entropy values. For one-dimensional systems, the transmission time plays the role of the volume, and the entropy of a waveform of fixed energy grows with the transmission time. In comparison, black body radiation occurring over an infinite bandwidth achieves the same entropy scaling with the volume of the system, but requires an energy expenditure proportional to the volume of the expanding system.

When the energy of the system becomes so large that gravitational constraints must be taken into account, universal entropy bounds must be modified accordingly. By imposing a quantization of the error per signal dimension, the ϵ-balls must grow with the number of dimensions, as indicated in (13.23). It follows that to have a signal-to-quantization ratio $\mathsf{SNR}_K \geq 1$, the energy of the signal must also grow at least as the square of the radius of the system, as indicated in (13.21). This scaling eventually conflicts with gravitational constraints.

As discussed in Chapter 11, general relativity imposes that the energy can be at most a linear function of the radius, to avoid collapse of the source of radiation and formation of a black hole. In practice, however, the quadratic lower bound (13.21) and the linear upper bound (11.78) from the theory of gravity can be satisfied with orders of magnitude to spare. To get an idea of the scales involved, assuming a signal-to-quantization ratio $\mathsf{SNR}_K = e$, and unit bandwidth, for lower and upper bounds to be comparable r must be a huge number, orders of magnitude larger than the size of the observable universe. It follows that at any reasonable scale the universal entropy bound provides a good indication of the maximum entropy growth.

One may also ask what happens if for a reasonable scale of r the energy approaches the critical value for gravitational collapse. In this case of extremely high energy, substituting (11.78) for E into (13.15) and (13.37), using the definition of Planck length (11.76) and ignoring second-order terms, we obtain the black hole entropy bound for the ϵ-entropy of

$$H_\epsilon \leq \frac{4\pi r^2}{4\ell_p^2} \frac{1}{2\pi e} \log e \text{ bits}, \tag{13.49}$$

and for the Shannon entropy of

$$H \leq \frac{4\pi r^2}{4\ell_p^2} \log e \text{ bits}, \tag{13.50}$$

which corresponds to (11.96). This simple substitution does not account for the complex physical processes that occur as E approaches the critical value for collapse, but it provides the correct result supported by more detailed physical arguments.

In comparison, the analogous gravitational bound for black body radiation (11.80) computed in Chapter 11 indicates that a lower entropy value is approached as the black body approaches gravitational collapse. Not surprisingly, black body radiation using a higher energy expenditure to obtain the same volumetric order of growth of the entropy reaches gravitational collapse at a lower entropy value.

13.5.2 Relation to Current Technologies

Since universal entropy bounds take into account all available degrees of freedom, when compared to current state-of-the-art engineering systems they are extremely loose. Let us consider a few examples. A flash memory card of the characteristic size of a centimeter and weight of a few grams can store about one terabyte of data, about 10^{12} bits. One gram of DNA of the same size can be used to store about 10^{15} bits. The internet connects some 3 billion users, and it is projected to grow to 8–10 billions of connected devices in the next 10 years. Assuming each connected device also stores one terabyte of information, the entire state of the internet could be described by 10^{22} bits. These may seem like large numbers, but they quickly disappear in comparison to nature's limits. If we apply the universal entropy bound to our little memory card using Einstein's equation for the energy, $E = mc^2$, we obtain a bound of the order of 10^{38} bits. Although this shows much room for technological improvement, by (13.27), saturating the universal entropy bound would require converting the entire mass–energy of our memory card into radiation carrying information at a rate of $(1/2)\log e$ bits per degree of freedom, and by doing so it considers all available degrees of freedom at the most fundamental level. Conventional devices use many more degrees of freedom than are available in nature to store their bits. Nature's limits, although important for matters of principle, are currently far from our reach.

Conceptually, the derivation of the universal entropy bounds provided in Sections 13.2 and 13.3 is very simple: we computed the maximum entropy and rate–distortion of bandlimited radiated signals when quantization is the minimum allowed by quantum indetermination constraints, and wrote the result in terms of the energy and radius of the system. This means that in order to exploit this amount of information, we need to be able to observe all the states of the system up to the quantum level. In this way, all degrees of freedom of the signal are saturated. Although we are far from reaching this capability, with the emergence of computing and communication devices using novel quantum architectures, degrees of freedom will be exploited more and more efficiently. For example, atomic-scale data storage has demonstrated storage of one bit of information using 12 iron atoms tightly packed over a copper substrate of size of the order of 1 nm^2. Ignoring the substrate over which the atoms are placed, the universal entropy bound for a nanometer ball filled by the same atomic mass gives about 10^{10} bits. Subatomic-scale storage, exploiting degrees of freedom of quantum wave functions representing subatomic particles, promises to shrink the gap even further. The possibility of storing information using the degrees of freedom of waves at the quantum level is becoming not so far-fetched. It seems reasonable to expect that progress will eventually meet nature's ultimate information limits, and that in the far future universal entropy bounds will play an important role in assessing the technological capabilities of human-made systems.

13.6 On Models and Reality

Information-theoretic bounds are aimed at describing information in physical systems at the most fundamental level. For continuous signals they arise by imposing the natural constraints of information theory on energy, dimensionality, and resolution. The energy determines the size of the signals' space, and its increase leads to a corresponding increase in the amount of information required to specify any arbitrary signal in the space. The dimensionality determines the number of coordinates required to identify the signal, and corresponds to the degrees of freedom of the space. The resolution determines the inherent uncertainty at which these coordinates can be observed. Physically, these constraints are not independent of each other. The energy depends on the frequency of radiation, and this influences the number of degrees of freedom and the granularity at which each signal can be observed. The size of the system also plays a role in determining the number of degrees of freedom and the resolution.

Of course, a physical system can in principle contain much more information than can be externally observed and imposed by the above constraints. As noted in Chapter 1, without any quantization constraint on the observation, the amount of information, measured in bits, is infinite. In this case, a continuum information model would consider the whole volume of the signals' space, containing infinitely many points, as a relevant information measure. In a stochastic setting, by the asymptotic equipartition property, this depends on the differential entropy and on the number of degrees of freedom of the signals' space. On the other hand, physics dictates that continuous signals are always observed in a quantized fashion, so discrete measures of information may be more appropriate in practical settings, where physical quantities are observed.

The question of whether the amount of information "observed" and the amount "contained" in a system differ from each other takes us into the realm of the philosophy of science. Much has been written on whether physics is inherently digital or not, namely whether unobservable information should be considered equivalent to the absence of information and reality should be considered "digital." In short: should we consider a continuum volume of information, or a discrete set of bits?

Proponents of the digital hypothesis believe that physical systems can only have a finite number of discrete states, and therefore physical processes can be reduced to digital computations. This leads to a strong positivist interpretation of universal entropy bounds as ultimate physical limits. If instead one embraces the point of view that the digital hypothesis is an unscientific ansatz in the sense of Popper, then a weaker interpretation of universal bounds would simply be that no measurement can ever prove that a physical system must be described by more than a finite number of bits. Nevertheless, a continuum of information hidden to the observer remains a physical possibility. Both points of view are consistent with information-theoretic derivations, but each requires a radically different philosophical stand.

A more relaxed position is the instrumentalist point of view, summarized by the phrase: "shut up and calculate!" This advocates refraining from any philosophical interpretation of the results of a mathematical model. According to this school of thought, scientific models are not meant to represent reality but are only instruments that allow us to describe and predict observations. In this case, whether reality is continuous or discrete does not matter at all. Maxwell's equations and quantum mechanics are regarded as mere inventions, each appropriate for describing phenomena observed within a certain range of conditions. A profound consequence of this point of view is that physics, like mathematics, is an evolving science that cannot be reduced to a finite set of rules. Every model is inherently incomplete and subject to paradigm changes. Physics, much like mathematics, is subject to ambiguity and paradox, and these naturally lead to its continuous evolution. There may be no ultimate physical laws, and scientific crises are natural and unavoidable.

In this book, we refrained from defending any of the above points of view. We argued that information-theoretic limits, including universal entropy bounds, arise in a mathematical model of continuous signals subject to constraints. The identification of these constraints is required for any rigorous derivation, but whether they can be relaxed or are the pillars of an untouchable physical theory is not something we can, or wish to, address. The mathematical laws of spectral and probabilistic concentration that are at the basis of our arguments are also at the basis of quantum uncertainty relations and statistical mechanics. Whether "real" information obeys these rules or whether they are ad hoc inventions, is irrelevant for a *theory* of information that rigorously predicts limits within the context of a specific mathematical model. As we stated at the beginning of the book, information is described mathematically, and this description requires confinement to a model with constraints. Information also has a physical structure, in the sense that the model *applies* to the real world, or at least to the world that we are able to observe. Although in what sense the model is applicable is subject to interpretation, the theory itself should not be. Instead, a fundamental requirement is that the theory should be robust, insensible to nuisances and perturbations of second-order details of the model.

A robust theory is more likely to be widely accepted by positivists and less prone to the paradigm shifts required by instrumentalists. Throughout the book, we have shown that the basic results of wave theory of information are robust. They remain essentially unaltered whether one works in a probabilistic or a deterministic setting, and whether one adopts a continuum approach based on degrees of freedom, typical volumes, and differential entropies, or an approach based on discrete information measures like entropies and capacities. After all, if we could summarize in one sentence the point of view of this book, we do not need to run onto Scylla wishing to avoid Charybdis, but we can safely navigate one of the many paths that lead to a rigorous notion of information.

13.7 Summary and Further Reading

The universal entropy bound for arbitrary matter systems was first proposed by Bekenstein (1981a). Its original derivation was based on a thought experiment involving black hole thermodynamics. Consider an object much smaller than a black hole and drop it into the black hole. As a result, the area of the event horizon of the black hole increases. The area increase is also proportional to the increase in entropy of the black hole. The computation of the area increase and an application of the second law leads to the universal entropy bound on the entropy of the object. This derivation was criticized and defended on several occasions in the literature; see Bekenstein (2005) for a review. An earlier proposal of a similar bound for one-dimensional radiation was also made by Bremermann (1967, 1982), using an heuristic argument that treated quantization as Gaussian uncertainty. Physically, a linear energy bound on entropy is generally accepted for weakly self-gravitating isolated objects, and several mathematical derivations based solely on quantum state counting and not involving gravity appeared in the literature; see, for example, Bekenstein and Schiffer (1990) for a review, and Schiffer (1991). The derivation provided here is for the specific setting of bandlimited electromagnetic radiation; it is based on spectral concentration arguments, and appeared in Franceschetti (2017).

A more general entropy bound for high-energy systems is the holographic bound discussed in Chapter 11. A general conclusion to be drawn from this, as well as the universal entropy bound, is that the number of degrees of freedom available in nature is much larger than is exploited in current engineering systems. Full exploitation of nature's degrees of freedom would require conversion of all matter into radiation energy and manipulation of the resulting quantum wave functions. Clearly, as information-theoretic bounds become more general, their range of applicability becomes more limited. Universal bounds are useful, however, to understand physical limits and to have an idea of the orders of magnitude at which natural information systems can be described. For the encoding capabilities of current technologies that were briefly discussed in Section 13.5.2 see Moon *et al.* (2009), Church, Gao, and Kosuri (2012), and Loth *et al.* (2012).

Finally, the philosophical discussion of whether universal bounds reflect real limits or are the result of artificial constraints imposed in a mathematical model is ongoing. This debate is related to whether reality is discrete or continuous, and to whether a complete physical theory can ever be developed. Views here vary widely. Some interesting points on the philosophy of information theory are raised by Lee (2017), and for the role of ambiguity in mathematical theories see Byers (2007). Robustness of information-theoretic models has been prominently advocated by Slepian (1976).

13.8 Test Your Understanding

Problems

13.1 The lower bounds (13.33) and (13.34) are the fundamental ingredients of the derivation of the universal entropy bounds. Provide a justification for these bounds in terms of the uncertainty principle.

Solution

The standard deviation of the time spread of the quantum uncertainty multiplied by the frequency spread must be at least $1/2$. The time spread of the energy is roughly T, and the frequency spread of the energy is roughly Ω. Since $T = r/c$, it follows that

$$\Omega \geq c/(2r). \tag{13.51}$$

Since each photon carries a quantum of energy $\hbar\omega$, the energy uncertainty of the signal is roughly $\sigma^2 = \hbar\Omega$, and multiplying both sides of (13.51) by $\hbar$ we obtain the desired result.

13.2 Discuss the tightness of the universal bounds on the ϵ-entropy, rate–distortion function, and Shannon entropy.

13.3 Illustrate the difference between the entropy of black body radiation and the maximum entropy of bandlimited radiation in one- and three-dimensional settings.

13.4 Consider one-dimensional bandlimited radiation and recall that universal entropy bounds are tight when the signal carries $\log\sqrt{e}$ bits per degree of freedom. Show that this also corresponds to a number of bits per second proportional to $\sqrt{\bar{P}/\hbar}$, where $\bar{P}$ is the energy per unit time. (Hint: follow the argument given in Section 12.2.6. Note that each degree of freedom is represented by a mode of radiation of bandwidth Ω, compute the minimum detectable energy of one mode of radiation, and from this an upper bound on the number of possible detectable modes. Finally, write an equation analogous to (12.49) to compute the number of bits per second.)

Appendix A: Elements of Functional Analysis

A.1 Paley–Wiener Theorem

The Paley–Wiener theorem relates the decay properties of the Fourier transform at infinity and the analyticity properties of its corresponding inverse transform. The most extreme form of decay at infinity of a Fourier transform, which is of interest to us, is to be *bandlimited*, namely to have compact support in the spectral domain. This yields a strong regularity property of the function, namely the inverse transform is the restriction over the real line of an *entire* function of a complex variable, and thus it can be expanded in a power series that converges everywhere. Furthermore, the rate of growth at infinity of the entire function is bounded by an exponential of Ω.

Let $f \in L^2(-\infty,\infty) : \mathbb{R} \to \mathbb{R}$. We have that $f \in \mathscr{B}(\Omega)$ if and only if the complex extension

$$f(z) = \int_{-\infty}^{\infty} F(\omega)\exp(j\omega z)d\omega, \quad z = x + jy,\ y > 0,\ x, y \in \mathbb{R} \tag{A.1}$$

is an entire function of exponential type Ω, namely there exists a constant C such that

$$|f(z)| \leq C\exp(\Omega|z|). \tag{A.2}$$

A.2 Uncertainty Principle over Arbitrary Measurable Sets

Let $f \in L^2$, with Fourier transform

$$\mathcal{F}f = F(\omega) = \int_{-\infty}^{\infty} f(t)\exp(j\omega t)dt, \tag{A.3}$$

and norm

$$\|f\| = \left(\int_{-\infty}^{\infty} |f(t)|^2 dt\right)^{1/2} = 1. \tag{A.4}$$

By Parseval's theorem, we have

$$\|F\| = 2\pi. \tag{A.5}$$

Let S_T and S_Ω be sets of measure T and 2Ω, and f be ϵ_T-concentrated over S_T, namely

$$1 - \int_{S_T} f^2(t)dt \le \epsilon_T^2, \tag{A.6}$$

and ϵ_Ω-concentrated over S_Ω, namely

$$1 - \frac{1}{2\pi}\int_{S_\Omega} |F(\omega)|^2 d\omega \le \epsilon_\Omega^2. \tag{A.7}$$

The uncertainty principle poses a lower bound on the product of the measures of the concentration sets, namely

$$\frac{\Omega T}{\pi} \ge [1 - (\epsilon_T + \epsilon_\Omega)]^2. \tag{A.8}$$

To prove the uncertainty principle, we need some definitions and some properties of operators on Hilbert spaces. Let the timelimiting operator be

$$\mathcal{T}f = \begin{cases} f & t \in S_T, \\ 0 & \text{otherwise,} \end{cases}$$

and the bandlimiting operator be

$$\mathcal{B}f = \mathcal{F}^{-1}\widehat{\mathcal{B}}\mathcal{F}f, \tag{A.9}$$

where

$$\widehat{\mathcal{B}}F = \begin{cases} F & \omega \in S_\Omega, \\ 0 & \text{otherwise.} \end{cases}$$

It follows that f is ϵ_T-concentrated on S_T if and only if

$$\|f - \mathcal{T}f\| < \epsilon_T, \tag{A.10}$$

and f is ϵ_Ω-concentrated on S_ω if and only if

$$\|f - \mathcal{B}f\| < \epsilon_\Omega. \tag{A.11}$$

Consider the integral operator

$$\mathcal{P}f(t) = \int_{-\infty}^{\infty} P(s,t)f(s)ds. \tag{A.12}$$

The operator norm is defined as

$$\|\mathcal{P}\| = \sup_{f \in L^2, f \neq 0} \frac{\|\mathcal{P}f\|}{\|f\|}, \tag{A.13}$$

and the Hilbert–Schmidt norm as

$$\|\mathcal{P}\|_{\text{HS}} = \left(\int_{-\infty}^{\infty}\int_{-\infty}^{\infty} |P(s,t)|^2 dsdt\right)^{1/2}. \tag{A.14}$$

The operator norm is dominated by the Hilbert–Schmidt norm,

$$\|\mathcal{P}\| \leq \|\mathcal{P}\|_{\text{HS}}; \tag{A.15}$$

namely, for all f,

$$\|\mathcal{P}f\| \leq \|f\| \cdot \|\mathcal{P}\|_{\text{HS}}. \tag{A.16}$$

This domination result is a consequence of the Schwartz inequality,

$$\left| \int_{-\infty}^{\infty} P(s,t) f(s) ds \right|^2 \leq \int_{-\infty}^{\infty} |P(s,t)|^2 ds \cdot \int_{-\infty}^{\infty} |f(s)|^2 ds. \tag{A.17}$$

Integrating both sides of (A.17), we get

$$\begin{aligned} \int_{-\infty}^{\infty} \left| \int_{-\infty}^{\infty} P(s,t) f(s) ds \right|^2 dt &\leq \int_{-\infty}^{\infty} \left(\int_{-\infty}^{\infty} |P(s,t)|^2 ds \cdot \int_{-\infty}^{\infty} |f(s)|^2 ds \right) dt \\ &= \int_{-\infty}^{\infty} |f(s)|^2 ds \cdot \int_{-\infty}^{\infty} \int_{-\infty}^{\infty} |P(s,t)|^2 ds dt, \end{aligned} \tag{A.18}$$

which is equivalent to (A.16).

Using the definitions above, the Hilbert–Schmidt norm of the operator $\mathcal{BT}$ can easily be computed to be

$$\|\mathcal{BT}\|_{\text{HS}} = \left(\frac{\Omega T}{\pi} \right)^{1/2}. \tag{A.19}$$

It then follows from the domination result that

$$\|\mathcal{BT}\| \leq \left(\frac{\Omega T}{\pi} \right)^{1/2}. \tag{A.20}$$

The proof of (A.8) is completed by performing the following computation:

$$\begin{aligned} 1 - \|\mathcal{BT}f\| &= \|f\| - \|\mathcal{BT}f\| \\ &\leq \|f - \mathcal{BT}f\| \\ &= \|f - \mathcal{B}f + \mathcal{B}f - \mathcal{BT}f\| \\ &\leq \|f - \mathcal{B}f\| + \|\mathcal{B}f - \mathcal{BT}f\| \\ &\leq \|f - \mathcal{B}f\| + \|\mathcal{B}\| \cdot \|f - \mathcal{T}f\| \\ &\leq \epsilon_T + \epsilon_\Omega, \end{aligned} \tag{A.21}$$

where the last inequality follows from $\|\mathcal{B}\| = 1$, (A.10), and (A.11).

A.3 Generalized Fourier Series

Orthogonal basis representations of a normed space can be viewed as a generalization of the Fourier series representation. Consider the set of functions in $L^2[a,b]$

$$\Phi = \{\phi_n : [a,b] \to \mathbb{C}\} \tag{A.22}$$

that are pairwise orthogonal with respect to the inner product

$$\langle \phi_n, \phi_m \rangle_w = \int_a^b \phi_n(x)\phi_m^*(x)w(x)dx = \begin{cases} \|\phi_n\|_w & \text{if } n = m, \\ 0 & \text{otherwise,} \end{cases} \tag{A.23}$$

where $w(x)$ is the weight function.

The generalized Fourier series of $f \in L_2[a,b]$ with respect to the orthogonal set Φ is

$$f(x) = \sum_{n=0}^{\infty} c_n \phi_n(x), \tag{A.24}$$

where the coefficients are given by

$$c_n = \frac{\langle f, \phi_n \rangle_w}{\|\phi_n\|_w}. \tag{A.25}$$

If Φ is a complete set, then the equality in the representation holds for all $f \in L^2[a,b]$ in the given norm. Namely,

$$\lim_{N\to\infty} \|f(x) - f_N(x)\|_w = 0, \tag{A.26}$$

where

$$f_N(x) = \sum_{n=0}^{N-1} c_n \phi_n(x). \tag{A.27}$$

The Bessel inequality for any set of orthogonal functions Φ is

$$\sum_{n=0}^{\infty} |c_n|^2 \le \frac{1}{\|\phi_n\|_w} \int_a^b |f(x)|^2 dx, \tag{A.28}$$

and the Parseval identity for any complete set of orthogonal functions Φ is

$$\sum_{n=0}^{\infty} |c_n|^2 = \frac{1}{\|\phi_n\|_w} \int_a^b |f(x)|^2 dx. \tag{A.29}$$

Many orthogonal representations are commonly used to represent square-integrable functions. The corresponding basis functions arise as solutions of boundary value problems in physics. The complex exponentials of the Fourier series are one common example. The sinc$(\cdot)$ functions of the cardinal series and the prolate spheroidal wave functions are other notable examples. We summarize some common classes of special orthogonal functions in Table A.1.

A.4 Self-Adjoint Hilbert–Schmidt Operators

Some auxiliary results on self-adjoint Hilbert–Schmidt operators introduced in Chapter 3 are described below.

Table A.1 Orthogonal representations.

Name	$f(x)$	$[a,b]$	$w(x)$
Fourier	$\exp(jnx)$	$(-\pi,\pi)$	1
Cardinal	$\text{sinc}(x-n)$	$(-\infty,\infty)$	1
Prolate spheroidal	$\varphi_n(x)$	$(-\infty,\infty)$, and $(-1,1)$	1
Chebyshev polynomials, 1st kind	$T_n(x)$	$(-1,1)$	$(1-x^2)^{-1/2}$
Chebyshev polynomials, 2nd kind	$U_n(x)$	$(-1,1)$	$(1-x^2)^{-1/2}$
Hermite polynomials	$H_n(x)$	$(-\infty,\infty)$	$\exp(-x^2)$
Laguerre polynomials	$L_n^{\alpha}(x)$	$[0,\infty)$	$\exp(-x)$
Legendre polynomials	$P_n(x)$	$(-1,1)$	1
Jacobi polynomials	$P_n^{(\nu,\mu)}(x)$	$(-1,1)$	$(1-x)^{\nu}(1-x)^{\mu}$

A.4.1 Variational Characterization of the Eigenvalues

For a wide class of linear self-adjoint operators, the eigenvalues can be characterized by three basic variational principles. Here, we refer to Hilbert–Schmidt self-adjoint operators with scalar product

$$\langle u,v\rangle = \int_c^d u(x)v^*(x)dx, \tag{A.30}$$

but more general versions of these results are also available in the literature. Consider a self-adjoint Hilbert–Schmidt operator $\mathcal{K}: \mathscr{H} \to \mathscr{H}$. Let $\mathscr{X}_N \subset \mathscr{H}$ be an N-dimensional subspace of $\mathscr{H}$. Let the Rayleigh quotient be

$$R_{\mathcal{K}}(x) = \frac{\langle \mathcal{K}f,f\rangle}{\langle f,f\rangle}, \quad f \neq 0. \tag{A.31}$$

Let $\lambda_0 \geq \lambda_1 \geq \cdots$ be the positive eigenvalues of $\mathcal{K}$, counted with multiplicities, and $\psi_0, \psi_1, \ldots$ the corresponding eigenfunctions satisfying

$$\mathcal{K}\psi_n = \lambda_n \psi_n. \tag{A.32}$$

Assume that there are at least $N+1$ positive eigenvalues, and indicate by $f \perp \mathscr{X}_n$ if f is orthogonal to each element of $\mathscr{X}_n$. We have:

- Rayleigh's principle:

$$\lambda_N = \sup\{R_{\mathcal{K}}(f) : \langle f,\psi_n\rangle = 0, \quad n = 0,\ldots,n-1\}. \tag{A.33}$$

- Max–min principle of Poincaré:

$$\lambda_N = \sup_{\mathscr{X}_{N+1}} \inf_{f\in\mathscr{X}_{N+1}, f\neq 0} R_{\mathcal{K}}(f). \tag{A.34}$$

- Min–max principle of Courant, Fisher, and Weyl:

$$\lambda_N = \inf_{\mathscr{X}_N} \sup_{f\perp\mathscr{X}_N, f\neq 0} R_{\mathcal{K}}(f). \tag{A.35}$$

The negative eigenvalues of the operator can be characterized by the same principles as above by replacing the minimum by the maximum and vice versa.

A.4.2 Mercer's Theorem

Sometimes it is convenient to obtain a series representation of a self-adjoint Hilbert–Schmidt kernel with stronger convergence properties than L^2 convergence.

Mercer's theorem provides a general result for the absolute and uniform convergence of the series representation of a self-adjoint Hilbert–Schmidt kernel based on the spectral representation of the corresponding operator, provided that some regularity assumptions are made beside square-integrability.

Consider a Hilbert–Schmidt, non-negative, self-adjoint operator $\mathcal{K}'\mathcal{K} : L^2[c,d] \to L^2[c,d]$, with continuous kernel $K'K(x,y)$. If $\{\lambda_n\}$ and $\{\phi_n\}$ are the eigenvalues and corresponding eigenvectors of $K'K$, then for all $x,y \in [c,d]$ we have

$$K'K(x,y) = \sum_{n=0}^{\infty} \lambda_n \phi_n(x)\phi_n^*(y), \tag{A.36}$$

where convergence is absolute and uniform (and hence point-wise) on $[c,d] \times [c,d]$.

A.4.3 Young's Inequality

Let $p,q,r \geq 1$ be such that

$$1 + \frac{1}{r} = \frac{1}{p} + \frac{1}{q}. \tag{A.37}$$

Let $f \in L^p(\mathbb{R}^n)$ and $g \in L^q(\mathbb{R}^n)$. Young's inequality for convolutions is

$$\|f * g\|_{L^r} \leq \|f\|_{L^p} \cdot \|g\|_{L^q}, \tag{A.38}$$

where

$$\|f(x)\|_{L^p} = \left(\int_{\mathbb{R}^n} |f(x)|^p dx \right)^{1/p} \tag{A.39}$$

and

$$f * g(x) = \int_{\mathbb{R}^n} g(x-y)f(y)dy. \tag{A.40}$$

In Chapter 8, we encountered Young's inequality in the following context: let $p = 2$, $q = 1$, $r = 2$, $g \in L^2(\mathscr{P})$, and $f \in L^1(V)$, where

$$\mathscr{P} = \{x - y, x \in S, y \in V\}. \tag{A.41}$$

In this case, we have

$$\|f * g\|^2_{L^2(S)} = \int_S |f * g(x)|^2 dx = \int_S \left| \int_V g(x-y)f(y)dy \right|^2 dx, \tag{A.42}$$

and Young's inequality is modified as follows:

$$\int_S \left| \int_V g(x-y)f(y)dy \right|^2 dx \leq \sup_{y\in V} \int_S |g(x-y)|^2 dx \cdot \left(\int_V |f(y)|dy \right)^2. \tag{A.43}$$

This result follows by first noting that, by the Schwarz inequality,

$$\begin{aligned} \left| \int_V g(x-y)f(y)dy \right|^2 &\leq \left(\int_V |g(x-y)| \cdot |f(y)|dy \right)^2 \\ &\leq \left(\int_V |g(x-y)|^2 |f(y)|dy \right) \left(\int_V |f(y)|dy \right), \end{aligned} \tag{A.44}$$

and, integrating over S and using Fubini's theorem, we obtain

$$\begin{aligned} \int_S \left| \int_V g(x-y)f(y)dy \right|^2 dx &\leq \int_S \left(\int_V |g(x-y)|^2 |f(y)|dy \right) \left(\int_V |f(y)|dy \right) dx \\ &= \int_V |f(y)|dy \int_S \int_V |g(x-y)|^2 |f(y)|dydx \\ &= \int_V |f(y)|dy \int_V |f(y)| \int_S |g(x-y)|^2 dxdy \\ &\leq \int_V |f(y)|dy \int_V |f(y)| \left(\sup_{y\in V} \int_S |g(x-y)|^2 dx \right) dy \\ &= \sup_{y\in V} \int_S |g(x-y)|^2 dx \left(\int_V |f(y)|dy \right)^2. \end{aligned} \tag{A.45}$$

In the case $S = \mathbb{R}^n$, we have

$$\int_{\mathbb{R}^n} |g(x-y)|^2 dx = \int_{\mathbb{R}^n} |g(x)|^2 dx, \tag{A.46}$$

and we obtain (A.38) without having to take the supremum.

A.5 Fredholm Integral Equations

A.5.1 First Kind

A Fredholm integral equation of the first kind is of the form

$$g(x) = \int_a^b K(x,y)f(y)dy, \tag{A.47}$$

where g and K are given and f is the unknown function. If all the functions are in L^2, the limits are infinite, and the kernel K is of the special form $K(x-y)$, then the right-hand side is a convolution and the solution is immediately given by the inverse Fourier transform,

$$f(x) = \mathcal{F}^{-1} \left\{ \frac{\mathcal{F}[f(x)](y)}{\mathcal{F}[K(x)](y)} \right\}. \tag{A.48}$$

A.5.2 Second Kind

An inhomogeneous Fredholm integral equation of the second kind is of the form

$$f(x) = \alpha \int_a^b K(x,y)f(y)dy + h(x), \tag{A.49}$$

where h and K are given and f is the unknown function. Its homogeneous version is obtained by letting $h(x) = 0$, yielding

$$f(x) = \alpha \int_a^b K(x,y)f(y)dy. \tag{A.50}$$

The resolvent formalism for the inhomogeneous Fredholm integral equation of the second kind (A.49) is the Liouville–Neumann series

$$f(x) = \sum_{n=0}^{\infty} \alpha^n \phi_n(x), \tag{A.51}$$

where

$$\phi_0(x) = h(x), \tag{A.52}$$

$$\phi_n(x) = \int_a^b K_n(x,y)f(y)dy, \quad n \geq 1, \tag{A.53}$$

and

$$K_n(x,y) = \int_a^b \int_a^b \cdots \int_a^b K(x,y_1)K(y_1,y_2)\cdots K(y_{n-1},y)dy_1 \cdots y_{n-1}. \tag{A.54}$$

This is obtained by solving the equation by iterative approximations. Taking the function $f_{-1}(x) = 0$ as the first approximation, we obtain the successive approximations

$$\begin{cases} f_0(x) = h(x) \\ f_1(x) = h(x) + \alpha \displaystyle\int_a^b K(x,y_1)h(y_1)dy_1 \\ f_2(x) = h(x) + \alpha \displaystyle\int_a^b K(x,y_1)h(y_1)dy_1 + \alpha^2 \int_a^b \int_a^b K(x,y_1)K(y_1,y_2)h(y_2)dy_1dy_2 \\ \quad \vdots \end{cases}$$

We are then led to the series

$$\lim_{n\to\infty} f_n(x) = \sum_{n=0}^{\infty} \alpha^n \phi_n(x). \tag{A.56}$$

The resolvent formalism is indeed a solution if the series converges uniformly:

> If the Liouville–Neumann series converges uniformly, then its sum is a solution of the Fredholm integral equation of the second kind (A.49).

This result follows immediately: noting that in this case term by term integration is permissible, and substituting (A.51) into the integral in (A.49), we obtain $f(x) - h(x)$.

The resolvent kernel for the Fredholm integral equation of the second kind is then defined as

$$K(x,y;\alpha) = \sum_{n=0}^{\infty} \alpha^n K_{n+1}(x,y), \tag{A.57}$$

so that the solution is also written in integral form as

$$f(x) = \int_a^b K(x,y,\alpha) f(y) dy. \tag{A.58}$$

A.6 Spectral Concentration over Arbitrary Measurable Sets

A special case of a Hilbert–Schmidt self-adjoint operator is the duration frequency limiting one $\mathcal{T}_P \mathcal{B}_Q \mathcal{T}_P$ encountered in Section 3.5.3. A general result regarding the behavior of the eigenvalues of this operator is stated in Section 3.5.4 and proved here.

Let P and Q be measurable sets in $\mathbb{R}^N$. For any real matrix A of size $N \times N$, we indicate by AP the set of points of the form Ax, where x is the column vector composed of the elements of $\mathbf{x} \in P$. We also indicate with $|\mathrm{A}|$ the determinant of A, with A^{T} the transpose of A, and with U the unit ball in $L^2(\mathbb{R}^N)$. We let $\mathrm{A} = \mathrm{A}(\tau)$ and $\mathrm{B} = \mathrm{B}(\rho)$ for real parameters τ and ρ. The case in which either matrix is constant or depends on multiple parameters is completely analogous.

For any $0 < \epsilon < 1$, let $N_\epsilon(\mathrm{A},\mathrm{B})$ be the number of eigenvalues of $\mathcal{T}_{\mathrm{A}P} \mathcal{B}_{\mathrm{B}Q} \mathcal{T}_{\mathrm{A}P}$ not smaller than ϵ. If

$$\lim_{(\tau,\rho)\to\infty} \mathrm{B}^{\mathrm{T}} \mathrm{A} U = \mathbb{R}^N, \tag{A.59}$$

then we have

$$\lim_{(\tau,\rho)\to\infty} \frac{N_\epsilon(\mathrm{A},\mathrm{B})}{|\mathrm{A}|\,|\mathrm{B}|} = (2\pi)^{-N} m(P) m(Q). \tag{A.60}$$

We let $K_{\mathrm{A,B}}(\mathbf{x},\mathbf{y})$ be the kernel of the operator $\mathcal{U}_{\mathrm{A,B}} = \mathcal{T}_{\mathrm{A}P} \mathcal{B}_{\mathrm{B}Q} \mathcal{T}_{\mathrm{A}P}$ having eigenvalues $\{\lambda_k\}$. We have

$$K_{\mathrm{A,B}}(\mathbf{x},\mathbf{y}) = \mathbb{1}_{\mathrm{A}P}(\mathbf{x}) \mathbb{1}_{\mathrm{A}P}(\mathbf{y}) h_{\mathrm{B}}(\mathbf{x}-\mathbf{y}), \tag{A.61}$$

so that

$$K_{\mathrm{A,B}}(\mathbf{x},\mathbf{x}) = \mathbb{1}_{\mathrm{A}P}(\mathbf{x}) h_{\mathrm{B}}(0). \tag{A.62}$$

All that is required is to establish Claim A.1 and Claim A.2 below, as they imply the statement of the result. The two claims state that all eigenvalues must be either close to one or close to zero, since both their sum and the sum of their squares have the same scaling order. The result then follows by noticing that the sum of the eigenvalues essentially corresponds to the number of non-zero eigenvalues and is of the order of $|\mathrm{A}||\mathrm{B}|$. We prove the two claims first, and then justify the implication that yields the final result.

Claim A.1 $\sum_k \lambda_k = |\mathrm{A}|\,|\mathrm{B}|\,(2\pi)^{-N} m(P) m(Q)$.

Claim A.2 $\sum_k \lambda_k^2 = |\mathrm{A}|\,|\mathrm{B}|\,(2\pi)^{-N} m(P) m(Q) + o(|\mathrm{A}|\,|\mathrm{B}|)$.

Proof of Claim A.1
By Mercer's theorem, there exists an orthonormal basis set $\{\phi_k\}$ for $L^2(\mathbb{R}^N)$ such that

$$K_{\mathrm{A,B}}(\mathbf{x},\mathbf{y}) = \sum_k \lambda_k \phi_k(\mathbf{x}) \phi_k(\mathbf{y}). \tag{A.63}$$

By orthonormality and (A.62), we have

$$\int_{\mathbb{R}^N} \mathbb{1}_{AP}(\mathbf{x}) h_{\mathrm{B}}(0) d\mathbf{x} = \sum_k \lambda_k, \tag{A.64}$$

and performing the computation

$$\begin{aligned}\int_{\mathbb{R}^N} \mathbb{1}_{AP}(\mathbf{x}) h_{\mathrm{B}}(0) d\mathbf{x} &= h_{\mathrm{B}}(0) m(\mathrm{A}P) \\ &= (2\pi)^{-N} m(\mathrm{B}Q) m(\mathrm{A}P) \\ &= (2\pi)^{-N} |A|\,|B|\, m(Q) m(P) \\ &= \sum_k \lambda_k \end{aligned} \tag{A.65}$$

establishes Claim A.1. □

Proof of Claim A.2
We let $K^{(2)}_{\mathrm{A,B}}(\mathbf{x},\mathbf{y})$ be the kernel of the operator $\mathcal{U}^2_{\mathrm{A,B}} = (\mathcal{T}_{\mathrm{AP}} \mathcal{B}_{\mathrm{B}Q} \mathcal{T}_{\mathrm{AP}})^2$ having eigenvalues $\{\lambda_k^2\}$. We have

$$\begin{aligned} K^{(2)}_{\mathrm{A,B}}(\mathbf{x},\mathbf{y}) &= \int_{\mathbb{R}^N} K_{\mathrm{A,B}}(\mathbf{x},\mathbf{z}) K_{\mathrm{A,B}}(\mathbf{z},\mathbf{y}) d\mathbf{z} \\ &= \int_{\mathbb{R}^N} \mathbb{1}_{AP}(\mathbf{x}) \mathbb{1}_{AP}(\mathbf{z}) h_{\mathrm{B}}(\mathbf{x}-\mathbf{z}) \\ &\quad \mathbb{1}_{AP}(\mathbf{z}) \mathbb{1}_{AP}(\mathbf{y}) h_{\mathrm{B}}(\mathbf{z}-\mathbf{y}) d\mathbf{z} \\ &= \mathbb{1}_{AP}(\mathbf{x}) \mathbb{1}_{AP}(\mathbf{y}) \int_{\mathrm{A}P} h_{\mathrm{B}}(\mathbf{x}-\mathbf{z}) h_{\mathrm{B}}(\mathbf{z}-\mathbf{y}) d\mathbf{z}. \end{aligned} \tag{A.66}$$

By Mercer's theorem, there exists an orthonormal basis set $\{\psi_n\}$ for $L^2(\mathbb{R}^N)$ such that

$$K^{(2)}_{\mathrm{A,B}}(\mathbf{x},\mathbf{y}) = \sum_k \lambda_k^2 \psi_k(\mathbf{x}) \psi_k(\mathbf{y}). \tag{A.67}$$

By orthonormality, we have

$$\int_{\mathbb{R}^N} K^{(2)}_{\mathrm{A,B}}(\mathbf{x},\mathbf{x}) d\mathbf{x} = \sum_k \lambda_k^2. \tag{A.68}$$

By (A.66) it follows that

$$\int_{\mathbb{R}^N} K^{(2)}_{\mathrm{A,B}}(\mathbf{x},\mathbf{x})d\mathbf{x} = \int_{\mathrm{A}P}\int_{\mathrm{A}P} |h_{\mathrm{B}}(\mathbf{x}-\mathbf{y})|^2 d\mathbf{x}d\mathbf{y}. \tag{A.69}$$

We apply the change of variable $\mathbf{x} = \mathrm{A}\mathbf{p}$, obtaining

$$\begin{aligned}\int_{\mathbb{R}^N} K^{(2)}_{\mathrm{A,B}}(\mathbf{x},\mathbf{x})d\mathbf{x} &= |\mathrm{A}| \int_{\mathrm{A}P} d\mathbf{y} \int_P |h_{\mathrm{B}}(\mathrm{A}\mathbf{p}-\mathbf{y})|^2 d\mathbf{p} \\ &= |\mathrm{A}| \int_P d\mathbf{p} \int_{\mathrm{A}P} |h_{\mathrm{B}}(\mathrm{A}\mathbf{p}-\mathbf{y})|^2 d\mathbf{y}.\end{aligned} \tag{A.70}$$

We apply another change of variable $\mathbf{u} = \mathrm{A}\mathbf{p} - \mathbf{y}$, obtaining

$$\int_{\mathbb{R}^N} K^{(2)}_{\mathrm{A,B}}(\mathbf{x},\mathbf{x})d\mathbf{x} = |\mathrm{A}| \int_P d\mathbf{p} \int_{\mathrm{A}(\mathbf{p}-P)} |h_{\mathrm{B}}(\mathbf{u})|^2 d\mathbf{u}. \tag{A.71}$$

Substituting (A.71) into (A.68) and dividing by $|\mathrm{A}|\,|\mathrm{B}|$, we have

$$\frac{1}{|\mathrm{A}|\,|\mathrm{B}|}\sum_k \lambda_k^2 = \int_P F_{\mathrm{A,B}}(\mathbf{p})d\mathbf{p}, \tag{A.72}$$

where

$$F_{\mathrm{A,B}}(\mathbf{p}) = |\mathrm{B}|^{-1} \int_{\mathrm{A}(\mathbf{p}-P)} |h_{\mathrm{B}}(\mathbf{u})|^2 d\mathbf{u}. \tag{A.73}$$

The function $F_{\mathrm{A,B}}(\mathbf{p})$ is dominated as

$$\begin{aligned}F_{\mathrm{A,B}}(\mathbf{p}) &\le |\mathrm{B}|^{-1}\int_{\mathbb{R}^N} |h_{\mathrm{B}}(\mathbf{u})|^2 d\mathbf{u} \\ &= (2\pi)^{-N}|\mathrm{B}|^{-1} m(\mathrm{B}Q) \\ &= (2\pi)^{-N} m(Q),\end{aligned} \tag{A.74}$$

which is integrable over P. Next, we show that

$$\lim_{(\tau,\rho)\to\infty} F_{\mathrm{A,B}}(\mathbf{p}) = (2\pi)^{-N} m(Q), \tag{A.75}$$

so that by Lebesgue's dominated convergence theorem, we have

$$\begin{aligned}\lim_{(\tau,\rho)\to\infty} \frac{1}{|\mathrm{A}|\,|\mathrm{B}|}\sum_k \lambda_k^2 &= \int_P \lim_{(\tau,\rho)\to\infty} F_{\mathrm{A,B}}(\mathbf{p})d\mathbf{p} \\ &= (2\pi)^{-N} m(P) m(Q),\end{aligned} \tag{A.76}$$

establishing Claim A.2. □

What remains is to prove (A.75). Substituting the result in Claim A.3 below into (A.73) and performing the change of variable $\mathrm{B}^{\mathrm{T}}\mathbf{u} = \mathbf{v}$, we have

$$\begin{aligned}F_{\mathrm{A,B}}(\mathbf{p}) &= |\mathrm{B}|^{-1}\int_{\mathrm{A}(\mathbf{p}-P)} |\mathrm{B}|^2\,|h(\mathrm{B}^{\mathrm{T}}\mathbf{u})|^2 d\mathbf{u} \\ &= \int_{\mathrm{B}^{\mathrm{T}}\mathrm{A}(\mathbf{p}-P)} |h(\mathbf{v})|^2 d\mathbf{v}.\end{aligned} \tag{A.77}$$

Assuming the boundary of P has measure zero,[1] we can assume that $\mathbf{p}$ is an interior point of P, so that the set $\mathbf{p} - P$ contains a ball of non-zero measure centered at the origin. It then follows from Parseval's theorem that the integral (A.77) converges to $(2\pi)^{-N} m(Q)$ as $(\tau, \rho) \to \infty$, and the proof is complete. □

Claim A.3 $h_{\mathrm{B}}(\mathbf{u}) = |\mathrm{B}|\, h(\mathrm{B}^{\mathrm{T}}\mathbf{u})$.

Proof of Claim A.3
We have

$$\mathcal{F} h_{\mathrm{B}}(\mathbf{y}) = \mathbb{1}_{\mathrm{B}Q}(\mathbf{y}) = \mathbb{1}_{Q}(\mathrm{B}^{-1}\mathbf{y}), \tag{A.78}$$

and the proof of Claim A.3 follows by computing the inverse Fourier transform:

$$\begin{aligned}
h_{\mathrm{B}}(\mathbf{u}) &= \mathcal{F}^{-1}\mathbb{1}_{\mathrm{B}Q}(\mathbf{u}) \\
&= (2\pi)^{-N} \int_{\mathbb{R}^N} \mathbb{1}_{\mathrm{B}Q}(\mathbf{y}) \exp(j\mathbf{u} \cdot \mathbf{y}) d\mathbf{y} \\
&= (2\pi)^{-N} \int_{\mathbb{R}^N} \mathbb{1}_{Q}(\mathrm{B}^{-1}\mathbf{y}) \exp(j\mathbf{u} \cdot (\mathrm{B}\mathrm{B}^{-1}\mathbf{y})) d\mathbf{y} \\
&= (2\pi)^{-N} |\mathrm{B}| \int_{\mathbb{R}^N} \mathbb{1}_{Q}(\mathbf{z}) \exp(j\mathbf{u} \cdot (\mathrm{B}\mathbf{z})) d\mathbf{z} \\
&= (2\pi)^{-N} |\mathrm{B}| \int_{\mathbb{R}^N} \mathbb{1}_{Q}(\mathbf{z}) \exp(j\mathbf{z} \cdot (\mathrm{B}^{\mathrm{T}}\mathbf{u})) d\mathbf{z} \\
&= |\mathrm{B}| h(\mathrm{B}^{\mathrm{T}}\mathbf{u}).
\end{aligned}$$

□

Proof of (A.60)
We are now ready to prove that Claim A.1 and Claim A.2 imply the final result. We first show that the number of "intermediate eigenvalues" that are not close to one nor to zero is $o(|\mathrm{A}||\mathrm{B}|)$. Let

$$S = \sum_k \lambda_k (1 - \lambda_k) = \sum_k \lambda_k - \sum_k \lambda_k^2. \tag{A.79}$$

By Claim A.1 and Claim A.2, we have $S = o(|\mathrm{A}||\mathrm{B}|)$. For every $0 < \epsilon_1 < \epsilon_2 < 1$, we have that the eigenvalues in the range $\lambda_k \in [\epsilon_1, \epsilon_2]$ contribute to S by an amount no smaller than

$$\min\{\epsilon_1(1-\epsilon_1), \epsilon_2(1-\epsilon_2)\} \left[N_{\epsilon_1}(\mathrm{A},\mathrm{B}) - N_{\epsilon_2}(\mathrm{A},\mathrm{B})\right] \le S = o(|\mathrm{A}||\mathrm{B}|). \tag{A.80}$$

It follows that

$$\lim_{(\tau,\rho)\to\infty} \frac{N_{\epsilon_1}(\mathrm{A},\mathrm{B}) - N_{\epsilon_2}(\mathrm{A},\mathrm{B})}{|\mathrm{A}||\mathrm{B}|} = 0. \tag{A.81}$$

Next, we show that

$$\limsup_{(\tau,\rho)\to\infty} \frac{N_{\epsilon_2}(\mathrm{A},\mathrm{B})}{|\mathrm{A}||\mathrm{B}|} \le \frac{1}{\epsilon_2} (2\pi)^{-N} m(p) m(Q), \tag{A.82}$$

[1] If the boundary has positive measure, one can obtain the same result using an approximation argument.

and

$$\liminf_{(\tau,\rho)\to\infty} \frac{N_{\epsilon_1}(\mathrm{A},\mathrm{B})}{|\mathrm{A}||\mathrm{B}|} \geq (1-\epsilon_1)(2\pi)^{-N} m(p)m(Q). \tag{A.83}$$

Combining (A.81), (A.82), and (A.83), the result follows.

We have

$$\sum_k \lambda_k \geq \epsilon_2 N_{\epsilon_2}(\mathrm{A},\mathrm{B}), \tag{A.84}$$

from which it follows from Claim A.1 that

$$N_{\epsilon_2}(\mathrm{A},\mathrm{B}) \leq \frac{1}{\epsilon_2}|\mathrm{A}||\mathrm{B}|(2\pi)^{-N} m(P)m(Q). \tag{A.85}$$

Dividing by ($|\mathrm{A}||\mathrm{B}|$) and taking the limit for $(\tau,\rho)\to\infty$, (A.82) follows.

We have

$$\begin{aligned}\sum_k \lambda_k^2 &= \sum_{k=0}^{N_{\epsilon_1}-1} \lambda_k^2 + \sum_{k=N_{\epsilon_1}}^{\infty} \lambda_k^2 \\ &\leq N_{\epsilon_1}(\mathrm{A},\mathrm{B}) + \epsilon_1 \sum_{k=N_{\epsilon_1}}^{\infty} \lambda_k \\ &\leq N_{\epsilon_1}(\mathrm{A},\mathrm{B}) + \epsilon_1 \sum_{k=0}^{\infty} \lambda_k. \end{aligned} \tag{A.86}$$

From this, it follows from Claim A.2 that

$$\begin{aligned} N_{\epsilon_1}(\mathrm{A},\mathrm{B}) &\geq \sum_k \lambda_k^2 - \epsilon_1 \sum_k \lambda_k \\ &= |\mathrm{A}\|\mathrm{B}|(2\pi)^{-N} m(P)m(Q) + o(|\mathrm{A}\|\mathrm{B}|) - \epsilon_1 |\mathrm{A}\|\mathrm{B}|(2\pi)^{-N} m(P)m(Q) \\ &= (1-\epsilon_1)|\mathrm{A}\|\mathrm{B}|(2\pi)^{-N} m(P)m(Q) + o((|\mathrm{A}\|\mathrm{B}|). \end{aligned} \tag{A.87}$$

Dividing by ($|\mathrm{A}\|\mathrm{B}|$) and taking the limit for $(\tau,\rho)\to\infty$, (A.83) follows. □

A.7 Kac–Murdock–Szegö Theorem

Let $f(t)$ be an even, continuous almost everywhere, uniformly bounded, and square-integrable function, and let its Fourier transform be $F(\omega)$. Consider the eigenvalue equation

$$\int_{-T/2}^{T/2} f(t-\tau)\psi_n(\tau)d\tau = \lambda_n \psi_n(t), \quad t \in [-T/2, T/2]. \tag{A.88}$$

If (a,b) does not contain zero and the set where $F(\omega) = a$ or $F(\omega) = b$ has measure zero, then the number of eigenvalues that satisfy

$$a < \lambda_n < b \tag{A.89}$$

is

$$N_T(a,b) = \frac{T}{2\pi}\int_{F(\omega)\in(a,b)} d\omega + o(T). \tag{A.90}$$

This result is useful in the following context encountered in Chapter 6. If $f(t)$ is the autocorrelation function of a wide-sense stationary stochastic process, we can use the theorem to determine the number of terms in its Karhunen–Loève representation that are sufficient to obtain a satisfactory approximation.

Recall that by (6.71), the error of an N-term Karhunen–Loève representation of the process is given by the tail sum of the eigenvalues of (A.88). From Claim A.1 of Section A.6, we have

$$\sum_{n=0}^{N-1}\lambda_n + \sum_{n=N}^{\infty}\lambda_n = \frac{T}{2\pi}\int_{-\infty}^{\infty} F(\omega)\, d\omega. \tag{A.91}$$

Letting

$$N = \frac{T}{2\pi}\int_{F(\omega)\in(a,b)} d\omega, \tag{A.92}$$

by (A.90) we have that, as $T \to \infty$,

$$\sum_{n=N}^{\infty}\lambda_n \le \frac{T}{2\pi}\left(\int_{-\infty}^{\infty} F(\omega)\, d\omega - a\int_{F(\omega)\in(a,b)} d\omega\right) + o(T). \tag{A.93}$$

This provides a bound on the error (6.71) in terms of the power spectral density of the process.

A.7.1 Constant Power Spectral Density

In the case that the power spectral density is $F(\omega) = 1$ for $\omega \in [-\Omega, \Omega]$ and zero otherwise, letting $a = (1-\epsilon)$ and $b > 1$ in (A.93) and (A.92), we obtain

$$\sum_{n=N}^{\infty}\lambda_n \le \epsilon\frac{\Omega T}{\pi} + o(T), \tag{A.94}$$

$$N = \frac{\Omega T}{\pi}. \tag{A.95}$$

Recalling that the total energy of the process over the observation interval is

$$\int_{-T/2}^{T/2}\frac{1}{2\pi}\int_{-\infty}^{\infty} F(\omega)d\omega dt = \frac{\Omega T}{\pi}, \tag{A.96}$$

by dividing (A.94) by (A.96) it follows that the energy error normalized to the total energy of the process is at most ϵ.

This result should come as no surprise. By (A.90), the number of eigenvalues that are within ϵ from one is asymptotically equal to N_0, and by (A.96) the total energy of the process is N_0. It then follows that retaining N_0 terms in the approximation is enough to capture most of the energy of the process. A tighter result is described next, and is used in Section 6.4.4 to determine the stochastic diversity.

A.7.2 Evaluation of the Stochastic Diversity

We evaluate the tail sum of the eigenvalues in (A.88) when $F(\omega) = 1$ for $\omega \in [-\Omega, \Omega]$ and zero otherwise. This result is useful to determine the stochastic diversity of the class of processes studied in Section 6.4.4.

From Claim A.1 of Section A.6, we have

$$\sum_{n=0}^{N-1} \lambda_n + \sum_{n=N}^{\infty} \lambda_n = N_0. \tag{A.97}$$

Letting $N = (1-\epsilon)N_0$, dividing both sides of the equation by N_0, taking the limit for $N_0 \to \infty$, and using (2.131), we have that

$$(1-\epsilon) + \lim_{N_0 \to \infty} \frac{1}{N_0} \sum_{n=N}^{\infty} \lambda_n = 1, \tag{A.98}$$

from which it follows that

$$\sum_{n=N}^{\infty} \lambda_n = \epsilon N_0 + o(N_0). \tag{A.99}$$

This result should be compared with (A.94). In the case of the upper bound (A.94) we have used the Kac–Murdock–Szegö theorem, which tells us that the number of eigenvalues close to one is essentially N_0. In the case of (A.99), we have used Slepian's concentration result, which tells us even more: there are only slightly less than N_0 eigenvalues having value one. It then follows that $(1-\epsilon)N_0$ terms can capture most of the energy of the process, and we obtain a tight expression for the energy of the error of the N-dimensional Karhunen–Loève approximation of the process.

Appendix B: Vector Calculus

B.1 Coordinate Systems

Figure B.1 displays the notation and conventions for the Cartesian coordinate system. Figure B.2 does the same for cylindrical coordinates, while Figure B.3 illustrates spherical coordinates.

B.2 Coordinate Transformations

B.2.1 Rectangular–Spherical

$$
\begin{aligned}
x &= r \sin\theta \cos\phi \\
y &= r \sin\theta \sin\phi \\
z &= r \cos\theta
\end{aligned}
\tag{B.1}
$$

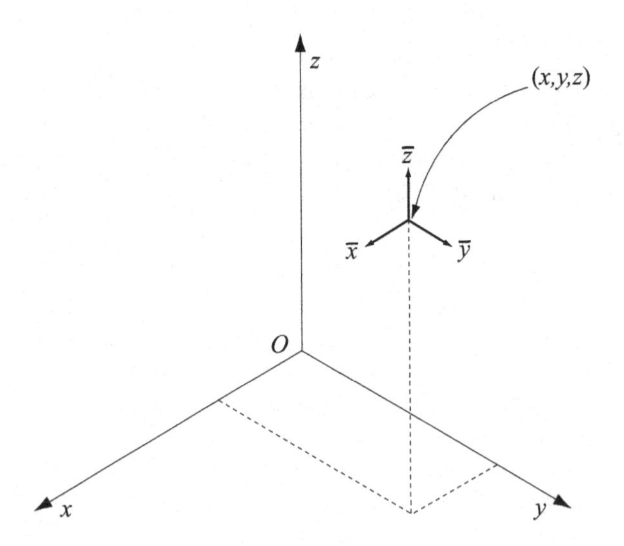

Fig. B.1 Cartesian coordinates.

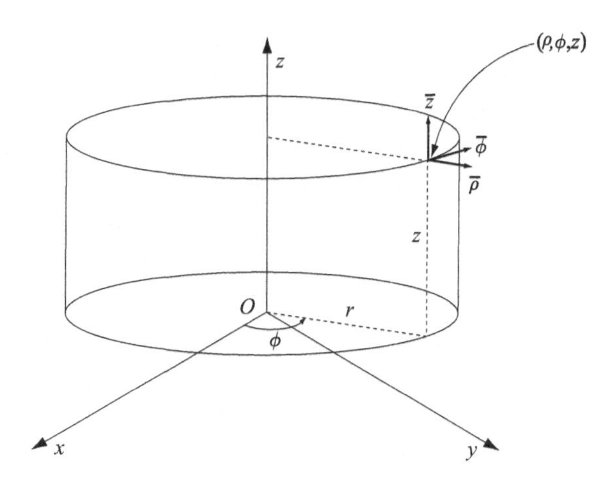

Fig. B.2 Cylindrical coordinates.

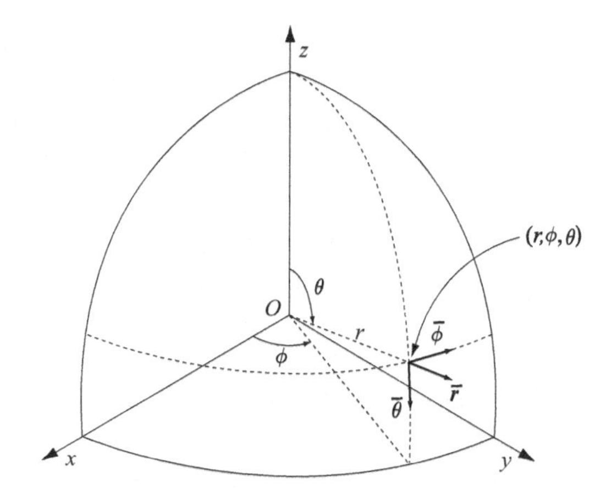

Fig. B.3 Spherical coordinates.

B.2.2 Rectangular–Cylindrical

$$
\begin{aligned}
x &= r\cos\phi \\
y &= r\sin\phi \\
z &= z
\end{aligned} \tag{B.2}
$$

B.3 Differential Operators

B.3.1 Rectangular Coordinates

$$
\nabla f = \frac{\partial f}{\partial x}\bar{\mathbf{x}} + \frac{\partial f}{\partial y}\bar{\mathbf{y}} + \frac{\partial f}{\partial z}\bar{\mathbf{z}} \tag{B.3}
$$

$$\nabla \cdot \mathbf{f} = \frac{\partial f_x}{\partial x} + \frac{\partial f_y}{\partial y} + \frac{\partial f_z}{\partial z} \tag{B.4}$$

$$\nabla \times \mathbf{f} = \left(\frac{\partial f_z}{\partial y} - \frac{\partial f_y}{\partial z}\right)\bar{\mathbf{x}} + \left(\frac{\partial f_x}{\partial z} - \frac{\partial f_z}{\partial x}\right)\bar{\mathbf{y}} + \left(\frac{\partial f_y}{\partial x} - \frac{\partial f_x}{\partial y}\right)\bar{\mathbf{z}} \tag{B.5}$$

$$\nabla^2 f = \nabla \cdot \nabla f = \frac{\partial^2 f}{\partial x^2} + \frac{\partial^2 f}{\partial y^2} + \frac{\partial^2 f}{\partial z^2} \tag{B.6}$$

$$\nabla^2 \mathbf{f} = \nabla^2 f_x \bar{\mathbf{x}} + \nabla^2 f_y \bar{\mathbf{y}} + \nabla^2 f_z \bar{\mathbf{z}} \tag{B.7}$$

B.3.2 Cylindrical Coordinates

$$\nabla f = \frac{\partial f}{\partial r}\bar{\mathbf{r}} + \frac{1}{r}\frac{\partial f}{\partial \phi}\bar{\boldsymbol{\phi}} + \frac{\partial f}{\partial z}\bar{\mathbf{z}} \tag{B.8}$$

$$\nabla \cdot \mathbf{f} = \frac{1}{r}\frac{\partial (r f_r)}{\partial r} + \frac{1}{r}\frac{\partial f_\phi}{\partial \phi} + \frac{\partial f_z}{\partial z} \tag{B.9}$$

$$\nabla \times \mathbf{f} = \left(\frac{1}{r}\frac{\partial f_z}{\partial \phi} - \frac{\partial f_\phi}{\partial z}\right)\bar{\mathbf{r}} + \left(\frac{\partial f_r}{\partial z} - \frac{\partial f_z}{\partial r}\right)\bar{\phi} + \frac{1}{r}\left(\frac{\partial (r f_\phi)}{\partial r} - \frac{\partial f_r}{\partial \phi}\right)\bar{\mathbf{z}} \tag{B.10}$$

$$\nabla^2 f = \frac{1}{r}\frac{\partial}{\partial r}\left(r\frac{\partial f}{\partial r}\right) + \frac{1}{r^2}\frac{\partial^2 f}{\partial \phi^2} + \frac{\partial^2 f}{\partial z^2} \tag{B.11}$$

$$\nabla^2 \mathbf{f} = \left(\nabla^2 f_r - \frac{2}{r^2}\frac{\partial f_\phi}{\partial \phi} - \frac{f_r}{r^2}\right)\bar{\mathbf{r}} + \left(\nabla^2 f_\phi + \frac{2}{r^2}\frac{\partial f_r}{\partial \phi} - \frac{f_\phi}{r^2}\right)\bar{\mathbf{y}} + (\nabla^2 f_z)\bar{\mathbf{z}} \tag{B.12}$$

B.3.3 Spherical Coordinates

$$\nabla f = \frac{\partial f}{\partial r}\bar{\mathbf{r}} + \frac{1}{r}\frac{\partial f}{\partial \theta}\bar{\theta} + \frac{1}{r\sin\theta}\frac{\partial f}{\partial \phi}\bar{\phi} \tag{B.13}$$

$$\nabla \cdot \mathbf{f} = \frac{1}{r^2}\frac{\partial (r^2 f_r)}{\partial r} + \frac{1}{r\sin\theta}\frac{\partial f_\theta \sin\theta}{\partial \theta} + \frac{1}{r\sin\theta}\frac{\partial f_\phi}{\partial \phi} \tag{B.14}$$

$$\nabla \times \mathbf{f} = \left(\frac{1}{r\sin\theta}\right)\left(\frac{\partial(f_\phi \sin\theta)}{\partial\theta} - \frac{\partial f_\theta}{\partial\phi}\right)\bar{\mathbf{r}} + \frac{1}{r}\left(\frac{1}{\sin\theta}\frac{\partial f_r}{\partial\phi} - \frac{(r\partial f_\phi)}{\partial r}\right)\bar{\boldsymbol{\theta}}$$
$$+\frac{1}{r}\left(\frac{\partial(rf_\theta)}{\partial r} - \frac{\partial f_r}{\partial\theta}\right)\bar{\boldsymbol{\phi}} \tag{B.15}$$

$$\nabla^2 f = \frac{1}{r^2}\frac{\partial}{\partial r}\left(r^2\frac{\partial f}{\partial r}\right) + \frac{1}{r^2\sin\theta}\frac{\partial}{\partial\theta}\left(\sin\theta\frac{\partial f}{\partial\theta}\right) + \frac{1}{r^2\sin^2\theta}\frac{\partial^2 f}{\partial\phi^2} \tag{B.16}$$

$$\nabla^2\mathbf{f} = \left[\nabla^2 f_r - \frac{2}{r^2}\left(f_r + f_\theta\cot\theta + \csc\theta\frac{\partial f_\phi}{\partial\phi} + \frac{\partial f_\theta}{\partial\theta}\right)\right]\bar{\mathbf{r}}$$
$$+\left[\nabla^2 f_\theta - \frac{1}{r^2}\left(f_\theta\csc^2\theta - 2\frac{\partial f_r}{\partial\theta} + 2\cot\theta\csc\theta\frac{\partial f_\phi}{\partial\phi}\right)\right]\bar{\boldsymbol{\theta}}$$
$$+\left[\nabla^2 f_\phi - \frac{1}{r^2}\left(f_\phi\csc^2\theta - 2\csc\theta\frac{\partial f_r}{\partial\phi} - 2\cot\theta\csc\theta\frac{\partial f_\theta}{\partial\phi}\right)\right]\bar{\boldsymbol{\phi}} \tag{B.17}$$

B.4 Differential Relationships

$$\nabla\cdot\nabla f = \nabla^2 f \tag{B.18}$$

$$\nabla(fg) = f\nabla g + g\nabla f \tag{B.19}$$

$$\nabla\cdot(f\mathbf{g}) = \mathbf{g}\cdot\nabla f + f\nabla\cdot\mathbf{g} \tag{B.20}$$

$$\nabla\times(f\mathbf{g}) = f\nabla\times\mathbf{g} - \mathbf{g}\times\nabla f \tag{B.21}$$

$$\nabla\cdot(\mathbf{f}\times\mathbf{g}) = \mathbf{g}\cdot(\nabla\times\mathbf{f}) - \mathbf{f}\cdot(\nabla\times\mathbf{g}) \tag{B.22}$$

$$\nabla\times(\mathbf{f}\times\mathbf{g}) = \mathbf{f}(\nabla\cdot\mathbf{g}) - \mathbf{g}(\nabla\cdot\mathbf{f}) + (\mathbf{g}\cdot\nabla)\mathbf{f} - (\mathbf{f}\cdot\nabla)\mathbf{g} \tag{B.23}$$

$$\nabla(\mathbf{f}\cdot\mathbf{g}) = \mathbf{f}\times(\nabla\times\mathbf{g}) + \mathbf{g}\times(\nabla\times\mathbf{f}) + (\mathbf{g}\cdot\nabla)\mathbf{f} + (\mathbf{f}\cdot\nabla)\mathbf{g} \tag{B.24}$$

$$\nabla\times\nabla f = 0 \tag{B.25}$$

$$\nabla\cdot(\nabla\times\mathbf{f}) = 0 \tag{B.26}$$

$$\nabla\times\nabla\times\mathbf{f} = \nabla(\nabla\cdot\mathbf{f}) - \nabla^2\mathbf{f} \tag{B.27}$$

B.5 Dyadic Analysis

In vector analysis, the scalar product transform $\mathbf{A} = c\mathbf{B}$ lets the vector $\mathbf{A}$ retain the same direction as $\mathbf{B}$. This is written in matrix form as

$$\begin{pmatrix} A_x \\ A_y \\ A_z \end{pmatrix} = c \begin{pmatrix} B_x \\ B_y \\ B_z \end{pmatrix}. \tag{B.28}$$

A more general linear transformation allows each component of $\mathbf{B}$ to influence each component of $\mathbf{A}$, so that the transformation changes the direction as well as the magnitude of the original vector. This *dyadic* transformation is written as

$$\mathbf{A} = \underline{\mathbf{C}} \cdot \mathbf{B}, \tag{B.29}$$

where the dyad $\underline{\mathbf{C}}$ is defined as

$$\underline{\mathbf{C}} = c_{xx}\bar{\mathbf{x}}\bar{\mathbf{x}} + c_{xy}\bar{\mathbf{x}}\bar{\mathbf{y}} + c_{xz}\bar{\mathbf{x}}\bar{\mathbf{z}} + \tag{B.30}$$

$$c_{yx}\bar{\mathbf{y}}\bar{\mathbf{x}} + c_{yy}\bar{\mathbf{y}}\bar{\mathbf{y}} + c_{yz}\bar{\mathbf{y}}\bar{\mathbf{z}} + \tag{B.31}$$

$$c_{zx}\bar{\mathbf{z}}\bar{\mathbf{x}} + c_{zy}\bar{\mathbf{z}}\bar{\mathbf{y}} + c_{zz}\bar{\mathbf{z}}\bar{\mathbf{z}}. \tag{B.32}$$

The dyadic transformation can be written in convenient matrix form as

$$\begin{pmatrix} A_x \\ A_y \\ A_z \end{pmatrix} = \begin{pmatrix} c_{xx} & c_{xy} & c_{xz} \\ c_{yx} & c_{yy} & c_{yz} \\ c_{zx} & c_{zy} & c_{zz} \end{pmatrix} \begin{pmatrix} B_x \\ B_y \\ B_z \end{pmatrix}. \tag{B.33}$$

The matrix C is a *second-rank tensor* and each element describes the influence of one field quantity on another. For example, c_{xy} describes the $\bar{\mathbf{x}}$ component of field $\mathbf{B}$ due to the $\bar{\mathbf{y}}$ component of field $\mathbf{A}$.

The dyad $\underline{\mathbf{C}}$ obeys the usual rules for dot product, but care should be taken in observing the order of the unit vectors and the dot product. For example, $\bar{\mathbf{x}}\bar{\mathbf{y}} \cdot \mathbf{B} = \bar{\mathbf{x}}B_y$, but interchanging the order of the unit vectors gives a different result, namely $\bar{\mathbf{y}}\bar{\mathbf{x}} \cdot \mathbf{B} = \bar{\mathbf{y}}B_x$. Similarly, $\underline{\mathbf{C}} \cdot \mathbf{B} \neq \mathbf{B} \cdot \underline{\mathbf{C}}$. This should not be surprising given the matrix interpretation.

B.6 Flux Theorems

Volumes V, surfaces S, lines C, and unit vectors $\bar{\mathbf{n}}$ and $\bar{\mathbf{c}}$ are defined in Figure B.4. All surfaces and curves are sufficiently regular that tangent and normal vectors can be defined unambiguously.

Gauss' theorem relates the integral of the divergence of a vector field $\mathbf{A}$ over a volume V to the outgoing flux of the vector field through its bounding surface S, namely

$$\iiint_V \nabla \cdot \mathbf{A}\, dV = \oiint_S \mathbf{A} \cdot \bar{\mathbf{n}}\, dS. \tag{B.34}$$

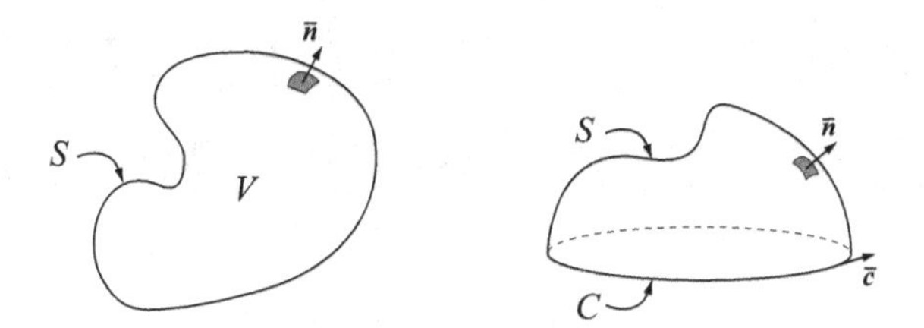

Fig. B.4 Volume, surfaces, and lines of integration.

Stokes' theorem relates the flux of the curl of a vector field **A** through an open surface S to the line integral of the vector field over its boundary C, namely

$$\iint_S \nabla \times \mathbf{A} \cdot \bar{\mathbf{n}} dS = \oint_C A \cdot \bar{\mathbf{c}} dc. \tag{B.35}$$

Appendix C: Methods for Asymptotic Evaluation of Integrals

C.1 Exponential-Type Integrals: Laplace's Method

Consider the integral

$$I(W) = \int_a^b f(x) \exp[-W\Phi(x)] dx, \tag{C.1}$$

where $\Phi(x)$ is a real-valued function and $f(x)$ and $\Phi(x)$ are smooth enough to be replaced by local Taylor approximations of appropriate degree. The idea of the method to approximate this integral as $W \to \infty$ goes back to Laplace (1774). According to Laplace, the major contribution to the integral arises from the immediate vicinity of those points of the interval $[a,b]$ at which $\Phi(x)$ assumes its smallest value.

Suppose the exponent has a single minimum at a critical point x_0 interior to $[a,b]$ so that $\dot{\Phi}(x_0) = 0$, $\ddot{\Phi}(x_0) > 0$. We expand the integrand's functions as

$$f(x) \simeq f(x_0), \quad \Phi(x) \simeq \Phi(x_0) + \frac{1}{2}\ddot{\Phi}(x_0)(x-x_0)^2, \tag{C.2}$$

obtaining

$$I(W) \sim \int_{x_0-\epsilon}^{x_0+\epsilon} f(x_0) \exp\left[W\Phi(x_0) + \frac{W}{2}\ddot{\Phi}(x_0)(x-x_0)^2\right] dx. \tag{C.3}$$

Since $\ddot{\Phi}(x_0) > 0$, this integral vanishes as $W \to \infty$ for all $x \neq x_0$. We can then extend the integration limits to the whole interval $(-\infty, \infty)$, obtaining, as $W \to \infty$,

$$I(W) \sim f(x_0) \exp[-W\Phi(x_0)] \int_{-\infty}^{\infty} \exp\left[-\frac{W}{2}\ddot{\Phi}(x_0)(x-x_0)^2\right] dx, \tag{C.4}$$

which leads to the final result:

$$I(W) \sim f(x_0) \exp[-W\Phi(x_0)] \sqrt{\frac{2\pi}{W\ddot{\Phi}(x_0)}} \quad \text{as } W \to \infty. \tag{C.5}$$

A factor of $1/2$ needs to be introduced if the minimum occurs at one of the end points, $x_0 = a$ or $x_0 = b$.

C.2 Fourier-Type Integrals: Stationary Phase Method

Consider the integral

$$I(W) = \int_a^b f(x) \exp[jW\Phi(x)]dx, \tag{C.6}$$

where the phase $\phi(x)$ is a real-valued function and $f(x)$ and $\Phi(x)$ are smooth enough to be replaced by local Taylor approximations of appropriate degree.

The idea is that as $W \to \infty$ the exponential becomes rapidly oscillating, leading to cancellations between positive and negative contributions to the integral. It follows that the integral tends to zero except at points where the phase is stationary, namely $\dot{\Phi}(x) = 0$.

Suppose the phase has only one stationary point x_0, with $\dot{\Phi}(x_0) = 0$, $\ddot{\Phi}(x_0) \neq 0$. We expand the integrand's functions as

$$f(x) \simeq f(x_0), \quad \Phi(x) \simeq \Phi(x_0) + \frac{1}{2}\ddot{\Phi}(x_0)(x - x_0)^2, \tag{C.7}$$

obtaining

$$I(W) \sim \int_{x_0-\epsilon}^{x_0+\epsilon} f(x_0) \exp\left[jW\Phi(x_0) + j\frac{W}{2}\ddot{\Phi}(x_0)(x - x_0)^2\right] dx. \tag{C.8}$$

This integral vanishes as $W \to \infty$ for all $x \neq x_0$. We can then extend the integration limits to the whole interval $(-\infty, \infty)$, obtaining, as $W \to \infty$,

$$I(W) \sim f(x_0) \exp[Wj\Phi(x_0)] \int_{-\infty}^{\infty} \exp\left[j\frac{W}{2}\ddot{\Phi}(x_0)(x - x_0)^2\right] dx, \tag{C.9}$$

which leads to the final result:

$$I(W) \sim f(x_0) \exp[Wj\Phi(x_0)] \sqrt{\frac{2\pi}{W|\ddot{\Phi}(x_0)|}} \exp[\pm\pi/4] \text{ as } W \to \infty. \tag{C.10}$$

The presence of the plus or minus depends on whether the stationary point is a minimum, $\ddot{\Phi}(x_0) > 0$, or a maximum, $\ddot{\Phi}(x_0) < 0$.

In the case that $\ddot{\Phi}(x_0) = 0$, the method is appropriately modified by taking into account higher-order terms in the Taylor expansion.

C.3 Complex Integrals: Saddle Point Method

Consider the integral

$$I(W) = \int_{\rho} f(z) \exp[W\Phi(z)]dz, \tag{C.11}$$

where $\Phi(z) = u(x,y) + jv(x,y)$ is a complex-valued function of complex argument $z = (x,y)$, the integration is along the path ρ in the complex plane, and $f(z)$ and $\Phi(z)$ are analytic at z_0.

As in the stationary phase method, the idea is that as $W \to \infty$ the dominant contribution to the integral comes from the points of stationary phase along the integration path. In addition, the path is deformed so that it passes through the critical points along the direction for which the real part $u(x,y)$ is maximized while the imaginary part $v(x,y)$ is stationary. This allows us to capture the largest possible contribution to the integral at the stationary point.

We require, at the stationary point,

$$\frac{\partial u}{\partial x} = \frac{\partial u}{\partial y} = \frac{\partial v}{\partial x} = \frac{\partial v}{\partial y} = 0, \tag{C.12}$$

or, equivalently,

$$\frac{\partial \Phi}{\partial z} = 0. \tag{C.13}$$

At such a critical point, Φ has neither a maximum nor a minimum, but has a *saddle point*. This follows from the Cauchy–Riemann conditions that require u and v to satisfy the Laplace equations

$$\frac{\partial^2 u}{\partial x^2} + \frac{\partial u}{\partial^2 y^2} = 0, \tag{C.14}$$

$$\frac{\partial^2 v}{\partial x^2} + \frac{\partial^2 v}{\partial y^2} = 0. \tag{C.15}$$

If, for example,

$$\frac{\partial^2 u}{\partial x^2} > 0, \tag{C.16}$$

then

$$\frac{\partial^2 u}{\partial y^2} < 0, \tag{C.17}$$

and hence the saddle.

Suppose there is only one stationary point z_0. Since both $u(x,y)$ and $v(x,y)$ have saddle points at z_0, a trajectory through such a saddle point can exhibit a maximum or a minimum, depending on the angle from which the saddle point is approached.

To evaluate the integral, we deform the integration path, as we may by Cauchy's theorem, so that it goes through the saddle point in such a way that the angle approaching the critical point makes $u(x,y)$ exhibit the steepest maximum while keeping $v(x,y)$ stationary. This is possible because the Cauchy–Riemann conditions ensure that the level surfaces of $u(x,y)$ and $v(x,y)$ are perpendicular to each other – see Figure C.1.

Expanding the integrand's functions as

$$f(x) \simeq f(z_0), \quad \Phi(z) \simeq \Phi(z_0) + \frac{1}{2}\ddot{\Phi}(z_0)(z-z_0)^2, \tag{C.18}$$

and expressing in polar coordinates

$$\ddot{\Phi}(z_0) = |\ddot{\Phi}(z_0)|\exp(j\theta), \tag{C.19}$$

$$(z-z_0) = r\exp(j\phi), \tag{C.20}$$

we have

$$\Phi(z) \simeq \Phi(z_0) + \frac{1}{2}|\ddot{\Phi}(z_0)|r^2\exp[j(\theta+2\phi)]. \tag{C.21}$$

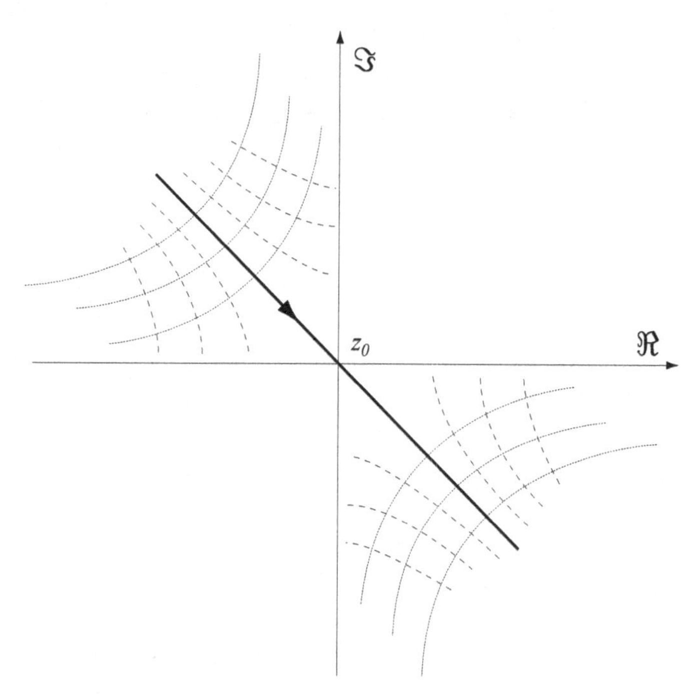

Fig. C.1 The dashed lines represent the locus of points of stationary phase, where $v(x,y)$ is constant. Perpendicular to these are the dotted lines representing the locus of points where $u(x,y)$ is constant. The directed line is the integration path of stationary phase and steepest ascent and descent through the critical point z_0.

It follows that moving away from z_0 by performing a step of length $|\ddot{\Phi}(z_0)|r^2/2$, the real part makes the maximum decrease if the direction of this step is such that

$$\exp[j(\theta+2\phi)] = -1. \tag{C.22}$$

This implies that choosing the angle $\phi = -\theta/2 \pm \pi/2$ makes the real part rise as fast as possible to a maximum at z_0 and subsequently decrease as rapidly as possible. The plus or minus ambiguity depends on the integration proceeding in the forward or backward direction through the critical point and thus depends on the order of the extrema in the integration path. The two directions obviously differ by an angle π.

The integration can now proceed in the usual way by extending the limits of integration to infinity:

$$\begin{aligned} I(W) &= \int_\rho f(z)\exp[W\Phi(z)]dz \\ &\approx f(z_0)\exp[W\Phi(z_0)]\exp(j\phi)\int_{-\infty}^{\infty}\exp\left[-W\frac{|\ddot{\Phi}(z_0)|}{2}r^2\right]dr \\ &\approx f(z_0)\exp[W\Phi(z_0)]\exp(j\phi)\sqrt{\frac{2\pi}{W|\ddot{\Phi}(z_0)|}}. \end{aligned} \tag{C.23}$$

In the case that $\ddot{\Phi}(x_0) = 0$, the method is appropriately modified by taking into account higher derivatives in the Taylor expansion. In addition, if singularities of $f(z)$ are crossed in the deformation of the path, their residual contribution should also be taken into account.

Appendix D: Stochastic Integration

In order to define the integral of a stochastic process rigorously, we may consider integration at every sample point of the process. In this case, the integral is defined as a linear functional of the process. Convergence issues should then be handled with care, because one needs to assume that the sample paths of the random process are well behaved, so that the integral can be defined for every sample function. A simpler approach is to consider mean square integration, which relies much less on the properties of the sample paths and is the typical interpretation when only statistical averages are of interest rather than individual sample functions.

D.1 Mean Square Integration

A sequence of random variables $\{\mathsf{X}_n\}$ converges to a random variable X in the mean square sense if all the random variables are defined on the same probability space,

$$\mathbb{E}(\mathsf{X}_n) < \infty \text{ for all } n, \tag{D.1}$$

and

$$\lim_{n\to\infty} \mathbb{E}[(\mathsf{X}_n - \mathsf{X})^2] = 0. \tag{D.2}$$

Let $\mathsf{G}(t)$ be a stochastic process and $x(t)$ be a deterministic function. We define the integral

$$I = \int_a^b \mathsf{G}(t)x(t)dt \tag{D.3}$$

as the mean-square limit of the corresponding Riemann sum.

Namely, given a partition of the interval $(a,b]$ of the form $(t_0,t_1],(t_1,t_2],\ldots,(t_{n-1},t_n]$, where $n \geq 0$, $a = t_0 < t_1 \cdots < t_n = b$, and a sampling point from each subinterval, $s_k \in (t_{k-1},t_k]$, for $1 \leq k \leq n$, the equality in (D.3) is defined if for any $\epsilon > 0$ there exists a $\delta > 0$ such that if

$$\max_k |t_k - t_{k-1}| < \delta, \tag{D.4}$$

then

$$\mathbb{E}\left[\sum_{k=1}^{n} \mathsf{G}(s_k)x(s_k)(t_k - t_{k-1}) - I\right] < \epsilon. \tag{D.5}$$

Appendix E: Special Functions

We give the definitions of some special functions. For integral representations and asymptotic expansions, the reader should consult Olver *et al.* (2010).

E.1 Gamma Function

The Gamma function is an extension of the factorial function to complex numbers:

$$\Gamma(n) = (n-1)!, \tag{E.1}$$

$$\Gamma(z) = \int_0^\infty \exp(-t)t^{z-1}dt. \tag{E.2}$$

E.2 Airy Function

The Airy function is the solution to the second-order linear differential equation

$$\frac{\partial^2 y}{\partial z^2} = zy, \tag{E.3}$$

$$y(z) = \mathrm{Ai}\,(z). \tag{E.4}$$

In integral form,

$$\frac{1}{2\pi j}\int_{-j\infty}^{j\infty} \exp(-t^3/3 + zt)dt. \tag{E.5}$$

For properties and asymptotic expansions, see Vallée and Soares (2010).

E.3 Bessel and Hankel Functions

Bessel functions of order n are solutions to the second-order linear differential equation

$$z^2\frac{\partial^2 B_n}{\partial z^2} + (z^2 + n^2)B_n = 0. \tag{E.6}$$

This exhibits two independent solutions: the Bessel functions of the first kind $J_n(z)$ and of the second kind $Y_n(z)$. Their linear combination gives the Hankel functions of the first kind,

$$H_n^{(1)}(z) = J_n(z) + jY_n(z), \tag{E.7}$$

and the Hankel functions of the second kind,

$$H_n^{(2)}(z) = J_n(z) - jY_n(z). \tag{E.8}$$

The modified Bessel functions are obtained with the substitution $z \to jz$, leading to the modified Bessel equation

$$z^2 \frac{\partial^2 B_n}{\partial z^2} - (z^2 + n^2) B_n = 0, \tag{E.9}$$

whose solutions are the modified Bessel function of the first kind,

$$I_n(z) = \exp\left(-jn\frac{\pi}{2}\right) J_n(jz), \tag{E.10}$$

and of the second kind,

$$K_n(z) = \frac{\pi j}{2} \exp\left(jn\frac{\pi}{2}\right) H_n^{(1)}(jz). \tag{E.11}$$

The spherical Bessel functions of the first, second, and third kinds are obtained when $n \to n + 1/2$, yielding

$$j_n(z) = \sqrt{\frac{\pi}{2x}} J_{n+1/2}(z) = x^n \left[-\frac{1}{z}\frac{\partial}{\partial z}\right]^n \frac{\sin z}{z}, \tag{E.12}$$

$$y_n(z) = \sqrt{\frac{\pi}{2z}} Y_{n+1/2}(z) = x^n \left[-\frac{1}{z}\frac{\partial}{\partial xz}\right]^n \frac{\cos z}{z}, \tag{E.13}$$

$$h_n^{(1,2)}(z) = \sqrt{\frac{\pi}{2z}} H_{n+1/2}^{(1,2)}(z). \tag{E.14}$$

E.4 Basic Connection Formulas

$$J_{-n}(z) = (-1)^n J_n(z) \tag{E.15}$$

$$Y_{-n}(z) = (-1)^n Y_n(z) \tag{E.16}$$

$$H_{-n}^{(1)}(z) = (-1)^n H_n^{(1)}(z) \tag{E.17}$$

$$H_{-n}^{(2)}(z) = (-1)^n H_n^{(2)}(z) \tag{E.18}$$

E.5 Addition Theorems

For more general forms, see Olver *et al.* (2010).

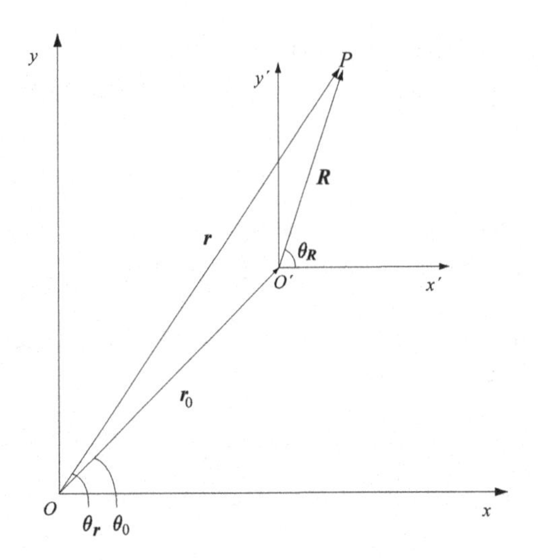

Fig. E.1 Geometry of the addition theorem.

E.5.1 Cylinder Functions

$$C_n(z \pm y) = \sum_{m=-\infty}^{\infty} C_{n \mp m}(z) J_m(y) \text{ for } |y| < |z|, \tag{E.19}$$

where $C_n(z)$ denotes $J_n(z)$, $Y_n(z)$, $H_n^{(1)}(z)$, $H_n^{(2)}(z)$, or any linear combination of these functions, the coefficients in which are independent of z and n. The restriction $|y| < |z|$ is unnecessary when $C = J$, and we have

$$J_n(z+y) = \sum_{m=-\infty}^{\infty} J_{n-m}(z) J_m(y). \tag{E.20}$$

E.5.2 Modified Bessel Functions

$$Z_n(z \pm y) = \sum_{m=-\infty}^{\infty} (\pm 1)^m Z_{n+m}(z) I_m(y) \text{ for } |y| < |z|, \tag{E.21}$$

where $Z_n(z)$ denotes $I_n(z)$, $\exp[jn\pi]K_n(z)$, or any linear combination of these functions, the coefficients in which are independent of z and n. The restriction $|y| < |z|$ is unnecessary when $Z = I$.

E.5.3 Hankel Functions

$$H_n(kR) \exp[jn(\theta_R - \theta_0)] = \sum_{m=-\infty}^{\infty} J_m(kr_0) H_{n+m}(kr) \exp[j(n+m)(\theta_r - \theta_0)]; \tag{E.22}$$

see Figure E.1.

Appendix F: Electromagnetic Spectrum

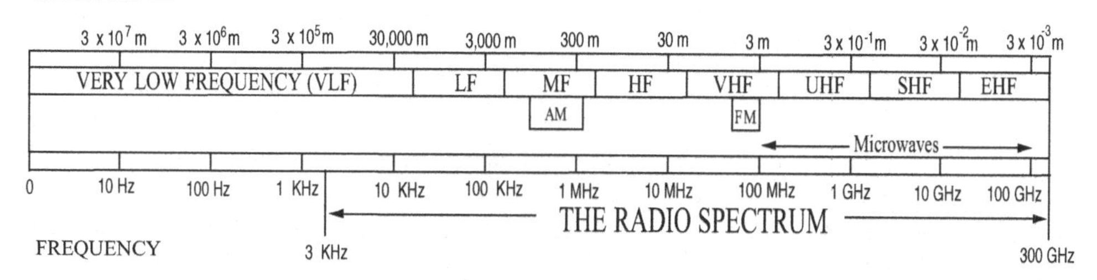

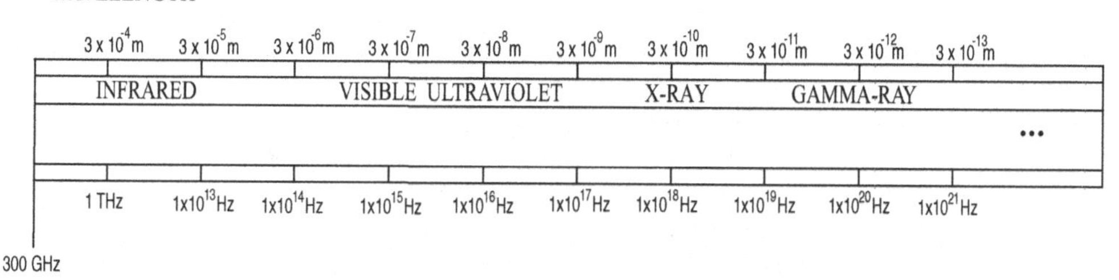

Fig. F.1 The electromagnetic spectral bands.

Bibliography

V. R. Algazi, D. J. Sakrison (1969). On the optimality of the Karhunen–Loève expansion. *IEEE Transactions on Information Theory*, 15(2), pp. 319–21.

B. C. Barber (1993). The non-isotropic two-dimensional random walk. *Waves in Random Media*, 3, pp. 243–56.

W. Beckner (1975). Inequalities in Fourier analysis. *Annals of Mathematics*, 102(6), pp. 159–82.

J. D. Bekenstein (1973). Black holes and entropy. *Physical Review D*, 7(8), pp. 2333–46.

J. D. Bekenstein (1981a). Universal upper bound on the entropy-to-energy ratio for bounded systems. *Physical Review D*, 23(2), pp. 287–98.

J. D. Bekenstein (1981b). Energy cost of information transfer. *Physical Review Letters*, 46(10), pp. 623–6.

J. D. Bekenstein (2005). How does the entropy/information bound work? *Foundations of Physics*, 35, pp. 1805–23.

J. D. Bekenstein, M. Schiffer (1990). Quantum limitations on the storage and transmission of information. *International Journal of Modern Physics C*, 1(4), pp. 355–422.

P. Bello (1963). Characterization of randomly time-variant linear channels. *IEEE Transactions on Communications*, 11(4), pp. 360–93.

E. Biglieri (2005). *Coding for Wireless Channels*. Springer.

L. Boltzmann (1872). Weitere Studien über das Wärmegleichgewicht unter Gasmolekülen. *Wiener Berichte*, 66, pp. 275–370. English translation: S. G. Brush (tr.) (2003). *The Kinetic Theory of Gases*. Imperial College Press.

L. Boltzmann (1896–8). *Vorlesungen über Gastheorie*. J. A. Barth. English translation: S. G. Brush (tr.) (1964). *Lectures on Gas Theory*. University of California Press.

E. Borel (1897). Sur l'interpolation. *Comptes rendus de l'Académie des sciences de Paris*, 124, pp. 673–6.

R. Bousso (2002). The holographic principle. *Reviews of Modern Physics*, 74(3), pp. 825–74.

J. Bowen (1967). On the capacity of a noiseless photon channel. *IEEE Transactions on Information Theory*, 13(2), pp. 230–6.

H. J. Bremermann (1967). Quantum noise and information. In L. M. Le Cam and J. Neyman *Proceedings of the Fifth Berkeley Symposium on Mathematical Statistics and Probability*. University of California Press.

H. J. Bremermann (1982). Minimum Energy Requirements of Information Transfer and Computing. *International Journal of Theoretical Physics*, 21(3–4), pp. 203–217.

J. L. Brown Jr (1960). Mean square truncation error in series expansions of random functions. *Journal of the Society of Industrial and Applied Mathematics*, 8(1), pp. 28–32.

O. M. Bucci, G. Di Massa (1988). The truncation error in the application of sampling series to electromagnetic problems. *IEEE Transactions on Antennas and Propagation*, 36(7), pp. 941–9.

O. M. Bucci, G. Franceschetti (1987). On the spatial bandwidth of scattered fields. *IEEE Transactions on Antennas and Propagation*, 35(12), pp. 1445–55.

O. M. Bucci, G. Franceschetti (1989). On the degrees of freedom of scattered fields. *IEEE Transactions on Antennas and Propagation*, 37(7), pp. 918–26.

O. M. Bucci, C. Gennarelli, C. Savarese (1998). Representation of electromagnetic fields over arbitrary surfaces by a finite and nonredundant number of samples. *IEEE Transactions on Antennas and Propagation*, 46(3), pp. 351–9.

W. Byers (2007). *How Mathematicians Think: Using Ambiguity, Contradiction, and Paradox to Create Mathematics*. Princeton University Press.

V. R. Cadambe, S. A. Jafar (2008). Interference alignment and degrees of freedom of the K-user interference channel. *IEEE Transactions on Information Theory*, 54(8), pp. 3425–41.

E. Candès (2006). Compressive sampling. In M. Sanz-Solé, J. Soria, J. L. Varona, J. Verdera (eds.) *Proceedings of the International Congress of Mathematicians*, Madrid, Spain.

E. Candès (2008). The restricted isometry property and its implications for compressed sensing. *Comptes rendus de l'Académie des sciences. Série I. Mathématique*, 346, pp. 589–92.

E. Candès, J. Romberg, T. Tao (2006). Robust uncertainty principles: Exact signal reconstruction from highly incomplete frequency information. *IEEE Transactions on Information Theory*, 52(2), pp. 489–509.

E. Candès, M. Wakin (2008). An introduction to compressive sampling. *IEEE Signal Processing Magazine*, 25(2), pp. 21–30.

C. M. Caves, P. D. Drummond (1994). Quantum limits on bosonic communication rates. *Reviews of Modern Physics*, 66(2), pp. 481–537.

S. Chandrasekhar (1960). *Radiative Transfer*. Dover.

G. M. Church, Y. Gao, S. Kosuri (2012). Next-generation digital information storage in DNA. *Science*, 337, p. 1628.

R. Clausius (1850–65). *The Mechanical Theory of Heat – with its Applications to the Steam Engine and to Physical Properties of Bodies*. John van Voorst.

J. B. Conway (1990). *A Course in Functional Analysis*, 2nd edn. Springer.

T. M. Cover (1994). Which processes satisfy the second law? In J. J. Halliwell, J. Perez-Mercader, W. H. Zurek (eds.), *Physical Origins of Time Asymmetry*. Cambridge University Press, pp. 98–107.

T. M. Cover, J. Thomas (2006). *Elements of Information Theory*, 2nd edn. John Wiley & Sons.

M. A. Davenport, M. B. Wakin (2012). Compressive sensing of analog signals using discrete prolate spheroidal sequences. *Applied Computational Harmonic Analysis*, 33, pp. 438–72.

A. De Gregorio (2012). On random flights with non-uniformly distributed directions. *Journal of Statistical Physics*, 147(2), pp. 382–411.

P. Dirac (1931). Quantised singularities in the electromagnetic field. *Proceedings of the Royal Society of London A*, 133, pp. 60–72.

D. L. Donoho (2000). *Wald Lecture I: Counting Bits with Shannon and Kolmogorov*. Technical report, Stanford University.

D. L. Donoho (2006). Compressed sensing. *IEEE Transactions on Information Theory*, 52(4), pp. 1289–1306.

D. L. Donoho, A. Javanmard, A. Montanari (2013). Information-theoretically optimal compressed sensing via spatial coupling and approximate message passing. *IEEE Transactions on Information Theory*, 59(11), pp. 7434–64.

D. L. Donoho, P. B. Stark (1989). Uncertainty principles and signal recovery. *SIAM Journal of Applied Mathematics*, 49, pp. 906–31.

O. El Ayach, S. W. Peters, R. W. Heath Jr (2013). The practical challenges of interference alignment. *IEEE Wireless Communications*, 20(1), pp. 35–42.

A. E. Gamal, Y. Kim (2011). *Network Information Theory*. Cambridge University Press.

K. Falconer (1990). *Fractal Geometry: Mathematical Foundations and Applications*. John Wiley & Sons.

P. Feng, Y. Bresler (1996a). Spectrum-blind minimum-rate sampling and reconstruction of multi-band signals. *Proceedings of the IEEE International Conference on Acoustics, Speech, and Signal Processing (ICASSP)*, 3, pp. 1688–91.

P. Feng, Y. Bresler (1996b). Spectrum-blind minimum-rate sampling and reconstruction of 2D multi-band signals. *Proceedings of the IEEE International Conference on Image Processing*, 1, pp. 701–4.

R. Feynman, R. Leighton, M. Sands (1964). *The Feynman Lectures on Physics*, vols. 1–3. Reprinted 2005. Addison Wesley.

C. Flammer (1957). *Spheroidal Wave Functions*. Stanford University Press.

G. B. Folland, A. Sitaram (1997). The uncertainty principle: A mathematical survey. *Journal of Fourier Analysis and Applications*, 3(3), pp. 207–38.

S. Foucart, H. Rauhut (2013). *A Mathematical Introduction to Compressive Sensing*. Springer.

G. Franceschetti (1997). *Electromagnetics: Theory, Techniques, and Engineering Paradigms*. Plenum Press.

M. Franceschetti (2004). Stochastic rays pulse propagation. *IEEE Transactions on Antennas and Propagation*, 52(10), pp. 2742–52.

M. Franceschetti (2007a). A note on Levéque and Telatar's upper bound on the capacity of wireless ad hoc networks. *IEEE Transactions on Information Theory*, 53(9), pp. 3207–11.

M. Franceschetti (2007b). When a random walk of fixed length can lead uniformly anywhere inside a hypersphere. *Journal of Statistical Physics*, 127, pp. 813–23.

M. Franceschetti (2015). On Landau's eigenvalue theorem and information cut-sets. *IEEE Transactions on Information Theory*, 61(9), pp. 5042–51.

M. Franceschetti (2017). Quantum limits on the entropy of bandlimited radiation. *Journal of Statistical Physics*, 169(2), pp. 374–94.

M. Franceschetti, J. Bruck, L. J. Schulman (2004). A random walk model of wave propagation. *IEEE Transactions on Antennas and Propagation*, 52(5), pp. 1304–17.

M. Franceschetti, O. Dousse, D. Tse, P. Thiran (2007). Closing the gap in the capacity of wireless networks via percolation theory. *IEEE Transactions on Information Theory*, 53(3), pp. 1009–18.

M. Franceschetti, R. Meester (2007). *Random Networks for Communication*. Cambridge University Press.

M. Franceschetti, M. D. Migliore, P. Minero (2009). The capacity of wireless networks: Information-theoretic and physical limits. *IEEE Transactions on Information Theory*, 55(8), pp. 3413–24.

M. Franceschetti, M. D. Migliore, P. Minero, F. Schettino (2011). The degrees of freedom of wireless networks via cut-set integrals. *IEEE Transactions on Information Theory*, 57(11), pp. 3067–79.

M. Franceschetti, M. D. Migliore, P. Minero, F. Schettino (2015). The information carried by scattered waves: Near-field and non-asymptotic regimes. *IEEE Transactions on Antennas and Propagation*, 63(7), pp. 3144–57.

D. Gabor (1946). Theory of communication. *Journal of the Institution of Electrical Engineers, Part III: Radio and Communication Engineering*, 93, pp. 429–57.

D. Gabor (1953). Communication theory and physics. *IRE Professional Group on Information Theory*, 1(1), pp. 48–59.

D. Gabor (1961). Light and information. In E. Wolf (ed.), *Progress in Optics*. Elsevier, vol. I, pp. 109–53.

R. G. Gallager (1968). *Information Theory and Reliable Communication*. John Wiley & Sons.

R. G. Gallager (2008). *Principles of Digital Communication*. Cambridge University Press.

A. G. Garcia (2000). Orthogonal sampling formulas: A unified approach. *SIAM Review*, 42(3), pp. 499–512.

J. Ghaderi, L.-L. Xie, X. Shen (2009). Hierarchical cooperation in ad hoc networks: Optimal clustering and achievable throughput. *IEEE Transactions on Information Theory*, 55(8), pp. 3425–36.

W. Gibbs (1902). *Elementary Principles of Statistical Mechanics Developed with Especial Reference to the Rational Foundation of Thermodynamics*. Reprinted 1960, Dover.

A. Goldsmith (2005). *Wireless Communications*. Cambridge University Press.

J. P. Gordon (1962). Quantum effects in communication systems. *Proceedings of the IRE*, 50(9), pp. 1898–908.

C. C. Grosjean (1953). Solution of the non-isotropic random flight problem in the k-dimensional space. *Physica*, 19, pp. 29–45.

P. Gupta, P. R. Kumar (2000). The capacity of wireless networks. *IEEE Transactions on Information Theory*, 42(2), pp. 388–404.

M. Haenggi (2013). *Stochastic Geometry for Wireless Networks*. Cambridge University Press.

M. Haenggi, J. Andrews, F. Baccelli, O. Dousse, M. Franceschetti (2009). Stochastic geometry and random graphs for the analysis and design of wireless networks. *IEEE Journal on Selected Areas in Communications*, 27(7), pp. 1029–46.

S. W. Hawking (1975). Particle creation by black holes. *Communications in Mathematical Physics*, 43, pp. 199–220.

W. Heisenberg (1927). Über den anschaulichen Inhalt der quantentheoretischen Kinematik und Mechanik. *Zeitschrift für Physik*, 43, pp. 172–98. English translation: In J. A. Wheeler, W. H. Zurek (eds.) (1983). *Quantum Theory and Measurement*. Princeton University Press, pp. 62–84.

W. Heitler (1954). *The Quantum Theory of Radiation*, 3rd edn. Reprinted 2000. Dover.

J. R. Higgins (1985). Five short stories about the cardinal series. *Bulletin of the American Mathematical Society*, 12(1), pp. 46–89.

D. Hilbert, R. Courant (1953). Methods of Mathematical Physics. Vols. 1, & 2, 2nd edn. Springer.

I. I. Hirschman, Jr (1957). A note on entropy. *American Journal of Mathematics*, 79, pp. 152–6.

J. A. Hogan, J. D. Lakey (2012). *Duration and Bandwidth Limiting: Prolate Functions, Sampling, and Applications*. Birkhäuser.

A. Holevo (1973). Bounds for the quantity of information transmitted by a quantum communication channel. *Problems of Information Transmission*, 9, pp. 177–83.

S. N. Hong, G. Caire (2015). Beyond scaling laws: On the rate performance of dense device-to-device wireless networks. *IEEE Transactions on Information Theory*, 61(9), pp. 4735–50.

B. D. Hughes (1995). *Random Walks and Random Environments. Volume I: Random Walks*. Oxford University Press.

A. Ishimaru (1978). *Wave Propagation and Scattering in Random Media*. IEEE Press.

S. Izu, J. Lakey (2009). Time–frequency localization and sampling of multiband signals. *Acta Applicandae Mathematicae*, 107(1), pp. 399–435.

J. D. Jackson (1962). *Classical Electrodynamics*. John Wiley & Sons.

S. A. Jafar (2011). Interference alignment: A new look at signal dimensions in a communication network. *Foundations and Trends in Communications and Information Theory*, 7(1), pp. 1–134.

D. Jagerman (1969) ϵ-entropy and approximation of bandlimited functions. *SIAM Journal on Applied Mathematics*, 17(2), pp. 362–77.

D. Jagerman (1970). Information theory and approximation of bandlimited functions. *Bell Systems Technical Journal*, 49(8), pp. 1911–41.

R. Janaswamy (2011). On the EM degrees of freedom in scattering environments. *IEEE Transanctions on Antennas and Propagation*, 59(10), pp. 3872–81.

E. T. Jaynes (1965). Gibbs vs. Boltzmann entropies. *Americal Journal of Physics*, 33(5), pp. 391–8.

E. T. Jaynes (1982). On the rationale of maximum entropy methods. *Proceedings of the IEEE*, 70, pp. 939–52.

A. Jerri (1977). The Shannon sampling theorem – its various extensions and applications: A tutorial review. *Proceedings of the IEEE*, 65(11), pp. 1565–96.

M. Kac, W. L. Murdock, G. Szegö. (1953). On the eigenvalues of certain Hermitian forms. *Journal of Rational Mechanics and Analysis*, 2, pp. 767–800.

T. Kawabata, A. Dembo (1994). The rate–distortion dimension of sets and measures. *IEEE Transactions on Information Theory*, 40(5), pp. 1564–72.

E. H. Kennard (1927). Zur Quantenmechanik einfacher Bewegungstypen. *Zeitschrift für Physik*, 44(4–5), pp. 326–52.

R. A. Kennedy, P. Sadeghi, T. D. Abhayapala, H. M. Jones (2007). Intrinsic limits of dimensionality and richness in random multipath fields. *IEEE Transactions on Signal Processing*, 55(6), pp. 2542–56.

C. Kittel, H. Kroemer (1980). *Thermal Physics*, 2nd edn. W. H. Freeman & Co.

J. J. Knab (1979). Interpolation of bandlimited functions using the approximate prolate series. *IEEE Transactions on Information Theory*, 25(6), pp. 717–19.

J. J. Knab (1983). The sampling window. *IEEE Transactions on Information Theory*, 29(1), pp. 157–9.

A. N. Kolmogorov (1936). Über die beste Annäherung von Funktionen einer gegebenen Funktionenklasse. *Annals of Mathematics*, 37(1), no. 1, pp. 107–10 (in German).

A. N. Kolmogorov (1956). On certain asymptotic characteristics of completely bounded metric spaces. *Uspekhi Matematicheskikh Nauk*, 108(3), pp. 385–8 (in Russian).

A. N. Kolmogorov, S. V. Formin (1954). *Elements of the Theory of Functions and Functional Analysis*, vols. 1, 2. Graylock.

A. N. Kolmogorov, V. M. Tikhomirov (1959). ϵ-entropy and ϵ-capacity of sets in functional spaces. *Uspekhi Matematicheskikh Nauk*, 14(2), pp. 3–86. English translation: (1961). *American Mathematical Society Translation Series*, 2(17), pp. 277–364.

V. A. Kotelnikov (1933). On the transmission capacity of "ether" and wire in electrocommunications. *Proceedings of the First All-Union Conference on Questions of Communication*, January 1933. English translation reprint in J. J. Benedetto, P. J. S. G. Ferreira (eds.) (2000), *Modern Sampling Theory: Mathematics and Applications*, Birkhauser.

M. Lachmann, M. E. Newman, C. Moore (2004). The physical limits of communication or why any sufficiently advanced technology is indistinguishable from noise. *American Journal of Physics*, 72(10), pp. 1290–3.

H. J. Landau (1975). On Szegö's eigenvalue distribution theorem and non-Hermitian kernels. *Journal d'Analyse Mathematique*, 28, pp. 335–57.

H. J. Landau (1985). An overview of time and frequency limiting. In J. F. Prince (ed.), *Fourier Techniques and Applications*. Plenum Press, pp. 201–20.

M. D. Landau, W. Jones (1983). A Hardy old problem. *Mathematics Magazine*, 56(4), pp. 230–2.

H. J. Landau, H. O. Pollak (1961). Prolate spheroidal wave functions, Fourier analysis and uncertainty, II. *Bell Systems Technical Journal*, 40, pp. 65–84.

H. J. Landau, H. O. Pollak (1962). Prolate spheroidal wave functions, Fourier analysis and uncertainty, III. *Bell Systems Technical Journal*, 41, pp. 1295–336.

H. J. Landau, H. Widom (1980). Eigenvalue distribution of time and frequency limiting. *Journal of Mathematical Analysis and Applications*, 77(2), pp. 469–81.

A. Lapidoth (2009). *A Foundation in Digital Communication*. Cambridge University Press.

P. S. Laplace (1774). *Mémoires de Mathématique et de Physique, Tome Sixiéme*. English translation: S. M. Stigler (tr.) (1986). Memoir on the probability of causes of events. *Statistical Science*, 1(19), pp. 364–78.

D. S. Lebedev, L. B. Levitin (1966). Information transmission by electromagnetic field. *Information and Control*, 9, pp. 1–22.

G. Le Caër (2010). A Pearson–Dirichlet random walk. *Journal of Statistical Physics*, 140, pp. 728–51.

G. Le Caër (2011). A new family of solvable Pearson–Dirichlet random walks. *Journal of Statistical Physics*, 144, pp. 23–45.

E. A. Lee (2017). *Plato and the Nerd. The Creative Partnership of Humans and Technology*. MIT Press.

S. H. Lee and S. Y. Chung (2012). Capacity scaling of wireless ad hoc networks: Shannon meets Maxwell. *IEEE Transactions on Information Theory*, 58(3), pp. 1702–15.

O. Lévêque, E. Telatar (2005). Information theoretic upper bounds on the capacity of large extended ad hoc wireless networks. *IEEE Transactions on Information Theory*, 51(3), pp. 858–65.

C. T. Li, A. Özgür (2016) Channel diversity needed for vector space interference alignment. *IEEE Transactions on Information Theory*, 62(4), pp. 1942–56.

T. J. Lim, M. Franceschetti (2017a). Deterministic coding theorems for blind sensing: Optimal measurement rate and fractal dimension. arXiv: 1708.05769.

T. J. Lim, M. Franceschetti (2017b). Information without rolling dice. *IEEE Transactions on Information Theory*, 63(3), pp. 1349–63.

G. Lorentz (1986). *Approximation of Functions*, 2nd edn. AMS Chelsea Publishing.

S. Loth, S. Baumann, C. P. Lutz, D. M. Eigler, A. J. Heinrich (2012). Bistability in atomic-scale antiferromagnets. *Science*, 335, pp. 196–9.

R. Loudon (2000). *The Quantum Theory of Light*, 3rd edn. Oxford University Press.

M. Masoliver, J. M. Porrá, G. H. Weiss (1993). Some two- and three-dimensional persistent random walks. *Physica A*, 193, pp. 469–82.

J. K. Maxwell (1873). A treatise on electricity and magnetism. Reprinted 1998, Oxford University Press.

N. Merhav (2010). Statistical physics and information theory. *Foundations and Trends in Communications and Information Theory*, 6(1–2), pp. 1–212.

M. Mézard, A. Montanari (2009). *Information, Physics, and Computation*. Oxford University Press.

D. A. B. Miller (2000). Communicating with waves between volumes: Evaluating orthogonal spatial channels and limits on coupling strengths. *Applied Optics*, 39(11), pp. 1681–99.

M. Mishali, Y. Eldar (2009). Blind multi-band signal reconstruction: Compressed sensing for analog signals. *IEEE Transactions on Signal Processing*, 57(3), pp. 993–1009.

C. R. Moon, L. S. Mattos, B. K. Foster, G. Zeltzer, H. C. Manoharan (2009). Quantum holographic encoding in a two-dimensional electron gas. *Nature Nanotechnology*, 4, pp. 167–72.

B. Nazer, M. Gastpar, S. A. Jafar, S. Vishwanath (2012). Ergodic interference alignment. *IEEE Transactions on Information Theory*, 58(10), pp. 6355–71.

H. Nyquist (1928). Thermal agitations of electric charges in conductors. *Physical Review*, 32, pp. 110–13.

B. M. Oliver (1965). Thermal and quantum noise. *Proceedings of the IEEE*, 53(5), pp. 436–54.

F. W. J. Olver, D. W. Lozier, R. F. Boisvert, C. W. Clark (eds.) (2010). *National Institute of Standards Handbook of Mathematical Functions*. Cambridge University Press.

A. Özgür, O. Lévêque, D. N. C. Tse (2007). Hierarchical cooperation achieves optimal capacity scaling in ad hoc networks. *IEEE Transactions on Information Theory*, 53(10), pp. 3549–72.

A. Özgür, O. Lévêque, D. N. C. Tse (2013). Spatial degrees of freedom of large distributed MIMO systems and wireless ad hoc networks. *IEEE Journal on Selected Areas in Communications*, 31(2), pp. 202–14.

C. H. Papas (1965). *Theory of Electromagnetic Wave Propagation*. Dover.

G. C. Papen, R. E. Blahut (2018). *Lightwave Communication Systems*. Preprint, to be published by Cambridge University Press.

J. B. Pendry (1983). Quantum limits to the flow of information and entropy. *Journal of Physics A: Mathematical and General*, 16, pp. 2161–71.

R. Piestun, D. A. B. Miller (2000). Electromagnetic degrees of freedom of an optical system. *Journal of the Optical Society America*, 17(5), pp. 892–902.

A. Pinkus (1985). *n-Widths in Approximation Theory*. Springer.

A. A. Pogorui, R. M. Rodriguez-Dagnino (2011). Isotropic random motion at finite speed with *k*-Erlang distributed direction alternations. *Journal of Statistical Physics*, 145, pp. 102–12.

A. S. Y. Poon, R. W. Brodersen, D. N. C. Tse (2005). Degrees of freedom in multiple-antenna channels: A signal space approach. *IEEE Transactions on Information Theory*, 51(2), pp. 523–36.

J. Proakis, M. Salehi (2007). *Digital Communications*. McGraw-Hill.

M. Reed, B. Simon (1980). *Functional Analysis*. Elsevier.

A. Rényi (1959). On the dimension and entropy of probability distributions. *Acta Mathematica Hungarica*, 10(1–2), pp. 193–215.

A. Rényi (1985). *A Diary on Information Theory*. John Wiley & Sons.

F. Riesz, B. Sz.-Nagy (1955). *Functional Analysis*. Ungar.

M. Schiffer (1991). Quantum limit for information transmission. *Physical Review A*, 43(10), pp. 5337–43.

E. Schmidt (1907). Zur Theorie der linearen und nichtlinearen Integralgleichungen. *Mathematische Annalen*, 63, pp. 433–76.

C. E. Shannon (1948). A mathematical theory of communication. *Bell System Technical Journal*, 27, pp. 379–423, 623–56.

C. E. Shannon (1949). Communication in the presence of noise. *Proceedings of the IRE*, 37, pp. 10–21.

D. Slepian (1964). Prolate spheroidal wave functions, Fourier analysis and uncertainty, IV. Extensions to many dimensions: Generalized prolate spheroidal functions. *Bell Systems Technical Journal*, 43, pp. 3009–58.

D. Slepian (1965). Some asymptotic expansions for prolate spheroidal wave functions. *Journal of Mathematics and Physics*, 44, pp. 99–140.

D. Slepian (1976). On bandwidth. *Proceedings of the IEEE*, 64(3), pp. 292–300.

D. Slepian (1978). Prolate spheroidal wave functions, Fourier analysis and uncertainty, V. The discrete case. *Bell Systems Technical Journal*, 57, pp. 1371–430.

D. Slepian (1983). Some comments on Fourier analysis, uncertainty and modeling. *SIAM Review*, 25(3), pp. 379–93.

D. Slepian, H. O. Pollak (1961). Prolate spheroidal wave functions, Fourier analysis and uncertainty, I. *Bell Systems Technical Journal*, 40, pp. 43–64.

W. Stadje (1987). The exact probability distribution of a two-dimensional random walk. *Journal of Statistical Physics*, 46, pp. 207–16.

T. E. Stern (1960). Some quantum effects in information channels. *IEEE Transactions on Information Theory*, 6, pp. 435–40.

G. W. Stewart (1993). On the early history of the singular value decomposition. *SIAM Review*, 35(4), pp. 551–66.

J. A. Stratton (1941). *Electromagnetic Theory*. McGraw-Hill.

A. Strominger, C. Vafa (1996). Microscopic origin of the Bekenterin–Hawking entropy. *Physics Letters B*, 379(1), pp. 99–104.

L. Susskind (1995). The world as a hologram. *Journal of Mathematical Physics*, 36(11), pp. 6377–96.

T. Tao (2012). *Topics in Random Matrix Theory*. Graduate Studies in Mathematics, vol. 132. American Mathematical Society.

G. 't Hooft (1993). Dimensional reduction in quantum gravity. In A. Ali, J. Ellis, S. Randjbar-Daemi (eds.), *Salamfestschrift: A Collection of Talks from the Conference on Highlights of Particle and Condensed Matter Physics*, World Scientific Series in 20th Century Physics, vol. 4. World Scientific.

G. Toraldo di Francia (1955). Resolving power and information. *Journal of the Optical Society of America*, 45(7), pp. 497–501.

G. Toraldo di Francia (1969). Degrees of freedom of an image. *Journal of the Optical Society of America*, 59(7), pp. 799–804.

D. N. C. Tse, P. Visvanath (2005). *Fundamentals of Wireless Communication*. Cambridge University Press.

A. Tulino, S. Verdú (2004). Random matrix theory and wireless communications. *Foundations and Trends in Communications and Information Theory*, 1(1) pp. 1–182.

V. Twersky (1957). On multiple scattering and reflection of waves by rough surfaces. *IRE Transactions on Antennas and Propagation*, 5, p. 81.

V. Twersky (1964). On propagation in random media of discrete scatterers. *Proceedings of the American Mathematical Society Symposium on Stochastic Processes in Mathematics, Physics, and Engineering*, 16, pp. 84–116.

J. Uffink (2008). Boltzmann's work in statistical physics. In E. N. Zalta (ed.), *The Stanford Encyclopedia of Philosophy*, Winter 2008 edn. Published online.

M. Unser (2000). Sampling – 50 years after Shannon. *Proceedings of the IEEE*, 88(4), pp. 569–87.

O. Vallée, M. Soares (2010). *Airy Functions and Applications to Physics*. World Scientific.

J. Van Bladel (1985). *Electromagnetic Fields*. Hemisphere.

R. Venkataramani, Y. Bresler (1998). Further results on spectrum blind sampling of 2D signals. *Proceedings of the IEEE International Conference on Image Processing*, 2, pp. 752–6.

M. Vetterli, J, Kovačević, V. Goyal (2014a). *Foundations of Signal Processing*. Cambridge University Press.

M. Vetterli, J, Kovačević, V. Goyal (2014b). *Fourier and Wavelet Signal Processing*. Cambridge University Press.

A. J. Viterbi (1995). *CDMA: Principles of Spread Spectrum Communication.* Addison Wesley.

H. Weyl (1928). *Gruppentheorie und Quantenmechanik.* S. Hirzel.

E. T. Whittaker (1915). On the functions which are represented by the expansions of the interpolation theory. *Proceedings of the Royal Society of Edinburgh*, 35, pp. 181–94.

H. Widom (1964). Asymptotic behavior of the eigenvalues of certain integral equations II. *Archive for Rational Mechanics and Analysis*, 17(3), pp. 215–29.

Y. Wu, S. Verdú (2010). Rényi information dimension: Fundamental limits of almost lossless analog compression. *IEEE Transactions on Information Theory*, 56(8), pp. 3721–48.

Y. Wu, S. Verdú (2012). Optimal phase transitions in compressed sensing. *IEEE Transactions on Information Theory*, 58(10), pp. 6241–63.

A. Wyner (1965). Capacity of the band-limited Gaussian channel. *Bell Systems Technical Journal*, 45, pp. 359–95.

A. Wyner (1973). A bound on the number of distinguishable functions which are time-limited and approximately band-limited. *SIAM Journal of Applied Mathematics*, 24(3), pp. 289–97.

L. L. Xie, P. R. Kumar (2004). A network information theory for wireless communication: Scaling laws and optimal operation. *IEEE Transactions on Information Theory*, 50(5), pp. 748–67.

H. Yuen, M. Ozawa (1993). Ultimate information carrying limit of quantum systems. *Physical Review Letters*, 70(4), pp. 363–6.

Index

achievable rate, 29, 345, 369, 385
addition theorems
  for Hankel functions, 247
  for special functions, 435
adjoint operator, 96
Airy function, 239, 243, 263
Airy integral, 262
anisotropic media, 133, 134
asymptotic equipartition, 18, 22, 319–321, 323, 337, 339, 404

backscattering, 258
bandlimitation, 5, 49
bandlimitation error, 236
bandlimited signals, 11, 14, 17, 33, 40, 41, 45, 49, 50, 52, 53, 55, 59–61, 63–66, 68, 69, 75, 79, 81, 82, 88–91, 95, 98, 104, 105, 116, 117, 125, 127, 192, 200, 214, 244, 248, 249, 278, 343, 344, 346, 350, 352, 358, 360–362, 368, 369, 378, 379, 381, 382, 392, 395
  approximate, 98
  of multiple variables, 108, 110, 169
bandlimiting operator, 69, 98, 106, 109
Bessel's inequality, 77
black body, 307, 311, 316, 318, 324, 326, 329, 330, 333, 337, 391, 396, 402
  average energy, 327
  entropy, 326, 327, 330
  radiation, 307, 330, 341
black hole, 38, 328, 329, 331–333, 401, 405
  entropy, 38, 39, 330, 331, 402
blind sensing, 42, 113, 114, 124
Boltzmann's constant, 303
Born approximation, 281
Bouligand dimension, 117

capacity
  AWGN channel, 30, 31, 347, 359, 362
  colored Gaussian channel, 357
  Kolmogorov, 14, 368, 369, 371–375, 377, 378, 380, 383
  Shannon, 26, 28, 344, 345, 347, 350–354, 356–366, 368, 369, 372, 373, 378, 381, 385–387
  zero-error, 383
capacity-achieving code, 347
cardinal series, 8, 50, 51, 59–63, 76, 78, 79, 81, 248
carrier frequency, 34–36, 355
carrier wavelength, 35
Cauchy's theorem, 235, 238
CDMA, 205–209, 212, 214, 216, 217, 219, 226
chips, 212–214
codebook, 15, 16, 28, 344, 345, 365–367, 378–380
codeword, 343, 345, 348
coding theorem, 18, 29, 350, 351
coherence
  bandwidth, 180–184, 196, 202, 227
  distance, 186, 227
  time, 181–184, 202, 212, 214, 227
coherent energy, 275
coherent response, 275, 283, 284, 295, 299
compact operator, 64, 96, 97, 100, 107, 109, 168
compressed sensing, 118, 124
compression rate, 121
compressor, 121
concentration, 2, 107, 124
  frequency, 63, 169
  geometric view, 68
  probabilistic, 3, 31
  Slepian's problem, 41, 63, 64, 68, 82, 87, 90, 95, 97, 100, 106, 109, 192, 193, 260, 368
  spatial, 169, 250
  spectral, 2, 41, 111, 265, 267
  time, 63, 169
conductivity, 133, 137
conductor, 134–137, 146
connection formulas, 247, 435
convolution
  space–time, 160
  spatial, 160, 161, 163, 166, 167, 169, 176, 233
convolution integral, 159, 162, 163, 166, 168, 171, 174
covering number, 14
critical bandwidth, 240, 243

cross section
absorption, 283, 284
scattering, 283, 284
cut-set, 5, 6, 9, 11, 34, 36, 40, 230, 256, 257, 265, 268, 269, 272–274
cut-set integral, 250, 251, 253–258, 261
linear, 251
surface, 253
cyclic prefix, 210

decoder, 121
decoding function, 29
decoding process, 343
decoding system, 4
decompressor, 121
degrees of freedom, 5, 7–13, 41, 87, 90, 91, 95, 104–112, 173, 187, 192, 200–202, 204, 205, 207–209, 212, 213, 217, 219, 222, 224–226, 230, 232, 233, 244–246, 248–257, 259–261, 265–270, 272–274, 327, 331, 358–360, 362, 363, 368, 369, 378, 379, 384–387, 391–393, 401–405
space–wavenumber, 11, 203, 207, 209, 217–222, 224–226, 265, 269
time–frequency, 11, 209, 219, 265, 269
total, 265
wide-band regime, 270, 271
density power, 140, 142, 147, 148, 150
deviation, 88, 104, 105, 234, 235, 244, 245
dielectric, 137
dielectric constant, 137
diffusion, 290
dimensionality reduction, 103
dipole, 147, 148, 151
moment, 147, 148
Dirac's delta, 121, 123, 304
Dirac's impulse, 132, 157, 159, 181, 293, 294
dispersive media, 132
distortion, 184, 201
Doppler
frequency, 182
frequency spread, 182, 183
power profile, 182–184
dyad, 160, 426

(ϵ,δ)-capacity, 378, 381
ϵ-capacity, 370, 372–376, 378–380
ϵ-covering, 369
ϵ-entropy, 369, 370, 372, 374–377, 379, 382, 392–395, 399, 402
ϵ-net, 369
effective dimension, 87, 91, 392
Einstein's equation, 38, 57, 402
electric dipole, 146
electric induction, 132
encoder, 121
encoding function, 28, 29
encoding process, 343
encoding system, 4, 347
energy constraint, 12, 14, 18, 20, 22, 23, 25, 27, 29, 30, 33, 40, 45, 61, 79, 84, 344, 346, 352, 353, 357, 359, 361, 379, 380, 383, 384, 387, 392, 397
entropy
Boltzmann, 20, 316, 321, 396
Clausius, 20, 318
conditional, 25
differential, 21, 23, 24, 30, 45–47, 317, 338, 339, 351, 367, 404
Gibbs, 20, 316
Kolmogorov, 14, 369, 371, 374, 375, 377, 378, 394, 399
relative, 322, 323
Shannon, 19, 316, 317, 320, 321, 364, 366, 368, 399, 402
statistical, 19, 316–323
thermodynamic, 317–322, 391, 398
equivalence principle, 153
evanescent waves, 143, 144, 156
event horizon, 38

fading, 179
far field, 150
Fourier series, 49
Fourier transform, 48
fractal dimension, 116, 117, 123, 124
Fraunhofer condition, 151
Fraunhofer region, 151
Fredholm integral equation, 92, 100, 104, 106, 109, 192, 198, 413–415
second kind, 64, 260, 276
Fredholm operator, 64
frequency diversity, 202, 204, 205, 216, 217
frequency response, 5, 164, 211, 216, 297
frequency spread, 203–205, 212
frequency-dispersive media, 137
Frobenius norm, 103, 196
Fubini's theorem, 413
functional dimension, 377

galactic noise, 324, 325
gauge invariance, 145, 146
Gaussian process, 345
Green's function, 42, 159–162, 164–168, 171, 174, 179, 200, 211, 214–216, 222, 225, 226, 232, 234, 236, 386
dyadic, 160
frequency-varying, 194
reduced, 233
spatially varying, 195
spectral, 159

stochastic, 178–181, 186, 188, 190, 193, 198
time-varying, 174, 187
Green's operator, 233, 246, 247, 261, 392
GSM, 206

Hardy's problem, 263
Helmholtz equation, 65, 72, 84
hierarchical cooperation, 221, 224–227
Hilbert dimension, 87
Hilbert space, 41, 69, 87, 95, 97, 248, 260, 408
Hilbert–Schmidt
decomposition, 41, 42, 95, 100, 102, 157, 167–170, 173, 187–189, 208, 211, 218, 226, 246–248, 261
kernel, 95, 96, 100, 246
operator, 95–98, 100, 188
Holevo's theorem, 388
holographic information bound, 38–40, 330–332, 337, 405
homogeneous media, 133, 134, 137, 139, 144, 146
Huygens' principle, 153

imaginary unit, 48
impulse response, 158, 164, 171
incoherent energy, 275
incoherent response, 275, 283, 284, 295–297, 299, 300
information dimension, 376
instantaneous power, 48
inter-symbol interference, 201
interference alignment, 222, 224, 225, 227
isotropic media, 133, 134, 137–139, 144, 146

Kac–Murdock–Szegö theorem, 193, 419
Karhunen–Loève representation, 42, 173, 186–191, 195, 196, 198, 420
keyhole effect, 186, 220
Kolmogorov capacity *see also* ϵ-capacity 370
Kolmogorov entropy *see also* ϵ-entropy 369

Liouville–Neumann series, 260, 414
logons, 83
lossless media, 144, 146

magnetic induction, 132
Maxwell's equations, 130, 132, 133, 135, 139, 152, 153, 157, 158, 160, 162, 163
Maxwell–Boltzmann collision equation, 276
measurement error, 115, 119, 120
measurement operator, 113, 115, 388
measurement rate, 115
Mercer's theorem, 102, 412, 416
metric entropy *see also* Kolmogorov entropy 369
metric order, 377
microwave window, 323
MIMO, 207–209, 218, 219, 221, 225, 226
Minkowski dimension, 117
Minkowski sum, 117
monopoles, 132
multi-band signals, 105, 376
multi-hop, 219, 222, 225, 226
multi-path, 176, 178–180, 182–184, 197, 277
multiple scattering theory, 42, 276, 278
mutual coherence function, 182–184
mutual information, 25, 26, 350, 351, 366, 367, 372, 373

N-width, 41, 81, 88–91, 95, 97, 100–105, 123, 368
narrow-band transmission, 201, 202
near field, 150
network information theory, 219
Newton's constant, 38, 328
noise, 54
additive, 344, 350, 383
colored Gaussian, 356
Gaussian, 29, 303–306, 344, 345, 350, 359
temperature, 323
thermal, 303–307, 312, 317, 325
non-dispersive media, 132–134, 137–139, 144, 146
Nyquist number, 51, 66, 82, 245, 252, 254, 255, 259, 260

OFDM, 205, 208–210, 212–217, 219

packing number, 15
Paley–Wiener theorem, 52
Parseval's theorem, 49, 70, 78, 92, 239, 418
path loss, 278, 291, 292, 299, 300
permeability, 133, 144
permittivity, 133, 144
phase transition, 8, 9, 11, 12, 50, 64–66, 79, 80, 83, 91, 107, 109, 110, 192, 236
photon, 35, 284–288, 290, 292, 294, 296, 298, 299, 306, 325, 326, 332, 333
energy, 36, 58, 313, 314, 324–326, 332
Planck's constant, 36, 58, 313
Planck's energy, 38
Planck's equation, 36, 325
Planck's length, 38, 328–330, 332, 333, 402
Planck's radiation formula, 330, 341, 391
Planck's scale, 40
plane wave, 139, 141, 142, 154, 155
Poisson distribution, 305
Poisson noise, 306, 355
Poisson process, 305, 317
power delay profile, 278
power density
full, 287–292
radiated, 287–291
Poynting vector, 155
Poynting–Umov theorem, 139, 154

Poynting–Umov vector, 138, 139
probabilistic channel, 25, 28
probabilistic method, 347
prolate spheroidal wave functions, 70, 74–76, 78, 79, 82, 87–89, 95, 98–100, 110, 114, 359
  angular, 73
  approximate, 82
  radial, 73
propagation velocity, 140–143, 155, 161, 164

quanta of energy, 35
quantum complementarity, 306, 325, 326
quantum fluctuations, 326
quantum noise, 306, 324, 326

radiation, 35, 130, 139, 146, 147, 151–153, 231, 275, 307, 393, 402, 403, 405
  entropy, 391
  noise, 306, 307, 311, 312, 323–325
  pressure, 155
  quantized, 394
  thermal, 307, 391
rake receiver, 216, 217, 227, 228
random coding, 347, 349
random walk, 276, 284, 286, 290–292, 296, 298–300
  recurrence, 290
rate gain, 173
rate–distortion, 364, 366–369, 373, 378, 379, 383, 384, 398–400
Rayleigh distribution, 178, 180, 334
Rayleigh–Jeans limit, 341
reconstruction error, 119, 120
reduced rank theorem, 103
Reed–Solomon decoding, 119
relaxation time, 140
reliability gain, 173
Riesz–Fisher theorem, 78

saddle point method, 237, 238
sampling, 8, 45, 50, 59, 76, 81, 82, 248–250, 261
  interval, 51, 76
scalar potential, 145
Schrödinger equation, 87
Schwarz inequality, 56, 69, 83, 97, 98
Schwarzschild radius, 38, 328
second law of thermodynamics, 42, 318
self-adjoint operator, 87, 95–97, 100, 102, 107, 109, 127, 168, 188, 192
Shannon's energy limit, 358
Shannon–McMillan–Breiman theorem, 18
shot noise, 304–306, 337
signal constellation, 54, 55
signal-to-noise ratio, 14, 16, 17, 23, 29, 31–36, 346, 352, 353, 363, 366, 378–380, 383, 384, 386, 387
singular value decomposition, 102, 166, 172, 187
singular values, 96, 97, 100–103, 124, 166, 167, 169, 211, 218, 246–248
Snell's law, 143, 155
solenoidal field, 131, 144
source codebook, 382
source coding, 121
sparsity number, 117
spatial bandwidth, 230, 233, 234, 236, 244, 250, 251, 255, 256, 260, 261
spatial spread, 203
spectral efficiency, 358
spectral theorem, 96
spectrum, 48
sphere covering, 17
sphere packing, 17, 31, 347
spheroidal coordinates
  oblate, 73
  prolate, 73
stochastic diversity, 42, 173, 187, 191–193, 197, 204, 216, 224–226, 275
Sturm–Liouville eigenvalue problem, 74
super-resolution, 11, 12, 363, 364

tap, 215, 217
tapped delay line, 215
TDMA, 206, 208, 209
time diversity, 202, 204–206
time spread, 203, 204, 206, 210–214
time-dispersive media, 133, 134
timelimited signals, 49, 51, 52, 55, 61, 63, 68, 69, 94, 108, 169, 234, 344, 359, 361, 362
timelimiting operator, 69, 98, 106, 109
transport theory, 276, 284, 290, 291
Twersky equation, 280, 281
Twersky's theory, 279
typical modes, 327
typical sequence, 320
typical set, 21, 320, 321, 323, 339
typical states, 319–321, 323, 326, 327, 339
typical waveforms, 22, 23

ultraviolet catastrophe, 312
uncertainty principle, 306
  over arbitrary measurable sets, 58, 82, 117, 407, 408
  converse, 59
  entropic, 58, 82
  Heisenberg, 41, 55, 56, 82, 325, 337
  in quantum mechanics, 56
  for signals, 55, 325, 406
  for a single photon, 328, 332

underspread channel, 214
universal entropy bound, 330, 337, 391, 394, 396, 398, 401–406

vector potential, 144

water filling, 357, 385
wave equation, 65, 71, 87
wavelet, 83
wavenumber bandwidth *see also* spatial bandwidth 230
wide-band transmission, 201, 202
Wiener–Khinchin theorem, 304

Young's inequality, 232, 234, 412

Printed in the United States
by Baker & Taylor Publisher Services